21 Jan. '86

chat, with best wishes and
pleasant memories of journeys
and conversations between
~~With Compliments~~
Northampton and Boston, from

Alistair Crombie

AUGUSTINE TO GALILEO

AUGUSTINE
TO
GALILEO

VOLUME I

SCIENCE IN THE MIDDLE AGES

5TH TO 13TH CENTURIES

VOLUME II

SCIENCE IN THE LATER MIDDLE AGES
AND EARLY MODERN TIMES

13TH TO 17TH CENTURIES

A. C. CROMBIE

Harvard University Press
Cambridge, Massachusetts

Library of Congress Catalog Card Number 61–16151
ISBN 674–05273–0

Printed in Great Britain

TO NANCY

AUGUSTINE
TO
GALILEO

VOLUME I

SCIENCE IN THE MIDDLE AGES
5TH TO 13TH CENTURIES

CONTENTS

9

Contents

Contents

PLATES

1 Aristotle's cosmology. From Petrus Apianus, *Cosmographia*, per Gemma Phrysius restituta, Antwerp, 1539.

2 The medieval mechanical model of solid spheres for the planet Saturn. From G. Reisch, *Margarita Philosophica*, Freiberg, 1503.

3 Drawing of an astrolabe. From Chaucer, *Treatise on the Astrolabe*, Cambridge University Library MS Dd. 3.53 (14th century).

4 A late Gothic astrolabe, *c.* 1430. In the Museum of the History of Science, Oxford.

5 An astrolabe in use. From an English MS, Bodley 614 (12th century) at Oxford.

6 Roger Bacon's geometrical diagrams showing the curvatures of the refracting media in the eye. From *Opus Majus*, British Museum MS Royal 7.F.viii (13th century).

7 Drawing from Theodoric of Freiberg, *De Iride*, Basel University Library MS F.iv.30 (14th century), showing an experiment with the refraction of light.

8 Drawing from Theodoric of Freiberg, *De Iride*, Basel University Library MS F.iv.30 (14th century), showing the paths of the rays inside a transparent sphere, to illustrate his explanation of the formation of the primary rainbow.

9 Drawing from Theodoric of Freiberg, *De Iride*, Basel University Library MS F.iv.30 (14th century), showing his explanation of the primary rainbow by double refraction and reflection within the spherical drops.

10 Diagram published in Jodocus Trutfetter, *Totius Philosophiæ Naturalis Summa* (Erfurt, 1514), to illustrate Theodoric of Freiberg's explanation of the rainbow.

11 Diagram from Petrus Peregrinus, *De Magnete*, Bodleian Library, Oxford, MS Ashmole 1522 (14th century), illustrating a chapter which contains the first known description of a pivoted magnet.

13

28 Ships showing construction, rig and rudder. From the *Luttrell Psalter.*

29 Knight firing a cannon against a castle. From Walter de Milemete, *De Nobilitatibus Sapientiis et Prudentiis Regum*, Christ Church, Oxford, MS 92.

30 Water-driven silk mill. From V. Zonca, *Novo Teatro di Machine et Edificii*, Padua, 1607.

31 Water-driven silk mill. From V. Zonca, *Novo Teatro di Machine et Edificii.*

32 Part of the so-called 'Gough Map' (1325–30). In the Bodleian Library, Oxford; showing S.E. England.

33 Part of a Portolan Chart. Showing Italy, Sicily and N. Africa. From British Museum MS Additional 25691 (*c.* 1327–30).

34 Ptolemy's map of the world, redrawn by Italian cartographers. From the second edition of his *Geographia* (Rome, 1478) to contain maps.

35 Screw-cutting lathe. From Jacques Besson, *Theatrum Instrumentorum et Machinarum*, Lyons, 1569 (1st ed. 1568).

36 Page from the *Album* of Villard de Honnecourt, showing the escapement mechanism in centre left. Above is a water-driven saw. From the Bibliothèque Nationale, Paris, MS français 19093 (13th century).

37 The Dover Castle clock, formerly dated 14th century but now believed to be later. Crown Copyright. Science Museum, London.

38 Glass making. From British Museum MS Additional 24189 (15th century).

39 Surgery. Sponging a patient, probably a leper, trephining, operating for hernia and treating fractures, from Roland of Parma, *Livre de Chirurgie*. From British Museum MS Sloane 1977 (13th century).

PREFACE TO THE SECOND EDITION

ONE of the most remarkable developments in the world of scholarly letters in the last generation, and especially since the end of the Second World War, has been the growth of the study of the history of science both as a professional historical discipline and among the interests of the general reader. Considering how science has quietly come to take a central position in our culture this is perhaps not surprising; some knowledge of the history of science has become unavoidably part of the acquisition of historical awareness. Certainly the considerable interest shown in the period covered by this book is not difficult to explain. It has long been a matter of curiosity to know something of the scientific thought of those medieval centuries in which so many other essential aspects of our civilization, ranging from the theory and practice of law and government to the character of feeling and execution in poetry and the plastic arts, had their genesis and formation. I hope that in these pages the reader curious to know something of the history of medieval science, not simply as the background to modern science but as interesting in itself, may find at least a general guide to his inquiries. The stories of science in antiquity and in modern times have been told more than once in recent works, both separately and as part of general histories of science, but there exists no adequate short history of science in this formative period that lies between. My purpose in writing this book has been to fill this gap.

The scholarship of the last half-century has long since banished the time when the rumours about medieval science, put about after the revival of interest in classical literature in the 15th century, could be regarded as an adequate substitute for the study of contemporary sources. In the pages that follow I have tried to use the results of recent research to tell, within the covers of a single general history, the story of Western science from its decay after the collapse of the Roman Empire in the West to its full reflowering in the 17th century. Especially I have tried to bring out, what I believe to be the most striking result of recent scholarship, the essential continuity of the Western scientific tradition from Greek times to the 17th century and, therefore, to our own day. Certainly the scientific thought of the period from Augustine to Galileo can often be deceptive both in its similarities to, and in its differences from modern science, but this is the inevitable result of its position as part of the great adventure of philosophical reformation undertaken by the barbarian invaders as they painfully educated themselves from the classical sources. If I have seemed to give too little attention to the originality of Arabic science in this period, that is not because I underrate the indispensable contribution made by medieval Arabic civilization in developing ancient science as well as in transmitting it to the West, but because it is specifically the history of science in the Latin civilization of the West that is the subject of this study. A broader treatment, perhaps too broad for a short work, would also include a full account of the history of science in both Islam and Byzantium.

My debt to those great pioneers whose documentary researches first let the light into medieval science in this century, Paul Tannery, Pierre Duhem, Charles Homer Haskins, Karl Sudhoff, to the bibliographical industry of George Sarton, and to the critical work of more recent scholars, especially of Lynn Thorndike, Alexandre Koyré and Anneliese Maier, will be obvious to anyone who turns these pages and indeed must be incurred by any student who enters this field. Since the war many specialized studies over the whole field have appeared and continue to do so in ever increasing numbers. I have tried to incorporate into this revision the relevant substance of those published since the first edition was completed, together with various changes in my own point of view.

Oxford, 6 January 1958   A.C.C.

I have corrected some details for this new edition.

Oxford, 1 July 1969   A.C.C.

ACKNOWLEDGEMENTS

VOLUME I

ACKNOWLEDGEMENTS are made to the following for supplying photographs for illustrations: the Chief Librarian of the Oeffentliche Bibliothek, University of Basel (Plates 7, 8, 9); the Librarian of Cambridge University (Figs. 12, 14 and Plates 1, 3, 30, 31, 35, 36); the Director of the Universitätsbibliothek, Erlangen (Plate 10); the Director of the British Museum, London (Fig. 4 and Plates 6, 12, 14, 21, 22, 24, 25, 26, 27, 28, 33, 38, 39); the Director of the Science Museum, London (Plate 37: Crown Copyright); the Director of the Wellcome Museum of Medical History, London (Plate 20); Bodley's Librarian, Oxford (Plates 2, 5, 11, 16, 32, 34); the Librarian of Christ Church, Oxford (Plate 29); the Curator of the Museum of the History of Science, Oxford (Plate 4); the Scriptor of the Vatican Library, Rome (Plates 13, 18, cover design); the Director of the Biblioteca Nazionale di S. Marco, Venice (Plate 17); the Director of the Oesterreichische National Bibliothek, Vienna (Plate 15). The following lent blocks for illustrations: Messrs William Heinemann, Ltd (Fig. 11); Oxford University Press (Fig. 4 and Plate 23); Penguin Books Ltd (Fig. 10).

I should like to thank Mr Stillman Drake and Dr Michael Hoskin for reading the proofs and suggesting a number of valuable improvements.

INTRODUCTION

T H E history of science is the history of systems of thought about the natural world. Though the most obvious characteristic of science in modern civilization is the control it has given over the physical world, even while such practical control was being acquired, and certainly for long periods before it became possible, men were trying to bring nature within the grasp of their understanding. The inventions and practical achievements of applied science are of great interest to the historian and so are the effects of natural science on the layman's view of the world as seen in literature, art, philosophy and theology; of even greater interest is the internal development of scientific thought itself. The chief problems before the historian of science are, therefore: What questions about the natural world were men asking at any particular time? What answers were they able to give? And why did these answers cease to satisfy human curiosity? What were the problems seen by the scientists of the period, and what were the problems they did not see? What were the limiting features, in philosophy of nature, in scientific method, in observational, experimental and mathematical technique, that characterized science in one period, and what changes shifted the point of view in another? An obsolete system of scientific thought, which may appear very strange to us looking back from the 20th century, becomes intelligible when we understand the questions it was designed to answer. The questions make sense of the answers, and one system has given place to another not simply because new facts made it obsolete, but more significantly because for some reason, sometimes the result of fresh observations, sometimes because of new theoretical conceptions, scientists began to rethink their whole position, to ask new questions, to make different assumptions, to look at long familiar evidence in a new way.

Introduction

The presentation of the thought of an age whose presuppositions and problems were not identical with our own must always involve delicate questions of both interpretation and evaluation. Many aspects of philosophy and science, especially in the period covered by this book, are fully understandable only within the whole context of thought and opportunity, metaphysical and theological as well as scientific and technical, and social and economic as well as intellectual, of which they formed part. An easy supposition that philosophies of different periods that look similar are in fact identical, and especially that a past philosophical opinion or method is identical with one current at the present time, is bound to be misleading. This is not to deny that it is legitimate to evaluate the contributions of past philosophers to present problems; but that is not the same as attempting to understand them in their contemporary setting. The dark metaphysical drama of the 17th-century enlightenment is here especially revealing.

The art of understanding the scientific thought of the past is for the same reason no less delicate, but its terms of reference are made somewhat different from those of philosophy because of a characteristic possessed pre-eminently by science, though also to some extent shared by history. Unlike other disciplines dealing with the world, the solutions to problems in science, past and present, can be judged by criteria that are in most cases objective, universally accepted, and stable from one period to the next. The historian of science would lose immensely if he failed to make use of superior modern knowledge to evaluate the discoveries and theories of the past. But it is just in doing so that he is exposed to the greatest danger. Because science does genuinely progress by making discoveries and detecting mistakes, the temptation is almost irresistible to regard the discoveries of the past as simply anticipating and contributing to the science of the present and to write off the mistakes as leading nowhere. It is precisely this temptation, belonging as it does to the essence of science, that can sometimes make it most difficult for us to understand how discoveries and theories were in fact made and were seen by their authors in their own day. It can lead to the most insidious form of the falsification of history.

Introduction

The aim of the historian of science in investigating the origins of a discovery or a new theory must in the first place be to find out what problems were puzzling scientists *before* the solution was reached, what questions they were asking, what were their assumptions and expectations, and what *they* regarded as an answer and an explanation. In pursuing his inquiries he must take account not only of the successful work, acclaimed in its own age and in our own, but also of unsuccessful theories and experiments, explanations that died stillborn or were killed in infancy or at least that have not survived, experiments that were, to our minds or even to contemporary minds, inept and misconceived. These may be even more revealing, because we are likely to prejudge them differently, than the great discoveries we have learned all too easily to accept. It is an interpretation of the aims, conceptions and solutions of the past, *as they occurred in the past*, that is the primary quarry of the historian of science. Of all man's activities thinking is the most human, and the famous phrase in Marc Bloch's *Métier d'historien* applies as forcefully to the historian of scientific thinking as to any other: '*L'historien ressemble à l'ogre de la fable. Là où il flaire la chair humaine, il sait que là est son gibier.*'

The period under review in this book is especially exposed to an unconscious temptation to falsification. Not only was it concluded by a genuine shift in the intellectual organization of science and by the beginnings of a massive increase in scientific knowledge; its history was first written by authors who used this scientific revolution to support other reforms in their own day. Led by Voltaire, the rationalist historians of the 18th century discounted any possibility of a connexion between medieval philosophy and the triumph of scientific reason which they located in the period of Galileo, Harvey, Descartes and Newton. Taking up the theme, Comte proposed the dangerous formula for his 19th-century followers of claiming precedent for the positivist enlightenment not in what Galileo or Newton may have *stated* their aims and methods to be, but in what these must *really* have been (although perhaps unknown to them) in order to have been as successful as they were. Certainly contemporary issues can be both a stimulus and a valuable guide to the study of the past. Certainly also Comte's distinction may be valid in a philosophical evaluation. It may even be true that in some cases a scientist thinks he is doing one thing when he is demonstrably doing another, as in Galileo's first formulation of the law of acceleration of falling bodies. It is certainly true that the relevant intentions and preconceptions of a scientist can rarely all be read directly in his writings; that he may indeed not be immediately aware of many of them; that what he says about them may be palpably influenced by an incomplete understanding of some contemporary philosophy or may be a crude rationalization of how he used them; that his *use* of his methods and conceptions may be even more revealing of his actual thought than what he says about them; that interpretation is an essential part of the historical analysis by which we reconstruct the past. But interpretation that eliminates as illusory all those elements of thought and usage that are unacceptable to a particular philosophy, or discounts those elements shown to be mistaken in the light of later scientific knowledge, can only succeed in hiding from us the indispensable evidence for the actual organization and development of scientific thought and the actual process of invention and discovery. And not only will this falsify history; with the same stroke the philosopher of science will be given so false an account of that 'natural history' of scientific thought which is his essential data, that he will be even more misled than by not studying the history of science at all.

It was the Greeks who invented science as we now know it. In ancient Babylonia, Assyria and Egypt, and in ancient India and China, technology had developed on a scale of sometimes astonishing effectiveness, but so far as we know it was unaccompanied by any conception of scientific explanation. Perhaps the most remarkable example of this ancient technology can be seen in the cuneiform texts of the Babylonians and Assyrians, setting out methods of predicting astronomical motions, which by the 3rd century B.C. were as accurate as the methods that had by then been developed in the Greece of Aristarchus of Samos. But the Babylonians and Assyrians offered no natural explanations of the phenomena they could predict with such skill. The texts in which they set out to 'explain' the world, as distinct from predicting its happenings, contain myths in which the visible order of things is attributed to a legal system obeyed by arbitrary choice by a society of gods personifying natural forces.

The Greeks invented natural science by searching for the intelligible impersonal permanence underlying the world of change and by hitting upon the brilliant idea of a generalized use of scientific theory; they proposed the idea of assuming a permanent, uniform, abstract order from which the changing world of observation could be deduced. The myths themselves were reduced to the status of theories. their entities tailored to the requirements of quantitative prediction. With this idea, of which their development of geometry became the paradigm, giving it its most precise expression, Greek science must be seen as the origin of all that has followed. It was the triumph of order brought by abstract thought into the chaos of immediate experience, and it remained characteristic of Greek scientific thought to be interested primarily in knowledge and understanding and only very secondarily in practical usefulness.

With the rise of Christianity, to this Greek rationalism was added the idea of nature as sacramental, symbolic of spiritual truths, and both attitudes are found in St Augustine. In Western Christendom during the early Middle Ages men were concerned more to preserve the facts which had been collected in classical times than to attempt original interpretations themselves. Yet, during this period, a new element was added from the social situation, an activist attitude which initiated a period of technical invention and was to have an important effect on the development of scientific apparatus. Early in the 12th century men asked how the facts recorded in the book of *Genesis* could best be explained in terms of rational causes. It was a 12th-century Byzantine writer, John Tzetzes, who in his versified *Book of Histories* (8. 973) was responsible for the phrase said to have been written by Plato over the door of the Academy : 'Let no one untrained in geometry enter my house.' (See Plate 4, Vol. II.) With the recovery of the full tradition of Greek and Arabic science in the 12th and early 13th centuries, and particularly of the works of Aristotle and Euclid, there was born, from the marriage of the empiricism of technics with the rationalism of philosophy and mathematics, a new conscious empirical science seeking to discover the rational structure of nature. At the same time a more or less complete system of scientific thought was provided by Aristotle's works. The rest of the history of medieval science consists of the working out of the consequences of this new approach to nature.

Gradually it was realized that the new science did not conflict with the idea of Divine Providence, though it led to a variety of attitudes towards the relation between reason and faith. Internal contradictions, contradictions with other authorities, and contradictions with observed facts eventually led to radical criticisms of the Aristotelian system. At the same time, extension of the use of experiment and mathematics produced an increase in positive knowledge. By the beginning of the 17th century the systematic use of the new methods of experiment and mathematical abstraction had produced results so striking that this movement has been given the name 'Scientific Revolution'. These new methods were expounded in the 13th century, but were first used with complete maturity and effectiveness by Galileo.

The origins of modern science are to be found at least as far back as the 13th century, but from the end of the 16th century the Scientific Revolution began to gather a breathtaking speed. The changes in scientific thought occurring then so altered the type of question asked by scientists that Kant said of them: 'a new light flashed on all students of nature'. The new science also profoundly affected man's idea of the world and of himself, and it was to have a position in relation to society unknown in earlier times. The effects of the new science on thought and life have, in fact, been so great and special that the Scientific Revolution has been compared in the history of civilization to the rise of ancient Greek philosophy in the 6th and 5th centuries B.C. and to the spread of Christianity throughout the Roman Empire in the 3rd and 4th centuries A.D. For this reason the study of the changes leading up to that revolution, the study of the history of science from the early Middle Ages to the 17th century, is of unique interest for the historian of science. The position of science in the modern world cannot be fully understood without a knowledge of the changes that occurred during that time.

The plan of this book is to start, in Chapter 1, with a brief account of ideas about the natural world in Western Christendom from the 5th to the 12th century and then, in Chapter 2, to show how the system of scientific thought accepted in the 13th century was introduced from Greek and Arabic sources. The purpose of Chapter 3 is to give a description of that system and to indicate the additions of fact and modifications in detail made to it during the century or more following its introduction. Chapter 4 is concerned with the relation of technical activity to science during the whole medieval period. In Chapter 1 of Volume II an account is given of the development of ideas on scientific method and criticism of the fundamental principles of the 13th-century system made from the end of the 13th to the end of the 15th century. This prepared the way for the more radical changes of the 16th and 17th centuries. The last chapter is devoted to the Scientific Revolution itself.

I

SCIENCE IN WESTERN CHRISTENDOM UNTIL THE TWELFTH-CENTURY RENAISSANCE

*'Our play leaps o'er the vaunt and firstlings
of these broils
Beginning in the middle'*
TROILUS AND CRESSIDA

THE contrast between the scientific ideas of the early Middle Ages, that is from about the 5th to the early 12th century, and those of the later Middle Ages, can best be seen in a conversation which is supposed to have taken place between the widely travelled 12th-century scholar and cleric Adelard of Bath and his stay-at-home nephew. Adelard's contribution to the discussion introduces the newly-recovered ideas of the ancient Greeks and the Arabs; that of his nephew represents the traditional view of Greek ideas as they had been preserved in Western Christendom since the fall of the Roman Empire.

The conversation is recorded in Adelard's *Quæstiones Naturales*, written, probably, after he had studied some Arabic science but before he had achieved the familiarity with it which is shown in his later translations, such as those of the Arabic text of Euclid's *Elements* and the astronomical tables of al-Khwarizmi. The topics covered range from meteorology to the transmission of light and sound, from the growth of plants to the cause of the tears which the nephew shed for joy at the safe return of his uncle.

·When not long ago, while Henry, son of William [Henry I, 1100–35], was on the throne, I returned to England after my long period of study abroad, it was very agreeable to meet my friends again. After

29

we had met and made the usual inquiries about one another's health and that of friends, I wanted to know something about the morals of our nation ... After this exchange, as we had most of the day before us and so lacked no time for conversation, a nephew of mine who was with the others – he was interested rather than expert in natural science – urged me to disclose something new from my Arab studies. To this, when the rest had agreed, I delivered myself as in the tract that follows.

The nephew declared himself delighted at such an opportunity of showing that he had kept his youthful promise to work hard at philosophy, by disputing the new ideas with his uncle, and declared:

if I were only to listen to you expounding a lot of Saracen theories, and many of them seemed to me to be foolish enough, I would get a little restless, and while you are explaining them I will oppose you wherever it seems fit. I am sure you praise them shamelessly and are too keen to point out our ignorance. So for you it will be the fruit of your labour if you acquit yourself well, while for me, if I oppose you plausibly, it will mean that I have kept my promise.

The scientific inheritance of the Latin West, represented by the nephew's contribution to the dialogue, was limited almost exclusively to fragments of Greco-Roman learning such as had been preserved in the compilations of the Latin encyclopædists. The Romans themselves had made hardly any original contributions to science. The emphasis of their education was upon oratory. But some of them were sufficiently interested in trying to understand the world of nature to make careful compilations of the learning and observations of Greek scholars. One of the most influential of these compilations, which survived throughout the early Middle Ages as a text-book, was the *Natural History* of Pliny (23–79 A.D.), which Gibbon described as an immense register in which the author has 'deposited the discoveries, the arts, and the errors of mankind'. It cited nearly 500 authorities. Beginning with the general system of cosmology it passed to geography, anthropology, physiology and zoology, botany, agriculture and horticulture, medicine, mineralogy and the fine arts. Until the 12th century, when translations of Greek and Arab works began to come into Western Europe, Pliny's was the largest known collection of natural facts, and it was drawn on by a succession of later writers.

The mathematics and logic of the Latin West rested on the work of the 6th-century Boethius, who did for those studies what Pliny had done for natural history. Not only did he compile elementary treatises on geometry, arithmetic, astronomy and music, based respectively on the work of Euclid, Nicomachus and Ptolemy, but he also translated the logical works of Aristotle into Latin. Of these translations only the *Categories* and the *De Interpretatione* were widely known before the 12th century, but until that time the translations and commentaries of Boethius were the main source for the study of logic as of mathematics. Knowledge of mathematics was largely confined to arithmetic. The only mathematical treatise remaining intact, the so-called 'Geometry of Boethius', which dates from no earlier than the 9th century, contained only fragments of Euclid and was concerned mostly with such practical operations as surveying. Cassiodorus (*c.* 490–580), in his popular writings on the liberal arts, gave only a very elementary treatment of mathematics.

Another of the compilers of the early Middle Ages who helped to keep alive the scientific learning of the Greeks in the Latin West was the Visigothic bishop, Isidore of Seville (560–636). His *Etymologies*, based on often fantastic derivations of various technical terms, remained popular for many centuries as a source of knowledge of all kinds from astronomy to medicine. For Isidore the universe was limited in size,* only a few thousand years old and soon to perish. The earth, he thought, was shaped like a

---

* The littleness of man in the universe was, however, a familiar theme for reflection and this passage from Boethius' *De Consolatione Philosophiæ* (2. 7) was well known throughout the Middle Ages : 'Thou has learnt from astronomical proofs that the whole earth compared with the universe is not greater than a point, that is, compared with the sphere of the heavens, it may be thought of as having no size at all. Then, of this tiny corner, it is only one-quarter that, according to Ptolemy, is habitable to living things. Take away from this quarter the seas, marshes, and other desert places, and the space left for man hardly even deserves the name of infinitesimal.'

wheel with its boundaries encircled by the ocean. Round the earth were the concentric spheres bearing the planets and stars, and beyond the last sphere was highest heaven, the abode of the blessed.

From the 7th century onwards the Latin West had to rely almost exclusively for scientific knowledge on these compilations, to which were added those of the Venerable Bede (673–735), Alcuin of York (735–804),...and the German Hrabanus Maurus (776–856), each of whom borrowed freely from his predecessors.

The gradual penetration of the barbarians into the Western Roman Empire from the 4th century had caused some material destruction and eventually serious political instability, but it was the eruption of the Mohammedan invaders into the Eastern Empire in the 7th century that gave the most serious blow to learning in Western Christendom. The conquest of much of the territory of the Eastern Empire by the Arabs meant that the main reservoir of Greek learning was cut off from Western scholars for centuries by the intolerance and mutual suspicion of opposing creeds, and by the dragon wing of the Mediterranean. In this intellectual isolation Western Christendom could hardly have been expected to make many original contributions to man's knowledge of the material universe. All the West was able to do was to preserve the collection of facts and interpretations already made by the encyclopædists. That so much was preserved in spite of the gradual collapse of Roman political organization and social structure under the impact, first, of Goths, Vandals and Franks, and then, in the 9th century, of Norsemen, was due to the appearance of monasteries with their attendant schools which began in Western Europe after the foundation of Monte Cassino by St Benedict in 529. The existence of such centres made possible the temporary revivals of learning in Ireland in the 6th and 7th centuries, in Northumbria in the time of Bede, and in Charlemagne's empire in the 9th century. Charlemagne invited Alcuin from Northumbria to become his minister of education, and one of Alcuin's essential reforms was to establish schools associated with the more important cathedrals. It was in such a school, at Laon, that the nephew of Adelard received his education in the 12th century, when the curriculum was still based on the work of the encyclopædists. Studies were limited to the seven liberal arts as defined by Varro in the first century B.C. and by Martianus Capella six hundred years later. Grammar, logic and rhetoric made up the first stage or *trivium*, and geometry, arithmetic, astronomy and music made up the more advanced *quadrivium*. The texts used were the works of Pliny, Boethius, Cassiodorus and Isidore.

One development of importance which had taken place in the studies of the Latin West between the days of Pliny and the time when Adelard's nephew pursued his studies at Laon was the assimilation of Neoplatonism. This was of cardinal importance for it determined men's views of cosmology until the second half of the 12th century. St Augustine (354–430) was the principal channel through which the traditions of Greek thought passed into the reflections of Latin Christianity, and St Augustine came profoundly under the influence of Plato and of Neoplatonists such as Plotinus (*c.* 203–70 A.D.). The chief aim of Augustine was to find a certain basis for knowledge and this he found in the conception of eternal ideas as expounded by the Neoplatonists and in the Pythagorean allegory, the *Timæus*, by Plato himself. According to this school of thought, eternal forms or ideas existed quite apart from any material object. The human mind was one of these eternal essences and had been formed to know the others if it would. In the process of knowing, the sense organs merely provided a stimulus spurring on the mind to grasp the universal forms which constituted the essence of the universe. An important class of such universal forms was mathematics. 'If I have perceived numbers by the sense of the body,' Augustine said in *De Libero Arbitrio* (book 2, chapter 8, section 21),

I have not thereby been able by the sense of the body to perceive also the nature of the separation and combination of numbers ... And I do not know how long anything I touch by a bodily sense will persist, as, for instance, this sky and this land, and whatever other bodies I perceive in them. But seven and three are ten and not only now but always; nor have seven and three in any way at any time not been

33

ten, nor will seven and three at any time not be ten. I have said, therefore, that this incorruptible truth of number is common to me and anyone at all who reasons.

In the 9th century such scholars as John Scot Erigena (d. 877) re-emphasized the importance of Plato. In addition to the work of the Latin encyclopædists and others, he began to use some original Greek works, some of the most important being the 4th-century translation by Chalcidius of Plato's *Timæus* and commentary by Macrobius, and the 5th-century commentary by Martianus Capella. Erigena himself showed little interest in the natural world and seems to have relied for his facts almost entirely on literary sources, but the fact that among his sources he included Plato, for whom St Augustine had also had so marked a preference, gave to men's interpretations of the universe a Platonic or Neoplatonic character for about 400 years, though it was not till the development of the school of Chartres in the 12th century that the more scientific parts of the *Timæus* were particularly emphasized.

In general the learning of Western Christendom as represented by the views of Adelard's nephew, the Latin encyclopædists, and the cathedral and the monastic schools was predominantly theological and moral. Even in classical times there had been very little attempt to pursue scientific inquiry for 'fruit', as Francis Bacon called the improvement of the material conditions of life. The object of Greek science had been understanding, and under the influence of later classical philosophers such as the Stoics, Epicureans and Neoplatonists natural curiosity had given way almost entirely to the desire for the untroubled peace which could only be won by a mind lifted above dependence on matter and the flesh. These pagan philosophers had asked the question: What is worth knowing and doing? To this Christian teachers also had an answer: That is worth knowing and doing which conduces to the love of God. The early Christians continued their neglect of natural curiosity and at first also tended to disparage the study of philosophy itself as likely to distract men from a life pleasing to God. St Clement of Alexandria in the 3rd century poked fun at this fear of pagan philosophy, which he compared to a child's fear of goblins. Both he and his pupil Origen claimed that all knowledge was good since it was a perfection of mind and that the study of philosophy and of natural science was in no way incompatible with a Christian life. St Augustine himself in his searching and comprehensive philosophical inquiries had invited men to examine the rational basis of their faith. But in spite of these writers natural knowledge continued to be considered of very secondary importance during the early Middle Ages. The primary interest in natural facts was to find illustrations for the truths of morality and religion. The study of nature was not expected to lead to hypotheses and generalizations of science but to provide vivid symbols of moral realities. The moon was the image of the Church reflecting the divine light, the wind an image of the spirit, the sapphire bore a resemblance to divine contemplation, and the number eleven, which 'transgressed' ten, representing the commandments, stood itself for sin.

This preoccupation with symbols is shown clearly in the bestiaries. Since the time of Aesop stories about animals had been used to illustrate various human virtues and vices. This tradition was continued in the 1st century A.D. by Seneca in his *Quæstiones Naturales*, and by later Greek works, culminating in the 2nd century with a work of Alexandrian origin known as the *Physiologus*, which was the model for all the medieval moralizing bestiaries. In these works facts of natural history collected from Pliny were mixed with entirely mythical legends to illustrate some point of Christian teaching. The phoenix was the symbol of the risen Christ. The ant-lion, born of the lion and the ant, had two natures and so was unable to eat either meat or seeds and perished miserably like every double-minded man who tried to follow both God and the Devil. The *Physiologus* had enormous popularity. It was translated into Latin in the 5th century and into many other languages, from Anglo-Saxon to Ethiopian. In the 4th century, when St Ambrose wrote a commentary on the Bible, he made liberal use of animals as moral symbols. As late as the early years of the 13th century Alexander Neckam could claim in his *De Naturis Rerum*, in which he showed very considerable interest in scientific fact, that he had written the book for purposes of moral instruction. In the 12th century there were many signs, as, for instance, in the illustrations to certain manuscripts and the descriptions of wild life by Giraldus Cambrensis (*c.* 1147–1223) and other travellers, that men were capable of observing nature very clearly, but their observations were usually simply interpolations in the course of a symbolic allegory which to their minds was all important. In the 13th century this passion for pointing out moral symbolism invaded even the lapidaries, which in the Ancient World, as represented in the works of Theophrastus (*c.* 372–288 B.C.), Dioscorides (1st century A.D.) and Pliny and even in the Christian works of 7th-century Isidore or 12th-century Marbode, Bishop of Rennes, had been concerned with the medical value of stones or with their magical properties.

This preoccupation with the magical and astrological properties of natural objects was, with the search for moral symbols, the chief characteristic of the scientific outlook of Western Christendom before the 13th century. There was a wealth of magic in the works of Pliny and one of its characteristic ideas, the doctrine of signatures according to which each animal, plant or mineral had some mark indicating its hidden virtues or uses, had a profound effect on popular natural history. St Augustine had to bring all the skill of his dialectic against the denial of free will which astrology implied, but had not been able to defeat this superstition. Isidore of Seville admitted that there were magical forces in nature, and though he distinguished between the part of astrology which was natural, since it led man to study the courses of the heavenly bodies, and the superstitious part which was concerned with horoscopes, he yet admitted that these heavenly bodies had an astrological influence on the human body and advised doctors to study the influence of the moon on plant and animal life. It was a very general belief during the whole of the Middle Ages and even into the 17th century that there was a close correspondence between the course of a disease and the phases of the moon and movements of other heavenly bodies, although throughout that time certain writers, as, for instance, the 14th-century Nicole Oresme and the 15th-century Pierre d'Ailly, had made fun of astrology and had limited celestial influence to heat, light and mechanical action. Indeed, medical and astronomical studies came to be closely associated.* Salerno and later Montpellier were famous for both and in a later age Padua welcomed both Galileo and Harvey.

An example of this astrological interpretation of the world of nature as a whole is the conception of the correspondence between the universe, or Macrocosm, and the individual man, or Microcosm. This theory had been expressed in the *Timæus* and had been elaborated in relation to astrology by the Stoics. The classical medieval expression of the belief was given in the 12th century by Hildegard of Bingen, who thought that various parts of the human body were linked with special parts of the Macrocosm so that the 'humours' were determined by the movements of the heavenly bodies.

Gilson has said of the world of the early Middle Ages, typified by the nephew of Adelard : 'To understand and explain anything consisted for a thinker of this time in showing that it was not what it appeared to be, but that it was the symbol or sign of a more profound reality, that it proclaimed or signified something else.' But this exclusively theological interest in the natural world had already begun to be modified even before the writings of the Greek and Arab natural philosophers became more fully and widely known in Western Christendom, as a result of increasing intellectual contact with the Arab and Byzantine worlds. One aspect of this change in outlook is to be seen in the increasing activity of the computists, doctors and writers of purely technical

* Cf. the Prologue to Chaucer's *Canterbury Tales* (ll. 411 *et seq.*) :
> With us ther was a Doctour of Phisyk;
> In al this world ne was ther noon hym lyk.
> To speke of phisik and of surgerye;
> For he was grounded in astronomye.
> He kepte his pacient a ful greet del
> In hourés, by his magik naturel.
> Wel coude he fortunen the ascendent
> Of his images for his pacient.
> He knew the cause of everich maladye,
> Were it of hoot or cold, or moiste, or drye,
> And where engendred, and of what humour;
> He was a verrey parfit practisour.

treatises of which there had been a continuous tradition through-out the early Middle Ages. In the 6th century Cassiodorus, when making arrangements for an infirmary in his monastery,* had in his *Institutio Divinarum Litterarum*, book 1, chapter 31, given some very precise and practical advice on the medical use of herbs:

Learn, therefore, the nature of herbs, and study diligently the way to combine various species ... and if you are not able to read Greek, read above all translations of the *Herbarium* of Dioscorides, who de-scribed and drew the herbs of the field with wonderful exactness. After this, read translations of Hippocrates and Galen, especially the *Thera-peutics* ... and Aurelius Celsus' *De Medicina* and Hippocrates' *De Herbis et Curis*, and divers other books written on the art of medicine, which by God's help I have been able to provide for you in our library.

A good example of the influence of practical problems in pre-serving the habit of observation, and a good illustration of the state of Latin scientific knowledge before the translations from Greek and Arabic, is provided by the writings of Bede. The main sources of Bede's ideas about the natural world were the Fathers, especially St Ambrose, St Augustine, St Basil the Great and St Gregory the Great; and Pliny, Isidore and some Latin writings on the calendar. Although he knew Greek, it was on Latin sources that he almost entirely drew. Based on these sources, Bede's writings on scientific subjects fall into two main classes: a largely derivative account of general cosmology, and a more independent treatment of some specific practical problems, in particular those connected with the calendar.

Bede's cosmology is interesting for showing how an educated person of the 8th century pictured the universe. He set out his views in *De Rerum Natura*, based largely on Isidore's book of the same title but also on Pliny's *Natural History*, which Isidore had not known. It was largely because of his knowledge of Pliny, as well as his more critical mind, that made Bede's book so greatly superior to Isidore's. Bede's universe is one ordered by ascertain-able cause and effect. Whereas Isidore had thought the earth

* At Monte Cassino St Benedict had also established an infirmary. The care of the sick was regarded as a Christian duty for all such foundations.

shaped like a wheel, Bede held that it was a static sphere, with five zones, of which only the two temperate were habitable and only that in the northern hemisphere actually inhabited. Surrounding the earth were seven heavens: air, ether, Olympus, fiery space, the firmament with the heavenly bodies, the heaven of the angels, and the heaven of the Trinity. The waters of the firmament separated the corporeal from the spiritual creation. The corporeal world was composed of the four elements, earth, water, air and fire, arranged in order of heaviness and lightness. At the creation these four elements, together with light and man's soul, were made by God *ex nihilo*; all other phenomena in the corporeal world were combinations. From Pliny, Bede got a much more detailed knowledge of Greek understanding of the daily and annual movements of the heavenly bodies than had been available to Isidore. He held that the firmament of stars revolved round the earth, and that within the firmament the planets circled in a system of epicycles. He gave clear accounts of the phases of the moon and of eclipses.

The problem of the calendar had been brought to Northumbria along with Christianity by the monks of Iona, but long before that time methods of computing the date of Easter had formed part of the school science of *computus*, which provided the first exercises of early medieval science.

The main problem connected with the Christian calendar arose from the fact that it was a combination of the Roman Julian calendar, based on the annual movement of the earth relative to the sun, and the Hebrew calendar, based on the monthly phases of the moon. The year and its divisions into months, weeks, and days belonged to the Julian solar calendar; but Easter was determined in the same way as the Hebrew Passover by the phases of the moon, and its date in the Julian year varied, within definite limits, from one year to the next. In order to calculate the date of Easter it was necessary to combine the length of the solar year with that of the lunar month. The basic difficulty in these calculations was that the lengths of the solar year, the lunar month and the day are incommensurable. No number of days can make an exact number of lunar months or solar years, and no number of lunar-months can make an exact number of solar years. So, in order to relate the phases of the moon accurately to the solar year in terms of whole days, it is necessary, in constructing a calendar, to make use of a system of *ad hoc* adjustments, following some definite cycle.

From as early as the 2nd century A.D. different dates of Easter, resulting from different methods of making the calculations, had given rise to controversy and had become a chronic problem for successive. Councils. Various cycles relating the lunar month to the solar year were tried at different times and places, until in the 4th century a 19-year cycle, according to which 19 solar years were considered equal to 235 lunar months, came into general use. But there was still the possibility of differences in the manner in which this same cycle was used to determine the date of Easter, and even when there was uniformity at the centre, sheer difficulty of communication could and did result in such outlying provinces as Africa, Spain and Ireland celebrating Easter at different dates from Rome and Alexandria.

Shortly before Bede's birth Northumbria had, at the Synod of Whitby, given up many practices, including the dating of Easter, introduced by the Irish-trained monks of Iona, and had come into uniformity with Rome. But there was still much confusion, by no means confined to Britain, as to how the date of Easter was to be calculated. Bede's main contribution, expounded in several treatises, beginning with *De Temporibus* written in 703, for his pupils at Jarrow, was to reduce the whole subject to order. Using largely Irish sources, themselves based upon a good knowledge of earlier Continental writings, he not only showed how to use the 19-year cycle to calculate Easter Tables for the future, but also discussed general problems of time measurement, arithmetical computation, cosmological and historical chronology, and astronomical and related phenomena. Though often relying on literary sources when he could have observed with his own eyes – as, for example, in his account of the Roman Wall not ten miles from his monk's cell – Bede never copied without understanding. He tried to reduce all observed occurrences to general laws, and, within the limits of his knowledge, to build up a consistent picture of the universe, tested against the evidence. His account of the tides in *De Temporum Ratione* (chapter 29), completed in 725 and the most important of his scientific writings, not only shows the practical curiosity shared by him and his Northum-brian compatriots but also contains the basic elements of natural science.

From his sources Bede learned the fact that the tides follow the phases of the moon, and the theory that tides were caused by the moon attracting the ocean. He discussed spring and neap tides, and, turning to things which 'we know, who live on the shore of the sea divided by Britain,' he described how the wind could advance or retard a tide and enunciated for the first time the important principle now known as 'the establishment of a port.' This states that the tides lag behind the moon by definite intervals which may be different at different points on the same shore, so that tides must be tabulated for each port separately. Bede wrote : 'Those who live on the same shore as we do, but to the north, see the ebb and flow of the tide well before us, whereas those to the south see it well after us. In every region the moon always keeps the rule of association which she has accepted once and for all.' On the basis of this, Bede suggested that the tides at any port could be predicted by means of the 19-year cycle, which he substituted for Pliny's less accurate 8-year cycle. Tidal tables were frequently attached to *computi* written after Bede's time.

Against the background of its time Bede's science was a re-markable achievement. It contributed substantially to the Carol-ingian Renaissance on the Continent, and found its way into the educational tradition dating from the cathedral schools estab-lished for Charlemagne by Alcuin of York. Bede's treatises on the calendar remained standard text-books for five centuries, and were used even after the Gregorian reform of 1582; *De Temporum Ratione* is still one of the clearest expositions of the principles of the Christian calendar.

Besides in Northumbria, Anglo-Saxon England saw some scientific developments in Wessex. In the 7th century, astronomy and medicine were taught in Kent; there is evidence that surgery was practised; and Aldhelm, Abbot of Malmesbury, wrote metrical riddles about animals and plants. But the most notable contribution came in the first half of the 10th century in the *Leech Book* of Bald, who was evidently a physician living during or shortly after the reign of King Alfred, to whom the book contains allusions. The *Leech Book* gives a good picture of the state of medicine at the time. The first part is mainly therapeutical, containing herbal prescriptions, based on a wide knowledge of native plants and garden herbs, for a large number of diseases, working downwards from those of the head. Tertian, quartan and quotidian fevers are distinguished, and reference is made to 'flying venom' or 'air-borne contagion,' that is, epidemic diseases generally, and to smallpox, elephantiasis, probably bubonic plague, various mental ailments, and the use of the vapour bath for colds. The second part of the *Leech Book* is different in character, dealing mainly with internal diseases and going into symptoms and pathology. It seems to be a compilation of Greek medicine, perhaps mainly derived from the Latin translation of the writings of Alexander of Tralles, together with some direct observation. A good example is the account of 'sore in the side,' or pleurisy, of which many of the 'tokens' or symptoms are described by Greek writers, but some are original. The Anglo-Saxon leech recognized the occurrence of traumatic pleurisy and the possibility of confusing it with the idiopathic disease, which the ancient writers did not. Treatment began with a mild vegetable laxative administered by mouth or enema, followed by a poultice applied to the painful spot, a cupping glass on the shoulders, and various herbs taken internally. Many other diseases are described, for example pulmonary consumption and abscesses on the liver, treatment here culminating in a surgical operation. But on the whole there is little evidence of clinical observation; no use was made of the pulse and little of the appearances of the urine, which were standard 'signs' for the Greeks and Romans. Anglo-Saxon surgery presents the same combination of empiricism with literary tradition as the medicine; treatments of broken limbs and dislocations, plastic surgery for harelip, and amputations for gangrene are described.

A remarkable work showing the intelligent interest of the Anglo-Saxon scholars in improving their knowledge of natural history in relation to medicine is the translation into Old English, probably made about 1000–1050 A.D., of the Latin *Herbarium* attributed apocryphally to Apuleius Barbarus, or Platonicus. As in most early herbals the text is confined to the name, locality found, and medical uses of each herb; there are no descriptions for identification, which was to be made by means of diagrammatic paintings, copied from the manuscript source and not from nature. About 500 English names are used in this herbal, showing an extensive knowledge of plants, many of them native plants which could not have been known from the Latin sources.

There are many other examples of the influence of practical interests on the scientific outlook of scholars. In the 8th century appeared in Italy the earliest known Latin manuscript on the preparation of pigments, gold-making, and other practical problems which might confront the artist or illuminator; one of Adelard's writings was to be on this subject. In the field of medicine, the traditional literary advice on the treatment of disease came under some criticism in Charlemagne's cathedral schools, and much sharper criticism in the light of practical experience is found in the *Practica* of Petrocellus, of the famous medical school of Salerno. The computists likewise continued to collect a body of experience and elementary mathematical techniques in their work on the calendar. It was this problem of calculating the date of Easter that was chiefly responsible for the continuous interest in arithmetic, and various improvements in technique were attempted from the beginning of the 8th century, when Bede produced his chronology and 'finger reckoning,' to the end of the 10th century when the monk Helperic produced his text-book on arithmetic, and down to the 11th and 12th centuries when there appeared numerous manuscripts on this subject. The calculation of dates led also to an interest in astronomical observations, and more accurate observations became possible when knowledge of the astrolabe was obtained from the Arabs by Gerbert and other scholars of the 10th century. The chief scientific centre at that time was Lotharingia, and Canute and later Earl Harold and William the Conqueror all encouraged Lotharingian astronomers and mathematicians to come to England, where they were given ecclesiastical positions.

Besides this persistent concern with practical problems, another tendency that was equally important in substituting a different approach to the world of nature for that of moralizing symbolism was a change in philosophical outlook, and especially that which is associated with the 11th-century nominalist, Roscelinus, and his pupil Peter Abelard (1079–1142). At the end of the 11th century the teaching of Roscelinus opened the great dispute over 'universals' which led men to take a greater interest in the individual, material object as such and not, as St Augustine had done, to regard it as simply the shadow of an eternal idea. The debate began over some remarks of Boethius concerning the relation of universal ideas such as 'man', 'rose' or 'seven' both to individual things and numbers and to the human minds that knew them. Did the universal 'rose' subsist with individual roses or as an eternal idea apart from physical things? Or had the universal no counterpart in the real world, was it a mere abstraction? One of the most vigorous attacks on St Augustine's point of view was made by Roscelinus' pupil Abelard, almost an exact contemporary of Adelard of Bath; his dialectical skill and violence won him the nickname of *Rhinocerus indomitus*. Abelard did not accept Roscelinus' view that universals were simply abstractions, mere names, but he pointed out that if the only reality were the eternal ideas then there could be no real difference between individual roses or men, so that in the end everything would be everything else. The outcome of this criticism of the extreme Augustinian view of the universal was to emphasize the importance of the individual, material thing and to encourage observation of the particular.

The effect of this changed philosophical outlook, of the increasing number of practical treatises, and of the rediscovery of Greek works through contact with the Arabs, is shown in the answers given by Adelard of Bath to the scientific questions put to him by his nephew. The first of the *Quæstiones Naturales* was:

Why do plants spring from the earth? What is the cause and how can it be explained? When at first the surface of the earth is smooth and still, what is it that is then moved, pushes up, grows and puts out branches? If you collect dry dust and put it finely sieved in an earthenware or bronze pot, after a while when you see plants springing up, to what else do you attribute this but to the marvellous effect of the wonderful divine will?

Adelard admitted that it was certainly the will of the Creator that plants should spring from earth, but he asserted his opinion that this process was 'not without a natural reason too.' He repeated this opinion in answer to a later question when his nephew asked him if it were not 'better to attribute all the operations of the universe to God,' since his uncle could not produce natural explanations for them all. To this Adelard replied:

I do not detract from God. Everything that is, is from him and because of him. But [nature] is not confused and without system and so far as human knowledge has progressed it should be given a hearing. Only when it fails utterly should there be recourse to God.

With this remark the medieval conception of nature began to cross the great watershed that divides the period when men looked to nature to provide illustrations for moralizing from that in which men began to study nature for its own sake. The realization of such a conception became possible when Adelard demanded 'natural causes' and declared that he could not discuss anything with someone who was 'led in a halter' by past writers.

Those who are now called authorities reached that position first by the exercise of their reason ... Wherefore, if you want to hear anything more from me, give and take reason.

The first explanation of the universe in terms of natural causes, after the dissatisfaction with the attempt to interpret it merely in terms of moral symbols, was associated with the school of Chartres and was deeply influenced by the teaching of Plato. Early in the 12th century Chartres had shown a renewed interest in the scientific ideas contained in the *Timæus*. Such scholars as Gilbert de la Porrée (*c.* 1076–1154), Thierry of Chartres (d. *c.* 1155) and Bernard Silvester (*fl.c.* 1150) studied Biblical questions with greater attention than before to the scientific matters involved, and all were deeply influenced by St Augustine. Like Adelard their attitude to earlier learned authorities was free and rational, and they believed in the progress of knowledge. As Bernard wrote:

We are like dwarfs standing on the shoulders of giants, so that we can see more things than them, and can see further, not because our vision is sharper or our stature higher, but because we can raise ourselves up thanks to their giant stature.

Thierry of Chartres in his *De Septem Diebus et Sex Operum Distinctionibus*, in which he attempted to give a rational explanation of the creation, declared that it was impossible to understand the story in *Genesis* without the intellectual training provided by the *quadrivium*, that is without the mastery of mathematics, for on mathematics all rational explanation of the universe depended. Thierry interpreted the story of the creation as meaning that in the beginning God created space or chaos, which for Plato had been pre-existing and had been shaped into the material world by a demiurge. In St Augustine's writings, the demiurge had been replaced by the Christian God, and the forms given to the material world were reflections of the eternal ideas existing in the mind of God.

According to Plato's *Timæus* the four elements out of which all things in the universe were made, earth, water, air and fire, were composed of small invisible particles, those of each element having a characteristic geometrical shape by which the demiurge had reduced to order the originally disorderly motions of chaos.*

* The conception of matter as being made up of small particles had been put forward by various Greek philosophers in an attempt to explain how change was possible in a world in which things still retained their identity. In the 5th century B.C. Parmenides had brought philosophers to an impasse by pointing out that the earlier Ionian school's conception of one, homogeneous substance such as water, air or fire as the identity persisting through change would in fact make change impossible, for one homogeneous substance could do nothing but remain one and homogeneous. Change would then involve the coming into being of something out of nothing, which was impossible. Change was therefore unintelligible. In order to overcome this difficulty other philosophers later in the 5th century assumed that there were several ultimate substances and that the re-

The elements were mutually transformable by breaking down each geometrical shape into others, but their main masses were arranged in concentric spheres with earth in the centre, water next to it, then air and finally fire, so as to form a finite spherical universe. The sphere of fire extended from the moon to the fixed stars, and contained within it the spheres of those heavenly bodies and of the other intermediate planets. Fire was the chief constituent of heavenly bodies.

In Thierry's view fire vaporized some of the waters on the earth and raised them to form the firmament dividing the waters which were under the firmament from the waters which were above the firmament. This reduction in the waters covering the central sphere of earth led to the appearance of dry land. The warmth of the air and the moisture of the earth engendered plants and trees. Next the stars were formed as conglomerations in the super-firmamental waters, and the heat developed by their sub-

---

arrangement of these produced the changes observed in the world. Anaxagoras said that each kind of body was divisible into homogeneous parts or 'seeds', each of which retained the properties of the whole, and was again divisible and so on to infinity. Empedocles, on the other hand, said that after a certain number of divisions of bodies there would be reached the four elements, earth, water, air and fire; all bodies were formed from combinations of these elements, each of which was itself permanent and unchanging. The Pythagorean school supposed that all objects were made up of points or units of existence, and that natural objects were made up of these points in combinations corresponding to the various geometrical figures. It should then have been possible for a line to be made up of a finite number of such points, and the Pythagorean theory broke down when faced with such facts as that the ratio of the diagonal to the side of a square could not be expressed in terms of an exact number but was $\sqrt{2}$, which to the Pythagoreans was 'irrational'. The Pythagoreans had in fact confused geometrical points with ultimate physical particles, and this seems to have been the point of Zeno's paradoxes. The atomists Leucippus and Democritus avoided this difficulty by admitting that geometrical points had no magnitude and that geometrical magnitudes were divisible to infinity, but held that the ultimate particles which made up the world were not geometrical points or figures but physical units which were indivisible, that is, atoms. According to the atomists the universe was made up of atoms moving continually at random in an infinite void. Atoms differed in size, shape, order and position, the

sequent motions hatched birds and fishes out of the terrestrial waters, and animals out of the earth itself. The animals included man – made in the image of God. After the sixth day nothing more was created, but Thierry adopted from St Augustine a theory to account for the appearance later of new creatures. Augustine had brought into agreement two apparently contradictory accounts in *Genesis*, in one of which all things were created at once, while in the other, creatures, including man, appeared in succession. He had accepted the idea, put forward in the 5th century B.C. by Anaxagoras and subsequently developed by the Stoics, of originative seeds or germs, and he had suggested that in the first stage of creation plants, animals and men had all been made simultaneously in germ or in their 'seminal causes,' and that in the second stage they had actually and successively appeared.

The falling and rising of bodies was explained by the Platonists of Chartres, following the *Timæus*, by supposing that bodies of like nature tended to come together. A detached part of any element would thus tend to rejoin its main mass: a stone fell to

---

number of different shapes being infinite. In their continual movements they formed vortices in which were produced first the four elements and then other bodies by mechanical attachments of like atoms, for instance, by a hook-and-eye mechanism. Since the number of atoms was limitless so was the number of worlds they might form in the infinite void. For the atomists the only 'truth' consisted in the properties of the atoms themselves, hardness, shape and size. All other properties such as taste, colour, heat or cold were simple sense impressions which did not correspond to anything in 'reality'. Both Pythagoreans and atomists agreed in thinking that the intelligible, persisting and real amid the changing variety of the physical world was something that could be expressed in terms of mathematics. This was also the view that Plato put forward in the *Timæus*, in which he was strongly influenced by the Pythagoreans. Down to the time of Plato the result of Greek efforts to explain change was thus to refine and make intelligible the idea of the identity persisting through change. This identity, which formed the 'being' or 'substance' of physical things, had been converted from something material into an intangible essence. For Plato this essence was the universal idea or 'form' which he held existed apart from physical things as the object of their aspiration. Change or 'becoming' was a process by which sensible likenesses of such eternal forms were produced in space and time (see Vol. II, p. 49 *et seq.*).

the earthy sphere at the centre of the universe, whereas fire shot upwards to reach the fiery sphere at the outermost limit of the universe. This Platonic theory of gravity had been known also to Erigena, who had held that heaviness and lightness varied with distance from the earth, the centre of gravity. Adelard of Bath had also accepted this theory of gravity and was able to satisfy his nephew's curiosity by saying that if a stone were dropped into a hole passing through the centre of the earth, it would fall only as far as the centre.

The movement of the heavenly bodies was explained by supposing that the universe, being spherical, had a proper motion of uniform eternal rotation in a circle about a fixed centre, as could be seen in the daily rotation of the fixed stars. The different spheres in which the seven 'planets,' Moon, Sun, Venus, Mercury, Mars, Jupiter and Saturn, were set, revolved with different uniform velocities such as would represent the observed movements of those bodies. Each of the spheres had its own Intelligence or 'soul' which was the source of its motion.

It was not only the cosmogony and cosmology of Thierry and his contemporaries that was influenced by the *Timæus*; it also coloured their physical and physiological conceptions. They followed Plato in holding that within the universe there was no void. Space was a *plenum*, that is, it was full. Movement could therefore take place only by each body pushing that next to it away and taking its place in a kind of vortex. Such functions as respiration and digestion Plato had explained as purely mechanical processes based on the movement of fiery and other particles. Sensations he supposed were produced by the motion of particles in the organs of the body. The particular quality of a member of any given class of sensations, for instance a particular colour or sound, he explained by the inherent qualities of the external object, depending on its structure, which in turn brought about particular physical processes in the special sense organ concerned. Vision he supposed to take place by means of a visual ray emitted from the eye to the object, colours being attributed to different sized fire particles which streamed off objects and interacted with this ray. Sounds he connected with the motion of air particles,

49

though he ignored the role of the ear drum. Different tastes and odours he related to the character of the particles composing or coming off the objects. Many of these views were taken over by the natural philosophers of the 12th century. The direct influence of the *Timæus* is seen in their belief in the indestructibility of matter and their explanation of the properties of the elements in terms of the motion of particles in which velocity and solidity were complementary, for no body could be set in motion without the corresponding reaction against a motionless body. One 12th-century philosopher, William of Conches, adopted a form of atomism based on a combination of Plato's ideas with those of Lucretius.

This Platonist conception of the universe continued to exert an important influence until the days of Roger Bacon, who, as a young man, sometime about 1245, lectured on physics from the point of view of the Chartres school. But Chartres itself was already in touch with the schools of translators who were working on Arabic and Greek texts at Toledo and in southern Italy, and it was in Chartres that the Ptolemaic astronomy and Aristotelian physics were first welcomed. Thus, because of developments within the thought of Western Christendom itself, the system of ideas represented by Adelard's nephew was beginning to appear a little antiquated by the middle of the 12th century. It was soon to be replaced by ideas developed by those who followed his uncle in the study of the Arabs and the Greeks and the pursuit of natural causes.

2

THE RECEPTION OF
GRECO-ARABIC SCIENCE IN
WESTERN CHRISTENDOM

THE new science which began to percolate into Western Christendom in the 12th century was largely Arabic in form, but it was founded on the works of the ancient Greeks. The Arabs preserved and transmitted a large body of Greek learning, and what they added to its content themselves was perhaps less important than the change they made in the conception of the purpose for which science ought to be studied.

The Arabs themselves acquired their knowledge of Greek science from two sources. Most of it they eventually learned directly from the Greeks of the Byzantine Empire, but their knowledge of it came also at second hand from the Syriac-speaking Nestorian Christians of Eastern Persia. During the 6th and 7th centuries Nestorian Christians at their centre of Jundishapur translated a number of important works of Greek science, chiefly on logic and medicine, into Syriac, which had replaced Greek as the literary language of Western Asia since the 3rd century. For a time after the Arab conquest Jundishapur continued to be the first scientific and medical centre of Islam, and there Christian, Jewish and other subjects of the Caliphs worked on the translation of texts from Syriac into Arabic. Damascus and Baghdad also became centres for this work, and at Baghdad in the early 9th century translations were also made direct from Greek. By the 10th century nearly all the texts of Greek science that were to become known to the Western world were available in Arabic.

Gradually the learning which had been amassed by the Arabs began to penetrate into Western Christendom as trading relations slowly revived between Christendom and Islam. By the 9th

century, towns such as Venice, Naples, Bari and Amalfi, later
joined by Pisa and Genoa, were carrying on trade with the Arabs
of Sicily and the eastern Mediterranean. In the 11th century a
Benedictine monk of Monte Cassino, Constantine the African,
was sufficiently familiar with Arab scientific work to be able to
produce a paraphrase of Galen and Hippocrates from the medical
encyclopædia of the Persian doctor Haly Abbas (d. 994). In the
12th century Adelard of Bath is known to have travelled in south
Italy and even in Syria and, at the beginning of the 13th century,
Leonardo Fibonacci of Pisa was in North Africa on business where
he acquired his knowledge of Arabic mathematics.

The chief centres from which the knowledge of Arabic and
ultimately of Greek science spread were Sicily and Spain. Toledo
fell to Alfonso VI in 1085 and towards the middle of the 12th
century became, under the patronage of its archbishop, the
Spanish centre of translation from Arabic into Latin. The very
great number of versions attributed to such a man as Gerard of
Cremona suggest the existence of some sort of school. The names
of known translators, Adelard of Bath, Robert of Chester, Alfred
of Sareshel (the Englishman), Gerard of Cremona, Plato of Tivoli,
Burgundio of Pisa, James of Venice, Eugenio of Palermo, Michael
Scot, Hermann of Carinthia, William of Moerbeke, bear witness
to the wide European character of the movement, as do their own
words, of which Adelard's are typical, to the feeling of excitement
with which the earlier scholars set out to gain Arab learning for
the Latin West. Many of the translations were works of collaboration, for example, the work of the Hispano-Jew John of Seville,
who translated the Arabic into vernacular Castilian which was
then rendered into Latin by Dominicus Gundissalinus. The earliest
known Latin-Arabic glossary is contained in a Spanish manuscript
dating, perhaps, from the 12th century, but the work of translat-
ing Greek and Arabic texts was severely hampered by the difficulty of mastering the languages involved, the intricacy of the
subject matter, and the complicated technical terminology. The
translations were often literal, and often words whose meanings
were imperfectly understood were simply transliterated from
their Arabic or Hebrew form. Many of these words have survived

down to the present day as, for example, alkali, zircon, alembic (the upper part of a distilling vessel), sherbet, camphor, borax, elixir, talc, the stars Aldebaran, Altair and Betelgeuse, nadir, zenith, azure, zero, cipher, algebra, algorism, lute, rebeck, artichoke, coffee, jasmine, saffron and taraxacum. Such new words went to enrich the vocabulary of medieval Latin, but it is not surprising that these literal translations sprinkled with strange words provoked complaints from other scholars. Many of the translations were revised in the 13th century either with a better knowledge of Arabic or directly from the Greek.

In Sicily, in addition to translations from the Arabic, there appeared some of the earliest translations to be made directly from Greek. Conditions in the island specially favoured the exchange of ideas between Arabic, Greek and Latin scholars. Until the fall of Syracuse in 878 it had been dominated by Byzantium. Then it passed under the control of Islam for nearly two hundred years until 1060, when a Norman adventurer with a small following captured Messina and was so successful in establishing his power that by 1090 the island had become a Norman kingdom in which Latin, Greek and Moslem subjects lived together in conditions even more favourable than those in Spain for the work of translation.

From the end of the 12th century to the end of the 13th the proportion of translations made direct from Greek to those made at second hand through Arabic gradually increased, and in the 14th century translation from Arabic practically ceased when Mesopotamia and Persia were overrun by the Mongols. It is said that from the end of the 12th century shiploads of Greek manuscripts came from Byzantium to Italy, though few can be definitely traced as having done so. When the Fourth Crusade was diverted against Byzantium, which was captured by the Westerners in 1204, one result was that many manuscripts passed to the Latin West. In 1205 Innocent III exhorted masters and scholars of Paris to go to Greece and revive the study of literature in the land of its birth, and Philip Augustus founded a college on the Seine for Greeks of Byzantium to learn Latin. Later in the 13th century Roger Bacon wrote a Greek grammar and, at the suggestion of St Thomas Aquinas, William of Moerbeke revised and completed the translation of almost all Aristotle's works in a literal version made direct from the Greek.

By the middle of the 12th century the number of new works added to the store of European learning included Aristotle's *logica nova*, that is, the *Analytics* and the other logical works not in the long familiar translations by Boethius which were included in the *logica vetus*, Euclid's *Elements*, *Optics* and *Catoptrics*, and Hero's *Pneumatica*. From the 12th century dates also the Latin version of the pseudo-Euclidean *De Ponderoso et Levi*, a work of Greek origin which provided both Islam and Western Christendom with their knowledge of specific gravity, the lever and the balance. In the third quarter of the century translations were made of the principal works of Ptolemy, Galen and Hippocrates, of which the popular versions came chiefly from Spain, and of Aristotle's *Physics* and *De Cœlo* and other *libri naturales* and the first four books of the *Metaphysics*. Early in the 13th century the complete *Metaphysics* was translated, and about 1217 appeared his *De Animalibus* comprising the *History*, *Parts* and *Generation of Animals*. At the same time was translated the pseudo-Aristotelian *Liber de Plantis* or *de Vegetabilibus*, which modern scholarship has attributed to the 1st-century B.C. Nicholas of Damascus and which, apart from the herbals deriving from Dioscorides and pseudo-Apuleius, was the most important single source of later medieval botany. By the middle of the 13th century nearly all the important works of Greek science were available in Latin translations (Table 1). Some works were also translated into vernacular languages, in particular into Italian, Castilian, French and, later, English. Of all these works the most influential were those of Aristotle, who had provided the basis for the natural philosophy of the Greeks and of the Arabs and was now to perform the same function for Western Christendom. The translations of his writings were chiefly responsible for the shift in educational interest that took place round about 1200 towards philosophy and science, which John of Salisbury (*c.* 1115–80) had complained were even in his time being preferred to the poetry and history of his youth.

TABLE I

THE PRINCIPAL SOURCES OF ANCIENT SCIENCE IN WESTERN CHRISTENDOM BETWEEN 500 AND 1300 A.D.

Author	Work	Latin translator and language of original of translation	Place and date of Latin translation
(i) EARLY GREEK AND LATIN SOURCES			
Plato (428–347 B.C.)	*Timaeus* (first 53 chapters)	Chalcidius from Greek	4th century
Aristotle (384–22 B.C.)	Some logical works (*logica vetus*)	Boethius from Greek	Italy 6th century
Dioscorides (1st century A.D.)	*Materia Medica*	from Greek	by 6th century
Anon.	*Physiologus* (2nd century A.D. Alexandria)	from Greek	5th century
Anon.	Various technical *Compositiones*	from Greek sources	earliest MSS 8th century
Lucretius (c. 95–55 B.C.)	*De Rerum Natura* (known in excerpts from 9th century; full text recovered 1417)		

Author	Work	Latin translator and language of original of translation	Place and date of Latin translation
Vitruvius (1st century B.C.)	De Architectura (known in 12th century)		
Seneca (4 B.C.–65 A.D.)	Quæstiones Naturales		
Pliny (23–79 A.D.)	Historia Naturalis		
Macrobius (fl. 395–423)	In Somnium Scipionis		
Martianus Capella (5th century)	Satyricon, sive De Nuptiis Philologiæ et Mercurii et de Septem Artibus Liberalibus		
Boethius (480–524)	Works on the liberal arts, particularly mathematics and astronomy, and commentaries on the logic of Aristotle and Porphyry		
Cassiodorus (c. 490–580)	Works on the liberal arts		
Isidore of Seville (560–636)	Etymologiarum sive Originum; De Natura Rerum		
Bede (673–735)	De Natura Rerum		

(2) ARABIC SOURCES FROM C. 1000

		from Arabic	12th and 13th centuries
Jabir ibn Hayyan corpus (written 9th–10th centuries)	Various chemical works		
Al-Khwarizmi (9th century)	*Liber Ysagogarum Alchorismi* (arithmetic)	Adelard of Bath from Arabic	early 12th century
	Astronomical tables (trigonometry)	Adelard of Bath from Arabic	1126
	Algebra	Robert of Chester from Arabic	Segovia 1145
Alkindi (d. c. 873)	*De Aspectibus; De Umbris et de Diversitate Aspectuum*	Gerard of Cremona from Arabic	Toledo 12th century
Thabit ibn Qurra (d. 901)	*Liber Charastonis* (on the Roman balance)	Gerard of Cremona from Arabic	Toledo 12th century
Rhazes (d. c. 924)	*De Aluminibus et Salibus* (chemical work)	Gerard of Cremona from Arabic	Toledo 12th century
	Liber Continens (medical encyclopædia)	Moses Farachi from Arabic	Sicily 1279
	Liber Almansoris (medical compilation based on Greek sources)	Gerard of Cremona from Arabic	Toledo 12th century

Author	Work	Latin translator and language of original of translation	Place and date of Latin translation
Alfarabi (d. 950)	*Distinctio super Librum Aristotelis de Naturdi Auditu*	Gerard of Cremona from Arabic	Toledo 12th century
Haly Abbas (d. 994)	Part of *Liber Regalis* (medical encyclopædia)	Constantine the African (d. 1087) and John the Saracen from Arabic	South Italy 11th century
	Liber Regalis	Stephen of Antioch from Arabic	c. 1127
pseudo-Aristotle	*De Proprietatibus Elementorum* (Arabic work on geology)	Gerard of Cremona from Arabic	Toledo 12th century
Alhazen (c. 965–1039)	*Opticæ Thesaurus*	from Arabic	end of 12th century
Avicenna (980–1037)	Physical and philosophical part of *Kitab al-Shifa* (commentary on Aristotle)	Dominicus Gundissalinus and John of Seville, abbreviated from Arabic	Toledo 12th century
	De Mineralibus (geological and alchemical part of *Kitab al-Shifa*)	Alfred of Sareshel from Arabic	Spain c. 1200
	Canon (medical encyclopædia)	Gerard of Cremona from Arabic	Toledo 12th century

		Michael Scot from Arabic	Toledo 1217
... (... century)	Liber Astronomie (Aristotelian concentric system)		
Averroës (1126–98)	Commentaries on Physica, De Cœlo et Mundo, De Anima and other works of Aristotle	Michael Scot from Arabic	early 13th century
Leonardo Fibonacci of Pisa	Liber Abaci (first complete account of Hindu numerals)	using Arabic knowledge	1202
(3) GREEK SOURCES FROM c. 1100			
Hippocrates and school (5th, 4th centuries B.C.)	Aphorisms	Burgundio of Pisa from Greek	12th century
	Various treatises	Gerard of Cremona and others from Arabic William of Moerbeke from Greek	Toledo 12th century after 1260
Aristotle (384–22 B.C.)	Posterior Analytics (part of logica nova)	Two versions from Greek	12th century
		from Arabic	
	Meteorologica (Book 4)	Henricus Aristippus from Greek	Toledo 12th century Sicily c. 1156

Author	Work	Latin translator and language of original of translation	Place and date of Latin translation
	Physica, De Generatione et Corruptione, Parva Naturalia, Metaphysica (1st 4 books), De Anima Meteorologica (Books 1-3), Physica, De Cælo et Mundo, De Generatione et Corruptione	from Greek	12th century
		Gerard of Cremona from Arabic	Toledo 12th century
Aristotle (384-22 B.C.)	De Animalibus (Historia animalium, De partibus animalium, De generatione animalium trans. into Arabic in 19 books by el-Batric, 9th century) Almost complete works	Michael Scot from Arabic	Spain c. 1217-20
		William of Moerbeke, new or revised translations from Greek	c. 1260-71
Euclid (c. 330-260 B.C.)	Elements (15 books, 13 genuine)	Adelard of Bath from Arabic	early 12th century
		Hermann of Carinthia from Arabic	12th century
		Gerard of Cremona from Arabic	Toledo 12th century

Author	Work	Translator / notes	Place	Date
	Optica and Catoptrica *Optica* *Data*	from Greek from Arabic from Greek several revisions; revision of Adelard's version by John Campanus of Novara	probably Sicily	c. 1254
Apollonius (3rd century B.C.)	*Conica*	perhaps Gerard of Cremona from Arabic (of this translation only a short fragment of Book 1 is now extant, as the introduction to Alhazen's *De Speculis Comburentibus*; but Book 2 was known to Witelo in the 13th century)		12th century
Archimedes (287–12 B.C.)	*De Mensura Circuli* Complete works (except for the *Sandreckoner*, the *Lemmata*, and the *Method*)	Gerard of Cremona from Arabic William of Moerbeke from Greek	Toledo	12th century 1269
Diocles (2nd century B.C.)	*De Speculis Comburentibus*	Gerard of Cremona from Arabic	Toledo	12th century

Author	Work	Latin translator and language of original of translation	Place and date of Latin translation
Hero of Alexandria (1st century B.C.?)	Pneumatica	from Greek	Sicily 12th century
	Catoptrica (attributed to Ptolemy in Middle Ages)	William of Moerbeke from Greek	after 1260
pseudo-Aristotle	Mechanica (Mechanical Problems) Problemata	from Greek Bartholomew of Messina from Greek	early 13th century Sicily c. 1260
	De Plantis or De Vegetabilibus (now attributed to Nicholas of Damascus, 1st century B.C.)	Alfred of Sareshel from Arabic	Spain, probably before 1200
pseudo-Euclid	Liber Euclidis de Ponderoso et Levi (statics)	from Arabic	12th century
Galen (129–200 A.D.)	Various treatises	Burgundio of Pisa from Greek	c. 1185
	Various treatises	Gerard of Cremona and others from Arabic	Toledo 12th century

	Anatomical treatises	from Greek	14th century
		from Greek	
Ptolemy (2nd century A.D.)	*Almagest*	from Greek Gerard of Cremona from Arabic	Sicily c. 1160 Toledo 1175
	Optica	Eugenius of Palermo from Arabic	c. 1154
Alexander of Aphrodisias (fl. 193–217 A.D.)	Commentary on the *Meteorologica*	William of Moerbeke from Greek	13th century
	De Motu et Tempore	Gerard of Cremona from Arabic	Toledo 12th century
Simplicius (6th century A.D.)	Part of commentary on *De Cælo et Mundo*	Robert Grosseteste from Greek	13th century
	Commentary on *Physica*	from Greek	13th century
	Commentary on *De Cælo et Mundo*	William of Moerbeke from Greek	1271
Proclus (410–85 A.D.)	*Physica Elementa* (*De motu*)	from Greek	Sicily 12th century

Of the actual knowledge from the stores of Greek learning which was transmitted to Western Christendom by the Arabs, together with some additional observations and comments of their own, some of the most important was the new Ptolemaic astronomy (below, pp. 93–103) and its associated trigonometry. This reached Europe through the translations of works by such writers as al-Khwarizmi, al-Battani (d. 929) and al-Fargani (9th century), but these authors had, in fact, added nothing new to the principles on which the astronomical system of Ptolemy had been founded. In the 12th century al-Bitruji, known in Latin as Alpetragius, revived the astronomical work of Aristotle, though here again the Arab did not advance much on the Greek. What the Arabs did do was to improve observing instruments and construct increasingly accurate tables for both astrological and nautical purposes. The most famous of these were prepared in Spain, which, from the time of the editing of the *Toledan Tables*, or *Canones Azarchelis*, by al-Zarqali (d. *c.* 1087) to their replacement under the direction of King Alfonso the Wise (d. 1284) by others compiled in the same town, had been a centre of astronomical observation. The meridian of Toledo was for a long time the standard of computation for the West and the *Alfonsine Tables* remained in use till the 16th century.

The second body of fact transmitted from Greek works to Western Christendom by way of Arabic translations and commentaries was the work on medicine and to this Arab scholars, though they did not modify the underlying principles much, added some valuable observations. Most of the information was derived from Hippocrates and Galen and became enshrined in the encyclopædias of Haly Abbas (d. 994), Avicenna (980–1037) and Rhazes (d. *c.* 924),* but the Arabs were able to add some new minerals such as mercury and a number of other drugs to the

* Cf. the Prologue to Chaucer's *Canterbury Tales* (ll. 429 *et seq.*).

Wel knew he the olde Esculapius
And Deiscorides, and eek Rufus,
Old Ypocras, Hal, and Galien;
Serapion, Razis, and Avicen;
Averrois, Damascien, and Constantyn;
Bernard, and Gatesden, and Gilbertyn.

predominantly herbal *materia medica* of the Greeks, and Rhazes was able to contribute original observations such as in his diagnoses of smallpox and measles.

The original Arabic contribution was more important in the study of optics and perspective for here, though the works of Euclid, Hero and Ptolemy had dealt with the subject, Alkindi (d. *c.* 873) and Alhazen (*c.* 965–1039) made a big advance on what had been known by the Greeks. Alhazen discussed, among other things, spherical and parabolic mirrors, the *camera obscura*, lenses and vision.

In the field of mathematics the Arabs transmitted to Western Christendom a body of most valuable knowledge which had never been available to the Greeks, though here the Arabs were not making an original contribution but simply making more widely known the developments in mathematical thought which had taken place among the Hindus. Unlike the Greeks, the Hindus had developed not so much geometry as arithmetic and algebra. The Hindu mathematicians, of whom Aryabhata (b. 476 A.D.), Brahmagupta (b. 598 A.D.) and later Bhaskara (b. 1114) were the most important, had developed a system of numerals in which the value of a digit was shown by its position. They knew the use of zero, they could extract square and cube roots, they understood fractions, problems of interest, the summation of arithmetical and geometrical series, the solution of determinate and indeterminate equations of the first and second degrees, permutations and combinations and other operations of simple arithmetic and algebra. They also developed the trigonometrical technique for expressing the motions of the heavenly bodies and introduced trigonometrical tables of sines.

The most important mathematical idea which the Arabs learnt from the Hindus was their system of numerals, and the adoption of this system in Christendom was one of the great advances in European science. The great merit of this system, which is the basis of the modern system, was that it contained the symbol for zero and that any number could be represented simply by arranging digits in order, the value of a digit being shown by its distance from zero or from the first digit on the left. It had very great advantages over the cumbrous Roman system. In the system which the Arabs learnt from the Hindus the first three numbers were represented by one, two and three strokes respectively, and after that 4, 5, 6, 7, 9 and possibly 8 were probably derived from the initial letters for the words representing those numbers in Hindu. The Arabs had learnt something of this system from the Indians, with whom they had considerable trading relations, as early as the 8th century, and a complete account of it was given by al-Khwarizmi in the 9th century. It was from a corruption of his name that the system became known in Latin as 'algorism'.

The Hindu numerals were introduced into Western Europe gradually from the 12th century onwards. It was symptomatic of the practical trend among mathematicians that al-Khwarizmi himself, whose work on algebra was translated by Adelard of Bath, said (as he is rendered by F. Rosen in his edition, *The Algebra of Mohammed ben Musa*, London, 1831, p. 3) that he had limited his activities

to what is easiest and most useful in arithmetic, such as men constantly require in cases of inheritance, legacies, partition, law-suits, and trade, and in all their dealings with one another, or where the measuring of lands, the digging of canals, geometrical computation, and other objects of various sorts and kinds are concerned.

Later in the same century Rabbi ben Ezra, by origin a Spanish Jew, fully explained the Arabic system of numeration and specially the use of the symbol O. Gerard of Cremona reinforced this exposition. But it was not till the 13th century that the Arabic system became widely known. This was due very largely to the work of Leonardo Fibonacci, or Leonardo of Pisa (d. after 1240). Leonardo's father was a Pisan merchant who was sent out to Bugia in Barbary to take charge of a factory, and there Leonardo seems to have learnt a great deal about the practical value of Arabic numerals and about the writings of al-Khwarizmi. In 1202 he published his *Liber Abaci* in which in spite of the name he fully explained the use of the Arabic numerals. He was not personally interested in commercial arithmetic and his work was highly theoretical, but after his time Italian merchants generally came gradually to adopt the Arabic, or Hindu, system of numeration.

During the 13th and 14th centuries the knowledge of Arabic numerals was spread through Western Christendom by the popular almanacs and calendars. As the dates of Easter and of the other festivals of the Church were of great importance in all religious houses, one almanac or calendar was usually found in these establishments. A calendar in the vernacular had been produced in France as early as 1116, and Icelandic calendars go back to about the same date. This knowledge was reinforced in the West by popular expositions of the new system by mathematical writers such as Alexander of Villedieu and John Holywood or, as he was called, Sacrobosco, and even in a surgical treatise by Henry of Mondeville. About the middle of the 13th century two Greek mathematicians explained the system to Byzantium. The Hindu numerals did not immediately drive out the Roman ones and in fact until the middle of the 16th century Roman numerals were widely used outside Italy, but by 1400 Arabic numerals were widely known and generally understood at least among men of learning.

*

A sphere in which the Arabs made a most important and original contribution to the history of European science was that of alchemy, magic and astrology, and this was partly because of the special approach to the problems of the world of nature that characterized a strong tradition of Arabic thought. The primary question in this tradition was not what aspects of nature most vividly illustrated the moral purposes of God nor what were the natural causes which would provide a rational explanation of the facts described in the Bible or observed in the world of everyday experience, but what knowledge would give power over nature. Inquirers wanted to find 'the Elixir of Life, the Philosopher's Stone, the Talisman, the Word of Power and the magic properties of plants and minerals', and the answer to their questions was alchemy. It was partly a desire to share this rumoured magical power that sent the first translators on their journeys from

Western Christendom to such centres of Arabic learning as Toledo or Sicily. Some scholars believed that the ancient Greeks had had such knowledge and had hidden it in cryptic writings and alchemical symbols.

Latin works written before the 12th century had been by no means free from magic and astrology (see above, pp. 35–7), but among the Arabs and those Latins who, after the 12th century, were influenced by their works magic and astrology fruited tropically. No sharp distinction was drawn between natural science and the magical or occult, for physical and occult causes were recognized as equally able to be responsible for physical phenomena. This point of view was expressed clearly by Alkindi, the 9th-century Arab Neoplatonist, in his work *On Stellar Rays* or *The Theory of the Magic Art*. The stars and terrestrial objects, and also the human mind through the potency of words suitably uttered, exerted 'influence' by means of rays whose ultimate cause was celestial harmony. The effects of the rays were supposed to vary with the configurations of the heavenly bodies. 'Celestial virtue' was admitted as a cause by nearly all the Latin writers of the 13th century, and Roger Bacon's famous discussion of the old theory of the 'multiplication of species' has been variously interpreted as a contribution to physics and as an account of astral influences going in straight lines. 'Marvels', when not the work of demons and therefore evil, might be produced by occult virtues resident in certain objects in nature, that is by 'natural magic'. The distinction between evil and natural magic was maintained by a number of scholastic natural philosophers, such as William of Auvergne, Albertus Magnus and Roger Bacon. The discovery of occult virtues was one of the principal objects of many medieval experimenters. The alchemists hoped to transmute metals, prolong human life, perhaps gain sufficient power over nature to discover the names of those who had committed theft or adultery.

Well down into the 16th century the connexion between magic and one side of experimentation was close. In the 17th century Bishop Wilkins, one of the founders of the Royal Society, was to include, in a book on mechanics called *Mathematicall*

Magick, being borne through the air by birds and by witches among recognized methods of human transport. But even in the 13th century many of the natural philosophers of Western Christendom were able to a large extent to keep magic out of their work. Albertus Magnus, Petrus Peregrinus and Rufinus are examples of observers and experimenters who did so. Roger Bacon (c. 1219–92), though he certainly derived the desire for power over nature as the object of his science, as well as his belief in the occult virtues of stones and herbs, from the ambitions and assumptions of magic, yet developed a view of scientific experiment which was perhaps the earliest explicit statement of the practical conception of the aims of science. With him the practical European genius was beginning to transform the magic of the Arabian Nights into the achievements of applied science.

In his *Opus Tertium,* chapter 12, Roger Bacon, having discussed speculative alchemy, goes on to say:

But there is another alchemy, operative and practical, which teaches how to make the noble metals and colours and many other things better and more abundantly by art than they are made in nature. And science of this kind is greater than all those preceding because it produces greater utilities. For not only can it yield wealth and very many other things for the public welfare, but it also teaches how to discover such things as are capable of prolonging human life for much longer periods than can be accomplished by nature ... Therefore this science has special utilities of that nature, while nevertheless it confirms theoretical alchemy through its works.

In his view of what could usefully be achieved by science Roger Bacon had the outlook common to his age: the future would be read more accurately than in the stars; the Church would overcome Antichrist and the Tartars. The ultimate value of science was to be in the service of the Church of God, the community of the faithful: to protect Christendom through power over nature and to assist the Church in her work of evangelizing mankind by leading the mind through scientific truth to the contemplation of the Creator already revealed in theology, a contemplation in which all truth was one. But in his conception of the immediate use of science he had almost the outlook of the 19th century.

'Next,' he says of agriculture in his *Communia Naturalium*,

comes the special science of the nature of plants and all animals, with the exception of man who by reason of his nobleness falls under a special science called medicine. But first in the order of teaching is the science of animals which precede man and are necessary for his use. This science descends first to the consideration of every kind of soil and the productions of the earth, distinguishing four kinds of soil, according to their crops; one soil is that wherein corn and legumina are sown; another is covered with woods; another with pastures and heaths; another is garden ground wherein are cultivated trees and vegetables, herbs and roots, as well for nutriment as for medicine. Now this science extends to the perfect study of all vegetables, the knowledge of which is very imperfectly delivered in Aristotle's treatise *De Vegetabilibus*; and therefore a special and sufficient science of plants is required, which should be taught in books on agriculture. But as agriculture cannot go on without an abundance of tame animals; nor the utility of different soils, as woods, pastures and heaths, be understood, except wild animals be nurtured; nor the pleasure of man be sufficiently enhanced, without such animals; therefore this science extends itself to the study of all animals.

Bacon did not develop this discussion of the sciences, but his appreciation of the potential usefulness of such studies is clear. His prophecies about the submarine and the motor car in the *Epistola de Secretis Operibus*, chapter 4, are well known and are another example of the extremely practical turn which he gave to scientific studies.

Machines for navigation can be made without rowers so that the largest ships on rivers or seas will be moved by a single man in charge with greater velocity than if they were full of men. Also cars can be made so that without animals they will move with unbelievable rapidity; such we opine were the scythe-bearing chariots with which the men of old fought. Also flying machines can be constructed so that a man sits in the midst of the machine revolving some engine by which artificial wings are made to beat the air like a flying bird. Also a machine small in size for raising or lowering enormous weights, than which nothing is more useful in emergencies. For by a machine three fingers high and wide and of less size a man could free himself and his friends from all danger of prison and rise and descend. Also a machine can easily be made by which one man can draw a thousand to himself by violence against their wills, and attract other things in like manner. Also machines can be made for walking in the sea and rivers, even to the bottom without danger. For Alexander the Great employed such, that he might see the secrets of the deep, as Ethicus the astronomer tells. These machines were made in antiquity and they have certainly been made in our times, except possibly the flying machine which I have not seen nor do I know anyone who has, but I know an expert who has thought out the way to make one. And such things can be made almost without limit, for instance, bridges across rivers without piers or supports, and mechanisms, and unheard of engines.

Bacon also urged the reform of the calendar, as had his master Robert Grosseteste, and described how this might be done, though, in fact, his suggestions had to wait until 1582 to be put into practice. In the later Middle Ages, however, scientific knowledge as distinguished from merely technical rule of thumb led to improvements in building and surgery and to the invention of spectacles, though in general the practical mastery over nature which the Arabs had sought through magic was not achieved for many centuries.

Most influential of all the contributions of Greco-Arabic learning to Western Christendom was the fact that the works of Aristotle, Ptolemy and Galen constituted a complete rational system explaining the universe as a whole in terms of natural causes. Aristotle's system included more than natural science as it is understood in the 20th century. It was a complete philosophy embracing all existence from 'first matter' to God. But just because of its completeness the Aristotelian system aroused much opposition in Western Christendom where scholars already had an equally comprehensive system based on the facts revealed in the Christian religion.

Moreover, some of Aristotle's theories were themselves directly contrary to Christian teaching. For instance, he held that the world was eternal and this obviously conflicted with the

Christian conception of God as creator. His opinions were doubly suspect because they reached the West accompanied by Arab commentaries which stressed their absolutely determinist character. The Arab interpretation of Aristotle was strongly coloured by the Neoplatonic conception of the chain of being stretching from first matter through inanimate and animate nature, man, the angels and Intelligences to God as the origin of all. When such commentators as Alkindi, Alfarabi, Avicenna and particularly Averroës (1126–98) introduced from the Mohammedan religion into the Aristotelian system the idea of creation, they interpreted this in such a way as to deny free will not only to man but even to God himself. According to them the world had been created not directly by God but by a hierarchy of necessary causes starting with God and descending through the various Intelligences which moved the celestial spheres, until the Intelligence moving the moon's sphere caused the existence of a separate Active Intellect which was common to all men and the sole cause of their knowledge. The form of the human soul already existed in this Active Intellect before the creation of man, and after death each human soul merged again into it. At the centre of the universe within the sphere of the moon, that is, in the sublunary region, were generated a common fundamental matter, *materia prima*, and then the four elements. From the four elements were produced, under the influence of the celestial spheres, plants, animals and man himself.

Several points in this system were entirely unacceptable to the philosophers of Western Christendom in the 13th century. It denied the immortality of the individual human soul. It denied human free will and gave scope for the interpretation of all human behaviour in terms of astrology. It was rigidly determinist, denying that God could have acted in any way except that indicated by Aristotle. This determinism was made even more repulsive to Christian thinkers by the attitude of the Arab commentators and especially of Averroës, who declared:

Aristotle's doctrine is the sum of truth because his was the summit of all human intelligence. It is therefore well said that he was created and given us by Divine Providence, so that we should know what it is possible to know.

Some allowance may be made here for oriental exaggeration, but this point of view came to be characteristic of the Latin Averroïsts. For them the world emanated from God as Aristotle had described it, and no other system of explanation was possible. Nor indeed did the extreme theological rationalism of this interpretation do violence to Aristotle's own thought. Aristotle had based his whole approach to natural science and metaphysics on the claim that it was possible to discover by reason the essence of things and of God, causing the regularities observed in the world. Plato's approach was the same, although differing over both the processes of reason involved and the nature of the essences discovered. In the brilliant *tour de force* in book 2, chapter 3 of *De Cælo*, Aristotle gave every support to the Averroïst interpretation of his cosmology. He set out to prove that his system was not only in fact true, but was necessarily true, for it alone followed from God's discovered essence and perfection. All things, he argued, existed for the ends they served and the perfection to which they tended. God's activity was eternal, and so therefore must be the motion of the heaven, which was a divine body. 'For that reason the heaven is given a circular body whose nature it is to move always in a circle . . .; and earth is needed because eternal movement in one body necessitates eternal rest in another.' Similarly he argued that the whole actual world was necessarily as he described and explained it, and could not in the nature of things be otherwise.

The situation that arose in the 13th century over Aristotelianism was not in fact the first experience that Christian thinkers had had of the encounter between Greek rationalism and the Christian revelation. Extensively discussed by both the Greek and the Latin Fathers, it was St Augustine's analysis of the relation between reason and faith that established the point of departure for the treatment of the problem in the medieval West. In a well-known passage in the *Confessions*, St Augustine described how he began, as a young man, following the method of

Greek philosophy, to search by reason alone for the intelligibility of existence, and how his conversion led him to believe that this could be grasped immediately through Christian faith. But he asserted emphatically that it was impossible to believe something that was not understood, and that to hold that it was sufficient to believe in Christian doctrine without aspiring to an understanding of it was to ignore the true end of belief. St Augustine thus added the content of revelation to that of experience as constituting the world given, the data whose nature and relations the Christian philosopher must try to elucidate by rational inquiry.

The most obvious and most influential problem arising from this programme was the relation that was to be understood between the two sources of data, revelation and experience, Scripture and science. This St Augustine tackled in his commentary *De Genesi ad Litteram*, whose exegetic methods Galileo was to expound. Beginning with the basic principle that truth is self-consistent, St Augustine ruled out *a priori* any real contradiction between the data of revelation, true by definition in the light of their source, and the equally true data of observation and conclusions of true reasoning. When there was an apparent contradiction, this must arise from our misunderstanding of the true meaning of the conflicting statements, and those, he said, may not be the literal meanings, whether in Scripture or in science. The problem of interpretation that arose in this way brought out in the first place the conflict between the Hebrew cosmology of Scripture, with its flat earth and domed sky, and the globe and spheres of the Greek astronomers. In dealing with such questions St Augustine insisted on clearly distinguishing the primary moral and spiritual purpose of Scripture from its accidental references to the physical world. The latter, as he agreed with St Jerome, were made according to the judgement of their time and not according to the literal truth. Though in no sense himself a natural scientist, St Augustine's writings show a competent knowledge of astronomy and other sciences, which he urged his fellow Christians to master. When they discussed natural questions, the shape and motion of the heavens or the earth, the elements, the nature of animals, plants and minerals, he was especially anxious that Christians should not jeopardize the acceptance of the fundamental doctrines of religion by making absurd statements, allegedly in according with Christian writings, about questions properly decidable by natural science alone. Certainly St Augustine enjoyed confirming Scripture from science, but his policy was to save Scripture from apparent falsification by observation and reason and, without prejudice, to hand over purely natural questions to scientific investigation. 'In points obscure and remote from our sight,' he wrote in book 1, chapter 18 of *De Genesi ad Litteram*, 'if we come to read anything in Holy Scripture that is, in keeping with the faith in which we are steeped, capable of several meanings, we must not, by obstinately rushing in, so commit ourselves to any one of them that, when perhaps the truth is more thoroughly investigated, it rightly falls to the ground and we with it.' Galileo was to quote this passage in urging the same reasonable policy upon his contemporaries, but the history of the problem, especially as it entered with medieval Aristotelianism, from the 13th century down to the time of Galileo himself, shows that while such a policy may help to reduce the area of conflict, it certainly does not provide automatic answers to all the questions arising between the cosmologies of reason and of revelation. Believing in the primary importance of the Christian apostolate, St Augustine himself went on to assert firmly in chapter 21 of *De Genesi ad Litteram* that should philosophers teach anything that is 'contrary to our Scriptures, that is to Catholic faith, we may without any doubt believe it to be completely false, and we may by some means be able to show this.'

It was the search for means of finding an accommodation between Aristotelian philosophy and Christian theology that gave rise to the most interesting and critical developments in philosophy and in the conception of science in the 13th and 14th centuries. After some initial hesitation and embarrassment, three main general lines of policy began to clarify. The first was that of the Latin Averroïsts who took their stand on the irrefutable rational truth of Aristotelian philosophy and accepted the consequence that Christian theology was irrational or even un-true. There seems little doubt that such a man as Jean de Jandun (d. 1328) was in fact an unbeliever, but that he lightly concealed his unbelief with an irony akin to Voltaire's. A threat equally to Christian theology and to empirical science, it was the Christian doctrine of the absolute freedom of God's will that formed the basis of the critique of Averroïst rationalism, although the criti-cism was pressed by means of logical arguments about the possibility of there being any necessary rational truths about the world. A moderate policy, for example that of Thomas Aquinas, was to accept the rationality of science but to deny that any necessity could be discovered in God. The extreme policy of the defenders of faith appeared in the 14th century, when for example William of Ockham eliminated the threat from reason by denying the rationality of the world altogether, and reduced its order to a dependence of fact on God's inscrutable will.

The 13th century saw first the categorical condemnation of Aristotle, but by the middle of the century he had been accepted as the most important of the philosophers. In 1210 in Paris, which by the end of the 12th century had already replaced Chartres as the greatest centre of learning in France, the provincial ecclesi-astical council prohibited the teaching of Aristotle's views on natural philosophy or of commentaries on them. In 1215 a similar decree was issued against reading his metaphysical and natural works, although this did not forbid private study and applied only to Paris; in fact, lectures on these works were announced in the University of Toulouse. Other prohibitions were issued subsequently, but it was not possible to enforce them. In 1231 Pope Gregory IX appointed a commission to revise some of the natural works and in 1260 William of Moerbeke began his translation from the Greek. Eminent teachers like Albertus Magnus (1193/1206–80) and his pupil Thomas Aquinas (1225–74) expounded the works of Aristotle, and by 1255 his most impor-tant metaphysical and natural works were set by the faculty of arts in Paris as a subject for examination. At Oxford the 'new Aristotle' made its first entry without attracting official opposi-tion. Lecturing on the new logical and physical treatises had begun by the first decade of the 13th century, but it was the influence of an inspiring philosopher and teacher, Robert Grosseteste (*c.* 1168–1253), that really established medieval Oxford's enduring interest in the new science, mathematics and logic, as well as in languages and Biblical scholarship. As *Magister Scholarum* or Chancellor of the University in 1214, perhaps the first to hold this office, as lecturer to the Franciscan house in Oxford, and from 1235 as Bishop of Lincoln, in which diocese Oxford lay, Grosseteste remained the chief ornament and guide of the University's early years (see Vol. II, p. 27 *et seq.*).

Throughout the Middle Ages there were various schools of thought about the Aristotelian system of the universe. In the 13th century in Oxford the Franciscan friars, who tended to remain loyal to the main features of Augustinianism, such as the theory of knowledge and of universals, accepted some important Aristotelian additions in the explanation of such natural phenomena as the movements of the heavenly bodies, but were often hostile to Aristotle's influence as a whole. At the same time in Oxford there was an interest characteristic of another aspect of Franciscan thought, exemplified by Roger Bacon, who was keenly alive to the mathematical, physical, astronomical and medical learning of Aristotle and the Arabs and less concerned with their metaphysical views. In the University of Paris black-habited Dominicans, such as Albertus Magnus and Thomas Aquinas, accepted the main principles of Aristotle's physics and philosophy of nature (see below, p. 80 *et seq.*) but rejected his absolute determinism. A fourth school of thought, represented by Siger of Brabant, who was a thorough-going Averroïst, accepted an entirely determinist interpretation of the universe. Yet a fifth group was in the Italian universities of Salerno, Padua and Bologna where theological matters counted for less than in England or France and where Aristotle and the Arabs were studied principally for their medical learning.

Those mainly responsible for making Aristotle acceptable to the Christian West were, besides Grosseteste, Albertus Magnus and Thomas Aquinas. The main problem confronting them was the relation between faith and reason. In his attempt to resolve this difficulty Albertus based himself, like St Augustine, on two certainties: the realities of revealed religion and the facts that had come within his own personal experience. Albertus and St Thomas did not regard Aristotle as an absolute authority as Averroës had done, but simply as a guide to reason. Where Aristotle, either explicitly or as interpreted by Arab commentators, conflicted with the facts either of revelation or of observation he must be wrong: that is, the world could not be eternal, the individual human soul must be immortal, both God and man must enjoy the exercise of free will. Albertus also corrected him on a number of points of zoology (see below, pp. 160–66 *et seq.*). But Albertus and more definitely St Thomas realized, as Adelard of Bath had done a century earlier, that theology and natural science often spoke of the same thing from a different point of view, that something could be both the work of Divine Providence and the result of a natural cause. In this way they established a distinction between theology and philosophy which assigned to each its appropriate methods and guaranteed to each its own sphere of action. There could be no real contradiction between truth as revealed by religion and truth as revealed by reason. Albertus said that it was better to follow the apostles and fathers rather than the philosophers in what concerned faith and morals. But in medical questions he would rather believe Hippocrates or Galen, and in physics Aristotle, for they knew more about nature.

The determinist interpretation of Aristotle's teaching associated with the commentaries of Averroës was condemned by the Bishop of Paris, Etienne Tempier, in 1277, and his example was followed in the same year by the Archbishop of Canterbury, John Pecham. In so far as this affected science it meant that in northern Christendom the Averroïst interpretation of Aristotle was banished. The Averroïsts retired to Padua where their views gave rise to the doctrine of the double truth, one for faith and another, perhaps contradictory, for reason. This condemnation of determinism has been taken by some modern scholars, notably by Duhem, as marking the beginning of modern science. The teaching of Aristotle was to dominate the thought of the later Middle Ages, but with the condemnation of the Averroïst view that Aristotle had said the last word on metaphysics and natural science, the bishops in 1277 left the way open for criticism which would, in turn, undermine his system. Not only had natural philosophers now through Aristotle a rational philosophy of nature, but because of the attitude of Christian theologians they were made free to form hypotheses regardless of Aristotle's authority, to develop the empirical habit of mind working within a rational framework, and to extend scientific discovery.

3

THE SYSTEM OF
SCIENTIFIC THOUGHT IN THE
THIRTEENTH CENTURY

I. EXPLANATION OF CHANGE AND
CONCEPTION OF SUBSTANCE

THE system of scientific thought that was made known to Western Christians in the 13th century came to them, in a collection of translations from Greek and Arabic, as a complete and for the most part coherent whole. This was a system of rational explanations in power and range quite beyond anything known earlier in the Latin West, and one the general principles of which in fact dominated European science until the 17th century. This Greco-Arabic scientific system was not, however, received merely passively in the 13th century. The activity of mind that had shown itself in the 12th century in the fields of philosophy and technology was applied in the 13th century to detect, and to endeavour to resolve, the contradictions that existed within the Aristotelian system itself, between Aristotle and other authorities such as Ptolemy, Galen, Averroës and Avicenna, and between the various authorities and observed facts. The Western scholars were trying to make the natural world intelligible and they seized upon the new knowledge as a wonderful, but not final, illumination of mind and as a starting-point for further investigation.

The object of this chapter is to describe this 13th-century scientific system, indicating the historical sources of each part of it, and to give a brief account of the additions of fact and modifications in detail made to it during the century or more after its introduction. These changes were made for the most part as a result of the gradual extension of observation, experiment and the use of mathematics, and they were made possible to a large extent as a result of habits acquired in technology. It will be necessary to mention some aspects of medieval technics in this chapter, but it is convenient to reserve a fuller discussion of the subject for Chapter 4. The experimental and mathematical methods were themselves the result of a definite theory of science, a theory postulating definite methods of investigation and explanation. Some indication of this theory of science will be necessary to make much of what follows in this chapter intelligible, and many of the additions of fact to be described were the results of its application. A fuller treatment of medieval scientific method will be reserved for Volume II, Chapter 1. Besides additions of fact, other important changes were made in the 13th-century scientific system as a result of criticism from a purely theoretical point of view. Those affecting the details of the system will be described in this chapter, but those involving criticism of its fundamental principles will also be reserved for Volume II, Chapter 1. These more radical criticisms derived for the most part from the change in the theory of science that began during the 13th century, a change which led to the conception that the experimental and mathematical methods should extend over the whole field of natural science. This was the conception that brought about the revolution in science that culminated in the 17th century, and so, while the present chapter is concerned with the 13th century scientific system itself, the two following chapters will give an account of the two traditions of scientific activity, the technical and the theoretical, that made possible the transition to the new scientific system of the 17th century.

For the system of scientific thought accepted in the 13th century to become fully intelligible to the 20th-century reader it is necessary to understand the nature of the questions it was designed to answer. The natural philosopher of the 13th century regarded the investigation of the physical world as part of a single philosophical activity concerned with the search for reality and truth. The purpose of his inquiry was to discover the enduring and intelligible reality behind the changes undergone by the world perceived through the senses. Exactly the same problem had, in fact, been the main preoccupation of the philosophers of ancient Greece, and their answer had been the conception of 'substance' as the identity persisting through change. This identity Plato had recognized as the universal idea or 'form' of a thing (see note on p. 46) and Aristotle had adopted this idea of form from Plato, though modifying it in various important ways. What makes sense of the general principles of 13th-century science, then, is the realization that the aim of scientific investigation was to define the substance underlying and causing observed effects.

It was Aristotle's conception of substance that dominated 13th-century science and this is best understood by starting with his conception of the methodological structure of science. According to Aristotle, scientific investigation and explanation was a twofold process, the first inductive and the second deductive. The investigator must begin with what was prior in the order of knowing, that is, with facts perceived through the senses, and he must proceed by induction to include his observations in a; generalization which would eventually lead him to the universal form. These forms were the intelligible and real identity persisting through and causing the changes observed; therefore, though most remote from sensory experience, they were 'prior in the order of nature.' The object of the first, inductive, process in natural science was to define these forms, for such a definition could then become the starting-point for the second process, that by which the observed effects were shown by deduction to follow from this definition and so were explained by being demonstrated from a prior and more general principle which was their cause. The definition of the form was necessary before demonstration could begin because all effects were considered to be attributes of some substance, and the cause of an effect was shown when the effect could be predicted as an attribute of a defined substance. This definition would include everything about a thing, its colour, size, shape, relations with other things, etc. No attribute, that is, no effect or event, could exist unless it inhered in some substance and, indeed, attributes and substance could be separated only in thought. In other words, it was essential to Aristotle's conception of scientific demonstration to reduce all science to subject-predicate propositions. This conception was to prove a most inconvenient framework in which to deal with many scientific problems that can be properly expressed only in terms of numerical relations, especially rates of change, as the early history of the modern science of motion was to show.

Aristotle described the process by which the form was discovered by induction as a process of abstraction from the data provided by the senses, and he held that there were three degrees of abstraction which revealed three different aspects of reality. These corresponded to the sciences of physics (or natural science), mathematics and metaphysics. The subject-matter of physics was change and motion as exemplified in material things; the subjects considered by mathematics were abstracted from change and from matter but could exist only as attributes of material things; metaphysics considered immaterial substances with an independent existence. This classification raised the important question of the role of mathematics in explaining physical events. The subjects considered by mathematics, Aristotle said, were abstract, quantitative aspects of material things. Therefore, different mathematical sciences had subordinate to them certain physical sciences, in the sense that a mathematical science could often provide the reason for facts observed in those material things, facts provided by physical science. Thus geometry could provide the reason for, or explain, facts provided by optics and astronomy, and the study of arithmetical proportions could explain the facts of musical harmony. Mathematics, being an abstraction from change, could provide no knowledge of the *cause* of the observed events. It could merely describe their mathematical aspects. In other words, mathematics alone could never prove an adequate definition of the substance, or, as it was called in the Middle Ages, 'substantial form,' causing the change, because it dealt only with mathematical attributes; an adequate definition of the causal substance could be reached only by considering all attributes, non-mathematical as well as mathematical. And, Aristotle held, qualitative differences as between flesh and bone, between one colour and another, and between motion up, down and in a circle could not be reduced simply to differences in geometry. This was a point on which Aristotle differed from Plato and the Greek atomists.

The science that considered the cause of change and motion, then, was physics. In putting forward an explanation that would account intelligibly for change as such, Aristotle attempted to avoid the defects which he considered had vitiated the explanations advanced by some of his predecessors (see note on p. 46). Thus, as he did not accept Plato's theory that the forms of physical things existed apart from them, he could not explain change by the aspiration of physical things to be like their eternal archetypes. Nor could he fall back on the atomist's explanation of change by the rearrangement of atoms in the void, for he could see no reason why there should be any limit to the division of physical bodies (or, indeed, of any other *continuum* whether of space, time or motion). For him the conception of void, which the atomists had considered as emptiness, or 'non-being,' between the atoms of substance, or 'being,' was untenable. 'Non-being' could not exist. His own explanation of change was to introduce between being and non-being a third state of potentiality, and to say that change was the actualization of attributes potential within any given physical thing because of that thing's nature. Attributes which were at any given time potential were as much part of a substance as those which were at that time actual.

Aristotle's conception of the cause of changes can be understood by considering his discussion of *phúsis* or 'nature'; the science of *physics* was in fact the science of 'nature' in a specific and technical sense. In a famous passage in the *Laws* (book 10), Plato had accused philosophers of leading youth away from the gods by teaching that this beautiful universe, the regularity of the heavenly movements, and the human soul arose 'not because of any mind, nor because of God, nor by art (*téchne*), but, as we may say, by nature (*phúsis*) and chance (*túche*).' Plato insisted that the material universe was the product of the *art* of God. In his *Physics* (book 2), Aristotle made this three-fold division of the causes of nature his starting-point for the rehabilitation of *phúsis* and of the naturalistic theories of the pre-Socratic philosophers.

84

The older philosophers, he said, had correctly applied the term *phúsis* to the matter out of which things are made, but by applying it only to the matter they had made it impossible to account for the cause of change. He therefore introduced the notion of *phúsis* as an active principle, whose spontaneous activity was the intrinsic source of the characteristic and regular behaviour of each natural thing; a natural spontaneity, he asserted with characteristic empiricism, that was directly observable in all the bodies we experienced. To *phúsis* or 'nature' as the spontaneous intrinsic source of change and of rest Aristotle applied the term 'form'; and for him 'matter' connoted the passive principle implying the *potentiality* to receive the attributes that became actual with the form. The 'nature' of a thing in both senses implied a substance in which it inhered; the form and the matter determined the 'nature' of the substance. A thing behaved 'naturally' when it behaved according to the nature of its intrinsic principle of change; otherwise its behaviour was forced upon it, as when a stone was thrown upwards against its natural tendency to go down. Such unnatural motion was known as forced or compulsory or violent.

This two-fold conception of nature as both active and passive involved further problems and distinctions, which were discussed and developed by the scholastics. In the first place, a 'natural' potentiality implied one that intrinsically tended towards perfect natural realization; in other words, it implied movement towards an end. The operation of final causality in this way was essential to the whole Aristotelian conception of nature. The substance or 'substantial form,' as not only the intelligible aspect of a thing but also as the active source of its behaviour, had a natural tendency or 'appetite' to fulfil its nature or form whether, as in living things, this might be the adult into which the embryo developed, or, as in the terrestrial elements, the 'natural' place in the universe (see below, pp. 89–91). To realize that end was to possess positively the natural potentialities in their full actuality, and so the 'nature' was the active source not only of natural change or motion but also of natural completion, or rest.

But it is clear that the passive potentialities could be actualized only by some active agent, a principle which Aristotle expressed in the well-known axiom of the *Physics*, book 7: 'Everything that is moved must be moved by something' (see below, pp. 91, 125–6, Vol. II, p. 61 *et seq.*).

This agent, according to Aristotle, might be an *intrinsic* source of activity, as in living things, which moved themselves, and in the spontaneous natural activity of inanimate substances, as when a stone fell naturally to the ground. And between these two intrinsic sources the important distinction (the subject of considerable discussion) had to be made between the movements that were actively initiated by the 'souls' of living things, and the movements of inanimate things, which were not initiated by them but simply took place given the necessary external conditions. The 'soul' of a living thing was thus the 'efficient cause' of its movement; the efficient cause of the spontaneous activity of an inanimate thing, on the other hand, was strictly speaking the agency that originally brought it into existence as that kind of thing.

Or the agent might be something *external* to the body in change, as in forced or 'violent' motion, for example when a boy threw a ball, or in natural change when potential attributes were made actual by contact with another substance in which they were actual, as wood burned when brought into contact with an already burning fire.

In accordance with these considerations, Aristotle distinguished four kinds of cause, of which two, the material and formal, defined the substance undergoing change, and two, the efficient and final, actually produced the movement. What he meant by each of these causes is clearly seen in his view of the generation of animals. He believed that the female contributed no germ or ovum but simply the passive matter out of which the embryo was made. This passive matter was the material cause. The efficient cause was the father whose seed acted as the instrument which started the process of growth. The male seed also carried to the female matter the specific form that determined what kind of animal the embryo would become. This form was the formal cause, and since it represented the final adult state to which development would proceed, it was also the final cause.

All changes of any kind whatever, of colour, growth, spatial relations or any other attribute, Aristotle explained on the same principle that attributes which had been potential became actual. Even the property of suffering eclipses was an attribute of the moon to be included in the definition of the moon's substance. And it is important to remember that the term 'motion' (*motus*) applied not only to change of place – local motion – but to change of any kind whatever.

Aristotle distinguished four different kinds of change: (1) local motion, (2) growth or decrease, (3) alteration or change of quality, and (4) substantial change which took place during the process of generation and corruption. In the first three the perceptible identity of the thing persisted throughout; in the fourth the changing thing lost all its old attributes and in fact became a new substance. This he explained by pushing the idea of substance as the persisting identity to its ideal limit and conceiving it to be pure potentiality, capable of determination by any form and having no independent existence. This pure potentiality was called by the medieval scholastics *materia prima*. Any given material thing could then be thought of as *materia prima* determined by a form.

An opinion of Aristotle's based on his idea of substance, which was to be the subject of some very important discussions in the 14th century, was his conception of infinity. He held that infinity, whether of the division or addition of time or of material things, was a potentiality implying that there was no assignable limit to the process concerned. Time, whether past or future, could have no limit assigned to it, so that the duration of the universe was infinite. But every material thing had a definite size determined by its form. In discussing the possibility of the existence of an infinitely small body he said that the division of material things could potentially go on to infinity, but this potentiality could never become actual. An infinitely large material body, that is, an infinite universe, was, however, not even a potentiality, for the universe was a sphere of finite size.

The conception of substance as developed by Aristotle was the basis of all natural explanation from the 13th to the 17th century, but even after Aristotle's ideas had become generally accepted they were still subject to criticism from Neoplatonists. The main difference between Aristotle's view of matter and that which had been put forward by such Neoplatonists as St Augustine and Erigena concerned the nature of the substance that persisted through substantial change. For these Neoplatonists this persisting substance was actual extension, that is pure potentiality (or *materia prima*) determined by spatial dimensions, and this underlay all other attributes of material things; for Aristotle it was simply pure potentiality. With some Arab philosophers such as Avicenna, al-Ghazzali and Averroës and the Spanish Rabbi Avicebron, the Neoplatonic theory of matter took the form that every material thing possessed a 'common corporeity' making it extended, and according to Avicebron, this corporeity was continuous through the universe. The importance of this theory was that it introduced the possibility of extending the use of mathematics to the whole of natural science, as is shown, for example, in the speculations of Robert Grosseteste (*c.* 1168–1253). He identified the common corporeity of the Neoplatonists with light, which had the property of dilating itself from a point in all directions and was thus the cause of all extension. He held that the universe arose from a point of light which by auto-diffusion generated the spheres of the four elements and the heavenly bodies and conferred on matter its form and dimensions. From this he concluded that the laws of geometrical optics were the foundation of physical reality and that mathematics was essential to the understanding of nature.

This problem of the use of mathematics in explaining the physical world remained, in fact, one of the central methodological problems, and was in many ways *the* central problem of natural science down to the 17th century. Even in the 12th century a prominent place had been given to mathematics in the teaching of the seven liberal arts. For example, Hugh of St Victor, author of one of the most important classifications of science relying on purely Latin sources, insisted that mathematics should be learnt before physics and was essential to it, even though mathematics was concerned with entities abstracted from physical things. Essentially the same view was taken by Dominicus Gundissalinus, author of the most influential 12th-century classification of science based on Arabic sources, most of his ideas being taken from Alfarabi. The mid-13th-century writer Robert Kilwardby (d. 1279), who used both Latin and Arabic sources in his classification of science, also paid special attention to the relation of the mathematical disciplines to physics, but maintained the Aristotelian distinction between them. Geometry, he said, abstracted from all aspects of physical bodies except the formal cause and considered that alone; the consideration of moving causes was the prerogative of physics. With the gradually increasing success of mathematics in solving concrete problems in physical science the reality of the sharp line that Aristotle had drawn between the two disciplines came slowly to be doubted. Indeed, from one point of view, the whole history of European science from the 12th to the 17th century can be regarded as a gradual penetration of mathematics (combined with the experimental method) into fields previously believed to be the exclusive preserve of 'physics.'

2. COSMOLOGY AND ASTRONOMY

Not only Aristotle's theory of substance and fundamental principles of scientific explanation, but also his ideas on the actual structure of the universe, dominated European thought in the 13th century. Aristotle's cosmology was founded on naïve observation and common sense and it had two fundamental principles: (1) that the behaviour of things was due to qualitatively determined forms or 'natures,' and (2) that the totality of these 'natures' was arranged to form a hierarchically ordered whole or cosmos. This cosmos or universe had many features in common with that of Plato and of the astronomers Eudoxus and Callippus (4th century B.C.); all had taught that the cosmos was spherical and consisted of a number of concentric spheres, the outermost being the sphere of the fixed stars, with the earth fixed in the centre, but Aristotle's system showed various refinements.

Aristotle's cosmos was a vast but finite sphere centred upon the centre of the earth and bounded by the sphere of the fixed stars, which was also the 'prime mover,' the *primum movens* of the scholastics, the originative source of all movement within the universe (Plate 1). Fixed in the centre of the universe was the spherical earth, and surrounding it concentrically were a series of spheres like the skins of an onion. First came the spherical envelopes of the other three terrestrial elements, water, air and fire, respectively. Surrounding the sphere of fire were the crystalline spheres in which were embedded and carried round, respectively, the moon, Mercury, Venus, the sun, Mars, Jupiter, and Saturn, which made up the seven 'planets.' Beyond the sphere of the last planet came that of the fixed stars and beyond this last sphere – nothing.

Thus each kind of body or substance in this universe had a place that was natural to it and a natural motion in relation to that place. Movement took place with reference to a fixed point, the centre of the earth as the centre of the universe, and there was a qualitative difference between the movements of a given body in one direction rather than in another in relation to that point. The natural behaviour of bodies depended, therefore, on their actual place within the universe as well as on the substance of which they were composed. The sphere of the moon divided the universe into two sharply distinct regions, the terrestrial and the celestial. Bodies in the former region were subject to all the four kinds of change, and the kind of motion that was natural to them was in a straight line towards their natural place in the sphere of the element of which they were composed. To be in that place was the fulfilment of their 'nature' and there they could be at rest. This was why to someone standing on the earth some substances, for example, fire, whose natural place was upwards, seemed light, while other substances, for example, earth, whose natural place was downwards seemed heavy. These directions represented an absolute up and down and the tendency to move up or down depended on the nature of the substance of which a particular body was composed. Plato had postulated the same kind of movement, but had explained it rather differently.

From the sphere of the moon outwards bodies were composed of a fifth element or 'quintessence' which was ingenerable and incorruptible and underwent only one kind of change, uniform motion in a circle, being a kind of motion that could persist eternally in a finite universe. This kind of motion Plato had said to be the most perfect of all, and his dictum that the motions of the heavenly bodies must be resolved into uniform circular motions was to dominate astronomy until the end of the 16th century. The spheres of the planets and stars composed of this celestial fifth element revolved round the central earth.

Motion as such Aristotle had regarded, as he did all other kinds of change, as a process of becoming from a state of privation and potentiality (in the case of motion this is rest) to actualization. Such a process of change required a cause and so every moving body required for its movement either an intrinsic principle of motion, as in the case of natural movement, or an external mover, as in the case of unnatural or forced movement (see above, pp. 84–7; below, pp. 125–6, Vol. II, p. 61 *et seq.*). As Aristotle expressed it in the *Physics*, book 8, chapter 4:

If then the motion of all things that are in motion is either natural or unnatural and violent, and all things whose motion is violent and unnatural are moved by something, and something other than themselves, and again all things whose motion is natural are moved by something – both those that are moved by themselves [*scil.* living things] and those that are not moved by themselves (*e.g.* light things and heavy things, which are moved either by that which brought the thing into existence as such and made it light and heavy, or by that which released what was hindering and preventing it); then all things that are in motion must be moved by something.

This conclusion, and the distinction between light and heavy elements, Aristotle justified by the direct observations that bodies do come to rest if nothing continues to push them and that when released on earth some bodies do rise while others fall.

The velocity of movement was supposed to be proportional to the moving force or power.

With the celestial spheres the original source of motion was the *primum movens* which moved itself, Aristotle said somewhat obscurely, by 'aspiring' to the eternal unmoved activity of God, eternal uniform circular motion being the nearest approach to that state possible for a physical body. In order that this 'aspiration' might be possible he had to suppose that this sphere had some sort of 'soul.' Indeed, he assigned 'souls' to all the spheres and this was the origin of the hierarchy of Intelligences or Motors that Arabic Neoplatonism was to attach to the spheres. Motion was communicated from the *primum movens* to the sphere inside it, the *primum mobile*, and so to the inner spheres.

With terrestrial bodies moving towards their natural place in the sublunary region the mover was their own 'nature' or 'substantial form,' whose fulfilment it was to be at rest in that place. There bodies would remain eternally were it not for two other agencies: the generation of substances outside their natural places by the transformation of one terrestrial element into another, and 'violence' due to an external mover. The ultimate cause of both these agencies was in fact the same, namely, the progress of the sun on its annual course round the ecliptic which, it was thought, produced periodic transformations of the elements into one another (see Fig. 3). The movement of these newly generated elements towards their natural place was the principal source of 'violence' in the regions through which they passed.

This generation of elements outside their natural places was also the reason why the actual bodies found in the terrestrial region were usually not pure but made up of a compound of four elements: for example, ordinary fire or water were compounds in which the pure elements with those names, respectively, dominated. And further, the annual motion of the sun was held to be the cause of the seasonal generation, growth and decay of plants and animals. Thus all change and motion in the universe! was ultimately caused by the *primum movens*. The remainder of this chapter will be devoted to a description of the explanations given, during the hundred years or more after the introduction of the Aristotelian system in the 13th century, of the different kinds of change observed in the different parts of the universe, beginning with astronomy and passing through the sciences concerned with the intermediate regions to finish with biology.

*

Thirteenth-century astronomy was, on its theoretical side, concerned mainly with a debate as to the relative merits of physical as compared with mathematical theories in accounting for the phenomena. The former were represented by Aristotle's explanations, the latter by Ptolemy's, and in fact the debate itself was an old one : it began in later Greek times and had passed through various vicissitudes among the Arabs. Both the Aristotelian and the Ptolemaic systems were known in the Latin West by the beginning of the 13th century. The controversy was opened by Michael Scot with his translation, in 1217, of the 12th-century Arab astronomer Alpetragius' *Liber Astronomiæ*, in which the Arab had tried to revive the waning fortunes of Aristotelian astronomy in face of the more accurate system of Ptolemy.

All ancient and medieval systems of astronomy were based on Plato's dictum that the observed movements of the heavenly bodies must be resolved into uniform circular motions. Aristotle had attempted to account for the facts by means of his system of concentric spheres. The geometrical refinements of this system he in fact took from Eudoxus and Callippus, but he tried to give physical reality to the geometrical devices with which they had accounted for the irregular movements, the 'stations' and 'retrogradations,' of the seven 'planets' as observed against the background of the fixed stars. Following Eudoxus and Callippus he postulated for each planet not one but a system of spheres (Fig 1). He supposed then that the axis of the sphere actually bearing the planet was itself attached to the inside of another rotating sphere, whose own axis was attached to a third, and so on. By postulating a sufficient number of spheres, arranging the axes at suitable angles and varying the rates of rotation, he was able to represent the observations to a fair approximation. The motion of the *primum movens* was communicated to the inner spheres mechanically by the contact of each sphere with that inside it, and this contact also prevented a void from occurring between the spheres. In order to prevent any sphere associated with a particular planet from imposing its motion on all the spheres beneath it, he introduced, between each planet's system and that of the next planet, compensating spheres which rotated about

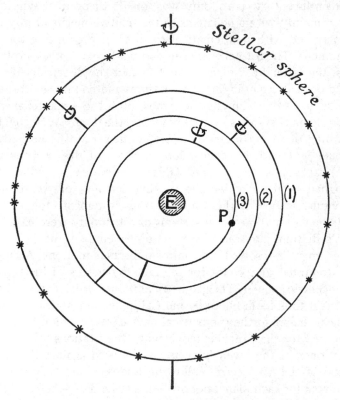

Fig. 1 The system of concentric spheres used by Eudoxus and Aristotle to explain the motion of a planet P, with the axes all placed to the plane of the paper. Supposing P to be Saturn, the outermost sphere is the stellar sphere, which rotates daily from East to West about a North-South axis passing through the centre of the stationary earth E, and accounts for the *daily* rising and setting of the 'fixed' stars and of the planet. Inside this sphere come three spheres which account for the *annual* motions of the

the same axis and with the same period as one of the planetary spheres of the external system, but in the opposite direction. In all there were 55 planetary and compensating spheres and one stellar sphere, making a total of 56. Further spheres were added after Aristotle's time: the *primum movens* was separated as a further sphere outside that of the fixed stars; and some medieval writers, such as William of Auvergne (*c.* 1180–1249), placed beyond the *primum movens* yet another sphere, an immobile Empyrean, the abode of the saints.

One weakness of all the systems postulating that the universe was made up of a series of concentric spheres was that they had to assume that the distance of each heavenly body from the earth was invariable. This assumption made it impossible to account, simply by means of orbits, for a number of obvious phenomena, in particular the variations in the apparent brightness of the planets and in the apparent diameter of the moon, and the fact that solar eclipses were sometimes total and sometimes annular. Later Greek astronomers had tried to account for these facts by devising different systems, and the most important of these was that devised by Hipparchus in the 2nd century B.C. and later adopted by Ptolemy in the 2nd century A.D. This was the most

---

planet against the background of the fixed stars on the stellar sphere. Sphere (1) accounts for the planet's annual motion from West to East in a great circle round the zodiac. Its axis is inclined to that of the stellar sphere at about the same angle as the zodiacal band makes with the celestial equator, which is the equator of the stellar sphere (cf. Plate I, and Fig. 3). Spheres (2) and (3) account for the annual stations and retrogradation of the planet and also for some change in latitude. The poles of sphere (2) lie in the zodiacal band; i.e., the equator of sphere (1). Spheres (2) and (3) rotate in opposite directions in equal times, with their speeds of rotation and the angle of inclination of the axis of (3) to that of (2) varying with different planets. The planet P is carried on equator of sphere (3). The combined motion of (2) and (3) causes P to describe a curve known in Greek as the 'hippopede' (or 'hobble'), in fact a spherical lemniscate, which bears a fair resemblance to the apparent looping motion of the planets. The spheres of the next planet, Jupiter, would come inside the sphere carrying Saturn, the outermost sphere in Jupiter's set repeating the daily rotation of the stellar sphere. Inside Jupiter would come the spheres of the remaining planets.

accurate and the most widely accepted astronomical system known in classical antiquity and in the Arab world. The treatise of Ptolemy's in which it was described, which in the Middle Ages went under the Latinized Arabic name of *Almagest*, was to dominate astronomical thinking on its mathematical side in the West until the time of Copernicus.

The astronomical system expounded by Ptolemy in the *Almagest* has often been interpreted, for example by Heath and Duhem, as merely a geometrical device by means of which to account for the observed phenomena, or 'save the appearances'. But it cannot be said without qualification that this was Ptolemy's own view. The assumptions with which he began, that the heavens are spherical in form and rotate as a sphere, that the earth is at the centre of this sphere and is motionless, that the heavenly bodies move in circles, were certainly not *arbitrary* assumptions, for without trying to prove them absolutely, he tried to make them as plausible as possible. In fact in his choice of assumptions and hypotheses Ptolemy was guided, it seems, by the reverse of arbitrary criteria, but rather by physical and metaphysical considerations which he regarded as empirically sound. In its physical conceptions his system was in fact basically Aristotelian, and Aristotle's influence may be read directly in the preface to the *Almagest*; but he supported it by empirical arguments showing as close a reliance on immediate direct observation as Aristotle himself. A good example is his discussion of the earth's immobility and his rejection of the hypothesis of Aristarchus of Samos, who had supposed that the earth spun on its axis and revolved round the sun, while the sun and the fixed stars remained at rest. This hypothesis, Ptolemy admitted, might allow the motions of the stars to be calculated with greater mathematical simplicity, but it was so completely contradicted by the immediate appearances that it had to be rejected. He never seems to have thought of explaining the immediate appearances away.

The mathematical aspect of his system Ptolemy based on the principle attributed to Plato, writing: 'We believe that it is the necessary purpose and aim of the mathematician to show forth all the appearances of the heavens as products of regular and circular motions.' This principle again he attempted to justify by an appeal to direct observation, for all the heavenly bodies do in fact return in their motions to their original positions. But it must be admitted that in his planetary theory Ptolemy used geometrical devices which subordinated questions of the actual physical paths of the planets, and the accepted principles of Aristotelian physics, to accuracy of calculation. This is the source of his reputation as a scientific 'conventionalist'.

Ptolemy used two different devices. The first, the device of the movable eccentric, was to suppose that the planets moved in a circle about a point, not at the centre of the earth, but somewhere on a line joining the centre of the earth with the sun. This eccentric point moved in a circle round the earth. The second device, that of the epicycle, which Ptolemy showed to be the geometrical equivalent of the movable eccentric, was to suppose that a planet moved in a circle about a centre, which itself moved in another circle of which the centre was stationary with respect to the earth, although not necessarily on it (Fig. 2). The inner circle was known as the *deferent* and the outer one, carrying the planet, as the *epicycle*. There was no limit to the number of circles that could be postulated in order to 'save the appearances'. In one point, in allowing that the linear velocity of the centre of the epicycle about the deferent might not be uniform, Ptolemy departed from Plato's dictum that only uniform circular motions could be used, if this is to be taken as applying to lateral velocities; but he made some attempt to preserve orthodoxy by making the angular velocity uniform about a point, the equant, inside the deferent though not necessarily at its centre.

By suitable arrangements of circles Ptolemy was able, in most respects, to give a very accurate description of the movements or 'appearances' of the planets. To account for another observed phenomenon, the precession of the equinoxes (that is, the steady increase in longitude of a star while its latitude remains unaltered), he supposed in a further work, his *Hypotheses of the Planets*, that outside the stellar sphere (the 8th in his system) there was a 9th sphere which imparted to the stellar sphere its diurnal motion from East to West, while the stellar sphere itself,

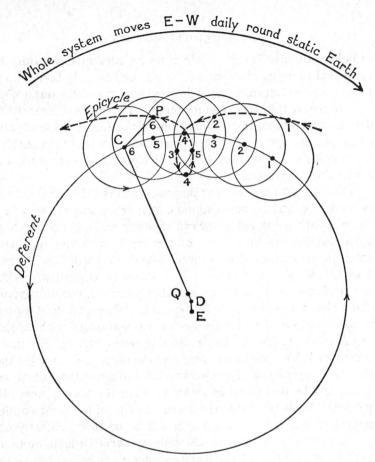

Fig. 2 The geometrical device of the epicycle in Ptolemy's system for the motion of a planet P. The daily motion of all the planets is produced by the whole system sharing in the daily rotation of the stellar sphere from East to West. The irregular journey of each planet round the ecliptic as viewed from the earth (cf. Fig. 3) is reproduced by supposing that while the planet travels round the epicycle centred at C, this centre itself travels round the deferent centred at D. The latter point does not coincide with the centre of the earth E; and C moves uniformly neither about D nor E but about a third point, the equant Q, chosen precisely to help to reproduce the apparently non-uniform velocity of the planet. The planet has a uniform angular velocity about Q, so that CQ sweeps out equal angles in equal times. The planet's irregular path through the fixed stars, as seen from the earth on successive days, is then described by the broken line, the positions of P on this path which correspond to those of C on the deferent being indicated by the numbers. The planet's 'stations', when to

98

together with the spheres of the planets, rotated slowly in the opposite direction with respect to the 9th sphere. When later the *primum movens* was separated from the stellar sphere, it became a distinct 10th sphere beyond this 9th. An erroneous theory that the equinoxes did not precess but oscillated, or 'trepidated', about an average position was advanced in the 9th century by an Arabic astronomer, Thabit ibn Qurra, and gave rise to considerable controversy in Europe from the 13th to the 16th century.

When the natural philosophers and astronomers of Western Christendom were confronted with the choice between the 'physical' system of Aristotle and the 'mathematical' system of Ptolemy, they at first hesitated, as in fact had the Greeks and Arabs before them. Ptolemy himself, after writing his mathematical *Almagest*, in which he was prepared to treat certain astronomical theories as convenient geometrical devices of which the simplest consistent with the appearances was to be used, later wrote another book, the *Hypotheses of the Planets*. In this he attempted to produce a system which would give a physical, mechanical explanation of the heavenly movements (Plate 2). The Ptolemaic system was quickly recognized early in the 13th century as being the best geometrical device for 'saving the appearances', and practical astronomers favoured it as being the only system capable of serving as the basis of numerical tables. But a desire was felt for a system that would both 'save the

---

an observer on the earth it seems to have stopped moving, occur about positions 3 and 5; and between 3 and 5 it seems to move backward, this being called a 'retrogradation'. With the upper planets, Mars, Jupiter, and Saturn, which are placed outside the sun, the centre C of the epicycle revolves round the deferent with the proper period of each planet's orbit round the ecliptic, while the planet revolves annually on its epicycle, this accounting for the annual irregularities (cf. Plate I, and Vol. II, Fig. 6). With the lower planets, Mercury and Venus, it is the epicycle that accounts for the proper period and the deferent that accounts for the annual irregularities. The sun itself revolves in an eccentric circle without an epicycle. The lower planets and the moon require somewhat more complex devices than the upper planets. In all cases accuracy could be increased by adding further spheres, giving additional components of motion to the deferent, or by adding further epicycles, the planet being carried on the outermost epicycle.

appearances' and also describe the 'real' paths of the heavenly bodies and account for the cause of their movements. Regarded in this light Ptolemy's eccentrics and epicycles were clearly inadequate themselves, and his system conflicted with a number of important principles of the only adequate system of physics known, that of Aristotle. In the first place, the theory of epicycles was not compatible with Aristotle's theory that circular motion required a solid fixed centre round which to revolve; and secondly, Ptolemy's explanation of precession would require that the stellar sphere had two different motions at the same time, and this was in conflict with Aristotle's principle that contradictory attributes cannot inhere in the same substance at the same time. Yet, although Ptolemy's system had these serious physical defects from which Aristotle's astronomical system was free, the latter was clearly inferior as a mathematical description of the observed facts.

The attitude taken to this dilemma in the second half of the 13th century seems to have been determined by that taken in the 6th century A.D. by the Greek philosopher Simplicius in his commentaries on Aristotle's *Physics* and *De Cælo*. A passage quoted by Simplicius in book 2, chapter 2 of his commentary on the *Physics* both expressed the failure of the Greeks after Aristotle to find a single system uniting astronomy with physics and dynamics, and clearly announced that the discovery of the true physical system was the ultimate objective of the science of motion in the heavens as on the earth. He wrote:

Alexander carefully quotes a certain passage by Geminus taken from his Summary of the *Meteorologica* of Posidonius; Geminus' account, which is inspired by the views of Aristotle, is as follows:

'It is the business of physical inquiry to consider the substance of the heaven and the stars, their force and quality, their coming into being and their destruction; it is in a position even to prove the facts about their size, shape and arrangement. Astronomy, on the other hand, does not attempt to speak of anything of this kind, but proves the arrangement of the heavenly bodies by considerations based on the view that the heaven is a real Cosmos, and, further, it tells us of the shapes and sizes and distances of the earth, sun, and moon, and of eclipses and conjunctions of the stars, as well as of the quality

and extent of their movements. Accordingly, as it is connected with
the investigation of quantity, size, and quality of form or shape, it
naturally stood in need, in this way, of arithmetic and geometry. The
things, then, of which alone astronomy claims to give an account it
is able to establish by means of arithmetic and geometry. Now in
many cases the astronomer and the physicist will propose to prove
the same point, e.g. that the sun is of great size, or that the earth is
spherical; but they will not proceed by the same road. The physicist
will prove each fact by considerations of essence or substance, of
force, of its being better that things should be as they are, or of
coming into being and change; the astronomer will prove them by the
properties of figures or magnitudes, or by the amount of movement
and the time that is appropriate to it. Again, the physicist will, in
many cases, reach the cause by looking to creative force; but the
astronomer when he proves facts from external conditions, is not
qualified to judge of the cause, as when, for instance, he declares the
earth or the stars to be spherical. Sometimes he does not even desire
to ascertain the cause, as when he discourses about an eclipse; at
other times he invents, by way of hypothesis, and states certain ex-
pedients by the assumption of which the phenomena will be saved.
For example, why do the sun, the moon, and the planets appear to
move irregularly? We may answer that, if we assume that their
orbits are eccentric circles, or that the stars describe an epicycle,
their apparent irregularity will be saved; and it will be necessary to
go further, and examine in how many different ways it is possible
for these phenomena to be brought about, so that we may bring our
theory concerning the planets into agreement with that explanation
of the causes which follows an admissible method. *Hence we actually
find a certain person* * [Heraclides of Pontus] *coming forward and
saying that, even on the assumption that the earth moves in a certain
way, while the sun is in a certain way at rest, the apparent irregularity with reference to the sun can be saved.* For it is no part of the
business of an astronomer to know what is by nature suited to a
position of rest, and what sort of bodies are apt to move, but he
introduces hypotheses under which some bodies remain fixed, while
others move, and then considers to which hypotheses the phenomena
actually observed in the heaven will correspond. But he must go to
the physicist for his first principles, namely that the movements of

* It is the theory of Aristarchus of Samos that is actually described. The
translation of the whole passage is by Heath.

the stars are simple, uniform, and ordered, and by means of these principles he will then prove that the rhythmic motion of all alike is in circles, some being turned in parallel circles, others in oblique circles.' Such is the account given by Geminus, or Posidonius in Geminus, of the distinction between physics and astronomy, wherein the commentator is inspired by the views of Aristotle.

The influence of these views is clearly seen in a distinction drawn by Thomas Aquinas, who pointed out in the *Summa Theologica*, part 1, question 32, article 1, that there was a difference between a hypothesis which must necessarily be true and one which merely fitted the facts. Physical (or metaphysical) hypotheses were of the first type, mathematical hypotheses of the second. He said:

For anything a system may be induced in a double fashion. One way is for proving some principle as in natural science where sufficient reason can be brought to show that the motions of the heavens are always of uniform velocity. In the other way, reasons may be adduced which do not sufficiently prove the principle, but which may show that the effects which follow agree with that principle, as in astronomy a system of eccentrics and epicycles is posited because this assumption enables the sensible phenomena of the celestial motions to be accounted for. But this is not sufficient proof, because possibly another hypothesis might also be able to account for them.

A few years later such writers as Bernard of Verdun and Giles of Rome (*c*. 1247–1316) were asserting that astronomical hypotheses must be constructed primarily with a view to explaining the observed facts, and that experimental evidence must settle the controversy between the Aristotelian 'physicists' and Ptolemaic 'mathematicians'. And according to Giles, when there were a number of equally possible hypotheses, the one to be chosen was the simplest. These two principles of 'saving the appearances' and of simplicity were to guide theoretical astronomy down to the time of Kepler and beyond.

By the end of the 13th century the concentric system of Aristotle had been discarded in Paris in the light of practical experience, and the Ptolemaic system became generally accepted. Some attempt was made to bring this astronomical system into line with physics by taking over the product of Ptolemy's later thought, and by considering the eccentric planetary spheres as solid spheres of the fifth element within each of which the epicycles might revolve.

The controversies between the different schools of astronomy by no means came to an end at once. Even in the 13th century at least one astronomer had shown a tendency to branch out with an entirely new hypothesis. Pietro d'Abano, in his *Lucidator Astronomiæ*, suggested that the stars were not borne on a sphere but were moving freely in space. In the 14th century, the even more radical innovation of considering the earth instead of the heavenly spheres to revolve was discussed by Jean Buridan and by Nicole Oresme, though the first references to this theory occurs at the end of the 13th century in the writings of the Franciscan François de Meyronnes. This and other new hypotheses discussed during the 14th and 15th centuries may have been suggested by ancient Greek speculations, in particular by the semi-heliocentric system postulated in the 4th century B.C. by Heraclides of Pontus, in which Venus and Mercury revolved around the sun while the sun itself revolved around the earth. This system was know in Western Christendom through the writings of Macrobius and Martianus Capella. (The completely heliocentric system of the 3rd century B.C. by Aristarchus of Samos was not known in the Middle Ages, although it was known, for example to Aquinas, that Aristarchus had taught such a system.) These innovations were based largely on the fundamental criticisms of Aristotle's physics that occurred in the 14th century and discussion of them will therefore be deferred to a later page (Vol. II, p. 49 *et seq.*).

As regards the practical astronomy of the 13th century, the observations were made largely for the purpose of constructing tables for calculating dates, in particular of Easter, for determining latitude and longitude, and for astrological prediction. The last specially occupied the Italians. At first the practical astronomy of medieval Christendom remained under Arab leadership. Omar Khayyam's calendar of 1079 was at least as accurate as anything produced until the Gregorian reform of the calendar in 1582, and Arab instruments, observations, tables and maps retained their superiority at least until the middle of the 13th century. From that time the astronomy of Western Christendom began to stand on its own feet. One of the earliest independent observations in Western Christendom was, in fact, made as early as 1091 or 1092, when Walcher of Malvern observed an eclipse of the moon in Italy and, by discovering the time it had been observed by a friend in eastern England, determined the difference in longitude between the two points. Another method for determining longitude which was suggested in the 12th century by Gerard of Cremona was to observe the position of the moon at noon and, from the difference between that and what would be expected from tables constructed for a standard place, for example Toledo, to calculate the difference in longitude between the two places. But the accurate determination of longitude required accurate measurement of time, and this only began to be possible in the 17th century. The determination of latitude, on the other hand, could be made with an astrolabe by observing the elevation of a star, or of the sun at noon. The Arabs had made accurate measurements of latitude, taking as the prime meridian (o) a point to the west of Toledo. Their tables were adapted for various towns in Christendom, for example London, Oxford and Hereford in England, and further observations were also made in the West itself.

The astrolabe was the chief astronomical instrument of both the Arab and the Latin astronomers of the Middle Ages and was known as 'the mathematical jewel'. The extent of its use by Greek astronomers is problematical. Hipparchus in the 2nd century B.C. knew the underlying theory of stereographic projection, but may not have known the instrument itself; but Ptolemy in the 2nd century A.D. certainly knew the astrolabe as an instrument. In later Hellenistic times it diffused both eastwards and westwards, possibly from Alexandria. Western astrolabes derive from the Moorish type found in Spain. The instrument was mentioned by Gerbert in the late 10th century (if the attribution of a work to him is correct) and by Radolf of Liège in the early 11th century, and it was described by Hermann Contractus (the

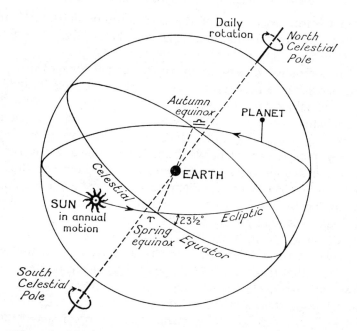

Fig. 3 The celestial sphere. The observer on the earth regards himself as
being at the centre of the stellar sphere. The position of a heavenly body
can then be determined by co-ordinates provided by systems of great
circles of which three were developed in antiquity. (1) The first system is
related to the celestial poles. These are the points on the stellar sphere
which are pierced by the axis about which this sphere appears to rotate
daily, which is the same as the axis of the earth. The celestial equator and
the tropics of Cancer and Capricorn all correspond to those of the earth,
and the circles of declination and right ascension provide co-ordinates for
determining the position of a point on the celestial sphere corresponding
respectively to latitude and longitude. The latter is measured in degrees
from the spring equinox, from West to East. (2) The second system is
related to the ecliptic. The sun and all the planets appear to move in one
great circle, although with different periods of revolution, when observed
against the background of the fixed stars on the stellar sphere. The sun,
having the most regular motion, is taken as defining this circle, which is
called the ecliptic, round the stellar sphere. The planets stray on their
orbits north and south of the solar circle in the course of their proper
periods. The ecliptic is inclined at an angle of approximately 23½ degrees
to the equator, the two intersections providing the fixed points of the
spring and autumn equinoxes (the equinoctial points). The celestial lati-

Lame) before 1048. One of the best accounts of this Western type was written in English in the second half of the 14th century by Geoffrey Chaucer, in his *Treatise on the Astrolabe.*

The astrolabe was essentially an instrument for measuring the angular distance between any two objects, and thus it could be used for taking the elevation of a heavenly body. It consisted of a graduated metal plate (usually of brass) with a datum line and a rotating pointer, called the *alidade,* on which were two sights (Plates 4, 5). The astrolabe was hung from a ring at the top of the diameter at right angles to the datum line, which was thus the horizon line, and, with this diameter always perpendicular to the earth, the *alidade* was rotated to point at a particular star whose altitude was read on the scale of degrees round the outer edge of the instrument. With this information it was possible to calculate time and determine north. The advantage of the astrolabe was that these values could be read off the instrument itself. For any particular latitude the Pole Star always has approximately a constant altitude and the other stars go round it. On the back of the *plate* of the astrolabe was a vertical stereographic projection of the celestial sphere on to a plane parallel to the equator as observed at a particular latitude on the earth, show-

---

tude of a point is then measured in degrees north or south of the ecliptic and the celestial longitude in degrees from the spring equinox, in the direction of the sun's apparent annual motion from West to East. The traditional division of the ecliptic into twelve equal sections of 30 degrees makes the signs of the zodiac, beginning at the spring equinox with the first degree of Aries (cf. Plate I). (3) The third system is related to the observer's horizon and zenith. The observer on the earth can see only the half of the heavens that appears above the horizon, which forms a great circle on the stellar sphere. Related to this circle are the almuncantars, or circles of equal altitude parallel with the horizon, and the azimuths, which pass through the zenith, vertically above the observer's head, and cut the horizon at right angles. Clearly with this system there is a different set of co-ordinates for each position on the earth's surface, a consideration taken into account in the design of instruments such as the astrolabe and sundial (see p. 106 *et seq.*). In the diagram, if the circle labelled 'Ecliptic' were an observer's horizon, his zenith would be vertically above the earth, and the meridian, the great circle that passes through the celestial poles and the zenith, would be the circle shown as the boundary of the sphere.

ing the equinoctials, tropics of Cancer and Capricorn, meridian, azimuths and almucantars (Fig. 3). Thus a different plate was needed for each latitude. If the observed altitude of a particular star were set to the corresponding altitude as shown on the plate, every other star would be in its correct position. Above this plate was a second plate, the *rete*, which was elaborately cut away and formed a rotating star map. On the *rete* was marked a circle which represented the ecliptic and so showed the position of the sun relative to the stars for each day of the year. If the stars were in their correct position, the position of the sun could therefore be read off. The line connecting the position of the sun with the position of the Pole Star was given by turning the *label* (a pointer rotating about the point representing the Pole Star) to the position of the sun. This gave the direction of the sun in azimuth and marked off time.

The astrolabe was most convenient in tropical latitudes where the variation in the altitude of the sun is great, and for this reason it was much used by the Arabs; for example, for determining the hours of prayers in mosques and for finding azimuths of the Qibla, that is, the direction towards Mecca.* But in spite of this, Arabic astrolabes, with one exception, do not show much development compared, for example, with later Western ones, especially in the 16th century. The exception is the so-called *Saphœa Azarchelis*, named after the 11th-century astronomer al-Zarqali but to be attributed, according to Millàs Vallicrosa, to his contemporary at Toledo, Ali ben Khalaf. This instrument used a horizontal instead of a vertical projection, thus making it possible for a single plate to be used at any latitude. But it also had disadvantages, and in fact it never superseded the older type, although many instruments, both Hispano-Moorish and later Western, were constructed combining both projections. The horizontal projection was revived in the 16th century by the Flemish cartographer Gemma Frisius under the name *astrolabum* [sic] *catholicum*, and the Roias and de la Hire projections are modifications of it. The latest European astrolabes date from the

* I should like to thank Mr F. R. Maddison for supplying information about the history of the astrolabe.

17th century, but in Arabic lands they were still being made in the 19th century. Their great convenience as instruments for telling the time was that they were portable. Sundials, which are instruments for showing the change in azimuth angle and must, therefore, be aligned north and south, could not be made portable until they could be combined with a compass. Such a combination was not produced until the end of the 15th century.

Another instrument used in the 13th century was the quadrant, of which improved versions were made by an Italian, John Campanus of Novara (d. after 1292), and by two Montpellier astronomers who lived about the same time. Another instrument which came into use at about this time was the mural quadrant, which had been used by Alexandrian, Arabian and Persian astronomers. This was mounted so that one end was level with a hole in the wall of the observatory. A travelling sight was swung round until it and the hole were aligned with the heavenly body under observation, and the angle was read on a scale. Another instrument constructed by Campanus was a sort of armillary sphere for determining the positions of the planets. This consisted of an armilla, or ring, fixed in the plane of the equator, with other rings representing the horizon, the meridian and the ecliptic so that it was a sort of model of the celestial sphere.

It was with such instruments that Guillaume de St Cloud, a follower of Roger Bacon and the founder of the Paris school of astronomy, determined from the solstitial altitudes of the sun the obliquity of the ecliptic in 1290 and the latitude of his place of observation at Paris. The obliquity he calculated as being 23° 34′ and the latitude of Paris he made 48° 50′. The modern figure for the obliquity of 1290 is 23° 32′ and his value for the latitude of Paris is that which is now accepted. Another of his observations was to note the meridian altitude of the sun when he himself was in a dark room with a small aperture to admit a beam of light, and from this he determined the epoch of the spring equinox. Another Frenchman, Jean de Murs, used a graduated arc of 15 feet radius to make the same determination at Evreux on 13 March, 1318.

The reform of the calendar, which had been urged by Grosseteste and Roger Bacon, again aroused attention when the *Alfonsine Tables* reached Paris about 1293. Pope Clement VI bade Jean de Murs and Firmin de Belleval to Avignon to report on the project, which they did in 1345. Another report was drawn up by Cardinal Pierre d'Ailly for the Council of Constance, 1414–18. The accuracy of the *Alfonsine Tables* was still mistrusted and the reform had to wait for nearly two centuries, though when at last it was made it was on the basis of numerical values very much the same as those arrived at in the 14th century.

Other instruments were invented or improved in France and observations were extended during the 14th century. Jean de Linières produced a catalogue of positions of 47 stars, the first attempt in Christendom to correct some of the star places given in the 2nd-century catalogue of Ptolemy. In 1342 a Jew, Levi ben Gerson of Montpellier, introduced the Baculus Jacobi, a cross-staff which had apparently been invented in the 13th century by Jacob ben Makir. Levi applied a diagonal scale to the instrument. The cross-staff was used for measuring the angular distance between two stars or, as a navigational instrument, for taking the elevation of a star or the sun above the horizon. It consisted of a graduated rod or staff with a cross-bar attached at right angles to it. The cross-staff was held with the staff against the eye. The cross-bar was moved until the sight at one end was in line with the horizon and the sight at the other end in line with a star or the sun. From the reading on the graduated scale on the staff the angle of elevation of the star could be obtained from a table of angles.

During the first half of the 14th century an important school of astronomy grew up also at Oxford, in particular at Merton College. One of the results of the work there was the development of trigonometry. Tangents were used by John Maudith (1310) and Thomas Bradwardine (d. 1349), and by Richard of Wallingford (c. 1292–1335) who took the loose methods used in the trigonometry of al-Zarqali's *Toledan Tables* and applied to them Euclid's rigorous methods of demonstration. John Maudith and Richard of Wallingford are the initiators of Western trigonometry, though an important treatise on the subject was written in Hebrew about the same time in Provence by Levi ben Gerson (1288–1344) and translated into Latin in 1342. An important improvement in technique adopted by these writers was to use the Hindu-Arabic practice, already found in the tables of al-Zarqali and other astronomical tables in wide circulation, of basing plane trigonometry on sines instead of chords, as had been done in the old Greco-Roman tradition dating from Hipparchus. Richard also adapted the *Alfonsine Tables* to Oxford and invented certain instruments, for example, an elaborate *rectangulus* for measuring and comparing altitudes and an improved *equatorium* for showing the positions of the planets.

The lively interest in astronomy during the 13th and 14th centuries, of which this work was the outcome, is shown also by the astronomical models then constructed. In 1232 the Emperor Frederick II had received a planetarium from the Sultan of Damascus. About 1320 Richard of Wallingford constructed an elaborate astronomical clock, showing the positions of the sun, moon and stars and also the ebb and flow of the tides. He also left a handbook describing how this instrument was to be used. An elaborate planetarium driven by weights was also made by the clockmaker Giovanni de' Dondi (b. 1318), and such things became popular as scientific toys.

3. METEOROLOGY AND OPTICS

Meteorology and optics formed a single heterogeneous subject in the 13th century, because these sciences were concerned with phenomena supposed to occur in the regions of the elements fire and air lying between the sphere of the moon and the terraqueous globe. These topics had been discussed by Aristotle in his *Meteorologica*, which was the chief source of 13th-century 'meteorology', and in this work Aristotle had attributed all the changes seen in the sky, except the movements of the heavenly bodies, to changes in those regions. The element fire was a sort of principle of combustion rather than actual flame and so was not itself visible, but it was easily set alight by movement, and agitation, brought about by hot dry exhalations rising up from the earth on which the sun's rays were falling, caused a number of phenomena to occur in the sphere of fire, as, for instance, comets, shooting stars and auroras. All these phenomena must occur in the region beneath the moon, because beyond it the heavens were ingenerable and incorruptible and could suffer no change but circular motion. In the sphere of the element air these hot dry exhalations caused wind, thunder and lightning, and thunderbolts, while cold moist exhalations produced by the sun's rays falling on water caused cloud, rain, mist, dew, snow and hail. A special group of phenomena associated with the moist exhalations were rainbows, halos and mock suns.

Throughout the Middle Ages comets and similar apparent changes in the heavens continued to be classed as 'meteorological' rather than astronomical phenomena, that is, as phenomena occurring in the sublunary region. In the 16th century more accurate measurements of their positions and orbits were to provide some of the most telling evidence against the truth of Aristotle's ideas on the structure of the universe. Comets were described several times in the 13th and 14th centuries, one of the most interesting references being made by Grosseteste to what may have been Halley's comet, which would have been due to appear in 1222.* Another interesting reference was made by Roger Bacon, who held that the awesome comet of July, 1264 had been generated under the influence of the planet Mars and had produced an increase of jaundice leading to bad temper, the result of which was the wars and disturbances in England, Spain and Italy at that time and afterwards!

Observations on weather and attempts, partly with an agricultural interest, to predict it by astrology had been made from the 12th century onwards. A most remarkable series of monthly weather records was kept during 1337–44 for the Oxford district by William Merlee. He based attempts to forecast the weather partly on the state of the heavenly bodies, and also on inferior signs such as the deliquescing of salt, the carrying of sound from distant bells, and the activity of fleas and the extra pain of their bites, all of which indicated greater humidity.

* H. C. Plummer, *Nature*, 1942, vol. 150, p. 253.

A subject which was to see the most remarkable progress during the 13th and 14th centuries was optics. The study of light attracted the attention in particular of those who tended to Augustinian-Platonism in philosophy, and this was for two reasons: light had been for St Augustine and other Neoplatonists the analogy of divine grace and of the illumination of the human intellect by divine truth, and it was amenable to mathematical treatment. The first important medieval writer to take up the study of optics was Grosseteste, and he set the direction for future developments. Grosseteste gave particular importance to the study of optics because of his belief that light was the first 'corporeal form' of material things and was not only responsible for their dimensions in space but also was the first principle of motion and efficient causation. According to Grosseteste, all changes in the universe could be attributed ultimately to the activity of this fundamental corporeal form, and the action at a distance of one thing on another was brought about by the propagation of rays of force or, as he called it, the 'multiplication of species' or 'virtue'. By this he meant the transmission of any form of efficient causality through a medium, the influence emanating from the source of the causality corresponding to a quality of the source, as, for instance, light emanated from a luminous body as a 'species' which multiplied itself from point to point through the medium in a movement that went in straight lines. All forms of efficient causality, as for instance, heat, astrological influence and mechanical action, Grosseteste held to be due to this propagation of 'species', though the most convenient form in which to study it was through visible light.

Thus the study of optics was of particular significance for the understanding of the physical world. Grosseteste's theory of the multiplication of species was adopted by Roger Bacon, Witelo, Pecham and other writers and they all made contributions to optics in the hope of elucidating not only the action of light but also the nature of efficient causality in general. For this purpose the use of mathematics was essential, for, as Aristotle had put it, optics was subordinate to geometry, and the progress made in medieval optics would certainly have been impossible without the knowledge of Euclid's *Elements* and Apollonius' *Conics*. Throughout the whole Middle Ages, and indeed much later, the Aristotelian distinction was maintained between the mathematical and the physical aspects of optics. As Grosseteste put it in discussing the law of reflection, geometry could give an account of what happened, but it could not explain why it happened. The cause of the observed behaviour of light, of the equality of the angles of incidence and reflection, was to be sought, he said, in the nature of light itself. Only a knowledge of this physical nature would make it possible to understand the cause of the movement.

The chief sources of 13th-century optics were, besides Aristotle's *Meteorologica* and *De Anima*, the optical writings of Euclid, Ptolemy and Diocles (2nd century B.C.), and of the Arab writers Alkindi, Alhazen, Avicenna and Averroës. Aristotle, who was more concerned with the cause of vision than the laws by which it was exercised, had held that light (or colour) was not a movement but a state of transparency in a body and was produced by an instantaneous qualitative change in a potentially transparent medium. Other Greek philosophers had put forward other explanations, Empedocles asserting that light was a movement which took time to be transmitted and Plato that vision could be explained by a series of separate rays going out from the eye to the object seen (see above, p. 49). In contrast with this theory of extramission, the Stoics had suggested that vision was due to rays of light entering the eye from the object. It was one or other of these theories of rays, implying that light travelled in straight lines, that had been adopted by the Greek geometers, such as Euclid and Ptolemy, who had developed optics to a place equal to that of astronomy and mechanics among the most advanced physical sciences of antiquity. These men discovered that the angle of reflection of rays from a surface was equal to the angle of incidence. Ptolemy, who measured the amount of refraction in rays passing from air into glass and into water, observed that the angle of refraction was always less than the angle of incidence but wrongly supposed that this was by a constant proportion. He concluded from this that the apparent position of a star did not always correspond to its real position because of refraction by the atmosphere.

This Greek work on optics was further developed by the Arabs and particularly by Alhazen (965–1039), whose work became the main source for physical and physiological optics in the medieval West.* Alhazen's achievements in both fields were revolutionary, though he persisted in the erroneous belief that the lens of the eye was the sensitive part. He showed that the angle of refraction was not proportional to the angle of incidence and studied spherical and parabolic mirrors, spherical aberration, lenses and atmospheric refraction. He also held that the transmission of light was not instantaneous and rejected the theory of extramission, which had been upheld by Euclid and Ptolemy, in favour of the view that light came from the object to the eye where it formed an image on the lens. Knowledge of the anatomy of the eye was also improved by the Arabs, whose chief source of information had been Rufus of Ephesus in the 1st century A.D. Outstanding work on this was done by Rhazes and Avicenna.

Among 13th-century writers on optics, Grosseteste himself is remarkable chiefly for his attempt to explain the shape of the rainbow by means of a single phenomenon which he could study experimentally, namely, the refraction of light by a spherical lens. Aristotle had held that the rainbow was caused by reflection from drops of water in the cloud, but Grosseteste attributed it definitely to refraction, though he thought that this was caused by the whole cloud acting as a large lens. Though his contribution to optics was more to emphasize the value of the experimental and mathematical methods than to add much to positive knowledge, he did make a few important additions. In his *De Iride* he attempted to treat refraction quantitatively, showing some knowledge of Ptolemy's work. Holding that vision was effected by extramitted visual rays, he proposed a 'law' of refraction according to which, when a ray passed from a rarer into a denser medium (e.g. from air into glass), it was refracted towards the perpendicular drawn to the common surface on a line bisecting

* See Crombie, 'The mechanistic hypothesis and the scientific study of vision', *Proc. Roy. Microscopical Soc.*, ii (1967) 3-112.

the angle made with this by the projection of the visual ray. This, he said, agreed with the principle of economy and what 'experiments show us', though simple experiments would have disproved it. He also used a classical rule for locating the optical image after reflection to give a rule for locating the refracted image. According to Grosseteste's rule the image would be seen at the junction of a projection of the visual ray and the perpendicular drawn to the refracting surface from the object seen (cf. below, Fig. 4). When a visual ray passed from a denser into a rarer medium, it was bent in the opposite direction. Grosseteste used this analysis to try to explain the shape of the rainbow and the operation of a spherical lens or burning glass: the sunlight was refracted twice, once on entering the lens and again on passing out on the farther side, the combined refractions bringing the rays to a focus at a point. He also suggested using lenses to magnify small objects and to bring distant objects closer; spectacles were in fact invented in northern Italy at the end of the 13th century (see below, pp. 227 and 236–7).

Another contribution by Grosseteste that may be mentioned is his attempt to formulate a geometrical and almost mechanical conception of the rectilinear propagation of light and of sound by a series of waves or pulses. In his commentary on the *Posterior Analytics*, book 2, chapter 4, he described how, when a sounding body is struck violently, it is set in vibration for a time because its violent motion and a 'natural power' alternatively send the parts back and forth, each overshooting the natural position. These vibrations are transmitted to the fundamental light incorporated as the first 'corporeal form' in the sounding body. 'Hence, when the sounding body is struck and vibrating, a similar vibration and similar motion must take place in the surrounding contiguous air, and this generation progresses in every direction in straight lines.' If the propagation strikes an obstacle, it is forced to 'regenerate itself by turning back. For the expanding parts of the air colliding with the obstacle must necessarily expand in the reverse direction, and so this repercussion extending to the light which is in the most subtle air is the returning sound, and this is an echo.' Just as the echo was propagated by the fundamental light, the basic principle of motion incorporated in the air, so a reflected image was produced by the analogous 'repercussion' of visible light, and refraction was similarly explained.

Grosseteste's chief disciple, Roger Bacon, made a number of small contributions to knowledge of reflection and refraction, though many of the experiments he described were repetitions of those made by Alkindi and Alhazen. He continued Grosseteste's teaching about methodology. He made some original experimental determinations, for instance, of the focal length of a concave mirror illuminated by the sun, and he pointed out that the sun's rays reaching the earth might be treated as parallel instead of radiating from a point, thus making possible a better explanation of burning lenses and parabolic mirrors. He firmly adopted the theory that in vision material light, travelling with enormous though finite velocity, passed from the object seen to the eye, but he pointed out that in the act of looking something psychological 'went forth,' so to speak, from the eye. Of the propagation of material light he gave an explanation similar to Grosseteste's asserting that it was not a flow of body like water, but a kind of pulse, as in sound, propagated from part to part. In the 'multiplication of the species' of light there was nothing but this kind of succession. But he noted that light travelled much faster than sound, for if someone at a distance is banging with a hammer we see the blow before we hear the sound, and in the same way we see lightning before we hear the thunder.

Bacon gave a better description of the anatomy of the vertebrate eye (Plate 6) and optic nerves than any previous Latin writer, and recommended that those who wished to study the subject should dissect cows or pigs. He discussed in detail the conditions necessary for vision and the effects of various kinds and arrangements of single lenses deriving eight special rules from Grosseteste's general rule for determining the angle of refraction and locating the optical image (see above pp. 114–15; Fig. 4). Assuming that the apparent size of an object depends on the angle it subtends at the eye, he tried to improve vision by using plano-convex lenses, though he understood them only imperfectly. His scientific imagination played freely with further possibilities: with arrangements of lenses or curved mirrors to magnify small objects indefinitely and bring distant objects nearer; with mirrors he thought Julius Caesar had erected in Gaul to observe events in England; with lenses that would make the sun and moon appear to descend above the heads of enemies : the ignorant mob, he said, could not endure it.

Roger Bacon's attempt to discover the cause of the rainbow is a good example of his conception of the inductive method (see Vol. II, p. 38 *et seq.*). He began by collecting phenomena similar to the rainbow : colours in crystals, in dew on the grass, in spray from mill wheels or oars when lit by the sun, or as seen through a cloth or through the eyelashes. He then examined the rainbow itself, noting that it appeared always in cloud or mist. By a combination of observation, astronomical theory and measurements with the astrolabe he was able to show that the bow was always opposite the sun, that the centre of the bow, the observer's eye and the sun were always in a straight line, and that there was a definite connexion between the altitudes of the bow and of the sun. He showed that the rays returning from the rainbow to the eye made an angle of 42 degrees with the incident rays going from the sun to the bow. To explain these facts he then adopted the theory, put forward in Aristotle's *Meteorologica*, that the rainbow formed the base of a cone of which the apex is at the sun and the axis passed from the sun through the observer's eye to the centre of the bow. The base of the cone would become elevated and depressed, thus producing a larger or smaller rainbow, according to the altitude of the sun; if it could be sufficiently elevated the whole circle would appear above the horizon, as with rainbows in sprays. This theory he used to explain the height of the bow at different latitudes and different times of year. It would imply, among other things, that each observer would see a different bow and this he confirmed by the observation that when he moved towards, away from or parallel to the rainbow it moved with him relative to trees and houses; 1,000 men in a row, he asserted, would see 1,000 rainbows and the shadow of each man would bisect the arc of his bow. The colours and form of the rainbow thus bore to the observer a relation unlike those of fixed objects such as crystals. As to the colours, Bacon's discussion was as inconclusive as that of everyone else till Newton; the form he explained as due to the reflection of light from spherical water drops in the cloud, the rainbow of any particular observer appearing only in the drops from which the reflected rays went to his own eyes. This theory he extended to explain halos and mock suns; it was not in fact correct.

Among Grosseteste's successors later in the 13th century, the Silesian writer Witelo (b. *c.* 1230) described experiments similar to Ptolemy's determining the values of the angles of refraction of light passing between air, water and glass, with angles of incidence increasing by 10 degrees to a maximum of 80 degrees. Alhazen had not described such measurements, but Witelo seems to have adapted an apparatus which Alhazen had described for another purpose. This consisted of a cylindrical brass vessel, in-

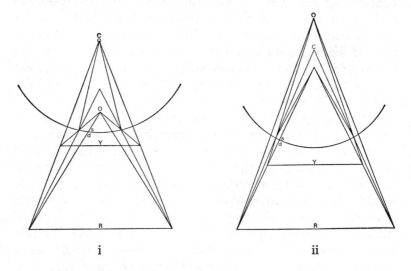

i ii

Fig. 4 Diagrams from British Museum MS Royal 7.F.viii (XIII cent.), illustrating Roger Bacon's classification of the properties of curved refracting surfaces, in the *Opus Majus*, V. Rays go from each end of the object (*res*, R), are bent at the curved surface separating the optically rarer (*subtilior*, s) and denser (*densior*, d) media, for example air and glass, and meet at the eye (*oculus*, O). The image (*ymago*, Y) is seen at the junction

side which was marked a circle in 360 degrees and minutes. The refracting media were suitably introduced into the cylinder, and the measurements were made by means of a sight and holes bored through each end of a diameter of the graduated circle.* Witelo's table setting out the concomitant variations in the angles of incidence and of refraction is remarkable for showing results with observations made in *both* directions across the refracting surface. These are revealing. For example, while the results obtained with light passing from air into water are reasonably accurate, those given for the reciprocal case are either very inaccurate or impossible. In fact it is clear that he never made these reciprocal measurements, but derived his values from a misapplication of the law that the amount of refraction is the same in both directions, not knowing also that at the higher angles of incidence there would be no values for refraction because all the light would be reflected at the under surface between the water and

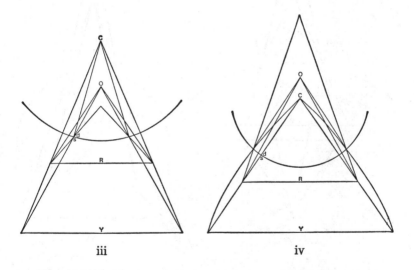

iii   iv

of the projection of these refracted rays and the perpendicular from the object, i.e. the line to the centre (C). It is magnified or diminished according

* For a full description see A. C. Crombie, *Robert Grosseteste and the Origins of Experimental Science* 1100–1700, Oxford, 1971, p. 220 *et seq.*

the air. Thus Witelo missed discovering the important phenomenon of total reflection at a critical angle. His work is nevertheless
interesting and he tried to express his results in a number of
mathematical generalizations. He pointed out that the amount of
refraction increased with the angle of incidence but that the
former increase was always less than the latter. He tried to
relate these generalizations to differences in the density of the
media. Witelo also carried out experiments in which he produced
the colours of the spectrum by passing white light through a
hexagonal crystal, and understood, at least by implication, that
the blue rays were refracted through a greater angle than the

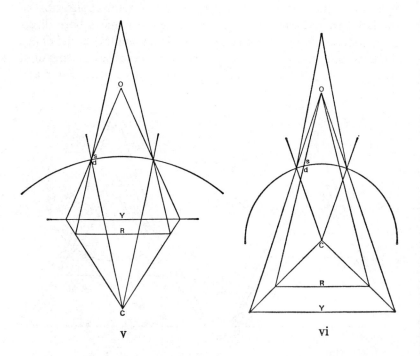

v   vi

to whether the concave (*i–iv*) or convex (*v–viii*) surface is towards the eye,
whether the eye is on the rarer (*i, ii, v, vi*) or denser (*iii, iv, vii, viii*) side
of the curvature, and whether the eye is on the side of the centre of
curvature (*centrum*, C) towards (*i, iii*) or away from (*ii, iv*) the object, or
the centre of curvature is on the side of the object towards (*vi, viii*) or

red. He supposed that the range of colours was produced by the progressive weakening of white light by refraction, thus allowing a progressively greater incorporation of darkness from the medium. The same explanation had been given in the so-called *Summa Philosophiæ* of pseudo-Grosseteste, a work by an English writer associated by Grosseteste's circle. Witelo used his optical investigations to offer an intelligent but erroneous explanation of the rainbow. Also of considerable interest is his discussion of the psychology of vision. Another English writer, John Pecham (d. 1292), made a useful contribution by writing a lucid little textbook on optics, though he made few original advances.

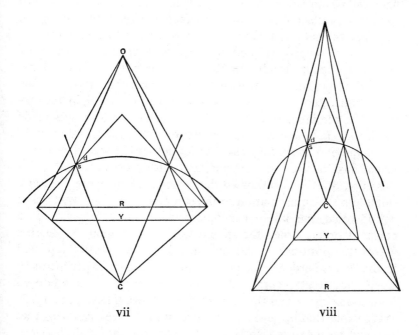

vii   viii

away from (*v, vii*) the eye. The rule which led Bacon to draw a diminished image in (*i*) and a magnified image in (*iii*) is corrected in a later section of the *Opus Majus*, where Bacon points out that 'the size of the visual angle is the prevailing factor in these appearances'; that is, the angles subtended by the object and the image at the eye. He recommended a convex lens forming a hemisphere (*vi*) or less than a hemisphere (*v*) to aid weak sight.

Some remarkable advances were made by the German writer Theodoric or Dietrich of Freiberg (d. 1311), whose work on refraction and on the rainbow is an outstanding example of the use of the experimental method in the Middle Ages.

Among those who had written on the rainbow before Theodoric composed his *De Iride et Radialibus Impressionibus*, Grosseteste had attributed the shape of the bow to refraction, and Albertus Magnus and Witelo, writing with much greater knowledge, had pointed out the need to consider the refraction as well as the reflection of the rays by *individual* raindrops. Theodoric himself advanced the theory that the primary bow was caused by light falling on spherical drops of rain becoming refracted into each drop, reflected at its inner surface and refracted out again; and that the secondary bow was caused by a further reflection before the second refraction. This is the explanation now accepted, though usually attributed to Descartes, whose mathematical exposition of it was certainly in every way superior. The important discovery on which it was based, that light was reflected at the inner concave surface of each raindrop, was made by Theodoric by means of experiments with a model raindrop in the shape of a spherical glass vessel filled with water, probably a urinal flask as used in medicine, and with a crystal ball.* With this apparatus he also showed that when such a sphere, held in a suitable position in relation to the sun and to the eye, was raised and lowered, the different colours of the rainbow appeared in a constant order. When the sphere was held at about 11 degrees above this position, he showed that the same colours appeared in the reverse order. Thus he was able in further experiments to trace with great accuracy the course of the rays that produced both the primary and the secondary rainbows (Plates 7, 8, 9, 10). It is curious that he gave a false value of 22 degrees, asserting that it could be measured with an astrolabe, for the angle between the rays going from the sun to the bow and from the bow to the eye of the observer. The approximately correct value of 42 degrees given by Roger Bacon was then well known.

* See Crombie, *Robert Grosseteste*, p. 232 *et seq.*, where Theodoric's drawings are reproduced in full.

The colours of the rainbow Theodoric also tried to investigate experimentally. He showed that the same colours as those seen in the rainbow could be produced by passing light through crystal balls or spherical glass vessels filled with water, and through hexagonal crystals, if either the eye were applied to the far side of the flask or crystal or the light were projected onto an opaque screen. The colours of this spectrum were always in the same order, red being nearest the line of incidence and being followed by yellow, green and blue, the four principal colours which he distinguished. His description shows that he understood that the colours were formed *inside* the refracting bodies after refraction at the first surface encountered and not simply on emerging from them. To explain the appearance of the spectrum Theodoric made use of the theory of colour which Averroës had developed in his commentaries on Aristotle, according to which colours were attributed to the presence in varying degrees of two pairs of opposite qualities; brightness and darkness, boundedness and unboundedness. The first pair were formal and the second material causes, and the reason why a spectrum could be produced was that the light stream consisted not of geometrical lines but of 'columns' with breadth and depth, so that different parts of it could be affected differently on passing through a suitable medium. Thus, when light fell perpendicularly on to the surface of a hexagonal or spherical crystal or of a spherical flask, it passed straight through without refraction and remained fully bright and unbounded. Such light therefore remained white. But light falling at an angle to the surface of the crystal or flask was refracted and weakened, its brightness was reduced by a positive amount of darkness, and it was affected by the boundedness of the surface of the refracting body. The different combinations of the qualities affecting the light stream then caused the range of colours emerging after refraction, from the brightest, red, to the darkest, blue, even though the crystal and the water in the flask were not themselves coloured as, for example, coloured glass was.

Theodoric carried out a number of experiments to demonstrate various points of his theory. He stated explicitly that in the rays refracted through a hexagonal crystal or glass flask filled with water, red appeared nearest the original line of incidence and blue farthest from it. He did not think of recombining the colours, so that they reformed white light, by passing them through a second crystal in a reverse position to the first, as Newton was to do. He did observe that if the screen were held very close to the crystal the light projected on to it showed no spectrum and appeared white, a fact which he explained by saying that at this distance the light was still too strong for the darkness and boundedness to produce their effects. Taken altogether, Theodoric made a remarkable advance both in optics and in the experimental method, and the technique of reducing a complicated phenomenon, like the shape and colours of the rainbow, to a series of simpler questions which could be investigated separately by specially-designed experiments was particularly pregnant for the future. Theodoric's theory was not forgotten; it was discussed by Themo Judæi later in the 14th century, by Regiomontanus in the 15th, and in the University of Erfurt and perhaps elsewhere in the 16th. At Erfurt, a certain Jodocus Trutfetter of Eisenach published in 1514 woodcuts of Theodoric's diagrams of the primary and secondary rainbows (Plate 10). An explanation of the rainbow similar to Theodoric's by Marc Antonio de Dominis was published in 1611 and this was almost certainly the basis of the much fuller explanation published by Descartes in 1637.

By a curious coincidence an explanation of the rainbow similar to Theodoric's was given also by the contemporary Arabic writers, Qutb al-din al-Shirazi (1236–1311) and Kamal al-din al-Farisi (d. c. 1320). Western and Eastern writers seem to have worked quite independently of each other but used the same ultimate sources, principally Aristotle and Alhazen. Al-Farisi gave also an interesting explanation of refraction, which he attributed to the reduction of the speed of light passing through different media in inverse proportion to the 'optical density', an explanation suggestive of that advanced in the 17th century by the supporters of the wave theory of light. By another interesting coincidence he was also improving the theory of the *camera obscura*, or pin-hole camera, at the same time as similar work was being done by Levi ben Gerson. Both showed that the images formed were not affected by the shape of the hole and that an accurate image was formed when the aperture was a mere point, but that a multitude of only partially superimposed images appeared with a larger hole. They used this instrument to observe eclipses and other astronomical phenomena, and the movements of birds and clouds.

Another important development in medieval optics was the geometrical study of perspective in connexion with painting. The beginnings of deliberate use of central projection dates from the paintings of Ambrogio Lorenzetti of Siena in the middle of the 14th century, and this was to revolutionize Italian painting in the 15th century.

4. MECHANICS AND MAGNETISM

Putting on one side the theory of the 'multiplication of species' of light, the only non-living causes of local motion in the terrestrial region considered in the 13th century were mechanical action and magnetism, and the only natural mechanical causes were gravity and levity. Mechanics was the part of physics, apart from astronomy and optics, to which mathematics was most effectively applied in the Middle Ages, and the chief sources of 13th-century mechanics were the most mathematical of all the treatises in the Aristotelian corpus, the *Mechanica* (*Mechanical Problems*), then generally but wrongly attributed to Aristotle himself, and a small number of late Greek and Arabic treatises. Aristotle's *Physics* was also important for mechanical ideas. Indeed the whole corpus of the mechanics that came down to the 13th century was based on the principle expounded in that work : the principle that local motion, like other kinds of change, was a process by which a potentiality towards motion was made actual. Such a process necessarily required the continued operation of a cause and when the cause ceased to operate, so did the effect. All moving bodies thus required for their motion either an intrinsic 'natural' principle, the 'nature' or 'form', which was responsible for the body's natural movement, or an external mover distinct from the body which necessarily accompanied the body it moved (see above,

pp. 84–5, 90–91). Further, the effect was proportional to the cause, so that the velocity of a moving body varied in direct proportion to the power or 'virtue'* of the intrinsic 'nature' or external motor and, for the same body and motive power in different media, in inverse proportion to the resistance offered by the medium. Movement, velocity, was thus determined by two forces, one, either internal or external, impelling the body, and the other, external to the body, resisting it. Aristotle had no conception of mass, the intrinsic resistance which is a property of the moving body itself, which was to be the basis of 17th-century mechanics.†
With falling bodies the force or power causing the movement was the weight, and so it followed from the above principles that in any given medium the velocity of a falling body was proportional to its weight and, further, that if a body were moving in a medium which offered no resistance its velocity would be infinite. Since this conclusion involved an impossibility Aristotle therefore saw in it an additional argument against the existence of void.

When Aristotle's mechanics became known in Western Christendom in the 13th century they were submitted, like the rest of his scientific ideas, to logical and empirical examination. This led, in the following century, to a radical criticism of his dynamical ideas and of their physical consequences, such as the impossibility of void, which prepared the way for the immense intellectual effort by which Galileo and his 17th-century followers escaped from Aristotelian principles and established the mathematical mechanics which was the central feature of the Scientific Revolution (see Vol. II, pp. 49–97 *et seq.*).

In the 13th century it was not dynamics but statics and to some extent kinematics, that is, the study of rates of motion, that underwent the most striking developments, particularly in the school of Jordanus Nemorarius. He is possibly to be identified with Jordanus Saxo (d. 1237), the second master-general of the

* This power was generally called *virtus*, meaning power or ability to do something.

† The concept of mass was deduced only in the 17th century from the supposition that in a vacuum, or a medium whose resistance was small in comparison with the body's weight, all bodies fell with equal velocity.

126

Order of Preachers, or Dominicans, but in fact his actual identity is still an unsolved problem. It followed from the Aristotelian principle that velocity was proportional to motive power, that motive power could be estimated as proportional to velocity. Aristotle had stated that if a certain motive power moved a certain body with a certain velocity, then twice the motive power would be necessary to move the same body with twice the same velocity. Motive power was measured, therefore, by the product of the weight of the body moved multiplied by the velocity impressed on it. This has been called 'Aristotle's axiom'. Dynamical and statical ideas were not clearly distinguished by Aristotle, by the author of the *Mechanica*, or by the writer of the Greek *Liber Euclidis de Ponderoso et Levi* and the derivative Arabic works which formed the basis of medieval Latin statics. But it would follow from the above dynamical statement, converted into statical terms, that motive power would be equal to the product of the weight of the body moved multiplied by the distance through which it was moved.

From these Aristotelian ideas and fragments of Alexandrian mechanics, containing only minor works of Archimedes, Jordanus Nemorarius and his school developed a number of important mechanical ideas which were to be taken over, in the 17th century, by Stevin, Galileo and Descartes. In the *Mechanica* it had been shown to follow from Aristotle's axiom that the two weights which balanced each other at opposite ends of a lever were inversely proportional to the velocities with which their points of attachment moved when the lever was displaced (Fig. 5).

In his *Elementa Jordani Super Demonstrationem Ponderis* Jordanus gave a formal geometrical proof, beginning with Aristotle's axiom, that equal weights at equal distances from the fulcrum were in equilibrium. In the course of this he made use of what has been called the 'axiom of Jordanus', that the motive power which can lift a given weight a certain height can lift a weight k times heavier to $1/k$ times the height. This is the germ of the pripiciple of virtual displacements.

The *Mechanica* also contained the idea of the composition of

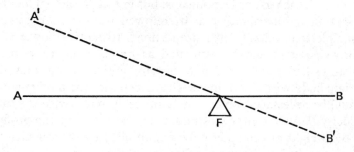

Fig. 5 The different weights A and B would balance if placed at such positions on the lever that when the lever turned on the fulcrum F the ratio of the velocities $\frac{A^1}{B^1}$ was proportional to the ratio of the weights $\frac{B}{A}$.

movements. It had been shown that a body moving with two simultaneous velocities (V_1 and V_2) bearing a constant ratio to each other would move along the diagonal (Vr) of the rectangle made by lines proportional to these velocities (Fig. 6); and also that if the ratio of the velocities varied, the resultant motion would not be a straight line but a curve (Fig. 7).

Jordanus applied this idea to the movement of a body falling

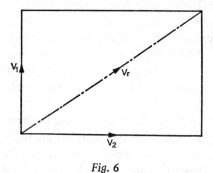

Fig. 6

along an oblique trajectory. He showed that the one effective force or motive power by which the body was moved at any given moment could be dissociated into two, the natural gravity downwards towards the centre of the earth and a 'violent' horizontal force of projection. The component of gravity acting along the trajectory he called *gravitas secundum situm*, or 'gravity relative to position'; he showed that the more oblique the trajectory, that is the nearer to the horizontal, the smaller was this component. The obliquity of two trajectories could be compared, he said, by measuring the distance fallen in a given horizontal distance.

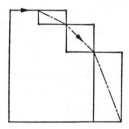

Fig 7 The vertical distances travelled increase in each successive unit of time while the horizontal distances travelled during the same interval remain constant.

In another treatise, *De Ratione Ponderis* or *De Ponderositate*, by tradition ascribed to Jordanus but possibly by another author whom Duhem, in his *Origines de la Statique*, has called 'the Forerunner of Leonardo', Jordanus' ideas were developed and applied to the study of the angular lever and of bodies on inclined planes. A faulty solution of the problem of the angular lever had been given in the *Mechanica*. The author of *De Ratione Ponderis*, considering the special case of equal weights hanging on the ends of the arms of the bent lever, showed, again using the principle of virtual displacements at least implicitly, that the weights will be in equilibrium when the horizontal distances from the vertical running through the fulcrum are equal. Presumably he knew also the more general principle, that *any* weights are in equilibrium when they are inversely proportioned to the horizontal distances, a principle involving the fundamental idea of the statical moment. Thus two weights E and F on a lever would be in equilibrium when they were inversely proportional to their effective distances BL and BR from the fulcrum (B), that is, $E/F = BR/BL$ (Fig. 8).

In fact Hero of Alexandria in his *Mechanica*, book 1, chapter 33, had already generalized the principle of the angular lever, but this work was not known to the author of the *De Ratione Ponderis*.

In discussing the component of gravity acting on bodies on inclined planes, the author of *De Ratione Ponderis* pointed out that the *gravitas secundum situm* of a body was the same at all

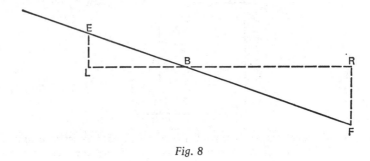

Fig. 8

points on the plane. He showed then, from the axiom of Jordanus, how to compare this value on planes of different inclination. He concluded:

if two weights descend on planes of different inclination and the weights are directly proportional to the lengths of the inclines, these two weights will have the same motive power in their descent. (Duhem, *Origines de la Statique*, 1905, p. 146.)

The same proposition was later proved by Stevin and Galileo, to whom *De Ratione Ponderis* would have been available in the printed text edited by Tartaglia and published posthumously in 1565. This treatise also contained the hydrodynamical principle, apparently coming from Strato (*fl.c.* 288 B.C.), that the smaller the section of a liquid flowing with a given fall, the greater its velocity of flow.

This work of Jordanus Nemorarius and his school became widely known in the 13th and 14th centuries. It was summarized by Blasius of Parma in the 15th century and, as Duhem has shown, it was used extensively by Leonardo da Vinci and was to become

1. Aristotle's cosmology. From Petrus Apianus, *Cosmographia per Gemma Phrysius restituta*, Antwerp, 1539

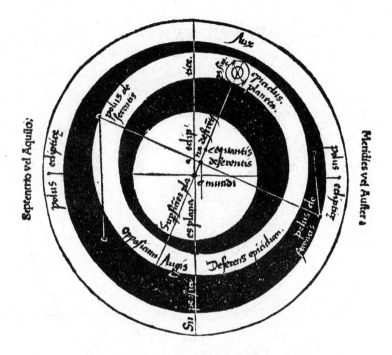

2. The medieval mechanical model of solid spheres for the planet
Saturn. From G. Reisch, *Margarita Philosophica*, Freiburg, 1503. The
outermost (white) sphere is the stellar sphere, centered on the
earth. The planet is shown on its epicycle, which is embedded in
the second of a system of three spheres which carry it round. This
second sphere (shown white) is the deferent and is eccentric, so
that the adjacent surfaces of the first and third spheres (shown
black) are also eccentric. The spheres are given the motions required
to make the planet's motion correspond with the observations.
Inside the innermost sphere of Saturn the systems for the other
planets would come in their proper order (cf Fig. 2)

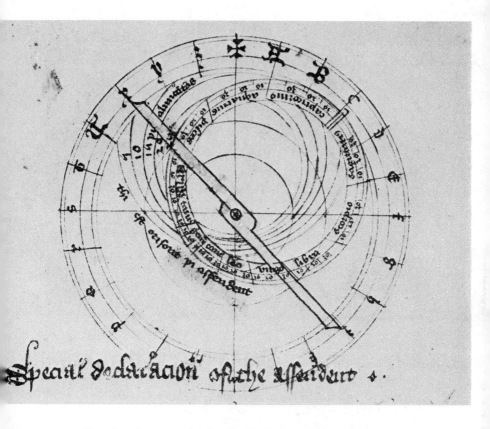

3. Drawing of an astrolabe showing the front with the *alidade*.
From Chaucer, *Treatise on the Astrolabe*, Cambridge University
Library MS Dd. 3.53 (14th century)

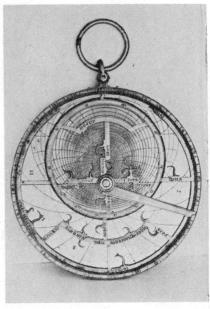

4. A late Gothic astrolabe, c. 1430, probably of French origin, showing back (left) and front (right). In the Museum of the History of Science, Oxford. See note on p. 245

5. (*opposite*) An astrolabe in use. From an English MS Bodley 614 (12th century) at Oxford

n aere uideni stelle aliquido cadere nulla cadente.
Cu eni sint nare ignee & ppe stellaru locʼ sit
ether: nunq̅. ad terrā descendunt. terrum cu sin̄t
maxime ꝙ si ex remotione parue uideani. si aliq̅
ex illis caderet. totā tam uʼ maximā eiusdē parcē
occuparet. Hon cade g̅: s̅, cade uideni. Sepe eni
in supioribʒ partibʒ aeris ÷ uent̅ & comotio. ꝗsi n̄
sit inferioribʒ: ex ea comotioē aer ignit̅ et spen
dens. p aera discurrit. Qui cu iuxta aliquam
stellam spende incipit: spendore suo uisum illi
stelle nobis aufēt. Uideecʒ ꝙ stella illa occidit.
S; dicee aliqs. Uide ÷ g̅ ꝙ stella illā n̄ uidemus:
hinc dicim. stellā eandem pea a nobʼ uidi s; eā esse
nescini. ꝗa cu ꝫ alio loco q̅. an eēt uidet̅. alia stella put

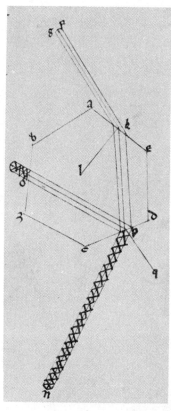

6. (*opposite*) Roger Bacon's geometrical diagrams showing the curvatures of the refracting media in the eye. From the *Opus Majus* British Museum MS Royal 7. F.viii (13th century). See note on p. 245

7. Drawing showing an experiment with the refraction of light. From Theodoric of Freiberg, *De Iride*, Basel University Library MS F. iv 30 (14th century). See note on p. 246

8. Drawing showing the paths of rays inside a transparent sphere, e.g. a spherical glass vessel of water or a raindrop, to illustrate the explanation of the formation of the primary rainbow. From *De Iride*. See note on p. 246

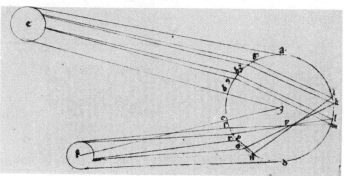

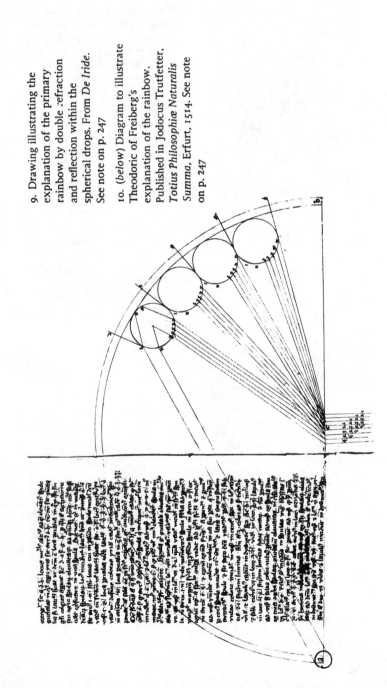

9. Drawing illustrating the explanation of the primary rainbow by double refraction and reflection within the spherical drops. From *De Iride*. See note on p. 247

10. *(below)* Diagram to illustrate Theodoric of Freiberg's explanation of the rainbow. Published in Jodocus Trutfetter, *Totius Philosophiae Naturalis Summa*, Erfurt, 1514. See note on p. 247

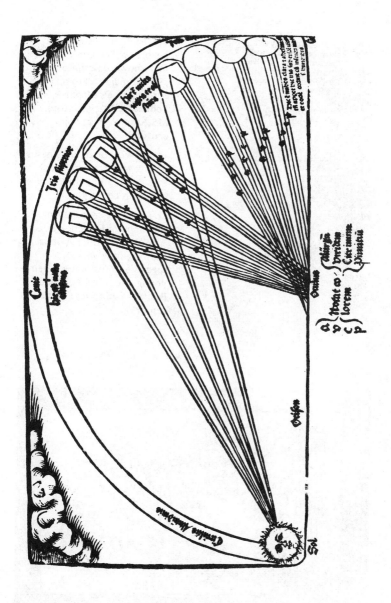

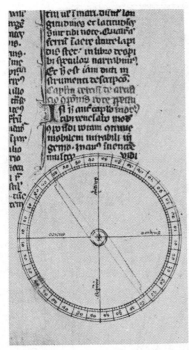

11. Diagram illustrating a chapter which contains the first known description of a pivoted magnet. From Petrus Peregrinus, *De Magnete*, Bodleian Library, Oxford, MS Ashmole 1522 (14th century)

12. Drawing of an ant's nest among wheat. From British Museum MS Royal 12. C. xix (late 12th century)

13. *(opposite)* A page from the Emperor Frederick II, *De Arte Venandi Cum Avibus*, showing how various species of birds protect their young. From MS *Vaticano Palatino Latino* 1071 (13th century)

...us capittee se aquis timet... aquas celestissime li... in capitulo se senili... aninim p̃haro tinq̃ a... ad ista confugiunt q̃... y celussione sui admit... y vnroq̃. Ille... q̃ nate sunt ut arboz... se consiguunt... modo cernimus plura... plura. Et die q̃ nate sii... alte se a ista eadem assig... te arbores confugiunt. qui... ad illis qui ad arbores ut un... cetomu. Si vero nate s... sup̃ petra fructiera. tur du... nios ad et consiguunt... nabis. thuribili e ameutari... plures q̃ nate sunt s̃ rupie... ad turbis consiguunt ut ta... partes sto q̃ nate sune sup̃... sonam e sunt coloris cinereus... sim lautame. e fit se ve... comentaunt ut pisces co... mumeie. cosuos. maris et... petres. caliandre e amouere... plures. se quib̃ multe sie... aveo trolite in cauela sui q̃... ardentes se esse cecuita y ter... ys cupiune euam manibus... homim. e qñ insequitur eos... rapit ad tram confugiune.

Perchices fastam e f̃ndi... q̃n sunt auri notarus e pr... obuetorum bon genus cesu...

sionem siue nubil louenr cerebar... longe ad o y apro cefendoni... sue celensionis. e impugna... tomb: s̃ouctis genuenti umi... maioz pure aurum con pin... obi. quia purisouno... ciprnales autem celensione... uslunt amoz e qua ut vie... neros e maros cumpiteros... est ppoa ascensio euatene lo... ge a se iteros sui in siueo... impaces que p̃requirur eas... Raurbis buiare e mares... cumpiteros cont uies rapa... ces compitant plures agm... sando se e cleuane alas trpo... neue capur ad mosd gallez... pugnamil quod p̃t amo... rem faciunt. se tame bitar... te e mares cumpiteros cont... ues impates. petuuue aues... e ptuce aues rapaces. Solr... e alie aues que refugiue ad... socieratem e congm̃atone... alas sue frmea ut me eas... femmees sue. e p eas cefen... oaut. siue siue cosibi gm... es thurueli. e alie sere osi... mo euam tota agmma in... se ista censius se courtrigile... auibs rapaub; facieutib; in... siltuum. Tea q̃ aures in sore... tare umeres siue siguum e... p̃ plres sunt s̃pis aunum e... concuruir ad cesentendum...

14. Drawings of spiders and insects. Formerly attributed to Cybo of Hyères. From British Museum MS Additional 28841 (14th century)

15. Water-colour painting of bramble (*Rubus fructicosus*). From the Juliana Anicia Codex of Dioscorides, *Codex Vindobonensis* (A.D. 512), in the National Bibliothek, Vienna

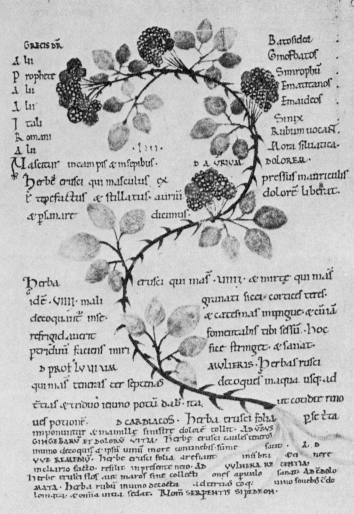

16. Painting of bramble (*Rubus fructicosus*). From the *Herbal of Apuleius Barbarus*, perhaps executed at Bury St Edmunds, Suffolk, Oxford MS Bodley 130 (12th century)

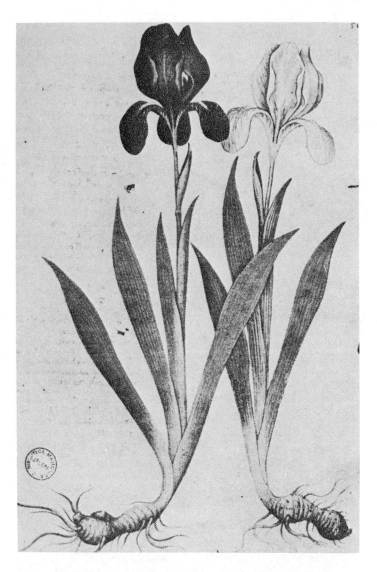

17. Painting of an iris, probably *I. chamæiris*. From Benedetto Rinio, *Liber de Simplicibus*, MS *Marciano Latino* vi. 59 (A.D. 1410), in the *Biblioteca Nazionale di S. Marco*, Venice

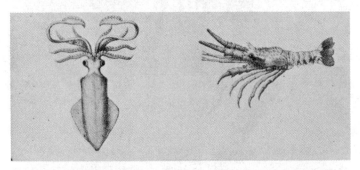

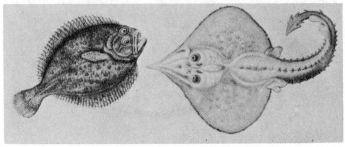

18. Zoological paintings from Petrus Candidus, *De Omnium Animantium Naturis*, MS Vaticano Urbinato Latino 276 (A.D. 1460). Reading from top to bottom, they illustrate: (A) *Formices* (unidentifiable ants); (B) Castor (beaver, *Castor fiber*); (C) *Loligines* (squid, *Loligio vulgaris*) and *Locusta maris* (lobster, *Palinurus vulgaris*); (D) *Rombus* (flounder, *Rhombus* spec.) and *Ratte vel rais* (sting ray, *Lophius piscatorius*). The first name in each case is that given by Petrus Candidus himself

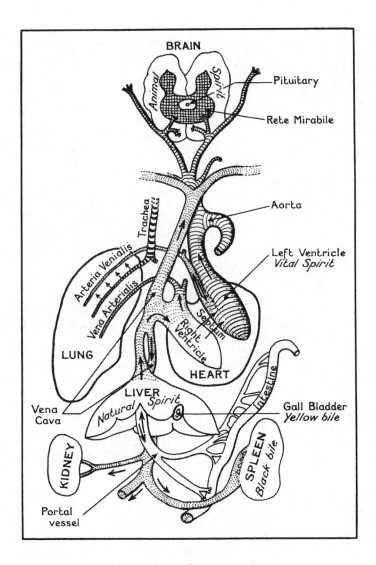

19. Galen's system of physiology. The arrows indicate the general direction of the movement of the blood and air

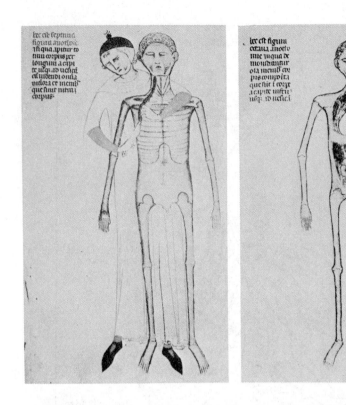

20. Two illustrations from Guido da Vigevano, *Anatomia*, showing respectively, a surgeon beginning a dissection, and the thoracic and abdominal viscera. The corpse is hung on a gibbet for dissection; cf Vol. II, Plate 21. From MS Chantilly 569 (14th century)

21. Richard of Wallingford measuring a circular instrument with a
pair of compasses. Note his abbot's crook and the mitre on the
floor, and the spots on his face, perhaps the leprosy he contracted
early in life and of which he died at the age of forty-three. From
British Museum MS Cotton Claudius E. iv (14th century)

22. Playing a stringed instrument with a bow. From British Museum MS Additional 11695 (12th century)

23. *(opposite)* Saxon ox-plough. From British Museum MS Julius A. vi (8th century)

24. *(opposite)* Harnessing with collar and lateral traces, and shoeing with nailed shoes. From the *Luttrell Psalter*, British Museum MS Additional 42130 (14th century)

25. *(opposite)* Watermill. From the *Luttrell Psalter*

26. *(opposite)* Spinning wheel. From British Museum MS Royal 10. E. iv (14th century)

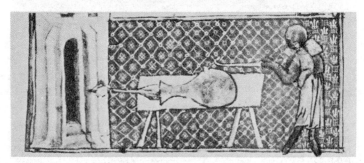

27. Windmill. From Oxford MS Bodley 264 (14th century)

28. Ships showing construction, rig, and rudder. From the *Luttrell Psalter*

29. Knight firing a cannon against a castle. From Walter de Milemete, *De Nobilitatibus Sapientiis et Prudentiis Regum*, Christ Church, Oxford, MS 92

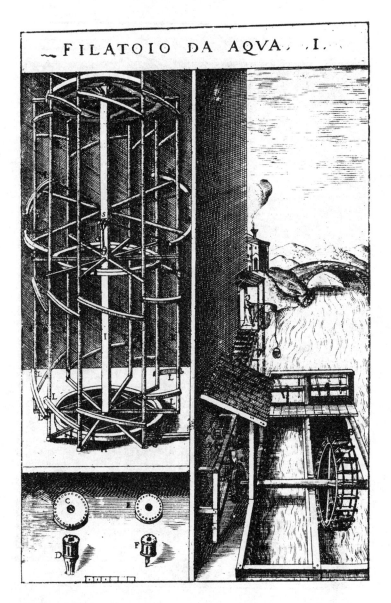

30. Water-driven silk mill. From V. Zonca, *Nova Teatro di Machine et Edificii*, Padua, 1607

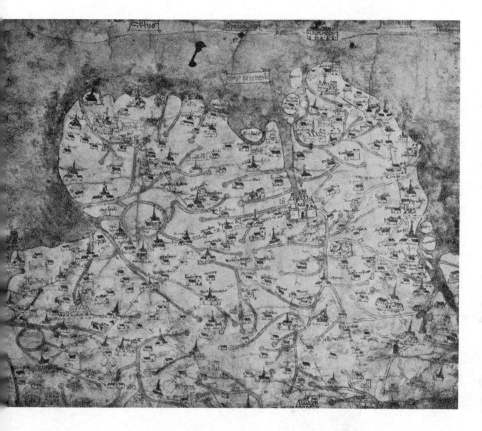

31. *(opposite)* Water-driven silk mill. From *Nova Teatro di Machine et Edificii*

32. Part of the so-called 'Gough Map' (1325–30), in the Bodleian Library, Oxford, showing S.E. England. The map is drawn with east at top

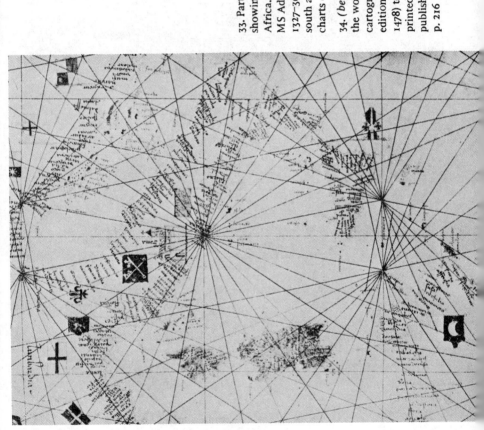

33. Part of a Portolan chart, showing Italy, Sicily, and N. Africa. From British Museum MS Additional 25691 (c. 1327–30). The map is drawn with south at the top, as usual with charts of this period

34. (*below*) Ptolemy's map of the world, redrawn by Italian cartographers. From the second edition of his *Geographia* (Rome, 1478) to contain maps. The first printed atlas was the edition published in Bologna, 1477. See p. 216

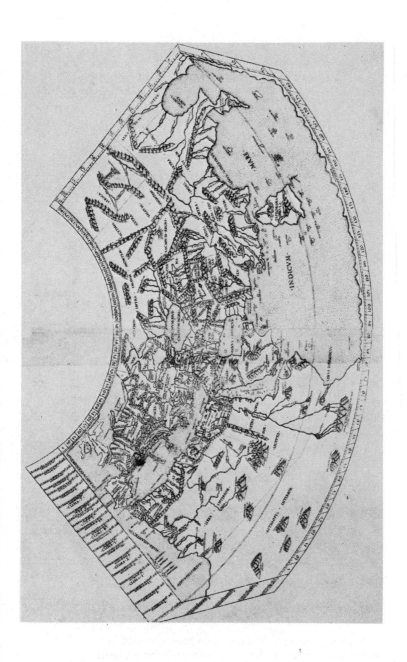

TERTIVM TORNI GENVS, SVBTILITATE NON CARENS, AD
INSCVLPENDAM PEDETENTIM-COCHLEAM CVIVS VIS
FORMÆ, IN AMBITVM CVIVSCVNQVÆ FIGVRÆ ROTVNDÆ
ET SOLIDÆ, VEL ETIAM OVALIS·

35. Screw-cutting lathe. From Jacques Besson, *Theatrum Instrumentorum et Machinarum*, Lyons, 1569 (1st ed. 1568)

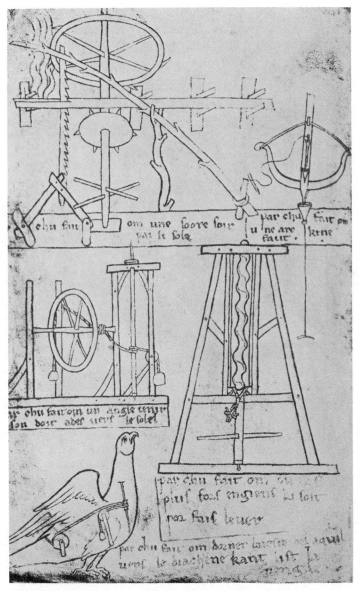

36. Page from the *Album* of Villard de Honnecourt, showing the escapement mechanism in centre left. Above is a water-driven saw. From the *Bibliothèque Nationale*, Paris, MS français 19093 (13th century)

37. The Dover Castle clock, formerly dated 14th century, but now believed to be later. On the left is the striking train, and on the right the going train regulated by the foliot. Crown Copyright. Science Museum, London

38. Glassmaking. From British Museum MS Additional 24189 (15th century)

39. Surgery. Sponging a patient, probably a leper, trephining, operating for hernia and treating fractures, from Roland of Parma, *Livre de Chirurgie*. From British Museum MS Sloane 1977 (13th century)

the starting-point for some of the striking developments in mechanics that took place in the late 16th and 17th centuries.

*

The other natural moving force or power besides gravity which chiefly engaged the attention of 13th-century physicists was magnetic attraction. This was the subject of what is one of the most striking examples of planned experimental research before the end of the 16th century, and William Gilbert, writing in 1600, acknowledged his debt to the little book completed on 8 August 1269. The *Epistola de Magnete* of Petrus Peregrinus of Maricourt, in which important sections of Gilbert's work were anticipated, was written as a letter to a fellow countryman of its author in Picardy, while Peregrinus waited in Charles of Anjou's besieging army outside the walls of the south Italian town of Lucera.

Certain properties of the lodestone were known before Petrus Peregrinus' researches. The fact that it attracted iron had been known to Thales and was later widely quoted as the classical example of occult 'virtue'. Its tendency to orientate itself north and south was known to the Chinese and adapted, perhaps by Moslems in maritime contact with them, for the invention of the compass. The earliest references to this instrument in medieval Latin literature occur in Alexander Neckam's *De Naturis Rerum* and other works round about 1200, but it is probable that its navigational use in the West preceded that date. Compasses with floating and later with pivoted needles were used from the end of the 13th century by both Arab and Christian sailors in the Mediterranean in conjunction with portolan maps or 'compass-charts' (see below, p. 214 *et seq.*). At the end of his treatise Petrus Peregrinus described improved instruments with both types of needle (Plate 11). His floating needle was used with a reference scale divided into 360 degrees.

Peregrinus opened his observations on magnets with the following injunction:

You must realize, dearest friend, that the investigator in this subject must know the nature of things and not be ignorant of the celestial motions; and he must also make ready use of his own hands, so that through the operation of this stone he may show remarkable effects. For by his carefulness he will then in a short time be able to correct an error which by means of natural philosophy and mathematics alone he would never do in eternity, if he did not carefully use his hands. For in hidden operations we greatly need manual industry, without which we can usually accomplish nothing perfectly. Yet there are many things subject to the rule of reason which cannot be completely investigated by the hand.

He then passed to the consideration of how to recognize lodestones, how to determine their poles and distinguish north from south, the repulsion of like poles, the induction in iron of the opposite pole to that of the lodestone with which it was rubbed, inversion of poles, the breaking of a magnetic needle into smaller ones, and the exertion of magnetic attraction through water and glass. One of the nicest experiments was made to determine the poles of a spherical lodestone or, as he called it, *magnes rotundus*, designed to illustrate the heavenly movements. A needle was held on the surface of the lodestone and a line drawn on the stone in the direction the needle took. The two points of junction of lines drawn from various positions would then be the poles of the lodestone.

The directive action on a magnet pointing north he attributed neither to the magnetic poles of the earth as Gilbert was to do in his theory that the earth was a large magnet, nor to the North Star as some of Petrus' contemporaries held. He pointed out that the lodestone did not always point directly at the North Star. Nor, he said, could its orientation be attributed to supposed deposits of lodestone in the northern regions of the earth, for lodestone was mined in many other places. He held that the magnet was directed towards the poles of the heavens on which the celestial sphere revolved, and he discussed the design of a *perpetuum mobile* based on this theory. But a contemporary, John of St Amand, at the end of his *Antidotarium Nicolai*, approached the modern conception of magnetism. He said :

Wherefore I say that in the magnet is a trace of the world, wherefore there is in it one part having in itself the property of the west, another of the east, another of the south, another of the north. And I say that in the direction north and south it attracts most strongly, little in the direction east and west.*

Petrus Peregrinus' explanation of the induction of magnetism in a piece of iron was based on Aristotelian principles of causation. The lodestone was an active agent which assimilated the passive iron to itself, actualizing its potential magnetism. This conception was elaborated by John of St Amand. He held that when a magnet pointed to the earth's poles,

the southern part attracts that which has the property and nature of the north, albeit they have the same specific form, and this is not except by some property existing more complete in the southern part which the northern part has potentially and thereby its potentiality is completed.

The action of magnetic attraction at a distance had been explained by Averroës as a form of 'multiplication of species'. The lodestone modified the parts of the medium touching it, for example air or water, and these then modified the parts next to them and so on until this *species magnetica* reached the iron, in which a motive virtue was produced causing it to approach the lodestone. The resemblance between this and Faraday's and Maxwell's tubes of force was brought even closer by John of St Amand's description of a 'current from the magnet through the entire needle placed directly above it'.

5. GEOLOGY

Geology in the 13th century was concerned mainly with the changes in the relative positions of the main masses of the elements earth and water forming the terraqueous globe in the centre of the universe, with the origin of continents and oceans and of mountains and rivers, and with the cause of the production of minerals and fossils. The three main sources of medieval geology were Aristotle's *Meteorologica* and two Arabic treatises, the pseudo-Aristotelian *De Proprietatibus Elementorum* or *De Elementis*, written probably in the 10th century, and Avicenna's

* L. Thorndike, *Isis*, 1946, vol. 36, pp. 156–57.

10th-century *De Mineralibus*. Aristotle did not fully discuss all the geological questions which later arose out of his cosmological theories, but he recognized that parts of the land had once been under the sea and parts of the sea floor once dry. He attributed this mainly to water erosion. He also offered explanations of rivers and minerals. He held that rivers originated in springs formed for the most part from water which, after being evaporated from the sea by the sun, rose to form clouds, and these, on cooling, fell again as rain and percolated into spongy rock. Thence the water ran out as springs and returned by rivers to the sea. He also believed that water was produced inside the earth by the transformation of other elements. Minerals he believed were formed by exhalations arising inside the earth under the action of the sun's rays. Moist exhalations produced metals, dry exhalations 'fossils'.

Some later Greek writers had used erosion by water as evidence for the temporal origin of the earth, for, they argued, if the earth had existed from eternity all mountains and other features would by now have disappeared. This view was opposed in other Greek works such as *On the Cosmos*, which some scholars have said was based on Theophrastus* (*c.* 372–287 B.C.). In this work it was maintained that there was a fluctuating balance between erosion by water and the elevation of new land caused by fire imprisoned in the earth trying to rise to its natural place. Against this a purely 'neptunic' theory was developed again from the *Meteorologica*, by late Greek commentators such as Alexander of Aphrodisias (*fl.* 193–217 A.D.), according to whom the earth had once been completely covered with water which the sun's heat had evaporated to expose the dry land. There was supposed to be a gradual destruction of the element water. This last conclusion had, in fact, been deduced by certain Greek philosophers of the 5th century B.C. from the presence of inland fossils; they alone in antiquity seem to have understood that fossils were the remains of animals which had lived under the waves once covering the places where they were found. The presence of inland shells had also been widely attributed by later Greek geographers to a partial

* Theophrastus' only surviving geological work is *Concerning Stones*.

withdrawal of the sea, such as that caused by the silting up of the Nile, but shells on mountains were believed to have been carried there by temporary deluges. The explanation of mountains, according to the theory contained in the late Greek commentaries on the *Meteorologica*, was that once the land had been exposed its perfectly spherical shape was then carved into valleys by water, leaving the mountains projecting above them.

Some time about the 10th century the author of the pseudo-Aristotelian *De Elementis* once more refuted this pure 'neptunism', and Avicenna in his *De Mineralibus* replaced it by a 'plutonic' explanation of mountains. He accepted the theory that the whole earth had once been covered with water and put forward the view that the emergence of dry land and the formation of mountains was due, sometimes to sedimentation under the sea, but more often to the eruption of the earth by earthquakes due to wind imprisoned under the earth. The mud thus raised was then transformed into rock partly by the hardening of clay in the sun and partly by the 'congelation' of water, either in the way stalactites and stalagmites are formed, or by some form of precipitation brought about by heat or by some unknown 'mineralizing virtue' generated in the petrifying clay. Plants and animals imprisoned in the clay were there turned into fossils. Once formed, mountains were eroded by wind and water and went on being gradually destroyed.

Avicenna's theory was adopted in his *De Mineralibus et Rebus Metallicis* (*c.* 1260) by Albertus Magnus, who quoted volcanoes as evidence for imprisoned subterranean wind and attributed the generation of the 'mineralizing virtue' to the influence of the sun and stars. The geology of Albertus was largely derived from the *Meteorologica, De Elementis*, perhaps *On the Cosmos*, and from Avicenna's *De Mineralibus*, but he worked his authorities into a coherent theory and made a number of observations of his own. He extended Avicenna's account of fossils, of which he said in his *De Mineralibus et Rebus Metallicis*, book 1, tract 2, chapter 8:

There is no-one who is not astonished to find stones which, both externally and internally, bear the impressions of animals. Externally they show their outline and when they are broken open there is found the shape of the internal parts of these animals. Avicenna teaches us that the cause of this phenomenon is that animals can be entirely transformed into stones and particularly into salt stones. Just as earth and water are the usual matter of stones, he says, so animals can become the matter of certain stones. If the bodies of these animals are in places where a mineralizing power (*vis lapidificativa*) is being exhaled, they are reduced to their elements and are seized by the qualities peculiar to those places. The elements which the bodies of these animals contained are transformed into the element which is the dominant element in them; that is the terrestrial element mixed with the aqueous element; then the mineralizing power converts the terrestial element into stone. The different external and internal parts of the animal keep the shape which they had beforehand.

He went on in another work, *De Causis Proprietatum Elementorum*, book 2, tract 3, chapter 5:

Evidence of this is that parts of aquatic animals and perhaps of naval gear are found in rock in hollows on mountains, which water no doubt deposited there enveloped in sticky mud, and which were prevented by the coldness and dryness of the stone from petrifying completely. Very striking evidence of this kind is found in the stones of Paris, in which one very often meets round shells the shape of the moon.

Albertus gave original descriptions of many precious stones and minerals, although he derived the substance of his mineralogy from Marbode. He accepted many of the magical properties ascribed to stones. He also described an explanation of rivers widely held until the 17th century. Some early Greek writers, such as Anaxagoras and Plato, had held that there was an immense reservoir in the earth from which springs and rivers came. This gave rise to the theory, supported by certain passages of the Bible, of the continuous circulation of water from the sea through underground caverns and up inside mountains, from which it flowed as rivers back again to the sea. Albertus accepted this. Among his own geological observations, those he made near Bruges led him to deny sudden universal overflowings of oceans and to reduce changes in the figures of continents and seas to slow modifications in limited areas.

Other writers in the 13th century made observations on various other geological phenomena. The tides had been correlated with the phases of the moon by Stoic Posidonius (b. *c.* 135 B.C.), and, like the menstruation of women, were commonly attributed to astrological influences. In the 12th century Giraldus Cambrensis had combined some observation with a discussion of this and other theories. Grosseteste in the next century attributed the tides to attraction by the moon's 'virtue', which went in straight lines with its light. He said that the ebb and flow of the tides was caused by the moon drawing up from the sea floor mist, which pushed up the water when the moon was rising and was not yet strong enough to pull the mist through the water. When the moon had reached its highest point the mist was pulled through and the tide fell. The second, smaller monthly tide he attributed to lunar rays reflected from the crystalline sphere back to the opposite side of the earth, these being weaker than the direct rays. Roger Bacon took over this explanation. In another work associated with Grosseteste's circle, the *Summa Philosophiæ* of pseudo-Grosseteste, a good account was given of contemporary thought about geology generally and many other related subjects. Another 13th-century work, the Norwegian encyclopædia, *Konungs Skuggsja* or *Speculum Regale*, contained descriptions of glaciers, icebergs, geysers and other phenomena. These, like Michael Scot's descriptions of hot sulphur springs and of the volcanic phenomena of the Lipari Islands, are evidence of a wide interest in local geology, which increased in the following centuries.

The most important Italian writer on geology in the 13th century was Ristoro d'Arezzo. It is probable that he knew the work of Albertus Magnus, though he may simply have used the same sources. But certainly Italian geology in general was dominated for the next two centuries by Albertus Magnus. In accordance with the Italian tradition Ristoro, in *La Composizione del Mondo* (1282), was very astrological. He attributed the elevation of dry land above the sea to attraction by the stars, as iron was attracted by magnets. He also recognized other influences, such as water erosion, sea waves throwing up sand and gravel, Noah's Flood depositing sediment, earthquakes, calcareous deposits from certain waters, and the activities of man. He made a number of observations, describing in the Apennines the eroded castellated strata containing iron which lay over the aqueous deposits of softer sandstones, shales and conglomerates. He recognized the marine origin of certain fossilized mollusc shells and discovered, apparently during a mountain expedition, a hot pool in which his hair became 'petrified' while bathing. He attributed the presence of these fossilized shells in mountains, not to their having been petrified where they had once lived, but to the Flood.

In the 14th century, the clockmaker Giovanni de' Dondi described the extraction of salt from hot springs and explained these as due to subterranean waters heated not, as Aristotle and Albertus Magnus had said, by flowing over sulphur, but by subterranean fire and gases produced by the heating action of celestial rays. The heating action of celestial virtue was also one explanation of the fire at the centre of the earth in which some alchemists believed and which they used to explain the presence of metallic ores, supposed to have been formed by condensation from metalligenous vapours, and also of volcanoes and similar phenomena. Geological matters were also discussed in Italy by such 14th-century writers as Dante (1265–1321), Boccaccio (1313–75) and Paulo Nicoletti of Venice (d. 1429), and in the 15th century by Leonardo Qualea (*c.* 1470) and Leon Battista Alberti (1404–72), who made observations on various local phenomena. All Italian writers who discussed the subject either accepted Ristoro's explanation of fossils in mountains as having been carried there by the Flood, or denied their organic origin altogether and regarded them either as having been spontaneously generated by a plastic or formative virtue produced by celestial influence or simply as accidents or 'sports' of nature.

In Paris in the 14th century, Jean Buridan (d. after 1358), in his *Quæstiones de Cælo et Mundo*, and Albert of Saxony or, as he was sometimes called, Albertus Parvus (*fl.c.* 1357), developed a new explanation of land and mountain formation. Albert based his conclusions on his theory of gravity (see Vol. II, p. 60). He held that the earth was in its natural place when its centre of gravity coincided with the centre of the universe. The centre of volume of the earth did not coincide with its centre of gravity, for the sun's heat caused part of the earth to expand and project above the enveloping water which, being fluid, remained with its centre of gravity at the centre of the universe. The shift of earth relative to water thus gave rise to dry land, leaving other parts submerged, and justified the hypothesis, later exploded by Christopher Columbus (1492), of a hemisphere of ocean balancing a hemisphere of land. The projecting land was then eroded by water into valleys, leaving the mountains. This was the only function Albert of Saxony ascribed to water and, together with the heat of the sun, it again displaced the centre of gravity of the earth, which thus underwent continuous little movements in order to coincide with the centre of the universe and caused continuous changes in the boundaries of land and sea. The erosion by water washed the land into the sea of which the floor, owing to the movements of the centre of gravity of the earth, gradually moved right through the middle of the earth eventually to reappear again as dry land on the other side. He used this theory of the shifting of the earth to explain the precession of the equinoxes. He made no mention of fossils.

Another northern successor of Albertus Magnus, Conrad von Megenburg (1309–74), put forward in his work *Das Buch der Natur* the view that springs and rivers were due to rain and rain alone. This had already been suggested by the Roman architect Vitruvius (1st century B.C.). This explanation, as well as Albert of Saxony's theory of mountains and Albertus Magnus' explanation of fossils, was accepted by Leonardo da Vinci and passed, via Cardano and Bernard Palissy, to the 17th century.

6. CHEMISTRY

Medieval chemistry began as an empirical art, but by the 13th century it had acquired a considerable body of theory, the purpose of which was to explain the particular kind of change with which chemistry was concerned, namely changes of quality and of substance in inanimate substances in the terrestrial region. This body of theory became inextricably interwoven with alchemy, and this association was to determine the character of chemical investigation for four centuries. Alchemy was empirical in spirit but was led up a blind theoretical alley by concentrating its attention rather on changes in colour and appearance than on changes in mass. So, while alchemical practice produced a large amount of useful information, alchemical theory had little to offer to the new chemistry that began to grow up in the 17th century.

The chief sources of practical chemistry in the 13th century were, apart from the practical experience handed down from generation to generation, the Latin translations of a number of Greek and Arabic treatises on dyeing, painting, glass making and other decorative processes, pyrotechnics, *materia medica*, mining and metallurgy, to which successive generations perhaps added one or two new recipes (see below, pp. 219–28 *et seq.*). The few Latin chemical manuscripts that remain from before the 12th century are entirely practical, but from about 1144, when Robert of Chester translated the *Liber de Compositione Alchemiæ*, Arabic alchemy began to enter western Europe.

The origin of alchemy seems to have been in the union of the practice of Egyptian metal workers with the theories of matter of Alexandrian Gnostics and Neoplatonists which, apart from a Timæan conception of *materia prima*, were fundamentally Aristotelian. The earliest alchemists, such as Zosimus and Synesius in the 3rd century A.D., who were Gnostics, thus combined descriptions of chemical apparatus and practical laboratory operations with an account of the visible universe as an expression of figures and symbols and a belief in sympathetic action, action at a distance, celestial influence, occult powers beneath manifest qualities, and the powers of numbers. These ideas permeated chemistry from the 3rd century A.D. to the 17th, and very often even practical laboratory operations were described in obscure symbolic language, perhaps to deceive others and keep the secrets hidden. It was Zosimus who first used the word *chemeia*, the Art of the Black Land, Egypt or *Khem*, which gave rise to the Arabic *alchemy* and the modern English *chemistry*. The main object of alchemy was the production of gold from the base metals. The possibility of doing so was based on the idea developed by Aristotle that one substance might be changed into another by changing its primary qualities.

Aristotle held that the generation and corruption of substantial forms in the sublunary region occurred at various levels in a hierarchy of substances. The simplest instances of perceptible matter were the four elements, but these were analysable in thought into *materia prima* determined by various combinations of the two pairs of primary contrary qualities or elementary principles acting as 'forms'. Perceptible substances differed from each other in many ways, for instance, in smell, taste or colour, but all, Aristotle said, were either hot or cold, wet or dry (fluid or solid). These four qualities were therefore primary and all others were secondary and derivative. The four elements were determined by the primary qualities as follows: Hot Dry = fire, Hot Wet = air, Cold Wet = water, Cold Dry = earth. The four elements of Empedocles had been unchangeable but with Aristotle, by interchanging members of the two pairs of primary contrary qualities, one element might be transformed into another. The old form (e.g., Cold Wet) was then said to have

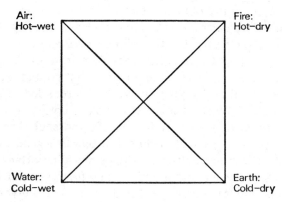

Fig. 9 The four elements.

been corrupted and the new one (e.g., Hot Wet) generated. Such substantial changes might involve a change of one or both qualities, or two elements might come together and interchange their qualities to produce the two others, as, for instance: Water (Cold Wet)+Fire (Hot Dry) $\rightleftarrows$ Earth (Cold Dry)+Air (Hot Wet) (Fig. 9). The second kind of change could not, of course, occur between consecutive elements, for this would bring together either two identical or two contrary qualities, which was *ipso facto* impossible. In chemical change and combination the combining substances thus disappeared with their properties, although they remained potentially regenerable, and new substances with new properties arose from their union. In a mixture, on the other hand, all the substances retained their properties and no new 'substantial form' arose. This Aristotelian idea that the elements might be transformed suggested that by depriving metals of certain attribuutes, or perhaps of all their attributes and thus reducing them to *materia prima*, they could subsequently be given the attributes of gold. For this purpose alchemists tried to discover an elixir, the 'Philosopher's Stone', which would act as a catalyst or as a ferment as yeast acted on dough.

By the 7th century, when the Arabs captured Alexandria, the magical element in Greek alchemy had gone far beyond the practical. Arabic alchemy derived mainly from Greek sources, but the leading exponents gave it once again a more practical turn. The first important Arabic alchemical documents are those that were traditionally attributed to Jabir ibn Hayyan, alleged to have lived in the 8th century, but the brilliant researches of Paul Kraus have left little doubt that in fact they date from the late 9th and early 10th centuries. Indeed the writings that go under the name of Jabir are in all probability the work of a sect dedicated to the pursuit of alchemy as a science with power both to give control over the forces of nature and to purify the soul. The Jabir to whom they are attributed is probably purely legendary. In the pursuit of their inquiries, these writings mark important developments in both theory and practice. 'The first essential in chemistry,' runs one passage as rendered by E. J. Holmyard in his *Makers of Chemistry* (Oxford, 1931, p. 60), is that thou shouldst perform practical work and conduct experiments, for he who performs not practical work nor makes experiments will never attain to the last degree of mastery. But thou, O my son, do thou experiment so that thou mayest acquire knowledge.

The Jabir corpus accepted the Aristotelian theory that minerals were generated from exhalations in the earth, but held that in the formation of metals the dry exhalations first produced sulphur and the moist exhalations mercury, and that metals were formed by the subsequent combination of these two substances. It contains the discovery, however, that ordinary sulphur and mercury combined to form not metals but a 'red stone' or cinnabar (mercuric sulphide), and it therefore concludes that it was not these which formed metals but hypothetical substances to which they were the nearest approach. The most perfect natural harmony and proportion of combination produced gold; other metals were the result of defects in either purity or proportion of the two ingredients. The object of alchemy was therefore to remove these defects. As regards practical chemistry, the Arabic manuscripts attributed to Jabir contain descriptions of such processes as distillation and the use of sand-baths and water-baths, crystallization, calcination, solution, sublimation, and reduction, and of such practical applications as the preparation of steel, dyes, varnishes and hair-dyes.

Among the other Arab alchemists who influenced Western Christendom, the most important were Rhazes (d. *c.* 924) and Avicenna (980–1037). Rhazes gave both a clear account of apparatus for melting metals, distilling and other operations, and a systematic classification of chemical substances and reactions. He also combined Aristotle's theory of *materia prima* with a form of atomism. Avicenna, in his *De Mineralibus*, the geological and alchemical part of the *Sanatio* (*Kitab al-Shifa*), made few fundamental chemical advances on his predecessors, but gave a clear account of the accepted theories. One aspect of chemical theory which caused difficulty was to explain how, in chemical combination, elements which no longer existed in the compound could be regenerated. Avicenna held that the elements were present in the compound not merely potentially but actually, but the question continued to trouble the medieval scholastics. Avicenna also made an attack on the makers of gold. Disbelief in transmutation had existed since the period of the Jabir writings and Avicenna, while accepting the theory of matter on which the claim was based, denied that alchemists had ever brought about more than accidental changes as, for instance, in colour. In spite of the practical spirit of Rhazes, through which the Arab chemists developed such processes as the refinement of metals by cupellation, that is, refining in a shallow vessel or cupel, and solution in acids and assays of gold and silver alloys by weighing and determining specific gravity, and in spite of Avicenna's criticism, the esoteric and magical art of alchemy continued to flourish vigorously. The earliest Arabic works translated into Latin thus included not only Rhazes' treatise on alums (or vitriols) and salts and Avicenna's *De Mineralibus*, but also the magical *Emerald Table*.

Both aspects of alchemy became popular in Western Christendom from the 13th century onwards, though such writers as Albertus Magnus usually adopted Avicenna's scepticism about transmutation. The encyclopædias of writers like Bartholomew the Englishman (*fl.c.* 1230–40), Vincent of Beauvais, Albertus Magnus and Roger Bacon contained a large amount of chemical information derived from both Latin and Arabic sources, and the last two seem to show some practical acquaintance with laboratory techniques. No fundamental advance was made on the Arabs in chemical theory before Paracelsus in the early 16th century, but in practical chemistry some important additions were made in the later Middle Ages.

Perhaps the most important Western contribution to practical chemistry was in methods of distillation. The traditional form of the still had been developed in Greco-Roman Egypt and was described by Zosimus and other early alchemical writers. This consisted of the curcurbite or vessel in which was placed the matter to be distilled, the alembic or still-head in which condensation occurred, and the receiver which received the distilled fraction after it had condensed (Fig. 10). The curcurbite was heated over a fire or in a sand-bath or water-bath. Modifications of this standard design were made for various purposes and were taken over

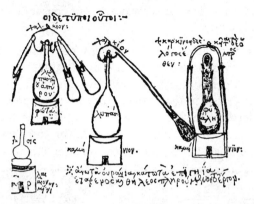

Fig. 10 Types of apparatus for distillation and sublimation (alembics), and for digestion, used by the Greek alchemists, *c.* A.D. 100–300. Similar types remained in use in the West till the end of the 18th century. From Bibliothèque Nationale, Paris, MS Grecque 2327.

by the Arabs, through whom they became known in the West, and some of these early designs, including the turk's-head type, in which the still-head was partly immersed in water to give more rapid condensation, remained in use as late as the 18th century. The Greco-Egyptian still was used at relatively high temperatures and was useful for distilling or sublimating substances like mercury, arsenic and sulphur. The Arabs improved it in various ways and introduced the gallery with several stills heated in one oven for producing substances like oil of roses and naphtha on a large scale, but neither the Greeks nor the Arabs developed efficient methods of cooling the alembic that would permit the condensation of volatile substances like alcohol. This seems to have been the contribution of the West (Fig. 11).

The earliest known account of the preparation of alcohol is described in the following paragraph translated from an early 12th-century manuscript of the technical treatise *Mappæ Clavicula* discussed by Berthelot in *La Chimie au Moyen Age*, vol. 1, p. 61:

On mixing a pure and very strong wine with three parts of salt, and heating it in vessels suitable for the purpose, an inflammable water is obtained which burns away without consuming the material [on which it is poured].

In the 13th century in Italy, *aqua ardens*, containing about 60 per cent alcohol, was prepared by one distillation, and *aqua vitæ*, with about 96 per cent alcohol, by redistillation. The method of cooling as described in the 13th century by the Florentine doctor, Taddeo Alderotti (1223–1303), was to extend the length of the tube leading from the alembic to the receiver and pass it horizontally through a vessel of water. The introduction of rectification

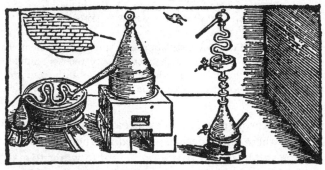

Fig. 11 Still with *canale serpentium* or *serpentes* condensing tube. From V. Biringuccio, *Pirotechnia*, Venice, 1558 (1st ed. 1540).

by distillation with limestone or calx is attributed to Raymond Lull (*c.* 1232–1315) and further improvements to the cooling apparatus in the 14th century are attributed to the Franciscan alchemist, John of Rupescissa (d. after 1356). Most of the early stills were probably of metal or pottery but in the early 15th century the Italian doctor, Michael Savonarola (1384–1464), speaks of distilling apparatus made of glass, which would be an obvious advantage in distilling substances like the mineral acids. By the end of the 13th century alcohol had become an important substance: it was used as a solvent in the preparation of perfumes and for extracting medicines, was prescribed as a medicine by doctors like Arnald of Villanova (*c.* 1235–1311), and spirits were beginning to take their place with wine and beer as a drink. By the 15th century distillers had become incorporated as a guild.

The still was used to prepare a number of other substances as well as alcohol. The earliest descriptions of the preparation of nitric and sulphuric acids are contained in a late 13th-century Latin manuscript of a work entitled *Liber de Investigatione Perfectionis*, which was attributed to Geber (the Latinized form of Jabir) and is probably based on Arabic sources but with Latin additions. In the 13th century a new type of still appeared for preparing concentrated acids, in which the neck of the curcurbite was extended and bent over to form a 'retort' and so prevent the distilling acids from attacking the *lutæ* or cements used in making the join between it and the alembic airtight. Mineral acids were prepared in fairly large quantities for assaying in metallurgy and good descriptions of their manufacture as well as that of sulphur, mercury and other substances obtained by distillation were given in the 16th century by metallurgical writers like Agricola and Biringuccio. The 'waters' or 'essences' of organic substances like plants and dried herbs and even ants and frogs were also obtained by steam-distillation, as well as by solution in alcohol, for use as medicines; and at least by the 16th century, with Hieronymus Brunschwig, it was recognized that these 'essences' were the active principles of drugs.

Some other improvements in practical chemistry appear in another late 13th-century Latin alchemical treatise attributed to Geber, the well-known *Summa Perfectionis*. This also was probably of Arabic origin but with Latin additions. It contains very clear and complete descriptions of chemical apparatus and practices used in trying to make gold. Beginning with a discussion of the arguments against transmutation and their refutation, it passes on to the theory that metals are composed of sulphur and mercury and to a description of the definition and qualities of each of the six metals, gold, silver, lead, tin, copper, iron. Then there is a description of chemical methods such as sublimation, distillation, calcination, solution, coagulation and fixation, of the nature of different substances, and of the preparation of each towards its transmutation by elixirs. Finally, there is a description of methods of analysis to ascertain whether the transmutation has succeeded. These include cupellation, ignition, fusion, exposition over vapours, admixture of burning sulphur, calcination and reduction. The *Summa Perfectionis* shows the considerable knowledge of chemical apparatus and processes in the possession of Western alchemists by the end of the 13th century and not least in interest is the evidence it gives of the use of the balance (Fig. 12), as in the observation that lead gains weight

Fig. 12 Chemical balance and furnace. From V. Biringuccio, *Pirotechnia*.

when calcined because 'spirit is united with the body.' Thus if alchemical theory went astray because it was based on too exclusive an attention to changes in colour and appearance, the alchemists' familiarity with the balance at least prepared the way for the concentration on mass on which modern chemistry is based.

The magical as well as the practical side of alchemy flourished vigorously in the West during the later Middle Ages. The search by alchemists for a formula that would give health and eternal youth, riches and power, is the origin of legends like that of Dr Faustus, and the wide publicity given to the more scientific alchemy by the great 13th-century encyclopædists gave rise, from the 13th to the 17th century, to an enormous number of manuscripts claiming the production of gold. These were at first written by men of some learning, but later, during the 14th and 15th centuries, by members of all classes; as Thomas Norton put it in *The Ordinall of Alchimy* (c. 1477) by 'Free *Masons* and *Tinkers* with poore *Parish Clerks*; *Tailors* and *Glasiers* ... And eke Sely Tinkers'; and often they were fathered on such names as Albertus Magnus, Roger Bacon, Arnald of Villanova and Raymond Lull,

Indeed, at times the practice became so common that it was condemned by princes and prelates alarmed for its effect on the value of money.

7. BIOLOGY

The common characteristic which distinguished all living from non-living things, according to Aristotle and to 13th-century ways of thought, was the ability to initiate movement and change without an external mover, that is, the power of self-movement or self-change. The kinds of movement or change common to all living things were growth, the assimilation of diverse matter under the form of the organism, and the continuation of this process in the reproduction of the species. These were the only kinds of living activity displayed by plants. Their substantial form was thus a 'nutritive soul' (or vital principle), which was not, of course, something separate and distinct from the material plant itself, but an inherent principle causing the observed behaviour. Animals added to nutrition the power of sensitivity, that is, the power to respond to environmental stimuli by local motion, and theirs was therefore a 'sensitive soul.' Men were distinguished again by the power of abstract reflection and the exercise of the will, which were the marks of a 'rational soul.' Men were also capable of sensation and nutrition, and animals of nutrition, the higher forms of soul including the activities of all those below them. Aristotle thus recognized a hierarchy of living forms stretching, as he said in the *Historia Animalium* (588 b 4), 'little by little from things lifeless to animal life,' from the first manifestations of life in the lower plants, through the plants to sponges and other sessile animals scarcely distinguishable from plants, and again through invertebrate and vertebrate animals, apes and pygmies to man. Each type was distinct and unchanging, its substantial form being both the efficient and final cause of its particular bodily activity, whether in nutrition, reproduction, locomotion, sensation or reasoning.

The subject of 13th-century biology, then, was these activities of the different beings making up the scale of living nature, and the way in which they were conceived of opened the way naturally for teleological as well as mechanical explanations. Aristotle and Galen had both taken a teleological view of the existence and functioning of organic structures, and this had led them to make valuable discoveries about the adaptation of the parts of organisms to each other and of the whole to the environment. Certainly, in the 13th century and later, the search for the purpose or function of organs often led to valuable conclusions. It was certainly also sometimes abused, as in what have been described as the wearisomely reiterated reasons for the existence of imperfectly described structures given by a writer like Guy de Chauliac.

Until the 13th century, the chief interest of the Latins in botany had been medical and in zoology moral and didactic. The same attitudes, in fact, characterized much of natural history down to the 17th century. When, in the 13th century, biology became a science combining observation with a system of natural explanations, this was largely due to the translations of Aristotle's own biological works, of the pseudo-Aristotelian *De Plantis* (a compilation from Aristotle and Theophrastus believed in the Middle Ages to be an original work of the former), and of various treatises by Galen. Robert of Cricklade's (Prior of St Frideswide's, Oxford, *c.* 1141–71) anthology of extracts from the *Natural History* witnesses to a revival of interest also in Pliny in the middle of the 12th century, and what the Arabs and, in particular, Avicenna and Averroës had to teach was quickly assimilated as it became available.

The early encyclopædias deriving from this movement included many incredible stories. Alexander Neckam (1157–1217) dismissed, as a ridiculous popular notion, the legend that the beaver, of which the testicles were the source of a certain medicine, castrated itself to escape its hunters, but he accepted the basilisk as the progeny of a cock's egg brooded by a toad and the common belief that an animal knew the medicinal value of herbs. For, as he said in his *De Naturis Rerum*, book 2, chapter 123:

educated by nature, it knows the virtues of herbs, although it has neither studied medicine at Salerno nor been drilled in the schools at Montpellier.

But Neckam made no claim to be a scientist. Like Hildegarde of Bingen (1098–1179), who, besides expounding mystical cosmology, in another work perhaps wrongly attributed to her named nearly a thousand plants and animals in German, he believed that man's Fall had had physical effects on nature, causing spots on the moon, wildness in animals, insect pests, animal venoms and disease, and his purpose was frankly didactic.

This didactic attitude was continued in many of the later encyclopædias, but other activities provided opportunities for observation. Some of these were associated with agriculture (see below, p. 196 *et seq.*) and produced the treatises on husbandry of Walter of Henley (*c.* 1250?) and Peter of Crescenzi (*c.* 1306) and the sections on agriculture in the encyclopædias of Albertus Magnus (*De Vegetabilibus et Plantis*) and Vincent of Beauvais (*Speculum Doctrinale*). Crescenzi's treatise remained the standard European work on the subject until the end of the 16th century. Also, Thomas of Cantimpré's *De Natura Rerum* (*c.* 1228–44) contains a description of herring fisheries, the *Konungs Skuggsia* of seals, walruses and whales, and Albertus Magnus, whose duties as provincial of the German Dominican province took him long distances on foot, gave an account in his *De Animalibus* of whaling and fishing and of German agricultural life. Travellers such as Marco Polo and William of Rubruck also brought back descriptions of new creatures, of the wild asses of Central Asia, and of rice, ginger and fat-tailed sheep.

The circle of natural philosophers and magicians which the Emperor Frederick II (1194–1250) kept at his court can claim a treatise on horse diseases, and the *De Arte Venandi cum Avibus* of Frederick himself is one of the most important medieval works on zoology. *The Art of Falconry*, based on Aristotle and various Moslem sources, began with a zoological introduction on the anatomy and habits of birds and went on to describe the rearing and feeding of falcons, the training of dogs for hunting with them, the various types of falcons, and the cranes, herons and other birds that were hunted. When Frederick made use of other practical treatises on falconry he did not hesitate to describe them as 'lying and inadequate,' nor did he hesitate to call Aristotle a man of books. The Emperor's book contains 900 pictures of individual birds, some of them possibly by Frederick himself, which are accurate even down to details of plumage, and the representations of birds in flight are obviously based on close and careful observation (Plate 13). He watched and questioned Saracen falconers, observed the nests of herons, cuckoos and vultures, and exploded the popular belief that barnacle geese were hatched from barnacles on trees. He had barnacles brought to him and, seeing that they contained nothing in any shape like a bird, concluded that the story had grown up simply because the geese bred in such remote parts that no-one had been there to see. He was interested in the mechanical conditions of flight and bird migrations, made experiments on the artificial incubation of eggs, and showed that vultures did not go for meat if their eyes were covered. He also noted various other points of bird behaviour, as, for example, how the mother falcon gave half-dead birds to her young to teach them how to hunt, and how the mother duck and other non-predatory birds feigned to be wounded and decoyed approaching strangers from their nests. He also described the air-cavities of the bones, the structures of the lungs, and other previously unrecorded facts of avian anatomy.

Other works on falconry, both in Latin and in the vernacular, witness to its wide popularity, but it was not the only sport that rendered service to zoology. The menageries which kings, princes and even towns kept, for amusements such as bear-baiting or out of curiosity, were in Italy and the East descended from those of antiquity. That which Frederick II carted about with him on his travels, even across the Alps, included elephants, dromedaries, camels, panthers, lions, leopards, falcons, bearded owls, monkeys and the first recorded giraffe to appear in Europe. The first large menagerie in the north was that established in the 11th century, at Woodstock, by the Norman kings. In the 14th century a large collection of exotic animals was kept by the Popes at Avignon. These forerunners of modern zoological gardens could satisfy the curiosity of the rich, and the charm exercised by animals over the minds of poorer people is shown by the well-known description of the domestic cat in Bartholomew the Englishman's

On the Properties of Things, the reputed source of Shakespeare's natural history.

A similar interest in nature is shown by the hounds, foxes, hares, and above all the foliage covering the capitals, bosses and misericords of York, Ely or Southwell cathedrals. There one may see, fresh and resilient, the leaves, flowers or fruit of the pine, oak, maple, buttercup, potentilla, hop, bryony, ivy and hawthorn. Émile Mâle, in his *Religious Art in France in the Thirteenth Century* (English translation, 1913, p. 52), has recognized in French Gothic cathedrals 'the plantain, arum, ranunculus, fern, clover, coladine, hepatica, columbine, cress, parsley, strawberry-plant, ivy, snapdragon, the flower of the broom and the leaf of the oak.' Even the conception of nature as symbolic of spiritual truths led, in the 12th and 13th centuries, to a special intensity of observation.

> The holly bears a bark
> As bitter as any gall
> And Mary bore sweet Jesus Christ
> For to redeem us all.

The same interest in nature is seen in the illustrations of certain manuscripts. Matthew Paris in his *Chronica Majora* (*c.* 1250) described an immigration of the crossbills (*cancellata*) and illustrated the bird. The borders of manuscripts from the 13th century onwards were frequently illuminated with naturalistic drawings of flowers and many kinds of animals, prawns, shells and insects. The 13th-century French architect, Villard de Honnecourt, interspersed his architectural drawings, studies in perspective and designs for engines of war and perpetual motion with illustrations of a lobster, a fly, a dragon-fly, a grasshopper, two parrots on a perch, two ostriches, a rabbit, a sheep, a cat, dogs, a bear, and a lion 'copied from life.' He also gave a recipe to preserve the natural colours of dried flowers (*d'un herbier*). The progress that was made in naturalistic illustration in the century after Villard de Honnecourt may be estimated by comparing his drawings with those in the late 14th-century Ligurian manuscript formerly attributed to a certain Cybo of Hyères. The borders of this manuscript contain illustrations of plants, quadrupeds, birds, molluscs and crustacea, spiders, butterflies and wasps, beetles and other insects, caterpillars as well as adults often being shown. A point of particular interest is the tendency to put together on the same page animals now classified as belonging to the same group (Plate 14).

In contrast with the naturalistic spirit of these manuscripts stands the conventional iconography of many of the encyclopædias and herbals. Singer has divided the illustrations of plants in the latter into what he calls the Naturalistic and the Romanesque traditions. Botanical iconography may be traced through the 6th-century Byzantine *Codex Aniciæ Julianæ* back to Dioscorides himself, whose own work was based on the herbal of Cratevas (1st century B.C.). He, according to Pliny, made coloured drawings of plants. The Benedictine monasteries not only cultivated extensive fields, but also planted kitchen and physic gardens, and the object of the herbal writer, who had little idea of the geographical distribution of plants, was usually to try to identify in his own garden the plants mentioned by Dioscorides and the *Herbarium* of pseudo-Apuleius (probably 5th century A.D.), the main text-books. Since the Mediterranean plants mentioned in these books were frequently absent or at best represented by other species of the same genus, neither the drawings nor the descriptions given in them corresponded to anything the Northern herbalist might see. In new herbals or new copies of the old texts, text and illustrations were usually made by a different hand, and in the Romanesque tradition the drawings made in the spaces left by the scribe became a matter of increasingly stylized copying. This tradition, which emanated from northern France and seems to have descended from a debased style of Roman art, reached its limit at the end of the 12th century.

Naturalistic representations of plants and animals were also made throughout the early Middle Ages, for example, in the mosaics in many churches in Rome, Ravenna and Venice. Some Latin herbals of the 11th and 12th centuries were also illustrated in this naturalistic tradition, of which the 12th-century Bury St Edmunds herbal is a striking example (Plate 16; cf. Plate 15).

From the 13th century onwards naturalistic illustrations steadily increased. Outside the herbals, naturalistic representations of plants and animals appear in the paintings of artists like Giotto (*c.* 1276–1336) and Spinello Aretino (*c.* 1333–1410) and, in the 15th century, herbal illustrators learnt from the three-dimensional realism of the art of Italy and Flanders, reaching perfection in the drawing of Leonardo da Vinci and Albrecht Dürer. An outstanding example is the herbal of Benedetto Rinio, completed in 1410, which was illustrated with 440 magnificent plates by the Venetian artist, Andrea Amodio (Plate 17). Both the naturalistic and the Romanesque traditions continued without a break into the early printed herbals with which histories of botany usually begin.

Considering the way they were composed it is not surprising that text and illustrations sometimes had little relation to each other, the former often describing a Mediterranean species known to the authority from whom it was copied and the latter being either purely formal or drawn from a native species known to the artist. But medical men relied on the herbals to identify plants with given pharmaceutical properties, and some attempt had to be made to improve verbal descriptions. These were almost always clumsy and frequently inaccurate and the synonyms given by writers of botanical lexicons or pandects, as, for instance, in the 13th century by Simon of Genoa and in the 14th century by Matthæus Sylvaticus (see below, p. 166), sometimes did not all correspond to the same object, even though considerable personal observation went into drawing them up. A clear, accurate and unambiguous nomenclature is, indeed, to be found nowhere before the 17th century and only imperfectly before Linnæus.

Not all medieval herbals restricted their interest wholly to pharmacy, nor were their descriptions all inaccurate. The *Herbal* (*c.* 1287) of Rufinus, which Thorndike has recently edited, was not only a medical herbal but a book of botany for plants' sake. Rufinus' authorities were Dioscorides, the *Macer Floridus* attributed to Odo of Meung who flourished at the end of the 11th century, the *Circa Instans* of the Salernitan doctor Matthæus

Platearius, the leading contribution to 12th-century botany, and several other works. As Thorndike has pointed out, Rufinus added to his authorities

careful, detailed description of the plant itself – its stalk, leaves, and flower – and an equally painstaking distinguishing of its different varieties or a comparison of it with, and differentiation of it from, other similar or related flora. He further takes care to inform us as to other names applied to a given herb or other plants indicated by the same name.

As in other herbals, the plants were nearly all in alphabetical order. Dioscorides had sometimes roughly grouped together plants of similar form and presented a series belonging to the Labiateæ, Componitæ, or Leguminosæ. The same tendency is seen in the Anglo-Saxon *Herbal* of about 1000 A.D., extracted from Dioscorides and pseudo-Apuleius; there was a real grouping of umbelliferous plants. Serious attempts at classification belonged to the natural scientific tradition of the North, whereas Rufinus, who had been brought up in the Italian medical tradition of Naples and Bologna, seems to have known nothing, in those days of expensive manuscripts, even of the *De Vegetabilibus et Plantis* of Albertus Magnus.

The botanical and zoological sections of the 13th-century encyclopædias of Bartholomew the Englishman, Thomas of Cantimpré and Vincent of Beauvais were by no means devoid of observation, but in this respect they cannot be compared with the digressions in which Albertus Magnus described his own personal researches when writing commentaries on Aristotle's works. The commentary, in which the text of the original might be either clearly distinguished from or included in the body of critical discussion, was the common medieval form of presentation of scientific work inherited by the 13th-century Latins from the Arabs. The *De Vegetabilibus et Plantis* (c. 1250) was a commentary on the pseudo-Aristotelian *De Plantis*, which in Alfred of Sareshel's translation was the chief source of botanical theory down to the 16th century. 'In this sixth book,' Albertus remarked at the beginning of a discussion of native plants known to him, we will satisfy the curiosity of the students rather than philosophy. For philosophy cannot discuss particulars ... Syllogisms cannot be made about particular natures, of which experience (*experimentum*) alone gives certainty.

Albertus' digressions show a sense of morphology and ecology unsurpassed from Aristotle and Theophrastus to Cesalpino and Jung. His comparative study of plants extended to all their parts, root, stem, leaf, flower, fruit, bark, pith, etc., and to their form. He observed that trees growing in the shade were taller, slimmer and had less branches than others, and that in cold and shady places the wood was harder. Both effects he attributed not to lack of light, but to lack of the warmth which favoured the activity of the roots in absorbing nourishment from the soil. The heat of the soil, which according to Aristotle served as the stomach of plants, was supposed to elaborate their food for them and therefore it was supposed that they needed to produce no excrement. Albertus claimed that the sap, potentially all parts of the plant because it supplied them with this nourishment, was carried in the veins which were like blood-vessels but had no pulse. The winter sleep of plants was caused by the retreat of the sap inwards.

He drew a distinction between thorns, which were of the nature of the stem, and prickles which were merely developed from the surface. Because in the vine a tendril sometimes grew in the place of a bunch of grapes, he inferred that a tendril was an imperfect form of a bunch of grapes. In the flower of the borage he distinguished, though without understanding their functions in reproduction, the green calyx, the corolla with its ligular outgrowths, the five stamens (*vingulæ*), and the central pistil. He classified floral forms into three types, bird-form as in the columbine, violet and dead nettle, pyramid- or bell-form as in the convolvulus, and star-form as in the rose. He also made an extensive comparative study of fruits, distinguishing between 'dry' and fleshy fruits, and described various types differing in the structure and relations of seed, pericarp and receptacle, in whether the pods burst or the flesh dried in ripening, and so on. He showed that in fleshy fruits the flesh did not nourish the seed, and in the seed he recognized the embryo. He also remarked in book 6, tract 1, chapter 31 :

On the leaves of the oak often form certain round ball-like objects called galls, which after remaining some time on the tree produce within themselves a small worm bred by the corruption of the leaf.

Theophrastus had suggested in his *Inquiry into Plants* that the vegetable kingdom should be classified into trees, shrubs, under-shrubs, and herbs, with further distinctions such as those between cultivated and wild, flowering and flowerless, fruit-bearing and fruitless, deciduous and evergreen, or terrestrial, marshy and aquatic plants within these groups. His suggestions were rather indefinite and tentative. Albertus' general classification follows the main outlines of this scheme. Though it is not set out in detail, Dr Agnes Arber, in her book on *Herbals*, has suggested that the following system might have been in his mind. His plants form a scale from the fungi to the flowering plants, though in the last group he did not explicitly recognize the distinction between monocotyledons and dicotyledons.

I. Leafless plants (mostly our cryptogams, that is, plants with no true flower).

II. Leafy plants (our phanerogams or flowering plants and certain cryptogams).

1. Corticate plants with stiff outer covering (our monocotyledons, having only one seed lobe).

2. Tunicate plants, with annular rings, *ex ligneis tunicis* (our dicotyledons, having two seed lobes).

a. Herbaceous.

b. Woody.

The appearance of new species had received an explanation from a number of natural philosophers before Albertus Magnus. In the cosmogonies of several of the early Greeks attempts were made to account for the origin of life and the variety of living things. Thus Anaximander held that all life had originated by spontaneous generation from water and that man had developed from fish. Xenophanes quoted fossil fish and seaweed as evidence that life arose from mud. Empedocles believed that life arose by spontaneous generation from earth: first plants appeared and then parts of animals (including man), heads, arms, eyes, etc., which united by chance and produced forms of all sorts, monstrous or proper. The proper forms extinguished the monstrous and, when the sexes had become differentiated, reproduced themselves, and the earth then ceased its generation. Similar views were adopted by Lucretius, and the notion of 'seeds' in the earth, to which Adelard of Bath alluded, received an explanation in the Stoic conception of *logoi spermatikoi*, which tended to produce new species of both animate and inanimate things from indeterminate matter. St Augustine's theory of the creation of things in their *rationes seminales*, or 'seminal causes' (see above, p. 48), which had a wide influence in the Middle Ages, was derived from this conception. It was paralleled among the Arabs by the 9th-century al-Nazzam and his pupil al-Jahiz, who speculated on adaptation and the struggle for existence.

Apart from Anaximander's, all these theories accounted for the succession of new species, not by modification from living ancestors, but by generation from a common source such as the earth. But some ancient writers, such as Theophrastus, had believed that existing types were sometimes mutable. Albertus accepted this belief and illustrated it by the domestication of wild plants and the running wild of cultivated plants. He described five ways of transforming one plant into another. Some of these did not involve a change of species but merely the actualization of potential attributes, such as when rye increased in size over three years and became wheat. Others involved the corruption of one substantial form and the generation of another, such as occurred when aspens and poplars sprang up in place of a felled oak or beech wood, or when mistletoe was generated from a sickening tree. Like Peter of Crescenzi later, he also believed that new species could be produced by grafting.

Speculations about the origin of new species and the mutation of those now existing continued in the next century with Henry of Hesse (1325–97), who referred to the appearance of new diseases and the new herbs which would be needed to cure them.

Later they entered the natural philosophies of Bruno, who was indebted also to the Stoics, and of Francis Bacon, Leibniz, and the evolutionists of the 18th century. The reflections of Albertus Magnus and of Henry of Hesse on the mutation of species were not related to any concept of an evolving, developing and progressing universe, animal kingdom or human race, an idea which is characteristically modern and had no place in medieval thought. Aristotle had described a scale of nature in his biological works, but in this there was no movement upwards, and when Albertus made this Aristotelian scale the basis of his botanical and zoological system he accepted, apart from accidents and the causes of mutation just mentioned, the continuance of breeding true to type.

Albertus' *De Animalibus*, and particularly the sections on reproduction and embryology, is one of the best examples of the way the system of facts and natural explanations provided by the translations of Aristotle's and other Greek works stimulated the natural philosophers of the 13th century to make similar observations of their own and to modify the explanations in their light. The first 19 books of the 26 books of the *De Animalibus* are a commentary embodying the text of Michael Scot's translation of Aristotle's *History of Animals, Parts of Animals* and *Generations of Animals*. In his commentary Albertus also made use of Avicenna's own commentary on these works, of Avicenna's *Canon*, which was based on Galen, and of Latin translations of some of Galen's own works. The remaining 7 of Albertus' 26 books consist of original discussions of various biological topics and descriptions of particular animals, taken partly from Thomas of Cantimpré.

For Aristotle, the reproduction of the specific form was an extension of growth, for, as growth was the realization of the form of one individual, reproduction was its realization in the new individual to which this gave rise. Albertus followed Aristotle in distinguishing four types of reproduction: sexual reproduction, in which male and female principles were either separate in different individuals, as in higher animals and in general those with local motion, or united as in plants and sessile animals and some others such as bees; reproduction by budding, as in some mussels; and spontaneous generation, as in some insects, eels and the lower creatures generally. The sexes of plants were clearly distinguished only by Camerarius (1694), though the point had been suggested by Theophrastus, Pliny and Thomas Aquinas. Like Aristotle, Albertus rejected the Hippocratic theory, also maintained by Galen, that both parents contributed to the form. Aristotle had held that the female provided merely the material (which he believed to be the catamenia – *menstruum* – in mammals and the yolk of the egg in birds) out of which the immaterial male form constructed the embryo. Albertus agreed with this but followed Avicenna in maintaining that the material produced by the female was a seed, or *humor seminalis*, separate from the catamenia or yolk, which he said was simply food. He incorrectly identified this seed with the white of the egg. The spermatozoon was not, of course, discovered until the invention of the microscope and he identified the cock's seed with the chalazea. The cause of the differentiation of sex, he held, was that the male 'vital heat' was able to 'concoct' the surplus blood into semen, informing it with the form of the species, while the female was too cold to effect this substantial change. All other differences between the sexes were secondary to this.

The efficacy of vital heat derived from the fact that, of the two pairs of primary qualities, hot-cold were active and dry-wet passive. The heart was the centre of vital heat and the central organ of the body. To it, and not to the brain, which Aristotle had said was a cooling organ, ran the nerves. Vital heat was the source of all vital activity. It was the cause of the ripening of fruit, of digestion which was a kind of cooking, and it determined the degree to which an animal would approach the adult form on being extruded from the parent. The facts of heredity Aristotle had explained by the degree of dominance of the male form over the female matter, female characteristics prevailing where the vital heat of the male was low. Monstrosities were produced where the female matter was defective for the purpose in hand and resisted the determining form. Vital heat, which Aristotle described in the *De Generatione Animalium* (736 b 36) as 'the spiritus [pneuma] included in the semen and the foam-like, and the natural principle in the spiritus, being analogous to the element of the stars', Albertus said was also the cause of spontaneous generation. The corruption of the form of a dead organism generated the forms of lower creatures which then organized the available matter, as worms generated in dung. The vital heat of the sun also caused spontaneous generation, and the Arabs and scholastics generally supposed that such forms were supplied by celestial 'virtue'.

Just as Aristotle was in opposition to Hippocrates and Galen over the question whether the male seed alone formed the embryo, so he was over the question whether in embryology any new characters arose or all were already preformed in the seed, which simply had to expand. Hippocrates had held a form of this preformation theory combined with pangenesis, that is, he held that the sperm was derived from all parts of the parent's body, and therefore gave rise to the same parts in the offspring. Aristotle showed that the theory that the embryo was an adult in miniature, which only had to unfold, implied that the parts developing later already existed in the earlier and all in the sperm, whose parts already existed in its parents and therefore in the sperm which produced the parent, and so on to infinity. He considered such *emboîtement*, or encasement, an absurd conclusion, and therefore maintained the epigenetic theory that the parts arose *de novo* as the immaterial form determined and differentiated the matter of the embryo. After the male seed had acted on the female matter by curdling it, he said that the embryo developed like a complicated machine whose wheels, once set going, followed their appointed motions. He described the development of a number of animals and made this comparative study the basis for a classification of animals. His observation that development was faster at the head end foreshadows the modern theory of axial gradients, and by showing that the more general preceded the more specific characters he anticipated von Baer. He also correctly understood the functions of the placenta and umbilical cord.

Albertus' own researches into embryology were guided by

Aristotle.* He never hesitated to accept the evidence of his eyes but, while he was ready to adopt the theories of alternative authorities and, for instance, like Avicenna combined epigenesis with a theory of pangenesis, he usually attributed errors of fact to copyists rather than to Aristotle. Following Aristotle's example, he opened hens' eggs after various intervals and added *per anathomyam*, and with considerable understanding, to Aristotle's description of what was going on, from the appearance of the pulsating red speck of the heart to hatching. He also studied the development of fish and mammals, of which he understood the fœtal nutrition. And while Aristotle had thought that the pupa was the egg of the insect, of which he supposed the life history to be from maternal female to larva to pupa (his egg) to adult, Albertus recognized the true insect egg, as well as that of the louse. In book 17, tract 2, chapter 1 of the *De Animalibus* he amplified Aristotle's text to say:

at first, eggs are something very small, and from them worms are generated, which in their turn are changed into the matter of ova [i.e., pupæ], and then from them the flying form emerges; and so there is a triple change from the egg, namely into the worm, and from the worm into a kind of egg, and from this into something that flies.

He said, in fact, that 'the generation of all animals is first from eggs.' At the same time he believed in spontaneous generation. He gave an excellent description of insect mating, and his description in book 5, tract 1, chapter 4 of the life history of a butterfly or moth represents a remarkable piece of sustained observation.

A certain kind of caterpillar is hidden in cracks after the sun has begun to recede from the summer tropic and it putrefies internally and becomes surrounded by a hard, horny, annular skin. In this is born a flying worm which has in front a long coiled tongue which it thrusts into flowers and it sucks out the nectar. It develops four wings, two in front and two behind, and flies and becomes multi-coloured and develops several legs, but not as many as it had when it

* In the text of Albertus' *De Animalibus* edited by H. Stadler it is possible to follow the original text with Albertus' amplifications.

was a caterpillar. The colours vary in two ways, either according to genus or in one individual. Some genera are white, some black and some of other intermediate colours. But there is a certain kind belonging to the last genus in which many different colours are found in the same individual. This animal, thus winged and generated from a caterpillar, is called by some people in Latin by the common name *verviscella*. It flies at the end of autumn and emits many eggs, for the whole lower part of its body below the thorax is converted into eggs, and in laying eggs it dies. And then again from these eggs caterpillars hatch next spring. But certain grubs do not become *verviscellæ* but gather at the ends of the branches of trees and there make nests and lay eggs, and from these arise grubs in the next spring. Those of this sort always extend the nest towards the sun at midday. But the sort that are generated from the flying forms place all their eggs in walls and cracks in wood and walls of houses near gardens.

Albertus' personal observations extended to many other zoological phenomena besides reproduction. Thomas of Cantimpré, though a good observer, had included a whole book of fabulous animals in his *De Natura Rerum* (c. 1228–44), but Albertus criticized the stories of the salamander, the beaver and the barnacle goose from personal observation. He said of the phœnix, the symbol of the resurrection, that it was studied more by mystical theologians than by natural philosophers. He gave excellent descriptions of a large number of northern animals unknown to Aristotle and noted the colour varieties of the squirrel (*pirolus*), which passed from red to grey as one went from Germany to Russia, and the lightening of colour in falcons (*falcones*), jackdaws (*monedulæ*) and ravens (*corvi*) in cold climates. He considered colour as compared to form as of little importance as a specific character. He noted the relation of build to method of locomotion and applied Aristotle's principle of 'homology' to the correspondence between the bones in the forefoot of the horse and the dog. He showed that ants whose antennæ had been removed lost their sense of direction, and he concluded (wrongly) that the antennæ bore eyes. His knowledge of internal anatomy was sometimes meagre, but he dissected crickets and observed the ovarian follicles and tracheæ. He seems to have recognized the brain and nerve cord of crabs and something of their function in movement. He observed that the moulting of crabs included their limbs, and showed that these regenerated if amputated. 'But,' he said in book 7, tract 3, chapter 4,

such animals are rarely regenerated in the abdomen, because in the bridge above which the soft parts are placed, the organs of their movement are fixed; and a motive virtue (*vis motiva*) goes down that bridge from the part of them which corresponds to the brain. Therefore, since it is the seat of a more noble power, it cannot be removed without danger.

The system by which Albertus classified the animals he described in books 23–26 followed the main lines of that suggested by Aristotle, which to some extent he elaborated. Aristotle had recognized three degrees of likeness within the animal kingdom: the 'species,' in which there was complete identity of type and in which differences between individuals were accidental and not perpetuated in reproduction; the 'genus', which consisted of such groups as fishes or birds; and the 'great genus,' which involved the morphological corresondence or homology between scale and feather, fish-bone and bone, hand and claw, nail and hoof, and of which the whole group of sanguineous animals (the modern vertebrates) was an example. Though no classification was actually set out by Aristotle, the main lines of his system are easily recognized, as they were by Albertus. As each species and genus had many differentiæ they might be grouped in many ways, and, again like Aristotle, Albertus did not keep to one system, but put animals sometimes into groups based on morphological or reproductive similarity, and sometimes into ecological groups such as flying (*volatilia*), swimming (*natatilia*), walking (*gressibilia*), and crawling (*reptilia*) animals. Here he advanced on Aristotle by proposing the division of water animals into ten genera: *malachye* (cephalopods), *animalia mollis testæ* (crabs), *animalia duris testæ* (shell fish), *yricii marini* (sea urchins), *mastuc* (sea anemones), *lignei* (sea-stars, sea-cucumbers), *veretrale* (pennatulide or gephyra?), *serpentini* (polychæte worms?), *flecmatici* (medusæ), and *spongia marina* (sponges). With some animals he repeated or aggravated Aristotle's mistakes, putting whales with fish and bats with birds, although he observed the bat's teeth and said in book 1 (tr. 2, c. 4) that 'she approaches the nature of quadrupeds.'

The main system of classification which Albertus derived from Aristotle was that based on the mode of generation, that is, on the degree of development, itself depending on the parents' vital heat and moisture, reached by the offspring at the time of extrusion from the parent's body. Thus, mammals were the hottest animals and produced viviparously young which were perfect likenesses to their parents, although smaller; vipers and cartilaginous fishes were internally oviparous, externally viviparous; birds and reptiles produced perfect eggs, that is eggs which did not increase in size after being laid; fish, cephalopods and crustacea produced imperfect eggs; insects produced a scolex (larva or premature 'egg') which then developed into the 'egg' (pupa); testacea produced generative slime or reproduced by budding; and in general members of the lower groups might be generated spontaneously. The complete 'Aristotelian' scale of living nature, as recognized and modified by Albertus, is set out in Table 2 (p. 167).

After the 13th century, descriptive botany and zoology was carried on by herbalists and naturalists having a variety of interests. Of the herbalists, Matthæus Sylvaticus included in his dictionary of medical 'simples', or *Pandectæ*, in 1317, a large amount of information based on personal observation of plants in various places he had visited or in the collection of domestic and foreign plants he kept in his botanical garden at Salerno. This is the earliest known non-monastic botanical garden and from this time others appear, particularly in connexion with the medical faculties of universities, the first of this kind being established at Prague in 1350. A number of surgeons and physicians such as John of Milano in Italy, John Arderne in England, and Thomas of Sarepta in Silesia wrote herbals in the 14th century. John of Milano illustrated his herbal, the *Flos Medicinæ* completed before 1328, with 210 drawings of plants. Thomas of Sarepta, who died as a bishop about 1378, is of particular interest for having in his youth made a herbarium of dried plants collected

TABLE 2

THE ARISTOTELIAN SCALE OF NATURE

Man (rational soul)

Animals (sensitive soul)

Enaima (sanguineous: modern vertebrates)

  1. Man.

  2. Hairy quadrupeds (land mammals, classified by cloven hoof, teeth, etc.).

  3. Cetacea (sea mammals, included by Albertus with fish).

  4. Birds (classified by raptorial and webbed feet, etc.).

  5. Scaly quadrupeds and apoda (reptiles and amphibia).

  6. Fishes (bony and cartilaginous).

Anaima (bloodless: modern invertebrates)

  7. Malacia (cephalopods).

  8. Malacostraca (crustacea).

  9. Entoma (insects, millipedes, spiders, intestinal worms, etc., Albertus' *animalia corpora annulosa vel rugosa habentia*).

  10. Ostracoderma or testacea (molluscs, except cephalopods; sea-urchins, ascidians).

  11. Zoophyta (sea-cucumbers, sea-anemones, jellyfish, sponges).

Plants (nutritive soul)

Left-hand classification:

A. Viviparous

B. Oviparous (sometimes internally oviparous, externally viviparous, as in some vipers and cartilaginous fish)
- *a.* with perfect egg
- *b.* with imperfect egg

C.
- *c.* with imperfect egg
- Oviparous
- Vermiparous, with scales
- Produced by generative slime, budding or spontaneous generation
- Produced by spontaneous generation

Produced without sexual differentiation from seeds, or by budding (classified as above, p. 158).

(A to C represents the scale of egg or embryo types depending on vital heat.)

in various places, including England. An anonymous French herbal compiled in Vaud about 1380 is of interest for containing fresh information about Swiss plants, but the most outstanding herbal of this period was the *Liber de Simplicibus* completed by Benedetto Rinio in Venice in 1410 (see above, p. 155). Besides the magnificent paintings of 450 domestic and foreign plants, this herbal contained brief botanical notes indicating collecting seasons, the part of the plant containing the drug, the authorities used, and the name of each plant in Latin, Greek, Arabic, German, the various Italian dialects, and Slavonic. Venice at that time had a vigorous trade in drugs with both East and West, and Rinio's herbal was kept in one of the main apothecary's shops, where it could be used for the practical purpose of identifying plants. The same medical interest was responsible for the printed herbals that began to appear later in the 15th century (see Vol. II, p. 267 *et seq.*).

Of the other naturalists of the 14th century, Crescenzi included in his *Ruralia Commoda* a large amount of information about varieties of domestic plants and animals of all kinds and devoted a special section to gardens (see below, p. 198 *et seq.*). His main agricultural authorities were the Roman writers Cato the Elder, Varro, Pliny, and the part of the *Geoponica* dealing with vines, which had been translated by Burgundio of Pisa, while for scientific biology he went to Albertus Magnus and Avicenna. The German naturalist, Conrad von Megenburg, is distinguished for having written about 1350 the first important scientific work in German, *Das Buch der Natur*. This was basically a free translation of Thomas of Cantimpré's *De Rerum Natura*, but it contained some fresh observations on rainbows, plague, and various animals and plants. It was very popular, and the first printed edition of 1475 was the earliest work in which woodcuts representing plants were used with the definite intention of illustrating the text and not merely for decoration. These illustrations probably did not much antedate the printing, but a late 14th-century naturalist whose illustrations showed very great powers of observation was 'Cybo of Hyères' (see above, p. 153). Gaston de Foix, who in 1387 began to write his celebrated French treatise, *Le Miroir de Phœbus*, which did for hunting what the Emperor Frederick II had done for falconry, also showed himself to be an excellent naturalist. This work, which was very popular and was translated into English in the early 15th century, contained very good and practical descriptions of how to keep hounds, falcons and other hunting animals and also a large amount of information about the habits of the hunted animals such as the hart, wolf, badger, and otter. Another French writer, Jehan de Brie, in a book of 'shepherdry' written in 1379 for Charles V, showed that even in court circles there could be an interest in nature. In England a series of treatises on various country sports culminated with the *Boke of St Albans* in two editions in 1486 and 1496, the second containing one of the first full accounts in English of fishing; an earlier *Treatyse of Fysshynge with an Angle*, on which the *Boke*'s account is based, dates from the first twenty years of the fifteenth century. An Italian zoological writer of the 15th century was Pier Candido Decembrio (1399–1477), or Petrus Candidus, who, in 1460, wrote a series of descriptions of animals, to which some excellent illustrations of birds, ants and other creatures were added in the 16th century (see Plate 18).

A large number of theoretical works on biology were also written in the 14th and 15th centuries, mainly in the form of commentaries on various books by Aristotle, Galen, Averroës or Avicenna. In the 13th century Giles of Rome (c. 1247–1316) had written a treatise on embryology, *De Formatione Corporis Humani in Utero*, based largely on Averroës, in which he discussed the development of the fœtus and the time at which the soul entered. On this last point there was much controversy and among those to discuss it was Dante, who put forward the view of St Augustine and of Averroës that the soul was generated together with the body, but manifested itself only with the first movement of the fœtus. Another 14th century writer, the Italian doctor Dino del Garbo (d. 1327), ascribed the birth and development of plants and animals from seeds to a kind of fermentation and tried to prove that the seeds of hereditary diseases lay in the heart. His compatriot, Gentile da Foligno, tried to work out the mathematical relation between the times of formation and movement of the fœtus and of birth of the infant. Another subject that attracted the attention of scholastic writers of the 14th century was the origin and nature of the movement of animals, and writers like Walter Burley, Jean de Jandun and Jean Buridan discussed this question in commentaries on Aristotle's *De Motu Animalium*. Other parts of Aristotle's *De Animalibus* were commented on by writers from the early 14th-century John Dimsdale or Teasdale in England to the mid 15th-century Agostino Nifo in Padua. Another series of treatises, beginning with that of Alfred of Sareschel, was written under the titles *De Corde* or *De Motu Cordis*. The problem whether, in generation, seed was contributed by both sexes was also argued out by theoretical writers, particularly with the popularity in the 15th century of Lucretius (see Vol. II, p. 116), who had upheld the double seed theory. This discussion went on into the 17th and 18th centuries in the dispute between the animalculists and ovists. At the end of the 15th century Leonardo da Vinci tried to bring some of these theoretical questions within reach of experiment, but it was not

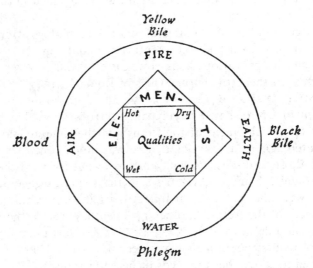

Fig. 13 The four humours. In the cycle of the seasons, the sequence of predominant humours is blood (spring), yellow bile (summer), black bile or melancholy (autumn), phlegm (winter).

until the 19th century that experimental embryology got properly under way.

The branch of biology in which the most interesting developments took place in the 14th and 15th centuries was neither botany, zoology nor embryology but human anatomy. The chief motive for the study of anatomy was its practical value for the surgeon and physician (see below, p. 237 *et seq.*, Vol. II, p. 275 *et seq.*). The chief sources of anatomical knowledge were Galen (129–200 A.D.) and Avicenna, the anatomical sections of whose *Canon of Medicine* were themselves based largely on Galen. Certain alternative ideas about anatomy were also known from Aristotle, as seen, for example, in the early 13th-century *Anatomia Vivorum* of Richard of Wendover, which was used by Albertus Magnus. By the end of the 13th century the preference was most commonly given to the usually more accurate Galen.

Galen's anatomical ideas, based on dissections of human and animal bodies, were intimately connected with a system of physiology. Both were avowedly derived in part from his great predecessors Herophilus and, more particularly, Erasistratus in the 3rd century B.C. According to Galen, the brain (and not the heart as Aristotle had said) was the centre of the nervous system, and the vital functions were explained by means of the three spirits (*spiritus* or pneuma) and the four Hippocratic humours, corresponding to the four elements (Fig. 13). The balance of these four humours – blood, phlegm (or *pituita*, found in the pituitary body), black bile (or *melancholia*, found in the spleen) and yellow bile (or *chole*, found in the gall bladder) – was necessary for the healthy functioning of the body, but the vital functions themselves were brought about by the production and movements of the three spirits, the 'natural spirit' of the liver, the 'vital spirit' of the heart, and the 'animal spirit'* of the brain (Plate 19). These were made ultimately from the food, and from the air drawn into the lungs by the act of respiration, when Galen held that the

---

* The term *spiritus animalis* refers to the *anima*, breath, the principle of animal life; the Greek equivalent is *pneuma psuchikon*. Contrasted with the *anima* in scholastic terminology is the *animus*, the spiritual principle of life, the rational soul.

principle of life entered the animal body. This physiological theory, with its three great systems, each associated with one of the three spirits and their functions, entirely dominated ideas of the significance of anatomical structures and connexions until it was overthrown by William Harvey (see Vol. II, p. 227 *et seq*.).

According to Galen, the food taken into the stomach was transformed first into the chyle by what was known as the first 'coction', a process activated by the innate heat of the animal body and analogous to domestic cooking. The useless parts of the food were at the same time absorbed by the spleen, there converted into black bile, and excreted through the bowel. The chyle itself, a white liquid, was carried from the stomach and intestines in the portal vein* to the liver. There, in the second coction, it was made into venous blood, the chief of the four humours, and imbued with a pneuma innate in all living substances, the 'natural spirit', the principle of nutrition and growth.

Although Aristotle had correctly related the veins as well as the arteries to the heart, Galen held that the veins formed a separate system, totally different in structure and function from the arteries, and that the venous system took its origin not from the heart but from the liver. The function of the venous system, he held, was to carry the venous blood, charged with natural spirit and nutriment, out from the liver to all parts of the body. He likened the *vena cava* to the trunk of a tree, with its roots in the soil, the liver, and its branches spreading out as the veins. It is this conception of the veins and arteries as belonging to two totally different physiological and anatomical systems that marks the fundamental opposition between Galen's theory of the motion of the blood and that which Harvey was to put in its place. For Galen, the function of the venous blood was to nourish the parts to which it flowed out from the liver. The process by which the

---

* It was only in this vessel that there was any reversal of flow, some of the venous blood returning from the liver to carry natural spirit and nutriment to the stomach and intestines. Owing to a misunderstanding, some recent historians have supposed that the blood 'ebbed and flowed' in the venous system as a whole. Cf. Donald Fleming, *Isis*, 1955, vol. 46, p. 14 *et seq*.

nutriment absorbed from the veins was converted into flesh was broadly speaking the third coction. The total amount of blood was not large, and it was continually, and slowly, being renewed from the liver.

Of the blood discharged by the liver into the *vena cava*, some, according to Galen only a small fraction, entered into the right side of the heart. This organ Galen held to have only two chambers, the ventricles; the auricles he regarded as simply dilatations of the great veins. He held that the heart was not a muscle, for, unlike true muscle, it could not be moved at will but beat involuntarily without ceasing because of a specific pulsific faculty, or *vis pulsifica*, possessed by its tissues. He held that the same faculty was possessed by the arteries and was manifested in the pulse.

Galen's views on the action of the heart and the arterial pulse differed from Aristotle's no less than his views on the venous system. Both held that the heart was the centre of the body's natural or innate heat, produced according to Galen by slow combustion, and both held that the active motion of the heart was its dilatation in diastole, not, as Harvey was to show, its contraction in systole. Aristotle attributed this dilatation to the cardiac heat itself, which boiled the blood and caused it to expand and erupt into the aorta and thence into the arteries and the body. Galen, by contrast, held that it was the heart's own *vis pulsifica* that caused it to dilate, *drawing the venous blood into it* from the *vena cava*; and that it was a similar active dilatation by the aorta and the arteries that *drew* the arterial blood and spirit *out* of the heart and into the body. As they dilated, he held that the left ventricle also drew in air from the lungs along the venous artery, and that the arteries similarly drew in air through the skin. In fact, he regarded the motions of the lungs in respiration, of the heart-beat, and of the arterial pulse, as all serving the same functions, the vitalizing and distributing of the arterial blood and the cooling and cleansing necessitated by the heart's beat.

Galen had an almost complete knowledge of the essential anatomy of the heart, and he knew that the course of the blood through it was governed by the presence of one-way valves at the four openings leading to and from its cavities. These valves had been discovered by Erasistratus (Vol. II, Plate 11). 'Nature,' Galen wrote in *De Naturalibus Facultatibus* (*On the Natural Faculties*), book 3, section xiii, 'provided the cardiac openings with membranous attachments to prevent their contents from being carried backwards.' This meant that the direction of the blood through the heart and lungs was in general forwards. The blood entering the right ventricle from the *vena cava* (apart from an insignificant amount which escaped back through the valve) then had two possible fates. The bulk of it passed through a valve opening outwards from the ventricle into the arterial vein (now called the pulmonary artery). On the contraction of the thorax this blood, its retreat cut off from behind by the valve, was forced into the lungs, to which it carried nutriment, and through fine channels into the venous artery (our pulmonary vein) with whose branches those from the arterial vein anastomosed. Whether he held that the venous artery then carried the blood to the left ventricle Galen does not make clear. Certainly the venous artery carried inspired air, or some quality derived from the air, drawn from the lungs into the left ventricle in diastole. In the opposite direction, 'sooty waste' derived from the combustion of the innate heat was carried from the left ventricle to the lungs, and thence expired. The effect of these actions was to cool and cleanse the heart, and it was these that he regarded as the principal functions of the lungs. The two-way traffic in the venous artery was made possible in Galen's view by the comparative inefficiency of the mitral valve opening into the heart, but it was to become one of the difficulties that led William Harvey to re-examine the whole Galenic system.

Besides this passage into the arterial vein, Galen held that a small amount of blood was squeezed from the right ventricle into the left ventricle through minute pores through the pits in the *septum* forming the dividing wall between them. In the left ventricle this blood encountered the pneuma brought from the lungs in the venous artery, and was there elaborated into the 'vital spirit', the principle of animal life as manifested in innate heat, and carried by the arterial blood. From the left ventricle the arterial blood was drawn out by the dilatation of the aorta through a valve opening outwards. Passing along the aorta it was distributed in the arteries throughout the body under the influence of the pulse, carrying the vital spirit into all the parts.

Some of the arteries went to the head, where, in the *rete mirabile** at the base of the brain, the blood was finely divided and charged with a third pneuma, the 'animal spirit'. These were contained in the ventricles of the brain and distributed to the sense organs and muscles by the nerves, which Galen supposed to be hollow. The animal spirit was the basis of sensation and voluntary muscular activity.

There were thus three principal organs of the body, each the centre of an anatomical system and a physiological function. The liver was the centre of the venous system and of the 'natural' or vegetative faculty concerned with nutrition; and this system Galen regarded (in complete contrast to the view since Harvey) as quite distinct in both structure and function from the arterial system, whose centre was the heart. The thick-walled arteries had quite a different appearance from the veins; the blood they contained was different from venous blood in colour and viscosity; and this agreed with the supposition that they served a different function. The arterial system served the 'vital' faculty whose seat was the heart, the origin of vital heat, cooled by the lungs. Finally there was the brain, the centre of the nervous system and of the 'animal' or psychic faculty, with the animal spirits corresponding to a venous material psyche (*anima*) and, at least in scholastic writings, serving as the liaison between the material body and the immaterial rational soul (*animus*).

Like the Alexandrian physiologists and anatomists going back to Herophilus and Erasistratus in whose teaching he had been brought up, Galen had been a good observer and experimenter. He studied the anatomy of bones and muscles, though in the latter, like Vesalius later, he sometimes drew conclusions about man from the dissection of such animals as the Barbary ape. He seems, in fact, to have worked largely with animals. He dis-

* A structure at the base of the brain which is well developed in some animals, for example the calf, but not in man.

tinguished between sensory ('soft') nerves entering the spinal cord from the body and motor ('hard') nerves going out from the spinal cord. He recognized many of the cranial nerves and made experiments on the spinal cord, showing that sectioning between different vertebræ of living animals had different effects: instantaneous death when the cut was made between first and second vertebræ, and arrest of respiration, paralysis of the thoracic muscles, and paralysis of the lower limbs, bladder and intestines when made at different points further down. He also had a fairly good idea of the general course of the veins and arteries, on the functions of which he made experiments. Erasistratus had thought that the latter contained only air, but Galen showed that when a length of artery was ligatured at both ends and punctured, blood came out of it. Thus, if his mistakes, such as his theory of the movements of blood, misled anatomists and physiologists until the 16th or 17th century, it was by his experimental method, with which he investigated problems ranging from the production of the voice by the larynx and the functioning of the kidney to the medicinal properties of herbs, that men learnt to correct them.

The medieval scholars who first read Galen's works were able to add little that was original, but, from the 12th century, it was recognized, as it was put in the Salernitan *Anatomia Ricardi*, that 'a knowledge of anatomy is necessary to physicians, in order that they may understand how the human body is constructed to perform different movements and operations.' The great 13th- and 14th-century surgeons insisted that some practical knowledge of anatomy was essential to their craft, Henry of Mondeville (d. c. 1325), for example, declaring that the mind must inform the hand in its operation and the hand in its turn instruct the mind to interpret the general proposition by the particular instance. In 12th-century Salerno the dissection of animal and human bodies seems to have been a part of medical training; the earliest Western work on anatomy is the early 12th-century *Anatomia Porci* attributed to a certain Copho of Salerno, which described the public dissection of a pig. This work was followed during the 12th century by four others from Salerno, the fourth of which, the *Anatomia Ricardi*, was the first to describe human anatomy. This was based largely on literary sources and contained descriptions of the eye, motor and sensory nerves, fœtal membranes and other structures similar to those given by Aristotle and Galen.

In the 13th century, the practice of dissection was continued at Bologna, where the first evidence of human dissection is found in the *Chirurgia* of the surgeon William of Saliceto, completed in 1275. This work was the first Western topographical anatomy and, though based largely on earlier Latin sources, it contained the observations of a practical surgeon such as on the thoracic organs of a man wounded in the chest and on the veins in joints and in the lower abdomen as seen in cases of hernia. Another Italian surgeon, Lanfranchi (d. before 1306), who worked in Paris, gave anatomical details associated with wounds in many different parts of the body. Further opportunities for human dissection were given at Bologna by the practice of making *post mortem* examinations to determine the cause of death for legal purposes. This practice was mentioned at the end of the 13th century by Taddeo Alderotti (d. 1303), who also attended dissections of animals, and the first formal account of a *post mortem* examination was given by Bartolommeo da Varignana in 1302. A manuscript of about the same date in the Bodleian Library at Oxford (MS Ashmole 399, *c.* 1290) has an illustration of a dissection scene and, later in the 14th century, many *post mortem* dissections were made during the Black Death. The same manuscript in the Bodleian Library contains stylized illustrations of the five systems, venous, arterial, skeletal, nervous, and muscular, and of the child in the womb. Similar illustrations are found in other manuscripts of the 14th and 15th centuries and have been published by Sudhoff.

The man who 'restored' anatomy by introducing the regular practice of public dissections of corpses for teaching purposes was Mondino of Luzzi (*c.* 1275–1326), who was a pupil of Alderotti and became a professor at Bologna. Mondino's *Anatomia*, completed in 1316, was the most popular text-book of anatomy before that of Vesalius in the 16th century and it exists in a large number of manuscripts and printed editions. Mondino himself dissected male and female human corpses and also, on one occasion, a pregnant sow. His book was the first work specifically devoted to anatomy and not merely an appendage to a work on surgery. It was, in fact, a practical manual of dissection in which the organs were described as they were to be opened : first those of the abdomen, then of the thorax and of the head, and finally the bones, spinal column, and extremities. This arrangement was imposed by the need, in the absence of good preservatives, to dissect the most perishable parts first and to complete the dissection within a few days. Mondino also used preparations dried in the sun to show the general structure of tendons and ligaments, and macerated bodies to trace the nerves to their extremities. A good account of the general procedure was given by Guy de Chauliac in his *Chirurgia Magna* completed in 1360.

In spite of his personal observations Mondino's *Anatomia* was very largely based on Galen, the 7th-century Byzantine writer Theophilus, and on Arabic authorities. The influence of the last can be seen in his Latinized Arabic terminology. Among the non-Arabic terms which he used two have survived to the present day, namely, matrix and mesentery. Mondino did not, in fact, dissect to make discoveries but, like a modern medical student, to gain some practical acquaintance with the teaching of the textbook authority. In his own manual he preserved both the mistakes and the correct observations of his authorities. He believed the stomach to be spherical, the liver five-lobed, the uterus seven-chambered, and the heart to have a middle ventricle in the septum. Yet he gave a good description of the muscles of the abdomen and may have been the first to describe the pancreatic duct. In at least one of his ideas, his attempt to establish the correspondence between the male and female generative organs, he was to be followed by Vesalius. Of particular interest are certain of his physiological ideas. He held that the production of urine was due to the filtering of the blood by the kidneys and attributed to the brain the old Aristotelian function of cooling the heart. In addition to this, the brain acted as the centre of the nervous system, and he held that its psychological functions

were localized in three ventricles, as follows: the anterior ventricle, which was double, was the seat of the *sensus communis*
or 'common sense' which, according to contemporary psychol-
₋ᵤ.', rep.   ₋ᵤ mᵤn's ability to make comparisons between
different senses; the middle ventricle was the seat of the imagination; the posterior, of memory. Mental operations were controlled
by the movement of the 'red worm' (that is the *choroid plexus*
of the third cerebral ventricle), which opened and closed the
passages between the ventricles and directed the flow of the
animal spirits (Vol. II, Plate 14 and p. 246).

After Mondino's time anatomical teaching, with public dissections of human bodies and even research, was carried on at
Bologna and elsewhere in northern Italy by a series of distinguished physicians, Guido da Vigevano, Niccolò Bertruccio,
Alberto de' Zancari, Pietro Torrigiano and Gentile da Foligno.
Guido da Vigevano, who worked both at Pavia and in France,
wrote a treatise in 1345 which was based partly on Mondino and
other authorities and partly on his own dissections. It is of
interest for its illustrations, which show a considerable advance
in the technique of dissection over that of the early 14th century
(Plate 20). One notable feature is that the corpse was slung from
a gibbet, as later in many of Vesalius' illustrations. Of the other
Paduan physicians, Gentile da Foligno is of particular interest for
being possibly the first to describe gallstones, and Niccolò
Bertruccio for his description of the brain. In 14th-century
France, Henry of Mondeville, a fellow student of Mondino at
Bologna, had already, by 1308, made systematic dissections and
used charts and a model of the skull for teaching at Montpellier.
In the anatomical section with his medical compendium he gave
a good account of the system of the portal vein. Mondeville's
definition of nerves includes tendons and ligaments, and it is of
interest that another famous Montpellier teacher, Bernard of
Gordon (d. *c.* 1320), seems to have suggested that nerves exerted
a mechanical pull on the muscles. Bernard followed Greek
authorities in believing that epilepsy was caused by the humours
blocking the passages of the brain and interfering with the supply
of air to the limbs. Guy de Chauliac, who had studied at Bologna

under Bertruccio, carried on the teaching by public dissections at Montpellier, and one manuscript of his surgical treatise contains some excellent illustrations showing dissections in progress. In the 15th century public dissections began in other centres, in Vienna in 1405 and in Paris in 1407. There are further anatomical illustrations in a manuscript of about 1420 of a treatise by the English physician John Arderne and in a German manuscript, written between 1452 and 1465, of the *Chirurgia* of the 13th-century Paduan physician, Bruno of Longoburgo. In the mid 15th century for about fifty years there seems to have been a decrease of interest in anatomy, possibly because of an over-concentration on purely practical and immediate surgical requirements, and possibly because of the custom, prevalent in the northern universities where surgery was held in low esteem and anatomy was taught by professors of medicine, of the anatomical teachers leaving dissection to a menial, while a demonstrator pointed out the parts, instead of the anatomist doing the dissection himself (see below, pp. 237–40). This slackening of progress did not last long, for already by the end of the 15th century Leonardo da Vinci had begun to make his brilliant anatomical drawings based on his own dissections and, early in the 16th century, Achillini made some fresh discoveries. By 1543, when Vesalius published his great work, the progress of anatomical research was already well under way (see Vol. II, pp. 273, 275–8 *et seq.*).

The position of man in the 13th-century universe was a special one: he was both the purpose and the final product of the material creation, and the centre of the complete scale of creatures. Man, 'who by reason of his nobleness falls under a special science called medicine', stood at the apex of the scale of material beings and at the base of the scale of spiritual beings: his body was the product of generation and fated to suffer corruption in the former realm: his soul was received at conception, or, according to some authorities, at some later period of gestation, direct from God who created it and destined it for eternal life. Thus man occupied a central position between two orders of being, the purely material order of the other animals, descending through plants to inanimate things, and the purely spiritual order of the angels, ascending to God.

One effect of this view of the special position of man in the universe was to emphasize the sacramental aspect of his scientific activities, to show that he, before all other creatures, was in a position to worship the Creator of this great chain of being stretching above and below him, in which each thing existed to fulfil its own nature in its special place and all to praise the Lord. The sentiment that was to inspire much of the 13th-century science had, in fact, been expressed at the beginning of the century by the founder of an order which was to give so many great innovators to Western scientific thought, particularly in England. St Francis of Assisi began his *Cantico del Sole*:

Be Thou praised, O Lord, for all Thy creatures, especially for our brother the sun, who brings the day and with it gives light. For he is glorious and splendid in his radiance and, Most High, signifies Thee.

This was certainly the sentiment of Grosseteste, Roger Bacon and Pecham in Oxford; and in Paris, Germany and Italy also, and among the other great order of friars to whom 13th-century science owes its chief progress, the belief was certainly not wanting that *amor intellectualis dei* included the study of nature, of the immense revolving spheres of the heavens and of the smallest living creatures, of the laws of astronomy, optics and mechanics, of the laws of biological reproduction and of chemical change. The feeling expressed by Vincent of Beauvais in his *Speculum Majus*, prologue, chapter 6, might equally well have come from the pen of Albertus Magnus or many another 13th-century scientific writer:

I am moved with spiritual sweetness towards the Creator and Ruler of this world, because I follow Him with greater veneration and reverence, when I behold the magnitude and beauty and permanence of His creation.

Another effect of the idea of man's nature held in 13th-century Christendom has already been touched upon, namely, the effect of the idea that man is rational and free-willed in leading to the rejection of Greek and Arabic determinism, and this was to be even more important in the sequel. Few people at the end of the 13th century, apart from the Averroïsts, believed that Aristotle had said the last word on philosophy and natural science, and though all would have admitted that he had provided them with the framework of their system of scientific thought, the theologians were careful to preserve both man and God from constraint within any particular system. The free speculation which resulted led to radical criticisms of many of the fundamental principles accepted in the 13th century, even of propositions whose acceptance then seemed necessary to the Christian religion itself (though most of these lay outside natural science); even, indeed, though radical views led to an occasional brush with ecclesiastical authority. Within natural science perhaps the most fundamental advance made as a result of these criticisms was in scientific method and the conception of scientific explanation, and this, together with the development of technology, formed the double track that led across the watershed of the 14th century and with many turns to the 16th- and 17th-century world.

4

TECHNICS AND SCIENCE
IN THE MIDDLE AGES

I. TECHNICS AND EDUCATION

IT has often been pointed out that science develops best when the speculative reasoning of the philosopher and mathematician is in closest touch with the manual skill of the craftsman. It has been said also that the absence of this association in the Greco-Roman world and in medieval Christendom was one reason for the supposed backwardness of science in those societies. The practical arts were certainly despised by the majority of the most highly educated people in classical antiquity, and were held to be work for slaves. In view of such evidence as the long series of Greek medical writings, stretching from the first members of the so-called Hippocratic corpus to the works of Galen, the military devices and the 'screw' attributed to Archimedes, the treatise on building, engineering and other branches of applied mechanics written during Hellenistic and Roman times by Ctesibius of Alexandria, Athenæus, Apollodorus, Hero of Alexandria, Vitruvius, Frontinus and Pappas of Alexandria, and the works on agriculture by the elder Cato, Varro and Columella, it may be doubted whether even in classical antiquity the separation of technics and science was as complete as has been sometimes supposed. In the Middle Ages there is much evidence to show that these two activities were at no period totally divorced and that their association became more intimate as time went on. This active, practical interest of educated people may be one reason why the Middle Ages was a period of technical innovation, though most of the advances were probably made by unlettered craftsmen. And certainly it was this interest of many theoretical

183

scientists in practical results that encouraged them to ask concrete and precise questions, to try to get answers by experiment and, with the aid of technics, to develop more accurate measuring instruments and special apparatus.

From the early Middle Ages, Western scholars showed an interest in getting certain kinds of results for which some technical knowledge was necessary. Medicine was studied in the earliest Benedictine monasteries, and the long series of medical works written during the Middle Ages, and continuing without a break into the 16th century and modern times, is one of the best examples of a tradition in which empirical observations were increasingly combined with attempts at rational and theoretical explanation, with the result that definite medical and surgical problems were solved. Another long series of treatises was written on astronomy by scholars from the time of Bede in the 7th century for purely practical purposes such as determining the date of Easter, fixing latitude, and showing how to determine true North and tell the time with an astrolabe. Even a poet such as Chaucer could write an excellent practical treatise on the astrolabe. Another series of practical treatises is that on the preparation of pigments and other chemical substances, which includes the 8th-century *Compositiones ad Tigenda* and *Mappæ Clavicula*, of which Adelard of Bath later produced an edition, the early 12th-century *Diversarum Artium Schedula* of Theophilus the Priest, who lived probably in Germany, the late 13th-century *Liber de Coloribus Faciendis* by Peter of Saint Omer, and the early 15th-century treatises of Cennino Cennini and John Alcherius. Technical treatises were among the first to be translated out of Arabic and Greek into Latin, and this was the work of educated men. It was, in fact, chiefly for their practical knowledge that Western scholars, from the time of Gerbert at the end of the 10th century, first began to take an interest in Arabic learning. The 13th-century encyclopædias of Alexander Neckam, Albertus Magnus and Roger Bacon contained a great deal of accurate information about the compass, chemistry, the calendar, agriculture, and other technical matters. Other contemporary writers composed special treatises on these subjects: Grosseteste and later writers on the calendar; Giles of Rome in *De Regimine Principum* on the art of war; Walter of Henley and Peter of Crescenzi on agriculture; Peregrinus, in the second part of *De Magnete*, on the determination of azimuths. It took a scholar to write about arithmetic, yet most of the advances that followed Fibonacci's treatise on the Hindu numerals were made in the interests of commerce.

In the 14th century, the Italian Dominican friar, Giovanni da San Gimignano (d. 1323), wrote an encyclopædia for preachers in which he gave for use as examples in sermons descriptions of numerous technical subjects: agriculture, fishing, cultivation of herbs, windmills and watermills, ships, painting and limning, fortifications, arms, Greek fire, smithing, glass making, and weights and measures. The names of two other Dominicans, Alessandro della Spina (d. 1313) and Salvino degl' Armati (d. 1317), are associated with the invention of spectacles. In the 15th century, a most interesting series of treatises was written on military technology. Beginning with Konrad Kyeser's *Bellifortis*, written between 1396 and 1405, this included a treatise by Giovanni de' Fontana (*c.* 1410–20), the *Feuerwerksbuch* (*c.* 1422), a treatise by an anonymous engineer in the Hussite wars (*c.* 1430), and the so-called 'Mittelalterliches Hausbuch' (*c.* 1480). The series went on in the 16th century with the treatises of Biringuccio and Tartaglia. These contained descriptions of how to make guns and gunpowder as well as problems of military engineering, which were discussed also by other contemporary writers such as Alberti and Leonardo da Vinci. Some of these treatises dealt also with general technical matters such as the construction of ships, dams and spinning-wheels. The series of practical chemical treatises which, in the earlier Middle Ages, had consisted mainly of recipes for pigments, continued in the 14th and 15th centuries with accounts of distillation and other practical techniques and went on in the 16th century with Hieronymus Brunschwig's books on distillation, the metallurgical *Probierbüchlein*, and Agricola's *De Re Metallica* (see above, p. 139 *et seq.*; below, p. 219 *et seq.*). Examples of the interest shown by medieval scholars in technics could, in fact, be multiplied considerably. They show not only that they had an abstract desire for power over nature such as Roger Bacon had expressed, but also that they were capable of getting the kind of knowledge that would lead to results useful in practice.

One reason for this interest of the learned in technics is to be found in the education they received. The popular handbook on the sciences by Hugh of St Victor (d. 1141), *Didascalicon de Studio Legendi*, shows that by the 12th century the seven liberal arts had been extended and specialized so as to include various kinds of technical knowledge. The mathematical subjects forming the *quadrivium* had, of course, had a practical object at least since the time of Bede, but from the early 12th century there was a tendency to increasing specialization. In the *Didascalicon* Hugh of St Victor followed a modified version of the classification of science in the tradition coming from Aristotle and Boethius; he divided knowledge in general into theory, practice, mechanics and logic. Giving a pseudo-historical account of the origin of the sciences, he said that they arose first in response to human needs as a set of customary practices, which were later reduced to formal rules. These practices began by man imitating nature: for example, he made his own clothes in imitation of the bark with which nature covered trees or the shell with which she covered shellfish. Each of the 'mechanical' arts, forming the 'adulterine' science of mechanics which provided for those things necessary because of the weakness of the human body, arose in this way. In mechanics Hugh included seven sciences: the manufacture of cloth and of arms, and navigation, which ministered to the extrinsic needs of the body; and agriculture, hunting, medicine and the science of theatrical performances, which ministered to intrinsic needs. He gave a brief description of each of these activities.

Later in the 12th century another popular classification of the sciences was written by Dominicus Gundissalinus, his *De Divisione Philosophiæ*. This was based partly on Arabic sources, in particular on Alfarabi, whereas Hugh had used only the traditional Latin sources. Gundissalinus, following another form of the Aristotelian tradition, classified the sciences into theoretical and practical. He subdivided the former into physics, mathematics and metaphysics and the latter into politics, or the art of civil government, the art of family government, which included giving instruction in the liberal and mechanical arts, and ethics or the art of self-government. The 'fabrile' or 'mechanical' arts were those concerned with making out of matter something useful to man, and the matter used could come either from living things, for example wood, wool, linen or bones, or from dead things, for example gold, silver, lead, iron, marble or precious stones. Through the mechanical arts resources were acquired which provided for the needs of the family. To each of the mechanical arts there corresponded a theoretical science which studied the basic principles which the mechanical art put into practice. Thus theoretical arithmetic studied the basic principles of numbers used in reckoning by the abacus, as in commerce; theoretical music studied in the abstract the harmonies produced by voices and instruments; theoretical geometry considered the basic principles put into practice in measuring bodies, in surveying, and in using the results of observing the motions of the heavenly bodies with the astrolabe and other astronomical instruments; the science of weights considered the basic principles of the balance and lever. Finally, the science of 'mathematical devices' turned the results of all the other mathematical sciences to useful purposes, for stone-masonry, for instruments for measuring and lifting bodies, for musical and optical instruments, and for carpentry.

In the 13th century, these ideas were taken up by a number of well-known writers, for example Roger Bacon, Thomas Aquinas and Giles of Rome. The treatises of Michael Scot and Robert Kilwardby are of special interest. Michael Scot held that each of the practical sciences was related to a theoretical science and was the practical manifestation of the corresponding theoretical science. Thus to different branches of theoretical 'physics' there corresponded such practical sciences as medicine, agriculture, alchemy, the study of mirrors, and navigation; to the different branches of theoretical mathematics there corresponded such practical arts as business concerned with money, carpentry,

smithing and stone-masonry, weaving, shoemaking. Robert Kilwardby's treatise, *De Ortu Scientiarum*, very widely read for
generations, expressed the same conviction of the importance of
the practical side of science concerned with getting useful results.
Of special significance is Kilwardby's pseudo-historical account
of the theoretical sciences as arising out of particular, concrete
problems encountered in attempting to satisfy the physical needs
of the body as, for example, his version of the ancient Greek
tradition that geometry arose first as a practical art among the
Egyptians because they had to survey the land after the flooding
of the Nile, and was transformed into a theoretical and demonstrative science by Pythagoras. Among the 'mechanical' sciences
he included agriculture, viticulture, medicine, cloth-making,
armouring, architecture and commerce. Roger Bacon gave elaborate descriptions of various practical sciences, asserted emphatically that the justification of the theoretical sciences was their
useful results, and stressed the need to include the study of the
practices of artisans and of practical alchemists in any scheme
of education.

Though it was only in guilds of artisans that any sort of practical training in the mechanical arts was received, the utilitarian
aims of medieval writers on education were reflected often to a
surprising extent, in the courses that might be taken at a university. This was the case, for example, in the 12th-century medical
school at Salerno, where the regulations of King Roger II of Sicily
and the Emperor Frederick II required that the medical student
should take a course lasting five years and including human
anatomy and surgery. After passing an examination at the end
of this course he was not allowed to practise until he had spent
a further year learning from a trained practitioner. From the end
of the 13th century, attendance at an 'anatomy' at least once a
year was prescribed for medical students at Bologna, and in the
14th century the medical school in the university devoted itself
increasingly to surgery. Some practical instruction in anatomy
seems, in fact, to have been required in most medical schools from
the end of the 13th century (see above, p. 171 *et seq.*; below, p.
237 *et seq.*).

In the 'arts' courses in most universities the mathematical subjects very often had some practical object in view. In 12th-century
Chartres a list of books recommended for study by Thierry of
Chartres included a high proportion of works on surveying,
measurement and practical astronomy; a list of text-books in use
at Paris at the end of the 12th century shows that the same
utilitarian tradition was continued there. At the beginning of the
13th century the arts course at Paris took six years, a Licence in
Arts not being granted before the age of 20, though at Paris
and most other universities the six years were later reduced,
sometimes to as little as four. The course usually consisted of
a study of the seven liberal arts followed by the 'three philosophies', natural philosophy (that is, natural science), ethics and
metaphysics. At Paris during the 13th century there was a tendency to reduce the time spent on mathematical subjects in favour
of the other arts subjects such as metaphysics. At Oxford a considerable emphasis was placed on the mathematical subjects, the
text-books prescribed including, for example, not only Boethius'
Arithmetic and Euclid's *Elements*, but also Alhazen's *Optica*,
Witelo's *Perspectiva*, and Ptolemy's *Almagest*. The arts course at
Oxford is also of interest for including the study of Aristotle's
De Animalibus as well as the more usual *Physica, Meteorologica,
De Cœlo*, and other works on 'natural philosophy'. A similar
emphasis on mathematics is seen in the arts course at Bologna,
where the subjects prescribed included a book on arithmetic
known as *Algorismi de Minutis et Integris*, Euclid, Ptolemy, the
Alfonsine Tables, a book of rules by Jean de Linières for using
astronomical tables to determine the motions of the heavenly
bodies, and a work on the use of the quadrant. Some of the
German universities seem also seriously to have cultivated the
study of arithmetic, algebra, astronomy, optics, music, and other
mathematical sciences. It seems unlikely that any actual practical
or laboratory instruction was included in the arts course at any
medieval university, but there is evidence that special courses in
astronomy were given at Oxford in the 14th century. According
to its preface, Chaucer wrote his treatises on the astrolabe to
explain to his son how to use the instrument which he gave him

when he went up to Oxford. Certainly the Fellows of Merton College made astronomical observations, and in at least one case, that of Richard of Wallingford and his planetarium, a scholar is known actually to have made his own instruments (Plate 21).

An important result of this mathematical training received in medieval education was that it encouraged the habit of expressing physical events in terms of abstract units and emphasized the need for the standardization of systems of measurement. Without this habit of thought mathematical physics would be impossible. Lewis Mumford has vividly described how it developed first in connexion with the purely practical regulation of affairs. The need to measure time for the orderly institutions of the Church and the routine of the monastery led to the sustained medieval interest in the calendar and to the division of the day into the unequal canonical hours, while the secular requirements of government and commerce led to the prevalence in civil life of the system of 24 equal hours in the day. The invention at the end of the 13th century of the mechanical clock, in which the hands translated time into units of space on the dial, completed the replacement of 'organic', growing, irreversible time as experienced, by the abstract mathematical time of units on a scale, belonging to the world of science. Space also underwent abstraction during the later Middle Ages. The symbolic arrangement of subjects in paintings according to their importance in the Christian hierarchy gave way, from the middle of the 14th century in Italy, to the division of the canvas into an abstract checkerboard according to the rules of perspective. Besides the symbolic maps, like the Hereford *Mappa Mundi* of 1314, there appeared maps by cartographers in which the traveller or mariner could find his position on an abstract system of co-ordinates of latitude and longitude. Commerce changed during the Middle Ages from a barter economy based on goods and services to a money economy based on abstract units, first of gold or silver coinage, and later also of letters of credit and bills of exchange. The problems arising from the dissolution of partnerships (some discussed in Italy as early as the 12th century) and in connexion with interest, discount and exchange were among the chief incentives to mathematical research. Problems of currency reform became the subject of treatises by academic mathematicians like Nicole Oresme in the 14th century and Copernicus two centuries later. This process of abstraction concentrated attention on the systems of units used. Attempts were made from as early as Anglo-Saxon times in England to standardize weights and measures and later (in legislation during the reign of Richard I) to replace units based on the human body like the foot and span by standard measures made of iron. Attempts were made also to establish the relationship between the different systems existing in different countries and even within the same country. A series of treatises was written by doctors interested in standardizing units of weight and volume for drugs.

One of the most interesting examples of a mathematical art developing an abstract language of its own in order to communicate knowledge of how to produce a precise practical effect is music. In the Middle Ages the theory of music was studied as part of the *quadrivium*, chants were sung and instruments played in church, secular music is known from about 1100, and some universities like 14th-century Salamanca and 15th-century Oxford even gave degrees in music: so for several centuries men of learning were closely acquainted with both the theoretical and practical aspects of the art. The basis of medieval music was the Greek system of modes, of which the major scale of C is the only one that sounds familiar in the 20th century. Greek music consisted entirely of melody. Though the Greeks used choirs of men's and boys' voices singing at intervals of an octave, a practice known as 'megadizing', and also harps with which they played in simultaneous octaves, this hardly amounted to harmony, of which they had no real conception. To write down a melodic line the Greeks had used letters to indicate the rise and fall of pitch and by the 7th century A.D., in Church music, this was done by strokes over the words, which themselves controlled rhythm. From these developed the system of 'neumes' written on a staff of parallel horizontal lines to indicate pitch, as in the *Micrologus de Disciplina Artis Musicæ*, written about 1030 by Guido d'Arezzo. He is of interest also as the originator of the system of designating the notes of the scale by the first syllables of six lines of a hymn to St John the Baptist: *ut, re, mi, fa, so, la.*

Early medieval church music was all plain chant, in which the notes had fluid time-values; mensural or measured music in which the duration of the notes had an exact ratio among themselves seems to have been invented in Islam. A number of Arab writers, of whom Alfarabi was one of the most distinguished, wrote on mensural music and, during the 11th and 12th centuries, knowledge of mensural music entered Christendom through Spain and through the translations of Arabic musical works by Christian scholars like Adelard of Bath and Gundissalinus. In the 12th century appeared in Christendom the system of notation in which the exact time-value of each note was indicated by black lozenges and diamonds on little poles, as explained in a treatise by John of Garland, who studied at Oxford early in the 13th century, and more fully in the *Ars Cantus Mensurabilis* attributed to Franco of Cologne, living during the second half of the 13th century. Hooks were attached to the black diamonds to serve the purpose of the modern crotchet, white notes were added, and eventually the so-called Franconian notation was developed into the modern system, completed with bar-lines about 1600 and key-signatures about 1700. The new system of mensural notation made it possible to have firmly defined rhythms, to sing, and, with the introduction of special notations for instruments, to play two different rhythms concurrently. With the last came also the beginning of the realization of the full potentialities of harmony.

Harmony began in the West with the practice of singing concurrently the same tune at two different pitches, usually in fourths or fifths. This system had been developed in Christendom by about 900 A.D. and was known as the *organum* or 'diaphony.' It is possible that something similar had been developed independently in Islam where, for example, the 10th-century Alfarabi recognized the major third and the minor third as concords. In the 10th century several Latin treatises were written on the *organum*, one of the best-known being written in the Low Countries by a certain Hucbald. About 1100 an Englishman, John

Cotton, and the author, probably French, of the anonymous treatise *Ad Organum Faciendum*, explained a new *organum*, in which the voices periodically changed from singing the same melody at different pitches to singing different melodies in such a way as to produce a carefully varied set of accepted concords. By the end of the 12th century the descant had appeared, then both parts began to be moved in counterpoint. About a century later the 'new art' had developed enough to see the appearance of the well-known English six-part round, 'Sumer is icumen in,' one of the earliest canons. By the mid-14th century quite complicated polyphony had been developed, as seen in the *Mass* for four voices composed by Guillaume de Machaut for the coronation of Charles V at Reims in 1364. Polyphony was elaborated still further by composers like John Dunstable and Josquin des Prés in the 15th century and Palestrina in the 16th. Besides developing vocal music these late medieval composers began to realize the possibilities of instruments. Pipes, trumpets and plucked stringed instruments were known from early times and the organ, which had been known to the Greeks, reappeared in the West in the 9th century, when it seems to have been tuned in the modern major scale with the keys named after the letters of the alphabet. About the same date the introduction of the bow made it possible to produce a sustained note on a stringed instrument (Plate 22), and in the 14th century stringed instruments began to be played with a fixed keyboard.

Throughout all these developments the musical theorist and composer worked closely together, and musicians were often distinguished in other branches of science. Typical results of this close contact between theorist and practitioner are the writings of the early-14th-century English mathematician and astronomer, Walter of Odington, who illustrated his important theoretical treatise on music with examples from his own composition. The contemporary mathematician, Jean de Murs, tried to order the mensural system according to a single rule relating the lengths of successive notes in the system, and experimented with new instruments foreshadowing the clavichord. The most outstanding 14th-century musical theorist was Philippe de Vitri (1291–1361), who made contributions to the methods and notation for establishing the relations between notes of varying length then recognized (*maxima* or *duplex longa, longa, brevis, semibrevis, minima* and *semiminima*), and to such notions as augmentation and diminution. Most of Philippe de Vitri's own compositions are now lost, but Guillaume de Machaut's *Mass* contains practical illustrations of many of his theoretical innovations. It was through this combination of theory and practice in the later Middle Ages that modern rhythmic and harmonic music realized the possibilities of the *organum* and the *Ars Cantus Mensurabilis*, and developed into an art which can be said to characterize the modern civilization of the West as much as the natural science developing at the same time.

Most of the fundamental techniques on which both classical and medieval economic life were based had been invented in prehistoric times. Prehistoric man discovered the use of fire, tools and agriculture, bred, domesticated, and harnessed animals, invented the plough, pottery, spinning and weaving and the use of organic and inorganic pigments, worked metals, made ships and wheeled carts, invented the arch in building, devised such machines as the windlass, pulley, lever, rotary quern, bow-drill and lathe, invented numbers, and laid the empirical fundations of astronomy and of medicine.

To this basic practical knowledge some important additions were made in the Greco-Roman world. Though the chief contribution of classical civilization to science was not in technics but in speculative thought, one of the most important contributions ever made to technology was made by the Greeks. This was their attempt to give rational explanations of the machines and other inventions and discoveries of their predecessors, which made it possible to generalize and extend their use. Thus it was the Greeks who first converted the practical, technological methods of reckoning and measuring, as developed in Mesopotamia and Egypt, into the abstract sciences of arithmetic and geometry, and who first attempted to give rational explanations of the facts observed in astronomy and medicine. By combining observation and theory they greatly extended the practical use of these sciences. Greek writers, from the author or authors of the Aristotelian *Mechanica* and Archimedes down to Hero of Alexandria, attempted to explain the lever and other mechanisms. Hero gave a full account of the five 'simple' machines by which a given weight might be moved by a given force and of some of their combinations: the wheel and axle, lever, pulley, wedge and endless screw. These were held to be the basis of all machinery until the 19th century. The Greeks developed also the elementary principles of hydrostatics. Some Hellenistic and Roman writers were the first to give descriptions of the various kinds of machinery then in practical use. Some of the most important of these were the crossbow, catapults and other ballistic devices, watermills involving the important method of transmitting power through geared wheels and perhaps a windmill, the screw press and trip hammer, syphons, vacuum pumps, force pumps and Archimedes' screw, the bellows organ and water organ, a steam turbine and a puppet theatre driven by falling weights, the water clock and such important measuring instruments as the cyclometer and hodometer, surveying instruments such as the dioptra (a theodolite without telescope described by Hero), and the cross-staff, astrolabe and quadrant which remained the basic astronomical instruments until the invention of the telescope in the 17th century. Most of these devices were, in fact, Greek inventions. In other technical fields as, for example, in medicine and in agriculture (where the Romans seem to have introduced legume rotation), important improvements were introduced in the classical world. But whether they were describing new techniques or simply ones inherited from the less expressive Egyptian, Babylonian and Assyrian civilizations, these Greco-Roman technical writings were to have a very important influence as a source of technical knowledge in both the Moslem and the Christian worlds in the Middle Ages. In Western Christendom these classical technical works exerted an influence right down into the 17th century.

In the period that followed the collapse of the Roman Empire in the West there was a considerable loss of technical knowledge, though this was compensated to a slight extent by some new techniques introduced by the invading Germanic tribes. From about the 10th century, however, there was a gradual improvement of technical knowledge in Western Christendom. This was brought about partly by learning from the practices and writings (often of classical origin) of the Byzantine and Arabic worlds, and partly by a slow but increasing activity of invention and innovation within Western Christendom itself. The gains thus made during the Middle Ages were never lost, and it is characteristic of medieval Christendom that it put to industrial use technical devices which in classical society had been known but left almost unused or regarded simply as toys. The result was that, as early as 1300, Western Christendom was using many techniques either unknown or undeveloped in the Roman Empire. By the year 1500 the most advanced countries of the West were in most aspects of technics distinctly superior to any earlier society.

2. AGRICULTURE

The basic occupation throughout the Middle Ages and, in fact, till the end of the 18th century was agriculture, and it was in agriculture that the first medieval improvements on classical techniques were introduced. Roman agriculture as described by Cato and Varro in the 2nd and 1st centuries B.C. had, in certain respects, reached a high level; such crops as vines and olives were intensively cultivated and the increased yields obtained by growing a leguminous crop alternately with a cereal were well understood. With the fall of the Western Empire agricultural methods at first declined, but from the 9th or 10th century improvement began and continued steadily into modern times. The first outstanding achievement of the medieval agricultural population was the great business of agricultural colonization. Rulers of the early Middle Ages like Theodoric the Great in Italy, the Lombard kings of the 7th and 8th centuries, Alfred the Great and Charlemagne had as their policy, in the words of Orosius, 'to turn the barbarians to the ploughshare,' to lead them 'to hate the sword.' The agricultural colonization of Europe sketched in the Carolingian period, the eastward felling of the German forests, the work of clearing, draining and cultivation from woody England and the flooded marshes of the Low Countries to the dry hills of Sicily and Christian Spain, which went on under the leadership of Cistercians and Carthusians, federal rulers and urban communes, had been practically completed by the 14th century. During that time, not only was Europe occupied and civilized, but agricultural productivity increased enormously as a result of improved methods. These maintained a steady increase in population and the growth of towns at least until the Black Death in the 14th century. As a result different regions became specialized for different crops and animals and for the production of wool and silk, hemp, flax and dye-plants and other materials to supply the growing needs of industry.

The first improvements in agriculture were brought about by the introduction of the heavy Saxon wheeled plough and a new system of crop rotation, both of which had come into operation in northwestern Europe by the 9th and 10th centuries. The use of the heavy wheeled plough equipped with coulter, horizontal share and mouldboard (Plate 23), instead of the light Roman plough, made it possible to cultivate heavier and richer soils, saved labour by making cross-ploughing superfluous, and thus gave rise to the strip system of land division in northern Europe as distinct from the older Mediterranean block system. Because it needed six or eight oxen to draw it, the use of this plough perhaps led to the grouping of the farming population in northwestern Europe into villages and the organization of agriculture on communal lines as seen in the manorial system. At the same time as the heavy plough was coming into use, the system of crop rotation in northwestern Europe was improved by having three fields, instead of two, of which one lay fallow. In the two-field system, one half of the land was left fallow while the other half was planted with grain. In the three-field system, one field was fallow, the second was planted with winter grain (wheat or rye), and the third with a spring crop (barley, oats, beans, peas, vetches). A complete rotation thus occurred every three years. The three-field system did not spread south of the Alps and the

Loire, apparently because it was only in the north that summers were wet enough to make spring sowing, the chief novelty of this system, profitable. Even in the north the two systems continued side by side till the end of the Middle Ages. The three-field system did distinctly increase productivity and, when combined with the superior plough, it may have been one of the reasons for the shift of the centre of European civilization to the northern plains in Charlemagne's time. Certainly one of its effects seems to have been to make possible the increasing use of the faster but more extravagant grain-fed horse, instead of the hay-fed ox, as the plough and draft animal.

Further improvements were introduced in methods of cultivation later in the Middle Ages. The ploughshare was made of iron, and the horse-drawn harrow with iron teeth replaced the older methods of breaking the clods with rakes or mattocks. Methods of draining low-lying land were also improved by the use of pumps and networks of sluices and canals; the lower reaches of the Rhine and the Rhône were confined to their courses by dykes; and along the coasts of the Netherlands large areas were reclaimed from the sea. Sand dunes were arrested by means of osier plantations along the North Sea coast, and forests of pines were planted on the dunes of Leiria in Portugal by King Dinis of Lavrador who ruled until 1325. In Spain and Italy hydraulic science was used to construct works for irrigation. The most remarkable of these were the dams and reservoirs of eastern Spain and the famous Lombard 'Naviglio Grande,' built between 1179 and 1258, which carried water from Lake Maggiore over 35,000 hectares to irrigate the lands on the banks of the Oglio, Adda and Po. Under the guidance of enlightened agriculturalists, monastic, royal or urban methods of restoring and enriching the soil were also improved. Thierry d'Hireçon, who managed the estates of Mahout, Countess of Artois and Burgundy, and who died in 1328 as Bishop of Arras, is an outstanding example.

A record of contemporary agricultural theory is found in the writings of Albertus Magnus, with his botanical approach, of Walter of Henley in England and Peter of Crescenzi in Italy, and of several other authors who attempted to reach rational methods by combining the study of ancient Roman sources and Arabic science with contemporary practice in Christendom. Thus Walter of Henley discussed marling and weeding, and Albertus Magnus discussed manuring. Walter of Henley's *Hosebondrie* (c. 1250) remained the standard work on the subject in England until the appearance of Sir Anthony Fitzherbert's *Husbandrie* in 1523. The best of the medieval treatises on agriculture was certainly Crescenzi's *Ruralia Commoda* (c. 1306). This was enormously popular on the Continent; it was translated into several European languages and it exists in a large number of manuscripts and was printed many times. Crescenzi had studied at Bologna logic, natural science, medicine and finally law. After holding a series of legal and political offices, he settled on his estate near Bologna and wrote his *Ruralia Commoda* late in life. This work was a critical compilation from books and observation, written with the object of giving the intelligent farmer a rational and practical account of all aspects of his occupation, from the biology of plants (taken from Albertus Magnus) to the arrangement of farm buildings and water supply. It treated such subjects as the cultivation of cereals, peas and beans; of vines and their wines, their varieties, their diseases and their remedies; of fruit trees, vegetables, medicinal plants and flowers; the care of woods; the rearing of all kinds of farm animals, large and small; horses and their ailments; and hunting and fishing. Perhaps the most original parts of his treatise were his elaborate discussion of the grafting of vines and trees and his account of the insect larvæ which destroy plants. His description of bee-culture shows that Roman methods had not been forgotten.

Of the methods of enriching the soil, the use of animal manures was fully appreciated in the Middle Ages: cattle were turned on to the stubble of arable fields; sheep were folded and the manure collected and spread. Lime, marl, ash, turf and calcareous sand were also used. And though, in most of Western Christendom, extensive cultivation with triennial rotation and fallow persisted, in the Netherlands, northern France and southern Italy it had become common by the 14th century to abandon the fallow year and plant instead root crops and legumes. Apart from the enrichment of exhausted soil, this had the advantage that it made it possible to maintain more animals through the winter; in the earlier Middle Ages most of the stock had to be killed off at the beginning of winter and the meat salted, the plough teams that were kept being fed on hay and straw. Yet in spite of these improvements, the expected yields in most parts of medieval Christendom remained low compared with those in the 20th century. For two bushels of wheat sown per acre in England the yield expected was 10 bushels; and for four bushels of oats sown the yield expected was 12 to 16 bushels. A marked improvement in yield came only with the 'scientific rotation' of the 18th-century agricultural revolution.

In other ways than in methods of cultivating and fertilizing the soil medieval agriculture made steady progress. Increasing attention was paid to the cultivation of fruit trees, vegetables and flowers in gardens, and new crops were introduced for special purposes: buckwheat or 'Saracen corn,' hops, rice and sugar cane were grown for food and drink; oil plants for food and lighting; hemp and flax, teazles, the dye-plants woad, madder and saffron, and in Sicily and Calabria even cotton and indigo, were grown for textile manufacture. Linen became the source of paper-making, which spread gradually northward for two centuries after entering southern Europe from the East in the 12th century. In the 13th century improvements were made in Italy over the method of manufacturing paper in Spain. By the 13th century mulberries were being cultivated and silkworms raised in industrial quantities in southern Italy and eastern Spain. From the 14th century large tracts of Italy, England and Spain were given over to raising sheep for the wool trade, so that already Prussia, Poland and Hungary were replacing them as grain growers. Sheep were in many ways the most important stock in the Middle Ages: they provided the most important raw material for textiles; they gave meat and were the most important source of animal manure for the fields. Different breeds were kept for different purposes and there was some attempt to improve breeds by crossing and the selection of rams. Of the other livestock, cattle were valued mainly as draft animals, though also for leather, meat and the

milk which was made into butter and cheese. With the introduction of fodder plants in the Netherlands in the 14th century
the first experiments were made in crossing. Pigs were the chief
source of meat, but were kept also for lard and tallow used for
candles. Poultry was abundant, the common guinea fowl or
Indian fowl having been introduced in the 13th century. Bees
were kept for honey, used in place of sugar, and for wax used for
lighting.

Another very important source of food in the Middle Ages was
fish, especially the herring, fished and marketed by the maritime
peoples living round the North Sea and in the Baltic. Herrings
were the staple food of poorer people. The herring industry was
much improved by a new method of preserving and packing
in kegs invented in the 14th century. By the 13th century
whales were being hunted by North Sea sailors and by Basques,
and on the shore beds of oysters and mussels were being
organized.

Of all the animals in which an interest was taken in the Middle
Ages, the horse was the one to whose breeding the greatest care
was devoted. The horse was one of the chief sources of non-human
power: it drew the plough; it was used with saddle or cart for
transport on land; it was ridden to the chase and for hawking;
and above all it was a primary engine of war. In classical times
cavalry had been of secondary importance because of inefficient
methods of harnessing, but the whole art of vigorous riding in
peace and war was transformed in early medieval times by the
introduction of stirrups. There is evidence that these were in use
in China in the 5th century A.D. and in Hungary in the 6th century, and shortly afterwards they were recommended for the
Byzantine cavalry. In northwestern Europe they are first found
in the 8th century in the graves of Vikings in Sweden. In the 9th
century stirrups are shown in the chessmen which are supposed
to have been sent to Charlemagne by Haroun al Raschid. By the
11th century stirrups were common, saddles were becoming
deeper, and prick spurs and the curb were coming into use. With
these methods of controlling the mount the cavalry charge with
lances became possible and remained the basis of tactics for

several centuries. Armour became heavier, and one of the chief points for breeding was to get a strong animal capable of carrying an enormous weight. Horse breeding was much influenced by Arab practices and the best works on the subject and on veterinary medicine relating to the horse were written in Arabic as late as the 14th century. Horse studs were set up in Christendom by rulers such as the counts of Flanders, the dukes of Normandy, and the kings of the two Sicilies. The kings of Castile introduced laws regulating stock breeding generally. The Arabs had traced pedigrees through the dam, but the Western practice from as early as the 12th century seems to have been to trace them through the sire, and certain Arab stallions were imported from time to time. In the 13th century several Spanish works were written on horse breeding and veterinary science; another was written by one of Frederick II's advisers in Sicily. In the 14th century Crescenzi's treatise had a section on the horse, and further veterinary works were written in Italy and Germany later in the century.

The value of the horse as a draft animal depended on the introduction of a new kind of harness which allowed the animal to take the weight on its shoulders by means of a rigid stuffed collar, instead of on its neck as hitherto (Plate 24). In Greek and Roman times, to judge by sculpture, vase paintings and medals, horses were harnessed in such a way that the pull was taken on a strap passing round the neck, so that the harder they pulled the closer they came to strangulation. The modern horse collar seems to have appeared in the West in the early 9th century, introduced perhaps from China. At the same time two other inventions appeared: the nailed horseshoe which improved traction, and the extension of the lateral traces for the tandem harness, allowing horses to be harnessed one in front of the other so that an indefinite number could be used to move heavy weights. This had not been possible with the classical method of harnessing horses side by side. Another improvement which came in during the same period was the invention of a multiple yoke for oxen. These inventions transformed life in the West in the 11th and 12th centuries, much as the steam engine did in the 19th. They made it possible to use the horse to pull the heavy wheeled plough, the first picture of a horse so engaged appearing in the Bayeux tapestry. Perhaps because of changed economic conditions, perhaps because of the opposition of the Church, slave labour, which was the basis of classical industry, had become increasingly scarce in the early Middle Ages. The new methods of harnessing animal power, and the increasing exploitation of water- and wind-power, came to make slavery unnecessary.

3. THE MECHANIZATION OF INDUSTRY

The great expansion of the use of watermills and windmills that took place during the later Middle Ages, in association with the growth of manufacturing, brought in an essentially new stage in mechanical technique. From this period must be dated that increasing mechanization of life and industry, based on the ever-increasing exploitation of new forms of mechanical power, which characterizes modern civilization. The initial stages of the industrial revolution, before the use of steam, were brought about by the power of the horse and ox, water and wind. The mechanical devices and instruments invented in classical times, pumps, presses and catapults, driving wheels, geared wheels and trip hammers, and the five kinematic 'chains' (screw, wheel, cam, ratchet and pulley) were applied in the later Middle Ages on a scale unknown in earlier societies. The remaining kinematic 'chain,' the crank, was known in its simple form in late classical times. It was used in such simple mechanisms as the rotary grindstone depicted in the mid-9th-century *Utrecht Psalter*. The combined crank and connecting-rod was a medieval invention. Though it is difficult to trace its later history, this mechanism had certainly come into general use by the 15th century. With the crank it became possible for the first time to convert reciprocating into rotary motion and *vice versa*, a technique without which modern machinery is inconceivable.

The earliest water-driven mills were used for grinding corn, though before them waterwheels operating chains of pots had been used in ancient Sumeria for raising water. These early cornmills were of three kinds. Horizontal millstones on a vertical shaft turned by water flowing past vanes attached to the bottom of the shaft are known from the 5th century A.D. from Ireland, Norway, Greece and other places, though there is no direct evidence for this kind of mill in antiquity. A second type of mill, in which a vertical undershot waterwheel operated a pestle by a trip hammer mechanism, is described by Pliny. An undershot waterwheel driving a millstone by means of geared wheels is described by Vitruvius. This is the first instance known of the use of geared wheels for transmitting power. Four centuries later Pappus of Alexandria described a toothed wheel rotating on a helix or worm gearing. There is evidence that the Romans used also overshot wheels, which had the mechanical advantage that they were driven by the weight of the water as well as the force of the current. From the Mediterranean, watermills spread northwestward and, by the 4th century A.D., they were in general use throughout Europe for grinding corn and pressing olives. In the 4th century Ausonius describes a water-driven saw in use on the Moselle for cutting marble. In the 11th century Domesday Book records 5,000 watermills in England alone. The first evidence as to the type of mill in use in medieval Christendom comes from the 12th century, by which time the vertical undershot wheel was the common type. Overshot wheels do not appear in illustrations before the 14th century (Plate 25) and even by the end of the 16th century they had by no means entirely displaced the undershot type.

With the spread of watermills came improvements in methods of transmitting power and of converting their rotary motion for special purposes. As early as the 12th century, illustrations show that the proportions of the crown and pinion wheels forming the gear were adjusted to give the millstone a high speed of rotation even in slow streams, and the general mechanism of the geared wheel was adapted to mills worked by other forms of power. Illustrations from the end of the 13th century to the 16th show such mechanisms in mills worked by horses or oxen or by hand, and 15th-century illustrations show them in windmills. By the end of the 12th century the rotary motion of the water wheel was being converted to operate trip hammers for fulling * and for crushing woad, oak bark, for tanning leather, and other substances, and by the 14th century the same mechanism was used for forge hammers. In the 14th century the treadle hammer, the English 'oliver,' appeared, and in the 15th century a stamping mill was described for crushing ore. In the late 13th century the waterwheel was adapted also to drive forge bellows (Fig. 14) and, if a device sketched by Villard de Honnecourt represents something actually used, sawmills for cutting wood. Water-driven sawmills certainly existed in the next century. By the 14th century waterwheels and also horse-driven wheels were used to drive grindstones for making edged tools; by the 15th century they were in use for pumping in mines and salt pits, for hoisting in mines with crank or windlass, and for driving iron-rolling mills and wire-drawing mills; by the 16th century they were used to drive silk mills.

Windmills came into use much later than watermills. The first certain knowledge of windmills comes from the writings of Arab geographers travelling in Persia in the 10th century, though mills may have existed before that time. These writings describe windmills with horizontal sails operating a vertical axle, to the lower end of which was attached a horizontal millstone. Windmills may have come to the West from Persia through the Arabs of Spain, through the Crusades, or through trade between Persia and the Baltic known to have passed through Russia. Certainly when windmills first appeared in Christendom in the 12th century it is in the northwest, though these had vertical sails driving a horizontal axle. But whatever its early history in the West, by the end of the 12th century the windmill was widespread in England, the Netherlands and northern France; it was used especially in those regions where there was no water. The chief mechanical problem introduced by the windmill arose out of the

---

* 'When a little farther, at the doubling of the Point of a Rock, they plainly discover'd ... Six huge Fulling-Mill Hammers, which, interchangeably thumping several Pieces of Cloth, made the terrible Noise that caus'd all *Don Quixote's* Anxieties and *Sancho's* Tribulation that night.' (*Don Quixote*, 1603, part i, book 3, section 6).

need to present the sails to the wind, and in the earlier mills the whole structure was rotated about a pivot or post (Plate 27). This meant that mills had to remain small, and only from the end of the 15th century did the windmill increase in size and develop in a really efficient form. The axle was then set at a slight angle to the ground, the sails adjusted to catch every breath of wind, a brake was fitted, and there were levers to adjust the position of the millstones. The 'turret' type of windmill with only the top section rotating, which was developed in Italy towards the end of the 15th century, was the last significant addition to the list of prime movers before the invention of the steam engine.

The development and application of these forms of power produced the same kind of social and economic changes and dislocations in the Middle Ages as were to occur again on a larger scale in the 18th and 19th centuries. As early as the 10th century the lords of the manor began to claim a monopoly for their cornmills, which were a source of money-income, and this led to a long struggle between the lords and the commune. The monks of Jumièges, as lords of the manor, destroyed the hand mills at Viville in 1207; the monks of St Albans carried on a campaign against hand mills from the end of the 13th century until the so-called Peasants' Revolt, the great rising of the English communities led by Wat Tyler in 1381. The mechanization of fulling in the 13th century led to a wholesale shift of the English cloth industry from the plains of the southeast into the hills of the northwest where water was available. Colonies of weavers settled round the fulling mills in the Lake District, the West Riding and the Stroud valley, and the cloth industry decayed in towns like York, Lincoln, London and Winchester which had provided the broadcloth that was the staple of English industry in the 12th century. The insistence of the landowners who erected these mills that cloth should be brought to them, and not fulled by hand or foot at home, led to a long struggle, aspects of which are vividly described in *Piers Plowman*, and this action by the owners of mills was certainly also one of the causes of the Peasants' Revolt.

Though the other processes involved in the manufacture of

cloth did not, until the 18th century, reach the complete mechanization achieved in fulling in the 13th, the first steps towards this also were made during the Middle Ages. The main stages in early cloth making were carding and combing by hand, spinning by hand from a distaff on to a loose spindle, and weaving of the yarn thus prepared into a loose 'web' on a loom worked by hand and foot. The 'web' was then fulled in water and so felted. After fulling the cloth went to the 'rower' who raised the nap with teasels and to the shearsman who cut off the loose threads, after which, when small blemishes had been repaired, it was ready for sale.*
The mechanization of spinning began in the 13th century when the spinning wheel turned by hand made its appearance (Plate 26). The processes of twisting silk and winding it on reels are said to have been mechanized in Bologna in 1272. Certainly various kinds of thread were being spun with wheels by the end of the 13th century and about the same time the quilling wheel came in, by means of which the spun yarn was wound regularly on to the quill or bobbin which was set in the shuttle for weaving. There are several 14th-century illustrations of this wheel in use. It is interesting from the mechanical point of view as one of the earliest attempts to use continuous rotary motion. At the end of the 15th century further improvements in mechanisms for spinning and winding were envisaged by Leonardo da Vinci, who drew a sketch of a 'flyer' by means of which these two processes could go on simultaneously; he seems to have had in mind large-scale machinery driven by water power or a horse winch. He designed also a power-driven gig mill for raising the nap on cloth with teasels. In fact no satisfactory substitute has ever been discovered for teasels, though unsuccessful attempts to use iron combs were made as early as the mid-15th century. The flyer actually came into use about 1530 in a wheel incorporating also another innovation, the drive by treadle and crank. Power-driven

---

* *Piers Plowman* (c. 1362) has a description of the cloth trade (ed. W. W. Skeat, Oxford, 1886, p. 466, B Text, Passus xv, ll. 444 *et* seq.): Cloth that cometh fro the wevying is nought comly to were, Tyl it is fulled under fote or in fullyng-strokkes, Wasshen wel with water and with taseles cracched, ytouked, and ytented and under tailloures hande.

spinning mills and gig mills seem to have been used on a considerable scale in the Italian silk industry from the end of the 16th century, and full descriptions of them are given by Zonca (1607) (Plates 30, 31).

In weaving, the improvements that took place between the end of the Roman Empire and the revival of Western silk manufacture in the 14th century occurred mainly outside the West in Byzantium, Egypt, Persia and China, though they were rapidly taken over in the West in the later Middle Ages. These improvements were introduced mainly to make possible the weaving of patterned silk materials, for which it was necessary to be able to select the particular threads of the warp to be moved. This was done by two distinct improvements to the loom: first, a loom worked by pedals with better heddles, and later with a reed frame to provide a runway for the shuttle; and secondly, the draw loom. Both these devices seem to have been in existence in Egypt by about the 6th century A.D. and they probably entered Christendom through Italy, perhaps as early as the 11th century. From the silk industry their use spread to other branches of textile manufacture. Some further minor improvements were made in weaving technique in Europe in the 14th and 15th centuries; a knitting machine or stocking frame was invented in the 16th century, hand knitting having been invented a century earlier; and a ribbon loom was introduced about 1621. But the major improvements in weaving were not to come until the invention of the flying shuttle and power loom which, with the contemporaneous advances in the mechanization of spinning, were to transform the textile industry, particularly in England, in the 18th and early 19th centuries.

Another industry which became rapidly mechanized at the end of the Middle Ages was the production of books. Of the different elements involved in printing, the manufacture of linen paper seems to have begun in China in the 1st century A.D., whence it spread westwards through the countries dominated by Islam, to enter Christendom through Spain and southern France in the 12th century. This was a more suitable material for printing than the older costly parchment and brittle papyrus. The inks with an oil base used in printing were developed first by painters rather than calligraphers. Presses were already known in the manufacture of wine and the printing of cloth. The most essential element, the type itself, was made possible by skills acquired by wood engravers and by goldsmiths who had developed a technique for casting metal. Type developed in three main stages, first in China and then in Europe, though since the techniques used in these two regions were very different it is difficult to say to what extent the one influenced the other. In China, printing from wooden blocks, a separate block being cut for each page, appeared in the 6th century A.D., printing from movable wooden characters in the 11th century, and from movable metal type (in Korea) in the 14th century. In Europe, the use of wood cuts for the elaborate initial letters of manuscripts first appeared in a monastery at Engelberg in 1147; block printing appeared at Ravenna in 1289 and was common throughout Europe by the 15th century; movable metal type came in at the end of the 14th century, appearing at Limoges in 1381, Antwerp in 1417, and Avignon in 1444. The advantage of cast metal type was that hundreds of copies of each letter could be cast from a single mould instead of having to be carved separately as with wooden type. Though the first record of it in Europe is in the Netherlands, it was at Mainz that the use of accurately set movable type was brought to perfection. At Mainz, between 1447 and 1455, Gutenberg and his associates introduced, in place of the older method of casting type in sand, first the adjustable metal type-mould for making lead type, then the improvement of punches and the preparation of copper type. These were the strategic inventions in printing and with them the multiplication of books on a large scale became possible.

Perhaps the most spectacular result of medieval mechanical techniques is to be seen in the buildings, and many of the devices employed by the medieval mason to solve the statical problems arising in the construction of large churches were altogether original. It is impossible to say to what extent the medieval builder was being purely empirical and to what extent he was able to use the results of theoretical work in statics, but it is significant that during the late 12th and 13th centuries, just when the erection of the great cathedrals was producing the most difficult practical problems, Jordanus Nemorarius and others were making important additions to theoretical statics; at least one 13th-century architect, Villard de Honnecourt, showed a knowledge of geometry. The original developments in Gothic architecture arose from the attempt to put a stone roof on the thin walls of the central aisle of a basilica, the usual form of Christian church since Roman times. The Romans never had to face the problems which arose for the medieval mason because they built the barrel or groined vaults over their baths in concrete, and domes, like that of the Pantheon, in horizontally coursed brickwork with mortar; when the concrete or mortar had set, the thrust of the roof on the wall was very small. This was not the case with medieval buildings, in which no such concrete or mortar was used.

The masons of 10th- and 11th-century Burgundy tried to roof their naves with barrel vaults in the Roman style, but they found that the enormous thrust on the side walls tended to push them out even though they were made very thick. The first attempt to overcome this difficulty was to make the side aisles nearly the same height as the nave and roof them by means of groined vaults formed by two barrel vaults intersecting at right angles. These groined vaults of the aisles counteracted the thrust of the barrel vault of the nave and themselves exerted very little thrust except at the corners, which could be supported by massive pillars. This arrangement had the disadvantage that it left the church lighted only by the aisle windows, and when, as in many Cluniac churches, the roof of the nave was raised to get windows above the aisles, the walls collapsed from lack of support. A solution was found at Vézelay and Langres by using groined vaults for the nave, two semi-circular wooden centrings being used on which to construct the diagonals of the vault. By this means the 11th-century builder could construct a vaulted roof to cover any space, square or oblong, with a separate vault over each bay resting on semicircular transverse arches separating the bays.

This arrangement still had serious defects. The form of the

semicircular arch, in which the height must be half the span, was quite inelastic and there was still a formidable outward thrust so that the transverse arches tended to drop. Considerable elasticity of design was introduced and the outward thrust reduced by adopting the pointed arch which appeared in Christendom first in Vézelay and other Cluniac churches in the late 11th century, and later in the Île de France. It is thought to have been brought to Europe from Asia Minor, where it had become common by the 9th century. Half-arches of this kind were used in the 12th century to buttress the walls of several French churches, flying buttresses, in fact, in all respects except that they were hidden under the triforium roof.

A further step which completed the change from the Roman to the Gothic vaulted roof was to build diagonal arches over the wooden centrings used in constructing the groins, and to use these as permanent ribs (sprung from columns) on which to build the vault surface. This seems to have been done in various parts of Europe during the 11th and early 12th centuries, and it was this invention which gave rise to the wonderful Gothic of the Île de France in the 12th century. It gave great elasticity of design to the vault and meant that any space of any shape could be vaulted with ease, so long as it could be divided up into triangles, and that the summits of all the arches and vaults could be kept at any level desired. This freedom was increased still further when it was realized that the diagonal ribs need not be complete arches, but that three or more half-ribs could be used butting against each other at the summit of a pointed roof. Following the introduction of the permanent rib, differences in the method of filling in the vault surface led to a striking divergence in roof design between France and England. The French method was to make each vaulting panel arched and self-supporting. The English, on the other hand, did not make their panels self-supporting, so that further ribs had to be added to keep them up and this led to the fan vaulting of which good examples are Exeter Cathedral and the Chapel of King's College, Cambridge.

Perhaps the most striking of all the devices invented to solve the problems created by stone vaulting was the flying buttress introduced in the Île de France in the 12th century. In contrast with English builders, who at first retained the Norman tradition of thick walls, the French reduced their walls to little else than frames for stained-glass windows, and in so doing they had to devise some means of counteracting the thrust of the nave roof. This they did, at Poissy in 1135 and later at Sens and St Germain des Près, by carrying up a half-arch above the roof of the side aisle to the junction of the roof and wall of the nave. Later it was realized that the roof thrust extended some way down the wall, and the flying buttress was doubled to meet this thrust as at Chartres and Amiens. This method of counteracting roof thrust created another problem, for it exposed the building to a considerable strain from east to west. To tie it together in this direction the wall arches and the gables over the windows were made specially strong. This gave the windows in French churches like La Sainte Chapelle in Paris a prominence they never had in England.

Probably many of the devices invented by the 12th- and 13th-century architects were based on rule of thumb, and the great period of medieval building is singularly lacking in treatises on the subject. But the notebook of Villard de Honnecourt, who designed parts of Laon, Reims, Chartres and other French cathedrals, shows that the 13th-century architect could possess a greater ability to generalize the problems of stress and weight lifting involved than the poverty of theoretical writings might suggest. The *Architettura* of Alberti shows that certainly by the 15th century architects had a good knowledge of mechanics. This knowledge becomes even more evident in the late 15th and early 16th centuries, when Leonardo da Vinci calculated the weight that a pillar or cluster of pillars of any given diameter could safely carry and tried also to determine the greatest weight that could be borne by a beam of any given span. By the 16th century, Vitruvius had begun to have a great influence on building, but his admirers, such as Palladio, whose *Architettura* was published in 1570, far surpassed him in scientific knowledge. By the 17th century, problems such as the strength of materials and the stability of arches had become a subject of research by professional mathematicians like Galileo, Wren and Hooke; Wren and Hooke were also employed as architects.

Another branch of construction in which considerable progress was made during the Middle Ages, with the object of making better use of wind power, was ship building. The two common types of medieval European ship derived, respectively, from the Roman galley and the Norse long ship. They had a number of features in common: both were long, narrow and flat-bottomed, with a single mast and a square sail; both were steered by an oar on the side at the stern of the ship. The first improvement on this arrangement was the fore-and-aft rig as seen in the lateen sail which appears suddenly in Greek miniatures in the 9th century. By the 12th century lateen sails were common in the Mediterranean, and from there they spread to northern Europe. At the same time ships grew larger, got higher out of the water, the number of masts increased, and in the 13th century the modern rudder fixed to the stern post, itself an extension of the keel, made its appearance (Plate 28). These improvements made it possible to tack effectively against the wind, made oarsmen unnecessary, and extended the range of exploration. An early exercise in the mechanization of ships, not necessarily representing anything actually built, appeared at the beginning of the 15th century with the drawing of ships with paddle-wheels by Konrad Kyeser and by the Sienese engineer, Jacopo Mariano Taccola. Ramelli also gave an illustration of a paddle boat in 1588 and another innovation, a submarine, was actually built and successfully used in the Thames in 1614.

An improvement to inland water transport was brought about in the 14th century by the introduction of lock gates on canals; new possibilities for transport on land were introduced by making roads of stone cubes set in a bed of loose earth or sand and by improvements in wheeled vehicles, including (in the 13th century) the invention of the wheelbarrow. Mechanization was attempted also with land vehicles as early as 1420, when Fontana described a velocipede. At the end of the 16th century wagons propelled by man-driven machinery and by sails were apparently constructed in the Low Countries. Flight had attracted attention in the West at least since the 11th century, when Oliver of Malmesbury is said to have broken his legs in an attempt to glide from a tower with wings fitted to his hands and feet. Roger Bacon was also interested in flight. Leonardo da Vinci actually designed a mechanical flying-machine which flapped its wings like a bird.

An important advance associated with these improvements in methods of transport was the appearance of the first good maps in the West since Roman times. When accurate maps were added to the rudder and the compass, which came into use in the 12th century (see above, p. 131 *et seq.*), ships could be navigated effectively away from sight of land and, as Mumford has put it, exploration was encouraged in an attempt to fill in the gaps suggested by the rational expectations of space. The first true medieval maps were the *portolani*, or compass-charts, for mariners. The earliest known *portolano* is the late 13th-century *Carte Pisane*, but its relative technical excellence suggests that others which have disappeared were made before it. Genoese sailors are said to have shown St Louis of France his position on a map when he was crossing to Tunis in 1270. Some of the evidence might seem to suggest a Scandinavian origin for *portolani*, but the Arabs certainly had charts from an early date, and charts were developed also by the Byzantines, Catalans and Genoese. The use of the Catalan *legua* for distances in all the known *portolani* perhaps supports the Catalan claim to primacy, but this use may have come in later as a matter of convenience. In fact, the question of the origin of *portolani* is undecided. The novel feature of the *portolani*, as compared with the old traditional symbolic *mappæ mundi*, is that they were made for use as guides to a specific area. Made by practical men and based on the direct determination of distances and azimuths by using log and compass, they were generally restricted to the coastline. They contained no indications of longitude and latitude. They were covered with networks of rhumb-lines giving the compass bearing of the places on the map. The rhumb-lines radiated from a number of points arranged in a circle, corresponding to the points marked on a compass-card.

Other accurate maps showing inland regions as well as the

coastline were produced by men of education from the 13th century, by which time scholars like Roger Bacon were taking an interest in real geography. Bacon himself made no practical contributions to cartography, though his belief that there was no great width of ocean between Europe and China is said to have influenced Columbus, who found it repeated in works by Pierre d'Ailly and Aeneas Sylvius. As early as about 1250 Matthew Paris drew four recognizable maps of Great Britain showing details such as the Roman Wall, roads and towns. Between 1325 and 1330 an unknown cartographer produced a remarkably detailed and accurate map of England, the so-called 'Gough Map', now in the Bodleian Library in Oxford, which shows roads with mileages probably as estimated by travellers (Plate 32). About the same time good maps showing northern Italy were made by Opicinus de Canistris, who died about 1352, and in 1375 the so-called Majorcan school of cartographers produced for Charles V of France the famous *Catalan Mappemonde*, which combined the virtues of the *portolani* and the land maps and included North Africa and parts of Asia (cf. Plate 33). This Majorcan centre had collected an enormous amount of marine and commercial information and was the forerunner of the colonial and naval institute founded by Prince Henry the Navigator at Sagres about 1437. These early maps showed no indication of latitude and longitude, though the latitude of many towns had been determined with the astrolabe (see above, p. 104 *et seq.*). But in his *Geographia* Ptolemy had drawn maps on a complete network of parallels and meridians. This work, as it has come down to us, seems to be at least partly a later compilation, and the maps in the existing manuscripts were probably made by Byzantine artists in the 13th and 14th centuries. It was recovered and translated into Latin by Giacomo d'Angelo, who dedicated his translation, with some excellent maps redrawn from the Greek original by a Florentine artist, to Pope Gregory XII in 1406 and Pope Alexander V in 1409. After this, cartographers began to adopt Ptolemy's practice. Good examples are Andrea Bianco's map of Europe in 1436, and the map of central Europe found among the manuscripts of Nicholas of Cusa (1401–64) and printed in 1491.

Ptolemy's own atlas of the world was printed in numerous editions from 1477, when the *Geographia* was published, at Bologna, for the first time with Ptolemy's maps. These were re-drawn by Italian cartographers (cf. Plate 34). It gradually trans-formed cartography by emphasizing the need for an accurate linear measure of the arc of the meridian, the essential require-ment for accurate terrestrial cartography.

Until the end of the 18th century the most important material for machinery and construction generally was wood. Most of the parts of watermills and windmills, spinning wheels, looms, presses, ships and vehicles were of wood, and wood was used for geared wheels in much machinery as late as the 19th century. Thus it was that the first machine tools were developed for work-ing wood, and even in the tools themselves only the cutting edge was of metal. Of the boring machines, the bow-drill known since Neolithic times, in which the drill was driven rapidly by a string wound round it and attached at each end to a bow which was moved back and forth, was replaced during the later Middle Ages by the brace and bit, and a machine for boring pump barrels from solid tree trunks was also known. The most important of the machine tools for accurate work, the lathe, may have been known in some form in antiquity, but the pole-lathe was prob-ably a medieval invention. The first known illustrations of pole-lathes appear only in sketches by Leonardo da Vinci, but they must have been in use before that time. The spindle was driven by a cord wound round it as in the bow-drill, and attached at the bottom to a treadle and at the top to a springy pole which flexed the cord back on taking the foot off the treadle. Leonardo shows also a rotary lathe driven by bands from a wheel, though rotary lathes with crank and treadle drive became common only from the 17th century. In these early lathes the work was turned be-tween fixed centres, but in the mid-16th century Besson designed a mandrel lathe in which the work was fixed to a chuck to which power was applied. Besson designed also a crude screw-cutting lathe (Plate 35), to which further improvements were made in the 17th century, in particular the change introduced by the clock-makers from traversing the work over a stationary tool to traversing the tool itself while the work merely rotated. Thus, from the early machine tools designed for working wood, were developed tools capable of accurate work with metals.

The earliest machines made entirely of metal were firearms and the mechanical clock, and the mechanical clock in particular is the prototype of modern automatic machinery in which all the parts are precisely designed to produce an accurately-controlled result. In the mechanical clock the use of geared wheels, the main point of interest in early machinery, was completely mastered.

Water clocks, like the clepsydra, measuring time by the amount of water dripping through a small hole, had been used by the ancient Egyptians, and the Greeks had improved them by fitting devices to indicate the hours by a pointer on a scale, and to regulate the movement. The water clocks developed by the Arabs and Latin Christians were based on these Greek devices and also on the devices of the automatic puppet theatre, which was popular in the Middle Ages. They were so successful that water clocks remained in use as late as the 18th century. These water clocks were worked by a float suspended in a basin filled and emptied by a regulating mechanism, and the motion of the float was communicated to the indicator, usually some kind of puppet show, by ropes and pulleys. In Islam, they were sometimes very large and were set up where the public could see them; and in Christendom smaller clocks were used in monasteries where they were looked after by a special keeper, one of whose duties was to adjust the clock at night by taking observations on a star. One such clock is said to have been made by Gerbert for the monastery at Magdeburg. Other early clocks were worked by a burning candle, and a mid-13th-century work prepared for Alfonso X of Castile describes a clock operated by a falling weight controlled by the resistance created by the passage of mercury through small apertures. Similar devices were developed in the long series of astronomical mechanisms – planetaria, mechanically rotated star-maps, etc. – which are as essential a part of the ancestry of the mechanical clock as time-keepers proper. In none of these devices were there any gears.

The essential features of the mechanical clock were a drive by a falling weight which set in motion a train of geared wheels, and an oscillatory escapement mechanism which prevented the weight accelerating as it fell, by stopping it at frequent intervals. The earliest illustration of an escapement mechanism, at least in the West, appears in the mid-13th century in a device drawn by Villard de Honnecourt for making an angel rotate slowly so that its finger always pointed towards the sun (Plate 36). It is possible that the first mechanical clocks were made shortly afterwards. There are references to what seem to have been mechanical clocks of some kind in London, Canterbury, Paris and other places during the second half of the 13th century and in Milan, St Albans, Glastonbury, Avignon, Padua and elsewhere during the first half of the 14th century. Some of these were planetaria, for showing the motions of the heavenly bodies, rather than clocks. Probably the earliest true clocks of which the mechanism is definitely known are the Dover Castle clock, usually dated 1348 but probably later (Plate 37), and Henri de Vick's clock set up in Paris at the Palais Royal, now the Palais de Justice, in 1370. These clocks were regulated by a verge escapement with a foliot balance. The essential components of this mechanism were a crown wheel with saw-like teeth, which were engaged alternatively by two small plates or pallets on a rod, so that the wheel was intermittently stopped and released. The foliot was a mechanism for regulating the speed of rotation of the crown or 'escape' wheel, and therefore of the whole train of wheels ending with the axle carrying the hands of the clock. The perfection of this verge escapement and foliot balance marks a limit in clock design on which, in point of accuracy, no real advance was made until the application of the pendulum to clocks in the 17th century, though before that time considerable refinements were made in construction. The early clocks were, in fact, mostly very large and the parts were made by a blacksmith. De Vick's clock was moved by a weight of 500 lb. which fell 32 feet in 24 hours and had a striking weight of nearly three-quarters of a ton. In the 15th century, clocks became smaller and were used in houses, screws were used to hold the parts together, and the end of the century saw the first 'clock-watches' driven by a spring.

These early clocks were reasonably accurate if set nightly by observing a star, and by 1500 most towns had public clocks on the outside walls of monasteries or cathedrals or on special towers. They either simply struck the hours or also showed them on a circular face marked in divisions of 12 or 24. The effect of placing them in public places was to bring about the complete replacement of the seven variable liturgical hours by the 24 equal hours of the clock. From an early date in antiquity astronomers had, in fact, divided the day into 24 equal hours, taking the hours of the equinox as standard, and throughout the early Middle Ages this system had existed, particularly in civil life, side by side with the ecclesiastical system. A decisive step was taken in 1370 by Charles V of France when he ordered all churches in Paris to ring the hours and quarters according to time by de Vick's clock, and from that time the equal hours became more common. The division of the hour into 60 minutes and of the minute into 60 seconds also came into general use in the 14th century and was fairly common as early as 1345. The adoption of this system of division completed the first stages in the scientific measurement of time, without which the later refinements of both physics and machinery would scarcely have been possible.

4. INDUSTRIAL CHEMISTRY

If wood, as Lewis Mumford has vividly pointed out, 'provided the finger excercises for the new industrialism,' the development of modern machinery and of precision instruments and scientific apparatus is inconceivable without the artificial products of the chemical industry, above all metals and glass (cf. above, p. 139 *et seq.*).

The metal in the working of which the greatest advances were made during the Middle Ages was iron. Already in Roman times the Gauls and Iberians had become efficient ironsmiths and their knowledge was never lost. By the 13th century iron was being worked in many of the main European fields, in Biscay, Northern France and the Low Countries, the Harz Mountains, Saxony and Bohemia, and in the Forest of Dean, the Weald of Sussex and

Kent, Derbyshire and Furness. The striking advances in iron working made during the Middle Ages were the result of using more efficient furnaces which gave higher temperatures for smelting. For this the chief fuel in medieval as in classical times was charcoal. Though 'sea coal' was mentioned by Neckam, and coal was being mined near Liège and Newcastle (whence it was transported to London in flat-bottomed boats) and in Scotland by the end of the 12th century and in most of the major European fields by the end of the 13th century, it was not until the 17th century that a method of using coal for iron working was introduced. This was invented by Dud Dudley about 1620. In the Middle Ages, one of the chief industrial uses of coal was for lime-burning, and already by 1307 the smoke had become such a nuisance in London that attempts were made to prohibit its use there. The improvements made in furnaces in the Middle Ages were due not to better fuel but to the introduction of mechanisms for producing blast air, and the production of charcoal for the ever-increasing needs of metallurgy to supply the demand for swords and armour, nails and horseshoes, ploughs and wheel-rims, bells and cannon remained a serious menace to the forests of Europe till the 18th century. In England it seems to have been the shortage of timber that brought the end of metallurgy in the Weald of Sussex and Kent.

From an early date the draught for furnaces was provided simply by wind tunnels, with hand bellows as auxiliaries. This was the method used in the so-called sponge iron process, in which the iron ore was heated with charcoal in small furnaces where the temperature was not high enough to melt the iron, but produced a spongy 'bloom' at the bottom of the furnace. By alternate heating and hammering, when the power-driven forge hammer came into play, the bloom was worked into wrought iron rods, which could be rolled and sheared or slit to form plates, or drawn through successively smaller holes in a tempered steel plate to form wire. Steel making was well understood in medieval Christendom, though the best steel came from Damascus, where it was made by a process apparently developed originally by the Hindus. Later, excellent steel was made at Toledo.

Improvements in the method of producing blast began with the introduction into the furnace of air under pressure from a head of water, a method that was used in Italy and Spain before the 14th century. Blast was produced also by steam issuing from the long neck of a vessel filled with water and heated, and by bellows operated by horse-driven treadles, but the most outstanding advance was the introduction of bellows driven by water power (Fig. 14). Such blast furnaces made their appearance in the Liège region in 1340 and quickly spread to the Lower Rhine, Sussex and Sweden. These new furnaces became much bigger than the old ones, and for the first time it was possible to produce temperatures that would melt the iron, so that it could be obtained

Fig. 14 Water-driven forge bellows. From V. Biringuccio, *Pirotechnia.*

directly instead of in the form of a bloom that had to be worked with hammers. Most important of all, the new furnaces made it possible for the first time to produce cast iron on a commercial scale.

Of the other metals, lead and silver, gold, tin and copper were mined in various parts of medieval Christendom. Cupellation furnaces with water-driven bellows for refining silver from lead appeared in Devon at the end of the 13th century. The lead was oxidized by heating to form litharge, which was skimmed off or absorbed by the porous hearth. Gold was mined in Bohemia, the Carpathians and Carinthia. Tin, of which the Cornish mines were the principal source, was used with copper for making bronze and with copper and calamine (hydrous zinc silicate) for making brass for bells, cannon, and monumental and ornamental 'dinan-derie', and with lead for making pewter for household ware. Specialized working of metals led to the development of separate guilds of silversmiths and goldsmiths, pewterers, blacksmiths, founders, bladesmiths, spurriers and armourers, and skill in weld-ing, hammering and grinding, chasing and embossing reached a very high level. Specialists also produced needles, scissors, shears, thimbles, forks, files, edged tools for the builder, nails, nuts and bolts, and spanners, clocks and locks, and some attempt was made at standardization. Brass wire was invented in the 11th century, and by the 14th century steel wire drawing was being done by water power. These specialist skills made possible the manufac-ture of articles of which the value depended on a precise finish. Attention to the finishing processes themselves made it possible to produce such instruments of precision as the astrolabe and the mechanical clock. Recognition of the need to control the con-tent of the alloy used led to the development of assaying, which laid the foundations of quantitative chemistry. Assaying famili-arized metallurgists with the use of the balance and led also to the development of other specialized branches of chemistry, of which the production of the mineral acids was one of the most important.

Of the medieval metallurgical processes from which an accurate product was required, the founding of bells and guns are perhaps the most interesting. The first European account of bell founding was given early in the 12th century by Theophilus the Priest, and from that time skill in casting bronze and brass de-veloped rapidly, to produce the monumental brasses of the 13th and 14th centuries, and such exquisite products as the southern door of the baptistery of Florence by Andrea Pisano in 1330 and the other even more wonderful doors by Ghiberti about a century later. Large bronze bells began to be made in the 13th century and became numerous in the 14th. The main problem was to produce bells that would ring in tune. The note of a bell varies with the proportions and the amount of metal used, and, though final tuning could be done by grinding down the rim if the note was too flat and by grinding down the inner surface of the sound bow if the note was too sharp, it was necessary for the founder to be able to calculate the exact size and proportions to give something near the right note before he began to cast the bell. For this each founder must have had his own empirical system, for example the system by which bells giving notes with intervals of the tonic, third, fifth and octave were produced by having diameters in the proportions of 30, 24, 20 and 15 and weights in the proportion of 80, 41, 24 and 10 respectively. The scientific temper of the time is shown by the attempt by Walter of Odington, in the late 13th or early 14th century, to devise a rational system according to which each bell would weigh eight-ninths of the bell next above it in weight. In practice this system was distinctly inferior to the empirical systems actually used by bell founders.

The earliest firearms appeared in the West during the first half of the 14th century, but they seem to have been made in China about a century earlier. In both regions considerable progress had previously been made in other forms of projectile-throwing weapons. In the West, by the end of the 12th century, the tre-buchet worked with counterweights had begun to drive out the older forms of torsion and tension engines of artillery coming from the Romans or Norsemen; by the early 14th century the crossbow had become a highly effective weapon with sights and a trigger mechanism and the longbow was no less powerful and accurate. The use of gunpowder as a propellant in an effective gun was simply the last of a number of improvements, and firearms did not immediately replace other projectile weapons, though they had become the chief weapon of artillery by the end of the 14th century. Cannon may have been used in the West as early as the siege of Berwick in 1319 and by the English at Crécy in 1346. The French fleet that was to invade England in 1338 is described as having '*un pot de fer à traire garros à feu,*' and cannon were certainly used the following year at the sieges of Cambrai and of Puy-Guillaume in Perigord. They were certainly also used by the English to capture Calais in 1347 and, according to Froissart, the English used 400 cannon, probably small mortars, to besiege St Malo in 1378.

Of the constituents of gunpowder, saltpetre seems to have been known in China before the first century B.C. and knowledge of the explosive properties of a mixture in the right proportions of saltpetre, sulphur and charcoal seems to have been perfected there by about 1000 A.D. In the West, other inflammable mixtures had been used in warfare much earlier. The Byzantines used an improved form, 'Greek fire', probably a mixture of quicklime, naphtha and pitch from petroleum and sulphur, against the Moslem fleet at the siege of Constantinople in 673, and later. Gunpowder itself became known in the West during the second half of the 13th century, perhaps introduced from China through the Mongols. Roger Bacon referred in his *Opus Majus* and *Opus Tertium* to an explosive powder, and pointed out that its power would be increased by enclosing it in an instrument of solid material. The earliest known Western recipe for gunpowder is in a Latin manuscript of about 1300 of the *Liber Ignium* attributed to a certain Marc the Greek, about whom nothing is known.

Having learnt the explosive and propulsive properties of gunpower, the West rapidly outstripped China in the manufacture of weapons. The earliest Western cannon were made from metal similar to that used for bells, often by the same founder, and the chief centres of manufacture were Flanders, Germany and, to a less extent, England. The earliest known illustration of a cannon in the West is of a small *vaso* or *pot de fer*, as they were called, in a manuscript of a work by Walter de Milemete dedicated to Edward III, in 1327 (Plate 29). In the middle of the 14th century cannon were cast from cuprous metal, and at the end of the century they were made also of wrought iron strips held together with iron bands. In the 15th century cannon, especially of wrought iron, reached a considerable size, the two largest known being 'Mad Meg', now in Ghent, which is 197 inches long, has a calibre of 25 inches, threw a stone ball of about 700 pounds and weighs approximately 13 tons, and 'Mons Meg', now in Edinburgh Castle, which is somewhat smaller. These early cannon were all muzzle-loaders, firing at first large round stones and later cast-iron balls. Lead shot was used from the 14th century with smaller guns. Breech loading was attempted quite early, but it was impossible to finish the metal surfaces with sufficient precision to produce gas-tight breech locks. A primitive form of rifling was introduced in the bronze guns, and during the 15th century standarization of guns and shot began to come in, culminating in the standard ordnance propagated by the artillery schools of Burgos and Venice in the early 16th century.

A great advance in gun making was made early in the 16th century by the introduction of a method of boring cast bronze or iron guns so that they could be given an accurate finish. Machines for boring wood had been known from an early date and, as early as 1496, the German mechanic, Philip Monch, had made an elaborate sketch of a gun borer worked by horse power. Leonardo da Vinci also sketched a boring machine for metal working, and Biringuccio described and illustrated one driven by a water-wheel in his *Pirotechnia* (1540). With the introduction of accurately-bored barrels began a new period in the history of gunnery which lasted until the 19th century.

The experience acquired in the production of metals in the later Middle Ages was transferred to other kinds of mining, and the great demand for minerals generally had some striking economic, political and industrial consequences. By the 14th century, apart from metals and coal, there was mining on a fairly large scale of sulphates in Hungary, rock salt in Transylvania, calamine and saltpetre in Poland, mercury in Spain, and in the 15th century of alum in Tuscany and the Papal states. Pumping, ventilation and haulage in ever-deepening seams made mining an expensive business that could be undertaken only by the man with capital, and as early as 1299 Edward I leased silver-lead mines in Devonshire to the Frescobaldi, a family of Florentine merchants and bankers, who in turn financed Edward I and II of England and also Philippe le Bel of France. Perhaps the most striking example of fortune and power acquired from mining is provided by the Fuggers. From small beginnings in the 14th century, the Fuggers had by the 16th century built up such capital from the silver-lead mines of Styria, the Tyrol and Spain that they were in a position to finance the big guns and mercenary troops on the scale required by a European ruler like the Emperor Charles V.

Of the industrial consequences of the growing demand for metals, perhaps the most striking are the improvements in pumps and eventually, at the end of the 17th century, the use of steam power to pump out the subsoil water, the experiments in the use of coal for metallurgy to overcome the increasing shortage of charcoal fuel, and the attempt to find substitutes for metals like tin which, before the exploitation of the mines of the New World and the Far East, was becoming ever scarcer. Of these substitutes the most important for science was glass, which, from the 14th century, was being produced as a substitute for pewter for household ware.

Glass making was well known in the Ancient World, and in various parts of the Roman Empire excellent dishes, bowls, beakers, bottles and other household objects were made from blown glass, and the art of engraving on glass was developed. In the early Middle Ages a high technique in glass making was carried on in Byzantium, in various Arabic centres and, more obscurely, also in the West. It was not until the 13th century that glass making began to revive generally in the West, though one of the best accounts of it is to be found in the early 12th-century treatise of Theophilus the Priest. The most famous Western centre was Venice. Though from the 13th century glass making made considerable progress also in Spain, France and England, it was not until the 16th century that glass was made on a large scale outside Italy.

Most medieval glass was blown (Plate 38). The materials, for example sand, carbonate of potash and red lead, were melted together in a furnace and, when the material had cooled enough to become viscous, a blob was picked up on the end of a long rod and rotated, or blown and worked with large tongs, until the required vessel or other object was formed. It might be reheated again to alter the shape. The essentials of the technique were dexterity, speed and the control of the temperature to which the cooling glass was exposed; on these depended its final strength. For plate glass the sand had to be free from iron oxide, and carbonate of lime, sulphate of soda and some form of carbon were required. The method of making plate glass was to blow a large bubble which was worked into a long, hollow cylinder hanging from the platform on which the blower stood, and was eventually slit open and worked flat. This method restricted the size of the sheet.

The chief use of glass in the Middle Ages was for windows and household vessels. Stained-glass windows for churches came in early in the 12th century and painted glass in the 14th. Glass vessels for household purposes were not common before the 16th century, pewter and glazed pottery being the usual materials for hardware, but from the 14th century glass was more commonly used. As early as the 13th century there are references to glass being used for scientific apparatus: Grosseteste and others mentioned optical experiments with a spherical urine flask, and by the early 15th century distilling apparatus was being made of glass. As Mumford has pointed out, the development of chemistry would have been greatly handicapped without glass vessels, which remain neutral in an experiment, are transparent, withstand relatively great heats, and are easy to clean and to seal. Optical instruments using lenses and the sciences which, from the early 17th century, developed with them would clearly have been impossible without glass. The Arabs had produced lenses as early as the 11th century, and lenses were discussed by the great Latin optical writers of the 13th century. Though medieval optical glass did not have the excellence of that produced since the 18th century, for which specially pure ingredients are used, it was good enough to make possible the invention of spectacles at the end of the 13th century (see below, pp. 236–7).

In other chemical industries as well as metallurgy and glass making medieval craftsmen acquired a considerable empirical knowledge. Considerable skill in controlling the processes involved was shown in pottery, in tile and brick making, in tanning and soap making, in the processes of malting, yeasting and fermentation involved in brewing, in the fermentation of wine and in the distilling of spirits. Salt making by dissolving the crude material from the mine in water, boiling the brine, and precipitating the crystals in open pans had been known to the Romans and was practised in the Middle Ages at various places, including Droitwich and Nantwich in England. Considerable skill was shown also in the dyeing of wool, silk and linen with vegetable dyes such as woad, madder, weld, lichens and a red dye obtained from 'greyne,' an insect resembling cochineal, and in fixing the dye with mordants, of which the most usual were alum, potash from wood ashes, tartar deposited by fermenting wine, iron sulphate and 'cineres' (possibly barilla or carbonate of soda). The treatises on the preparation of pigments, glues, siccatives and varnishes written from the eighth to the sixteenth centuries contained a large variety of recipes giving practical directions how to prepare chemical substances. At the beginning of the 12th century the treatise by Theophilus the Priest referred to oil paints, though it was not before the Van Eycks early in the 15th century that the siccative properties of oil paints were improved so that they dried quickly enough for several colours to be put on at the same time. Medieval painters and illustrators learnt how to prepare a large variety of colours of vegetable and mineral origin and new recipes were continually being added, for example that for 'mosaic gold,' a stannic sulphide, which was discovered about 1300. The ordinary black ink of medieval manuscripts was usually lampblack mixed with glue. The practical skill acquired in these industries helped to lay the foundations of modern chemistry.

5. MEDICINE

Perhaps of all the practical arts of the Middle Ages, medicine is the one in which hand and mind, experience and reason, combined to produce the most striking results. Of the higher faculties of theology, law and medicine in medieval universities, only in medicine was it possible to have further training in natural science after the arts degree, and many of the leaders of science from Grosseteste, in the 13th century, to William Gilbert, in the 16th, had studied medicine (see above, p. 169 *et seq.*; p. 186 *et seq.*), Medical men like Grosseteste, Petrus Hispanus and Pietro d'Abano, basing themselves on the logical writings of Galen, Ali ibn Rid-wan and Avicenna, as well as Aristotle, made some most import-ant contributions to the logic of induction and experiment which had a profound effect on science down to the time of Galileo, who himself began his university studies in medicine (see Vol. II, p. 25 *et seq.*). And certainly in practical medicine the medieval doctors found empirical solutions to some important problems and established the basic scientific attitude that characterizes modern medical practice.

After the decay of the Roman Empire, medicine in the West was largely folk-medicine, but some knowledge of Greek medi-cine was preserved by writers like Cassiodorus and Isidore of Seville and by the Benedictine monasteries. Latin summaries of parts of Hippocrates, Galen and Dioscorides were known, and something of the gynæcological tradition of the 2nd century A.D. Soranus survived in books for midwives. A revival of medical learning took place in Carolingian times at Chartres and other schools, in the 10th century the Leech-Books appeared in Anglo-Saxon England and in the 11th the writings of Hildegard of Bingen in Germany. The real revival of Western medicine began in the 11th century when the medical school at Salerno, which had come gradually into existence perhaps a century or two earlier, began its attested activity. Whether it was because of its Greek or Jewish population or because of its contacts with the Arabs in Sicily, certainly before 1050 Gariopontus was quoting freely from Hippocrates, and Petrocellus had written his *Practica*; about the same time Alphanus, Archbishop of Salerno, translated from Greek a physiological work by Nemesius under the title of *Premnon Fisicon*; and before 1087 Constantine the African had translated from Arabic Galen's *Art of Medicine* and *Therapeutics* and various works by Haly Abbas and the Jewish physician, Isaac Isræli. The school of Salerno acquired a considerable repu-tation, and Sudhoff has suggested that its teachers were practising doctors who taught medicine by dissecting animals. Certainly in the 12th century the *Anatomia Ricardi* emphasized the need for a knowledge of anatomy, and the *Anatomia Porci* attributed to Copho described the public dissection of a pig. At the end of the

12th century Salerno produced the first great Western surgeon, Roger of Salerno, whose work was carried on in the early 13th century by Roland of Parma (Plate 39). About the same time was composed the famous *Regimen Sanitatis Salernitanum*, which remained a classic of medical lore until the 16th century.

In the 12th century, Montpellier also began to rise as a medical centre and, in the 13th century, the university medical schools of Montpellier, Bologna, Padua and Paris gradually superseded Salerno. Medical teaching in these university schools was based on various works by Galen and Hippocrates and by Arab and Jewish doctors, the translation of which into Latin had been chiefly responsible for the revival of Western medicine in the 12th and 13th centuries. Of the Arabic and Hebrew works, the most important were Avicenna's encyclopædic *Canon of Medicine*, Isaac Isræli's classic work on fevers, and Rhazes' works in which were descriptions of diseases like smallpox and measles. The 10th-century Spanish Moor, Albucasis, provided the chief early text-book for surgery, and works by the 9th-century Hunain ibn Ishaq and by Haly Abbas were the chief sources through which Arabic ophthalmology became known. Other important works were those by the 7th-century Byzantine, Theophilus, on the pulse and the urine, the examination of which was the commonest method of diagnosis in the Middle Ages, and by Dioscorides' *De Materia Medica*.

Medical treatment in the Middle Ages, when not confined simply to the Hippocratic method of keeping the patient in bed and letting nature take its course, was based on herbs. In Greek medicine the physiological theory behind the use of herbs was that disease was due to an upset of the balance between the four humours, so that 'cooling' drugs were administered to counteract excessive heat in the patient, 'drying' drugs to counteract excessive moisture, and so on (Fig. 13). The supposed effects of drugs based on this theory were sometimes fanciful, but doctors in the ancient world from Egyptian times had accumulated an empirical knowledge of the effects of a considerable number of herbal drugs like mint, aniseed, fennel, castor oil, squill, poppy, henbane, mandragora, and also of a few mineral drugs like alum, nitre, hæmatite and copper sulphate. A common fumigant was prepared by burning horns with dung to produce ammonia. To the Greek list the Arabs added some herbs from India like hemp, senna and datura and mineral drugs like camphor, naphtha, borax, antimony, arsenic, sulphur and mercury. The Western doctors made further contributions. As early as the 12th century, the so-called *Antidotarium Nicolai*, a work on drugs composed at Salerno before 1150, recommended the use of the *spongia soporifera* to induce anæsthesia, and Michael Scot, who studied at Salerno, gave the recipe as equal parts of opium, mandragora and henbane pounded and mixed with water. 'When you want to saw or cut a man, dip a rag in this and put it to his nostrils.' Modern experiments suggest that this could not have been a very powerful anæsthetic, and various attempts to improve it were made during the Middle Ages, including, by the 16th century, the use of alcohol fumes. The extraction of the virtues of herbs with alcohol to make what is now known as a tincture was discovered by Arnald of Villanova (*c.* 1235–1311). Minerals like arsenious oxide, antimony and mercury salts were regularly used in drugs by the Bolognese doctors, Hugh (d. 1252–58) and Theodoric Borgognoni (1205–98), and also by Arnald of Villanova and others. Mercury ointments were especially popular as cures for various skin diseases and the salivation they produced was noticed.

A branch of medicine in which the empiricism of the medieval mind showed itself to good effect was observation of the effects of different diseases. A large number of diseases had been recognized and described by Greek and Arab physicians, and to this body of knowledge additions were made, particularly in the written *consilia* or case-histories that became common from the time of Taddeo Alderotti, of Bologna, in the 13th century. The practice of writing *consilia* was part of the general movement towards strictness in presenting evidence in theology as well as in the profane sciences, and sometimes it led to an emphasis on logical form to the detriment of observation, as when *consilia* were prepared and medical advice given from reports from unseen patients. When used properly and based on individual case-histories, as it was by doctors like Alderotti and Arnald of Villanova in the 13th century, Bernard of Gordon and Gentile da Foligno in the 14th, and Ugo Benzi in the 15th, this practice led to some excellent descriptions of the symptoms and courses of diseases such as bubonic and pneumonic plague, diphtheria, leprosy, phthisis, rabies, diabetes, gout, cancer, epilepsy, a skin disease known as *scabies grossa* or *scabies variola* which some historians have identified with syphilis, affliction with the stone, and numerous surgical cases. Many of these *consilia* were printed in the late 15th and 16th centuries. They are the origin of modern case-history books.

The chief limitation of medieval doctors was, in fact, not that they could not recognize diseases but that they could not often cure them. They had very little understanding of either normal or morbid physiology or of the causes of most diseases, and they were sometimes further misled by the habit, coming from Aristotelian philosophy, of regarding each separate symptom and even wounds as manifestations of a separate 'specific form.'

A good idea of the state of medical knowledge in the 14th century can be gathered from the tracts written by physicians at the time of the Black Death. This plague seems to have begun in India about 1332, where an Arab doctor gave an account of it, and to have spread westwards, reaching Constantinople, Naples and Genoa by 1347. It reached a climax in the Mediterranean in 1348, in the North in 1349, and in Russia in 1352. It died down then, but smaller plagues went on recurring in the West at fairly frequent intervals till the end of the 14th century and at less frequent intervals for another three centuries after that. More than twenty tracts written at various places during the years of the Black Death show the characteristics commonly found in late medieval medicine: an orderly approach to the problems of symptoms, progress, causes, transmission, prevention and cure, in which is seen the combination of intense speculation based on causes no longer accepted in the 20th century with some very sound ideas on which effective practical measures were based. The eastern origin of the epidemic was generally recognized and several of the tracts contain full descriptions of the symptoms, for example that written by Gentile da Foligno at the request of the University of Perugia in 1348 and the *Chirurgia Magna*, written in 1360 by Guy de Chauliac, an eminent product of Montpellier and Bologna and papal physician at Avignon. The symptoms included fever, pain in the side or chest, coughing, short breath and rapid pulse, vomiting of blood, and the appearance of buboes in the groin, under the armpit or behind the ears. Bubonic and pneumonic plague were distinguished. Some tracts gave, as early indications of the onset of the disease, pallor and an expression of anxiety, a bitter taste in the mouth, darkening of ruddy complexions, and a prickling of the skin above incipient abscesses which gave sharp pains on coughing.

Of the natural causes of the epidemic, considerable attention was devoted to astrological influences, and attempts were made to predict future plagues on the basis of planetary conjunctions. These remote causes were supposed to operate through near causes and in particular to cause the corruption of the air, though other causes of corruption were suggested, such as exhalations from the earthquake of 1347 and the unseasonable and very damp weather. Weather signs as well as astrological signs were watched as indications of the onset of plague, but some writers pointed out the lack of complete correlation of either with epidemics.

About prevention there was considerable uncertainty, most physicians advising flight as the only reliable precaution, and if that was impossible some form of protection against corrupt air, such as avoiding damp places, burning aromatic wood in the house, and abstaining from violent exercise which drew air into the body and from hot baths which opened the pores of the skin. Since the corrupt vapours were held to cause plague by acting as a poison in the body, one method of prevention was to take various antidotes against poison as, for example, theriac, mithridate, or powdered emerald. Bleeding to reduce the natural heat of the body was also advised. The usual methods of treatment were bleeding to remove the poison, administering purgative drugs, and lancing or cauterizing the buboes or the use of a strongly drawing plaster. Attention was given also to maintaining the strength of the heart.

Though the physicians who had to deal with the Black Death were in many ways poorly equipped for the task, their experience made them give serious thought to problems never before discussed. As John of Burgundy put it in a passage translated by A. M. Campbell in her *Black Death and Men of Learning* from his *Treatise on the Epidemic Sickness*, written about 1365:

modern masters everywhere in the world are more skilled in pestilential epidemic diseases than all the doctors of the art of medicine and the authorities from Hippocrates down, however many they are. For ... no one of them saw so general or lasting an epidemic, nor did they test their efforts by long experiment, but what most of them say and treat about epidemics they have drawn from the sayings of Hippocrates. Wherefore the masters of this time have had greater experience in these diseases than all who have preceded us, and it is truly said that from experience comes knowledge.

The most striking new ideas put forward by the physicians of the Black Death concerned the method of transmission of the epidemic by infection. Of this the Greeks seem to have had little notion, attributing all epidemics to a single general cause, *miasma*. In the Middle Ages, the idea that specific diseases could be caught by infection or contagion was worked out first in connexion with leprosy, and by the 13th century had been applied to other diseases like erysipelas, smallpox, influenza, diphtheria and typhoid fever. A dancing mania, St Vitus's Dance, that spread through the Germanies in the late 14th and 15th centuries, was also recognized as contagious. The segregation of lepers was based originally on the ritual of isolation described in the Bible and had been practised in Christendom at least from the 5th century. Leprosy was still a serious menace in the 12th century, when it seems to have increased somewhat and it is said that in France as many as one person in 200 was a leper, but from the end of the 13th century it began to decline. Physicians learnt to recognize the symptoms more accurately; in the mid 13th century Gilbert the Englishman described the local anæsthesia of the skin, which is one of the best diagnostic symptoms, and a century later Guy de Chauliac drew attention to the excessive greasiness of the skin.

So successful were the methods of diagnosis and segregation that by the 16th century Europe was almost entirely free from leprosy, and similar preventive measures were taken against other infectious diseases.

Among the tracts written during the Black Death, two by Spanish Moors contained the most remarkable statements about infection. Ibn Khatima of Almeria pointed out that people who came into contact with someone with plague tended to contract the same symptoms as the diseased person, and Ibn al-Khatib of Granada said that infection could take place through clothes and household objects, by ships coming from an infected place, and by people who carried the disease though they were themselves immune. Scarcely less remarkable was the slightly earlier *consilium* on the plague by Gentile da Foligno, who used the words 'seeds (*semina*) of disease' (found also in works by Galen and Haly Abbas) for what would now be called germs, and *reliquæ* for the infectious traces left by patients. Some of the methods of infection suggested by Black Death physicians appear rather strange in the 20th century, for example, one based on the optical theory of the 'multiplication of species' according to which plague could be caught from a glance from the eyes of the patient. When the sick man was in agony the poisonous 'species' was expelled from the brain through the concave optic nerves. But at a time long before the germ theory of disease was properly understood, physicians had learned enough about infection to advise governments on the precautions to be taken.

The first commission of public health was organized in 1343 in Venice, and in 1384 Lucca, Florence, Perugia, Pistoia and other towns made laws to prevent infected persons or goods from entering them. The first systematic efforts to isolate plague carriers date from the regulations made by Ragusa in Dalmatia, Avignon and Milan at this time. Ragusa, in 1377, issued a new law ordering the isolation of all travellers from infected regions for 30 days (called the *trentina*), and Marseilles, in 1383, extended this period to 40 days for ships entering the harbour, thus instituting the *quarantine*. Venice opened a quarantine hospital and brought in regulations about the airing of infected houses, washing and sunning of bedding, control of domestic animals, and other hygienic matters. Military hygiene had attracted attention since the early Crusades, when losses had been heavy because of ignorance of elementary sanitation, and in the 13th century several works had been written on precautions to be taken by soldiers and large bodies of pilgrims. The most outstanding were a work written by Adam of Cremona for the Emperor Frederick II, a short treatise on military hygiene by Arnald of Villanova, and the *Régime du Corps* by Aldobrandino of Siena. With the Venetian regulations began the interest of municipalities in hygiene.

A special branch of medicine in which some striking progress was made in the Middle Ages was ophthalmology. Operations like that for cataract had been known since classical times and the Arabs had acquired considerable skill in treating eye complaints, using zinc ointments and performing difficult operations like removing an opaque lens. The most popular Latin work on ophthalmology was written by a 12th-century Jew, Benvenutus Grassus, and based on Eastern sources. In the 13th century Petrus Hispanus described various cataract conditions in great detail and gave an account of the operation with gold needles.

The outstanding advance made in the West was the invention of spectacles. That weak sight and particularly the difficulty of reading in the evening was felt as a serious affliction is shown by the number of salves and lotions prescribed for this complaint, but although lenses had been known for some centuries in both Christendom and Islam, it is only at the end of the 13th century that there is evidence of spectacles with convex lenses being used to compensate for long sight. Roger Bacon had proposed this in 1266–7 in his *Opus Majus*. The invention of actual spectacles was associated traditionally with the names of certain north Italian Dominican friars, but it is more probable that the first spectacles were made, shortly after 1286, by an unknown inventor, and that the invention was made public by a friar, Alessandro della Spina of Pisa, who saw them being made and then constructed his own. Their manufacture was early associated with the Venetian glass and crystal industry, and spectacles were in fact sometimes made of crystal or *beryllus*. The earliest known occurrence of any term for spectacles is found in supplementary regulations for the Venetian guild of crystal-workers in 1300, which refers to *roidi da ogli* ('discs for the eyes'); and in the following year there is reference to making *vitreos ab oculis ad legendum* ('glasses for the eyes for reading'). In 1300 there is reference also to *lapides ad legendum*, which seem to be magnifying glasses. A little later there are further references in other Italian documents; for example in 1322 a Florentine bishop (quoted by E. Rosen in an article in the *Journal of the History of Medicine*, 1956, vol. 11, p. 204) bequeathed 'one pair of spectacles, framed in silver-gilt.' A statement by Bernard of Gordon in 1303 was formerly thought to refer to spectacles, but the earliest absolutely certain medical reference is much later, when Guy de Chauliac in 1363 prescribed spectacles as a remedy for poor sight after salves and lotions had failed. By this time spectacles had in fact become fairly common, and, for example, Petrarch (1304–74) wrote in his autobiographical *Letters to Posterity* : 'For a long time I had very keen sight which, contrary to my hopes, left me when I was over sixty years of age, so that to my annoyance I had to seek the help of spectacles.' These early spectacles were, it seems, all made with convex lenses; it is only in the 16th century that concave lenses are known to have been used for shortsightedness. From Christendom spectacles spread to the Arabs and China.

In surgery, progress began in the West with Roger of Salerno's *Practica Chirurgica*, written at the end of the 12th century. Roger seems to have been influenced more by Byzantine doctors such as the 6th-century Aëtius and Alexander of Tralles, and the 7th-century Paul of Aegina, than by the Arabs. He shows acute powers of observation and some sound clinical practice. He broke and reset badly-united bones, treated hæmorrhage with styptics and ligatures, had an efficient method of bandaging, and described a remarkable technique for operating for hernia. His early 13th-century follower, Roland of Parma, showed particular skill with head injuries and described trephining and the elevation of depressed fractures. He also recognized the need to keep the hands clean and the patient warm. Both these surgeons were in most of their work 'wound surgeons,' and in their treatment of wounds they followed Galen's advice and promoted suppuration by using greasy salves.

This treatment of wounds was opposed in the 13th century by the north Italian surgeons, Hugh and Theodoric Borgognoni, and in the early 14th century by the Frenchman Henry of Mondeville, all of whom had studied at Bologna. They said that it was not only unnecessary but also harmful to generate pus and that the wound should simply be cleaned with wine, the edges brought together with stitches, and then left for nature to heal. Another 13th-century Italian surgeon, Bruno of Longoburgo, repeated this insistence on keeping wounds dry and clean and spoke of healing 'by first and second intention.' Further advances were made by another Italian, Lanfranchi, who, in his *Chirurgia Magna* of 1296, said that the cut ends of nerves should be stitched together, and by the Fleming, Jan Yperman (d. *c.* 1330), like Mondeville an army surgeon, who described many different cases from personal experience and emphasized the importance of anæsthetics. Mondeville himself invented an instrument for extracting arrows and removed pieces of iron from the flesh by a magnet. Progress in these directions continued throughout the 14th and 15th centuries, but in the middle of the 14th century Guy de Chauliac unfortunately abandoned the antiseptic treatment of wounds and under the influence of his writings surgeons returned to the salves and suppurations of Galen.

Though in the Middle Ages surgery was concerned mainly with wounds and fractures, it was recognized that surgical treatment was necessary for certain other ailments and in some operations considerable skill was acquired. The operations for the stone and Cæsarian section had been known from classical times and specialized surgical instruments, scalpels, needles and thread, saws, ear syringes, levers, and forceps of all kinds had been developed by the Arabs. As early as the middle of the 13th century, Gilbert the Englishman, Chancellor of Montpellier in 1250, recognized the importance of surgical treatment for cancer, and at the end of the 13th century the Italian surgeon, William of Saliceto, described the treatment of hydrocephalic children by removing the fluid through a small hole made in the head with a cautery. Early in the 14th century Mondeville described the healing of wounds in the intestine by the antiseptic method and insisted on the necessity of binding arteries in cases of amputation. Mondino gave excellent descriptions of the operation for hernia both with and without castration, though the difficulty of this is shown by Bernard of Gordon's preference for the truss, of which he gave the first modern description. Gentile da Foligno noted that there was no ancient work on the rupture of the abdominal lining, for which physicians and surgeons had to rely on their own experience. Guy de Chauliac shows himself, in his *Chirurgia Magna* of 1360, also to have been a skilful surgeon and a good observer, and this treatise remained a standard work until the time of Ambrose Paré in the 16th century. He used the *spongia soporifera* and was particularly skilful with hernia and fractures, noting the escape of cerebrospinal fluid in fractures of the skull and the effect of pressure on respiration; he extended fractured limbs with pulleys and weights. A contemporary English surgeon, John Arderne (1307–77), who described the Black Death in England, gave an account of a new syringe and other instruments for use in the cure of fistula; his countryman, John Mirfeld (d. 1407), described a 'tornellus' for reducing certain dislocations. In 15th-century Italy the Brancas used plastic surgery to restore noses, lips and ears, the technique for which was suggested by the Roman doctor Celsus. For the nose, skin was taken in a loop from the upper arm, one end being left attached to the arm until the graft on the nose had become firmly attached. Plastic surgery was practised also by the German army surgeon, Heinrich von Pfolspeundt, who, in 1460, described the gunshot wounds; another German army surgeon, Hans von Gersdorff, in 1517, described some elaborate mechanical apparatus for treating fractures and dislocations.

A special branch of surgery in which progress was made in the Middle Ages was dentistry. The Byzantine and Arab physicians had recognized caries, treated and filled decayed teeth, and done extractions. The English surgeon, John of Gaddesden (d.

1361), described a new instrument for extracting teeth. Guy de Chauliac prescribed powder made from cuttle-bones and other substances for cleaning the teeth, and described the replacement of lost teeth by pieces of ox bone or by human teeth fastened to the sound teeth with gold wire. Later medieval dental writers described the removal of the decayed parts with a drill or file and the filling of the cavity with gold leaf.

This activity in surgery during the late Middle Ages concentrated attention on the need to study anatomy, and all the great surgeons from the 12th century onwards recognized that good surgery, and even good medicine, was impossible without a knowledge of anatomy (see above, pp. 176–80). For many years the Church had prohibited clerks from shedding blood and therefore from practising surgery; for this reason surgery was never recognized as a subject for study at medieval universities as medicine was. This meant that although some instruction was received in anatomy, the medieval student had to get his real knowledge of anatomy, as well as of surgery, as Mondeville advised, by working with a practising surgeon. The result of this exclusion of surgery from the universities, and particularly from the French and English universities, meant that surgery was sometimes relegated as a manual craft to itinerant barbers, who cut for the stone, hernia or cataract and had no training beyond apprenticeship to a barber. Only in Italy was surgery encouraged at the universities; at Bologna in particular *post mortem* examinations were carried out to determine the cause of death and, in the Black Death, to find out something about the effects of this disease. In the 15th century most of the best surgeons were Italians, and it was in Italy that the study of anatomy began to make rapid progress from the end of that century (see Vol. II, p. 273 *et seq.*).

A medieval institution which did much to help not only the care of the sick, but also the knowledge obtained from the observation of medical and surgical cases, was the hospital. In ancient times Greek doctors had kept patients in their houses and there had been temples of Aesculapius where the sick gathered for treatment, the Romans had built military hospitals, and the Jews had provided houses for the needy. The foundation of large numbers of charitable hospitals for the relief of the poor and treatment of the sick was a product of Christian civilization. The Emperor Constantine is credited with the first hospital of this kind, and they became very numerous in Byzantium, one particular hospital founded in the 11th century having a total of 50 beds in separate wards for different kinds of patients, with two doctors as well as other staff attached to each ward. These Byzantine hospitals were copied by the Arabs, who as early as the 10th century had a hospital in Baghdad with 24 doctors. In the 13th century there was a hospital in Cairo with four wings used respectively for patients with fevers, eye diseases, wounds and diarrhœa, and a separate wing for women, with each wing equipped for preparing medicines and supplied with running water from a fountain.

In the West, most monasteries had infirmaries and asylums, and hospitals were founded by special orders of hospitallers like the Order of St John of Jerusalem and the Brothers of the Holy Ghost. Many of these were leper hospitals, and a great impulse to the founding of hospitals was given by the Crusades, a movement which may have helped to spread this disease. When St Bartholomew's hospital was founded in London, in 1123, there were already 18 hospitals in England. By 1215, when St Thomas's was founded, there were about 170. In the 13th century 240 more hospitals were founded, in the 14th century 248, and in the 15th century 91. The same activity occurred in other countries. In 1145 the Brothers of the Holy Ghost founded a hospital at Montpellier which became famous, and from the early 13th century, under the inspiration of Pope Innocent III, Holy Ghost hospitals were founded in almost every town in Christendom. In 1225 Louis VIII of France made a gift of 100 sous to each of 2,000 houses for lepers located within his realm. The 13th-century hospitals were usually of one storey and had spacious wards with tiled floors, large windows, the beds in separate cubicles, an ample water supply, and arrangements for disposing of sewage. The earlier hospitals were simply for the care of the sick and feeble rather than for treatment, but in the later hospitals different diseases were isolated and specialized therapy introduced.

241

A notable feature of some medieval hospitals was that some attempt was made to understand and care for the insane and to give treatment to psychological disorders. As early as the 7th century Paul of Aegina had discussed at some length the causes and treatment of 'melancholia' and 'mania.' In 1203 *furiosi frenetici* were admitted to a hospital connected with the cathedral at Le Mans. Later, certain hospitals specialized in mental cases, as did the Royal Bethlehem or Bedlam in London at the end of the 13th century. Mental disorders were attributed to three classes of cause: physical as with rabies and alcoholism, mental as with melancholia and aphasia, and spiritual as with demonic possession. Treatment also fell into the same three classes, and in each case the method of trying to cure the patient involved at its best some attempt to bring the cause of his suffering into the light of his rational consciousness. But the effectiveness of medieval psychological medicine should not be exaggerated, and no doubt too common an attitude to the mental patient was blank incomprehension combined with brutality and pious despair. As late as 1671 René Bary tells us in his text-book *La physique divisé en trois tomes*, that fools are mostly so at full moon, that the English beat them on the 14th day of the moon in Nazareth church in London, and that 'les Mathurins de la Beausse' do the same and also strip the lunatics, pinch them, and commend them to God. No doubt the medieval doctor also was too often as far as this from the sympthetically scientific analysis of psychological cases practised by Bary's contemporary, a pioneer of modern psychiatry, Thomas Sydenham.

Taken as a whole, medieval medicine is a remarkable product of that empirical intelligence seen in Western technology generally in the Middle Ages. The medical knowledge and treatment, like the other techniques and devices which were introduced, gave Western man power to control nature and to improve the conditions of his own life such as was never possessed in any earlier society. Behind this inventivenes lay, without a doubt, the motive of physical and economic necessity; but, as Lynn White has pointed out in an article contributed to *Speculum* in 1940, 'this "necessity" is inherent in every society; yet has found inventive expression only in the Occident.' Necessity can be a motive only when it is recognized, and among the most important reasons for its recognition in the West must be included the activist tradition of Western theology. By asserting the infinite worth of responsibility of each person, this theology placed a value upon the care of each immortal soul and therefore upon the charitable relief of physical suffering, and gave dignity to labour and a motive for innovation. The inventiveness that resulted produced the practical skill and flexibility of mind in dealing with technical problems to which modern science is the heir.

NOTES TO PLATES

Plate 4. Left, the back, showing the *alidade* with a pair of sights. Round the outer edge is a scale of degrees, within which is a zodiac/ calendar scale by means of which the sun's position in the ecliptic can be found at any time of year. In the area inside this scale a diagram of unequal hours has been engraved in the top half, and a shadow square in the bottom half.

Right, the front, showing the *label*, and the elaborately cut-away, movable *rete*, lying inside a scale of equal hours round the outer edge. The points of the curly appendages on the *rete* represent 21 different fixed stars. The eccentrically placed circle above represents the ecliptic, the outer rim represents the Tropic of Capricorn, and the two segments inside this are parts of the equatorial circle. The meridian runs vertically through the pivot of the *label*, with North at the bottom. Underneath the *rete* is the *plate*, on which is marked a vertical stereographic projection of the celestial sphere. In the top section are shown the *almucantars* (circles of altitude) round the pole, with the horizon at the bottom. These are cut by the *azimuths*. Below are lines of unequal hours. See pp. 104–8 and Fig. 3.

Plate 6. To show how the eye focused the 'species' of light entering it, Bacon described the anatomical arrangement of its parts. Following Avicenna he said that the eye had three coats and three humours. The inner coat consisted of two parts, the *rete* or *retina*, an expansion of the nerve forming a concave net 'supplied with veins, arteries and slender nerves' (*Opus Majus*, V, i, ii, 2, ed. J. H. Bridges, ii, 15) and acting as a conveyor of nourishment; and outside this a second thicker part called the *uvea* (in modern terminology the choroid, including the iris). Outside the *uvea* were the *cornea*, which was transparent where it covered the opening of the pupil, and the *consolidativa* or *conjunctiva* [sclerotic]. Inside the inner coat were the three humours, and so for light entering the pupil: 'There will then be the *cornea*, the *humor albigineus* [aqueous humour], the *humor glacialis* [lens], and the *humor vitreus* [vitreous humour], and the extremity

245

of the nerve, so that the species of things will pass through the medium of them all to the brain. ... The crystalline humour [lens] is called the pupil, and in it is the visual power' (*Opus Majus*, V, i, ii, 3, ed. Bridges, ii, 17–18). Like everyone else at his time Bacon thought the lens was the sensitive part of the eye (cf. Vol. II, p. 257). In these figures Bacon's intention was to draw simply a geometrical diagram showing the various curvatures of the ocular media. 'I shall draw, therefore, a figure in which all these matters are made clear as far as is possible on a surface, but the full demonstration would require a body fashioned like the eye in all the particulars aforesaid. The eye of a cow, pig, and other animals can be used for illustration, if any-one wishes to experiment. I consider this figure better than the one that follows, although the following one is that of the ancients.' (*Opus Majus*, V, i, iii, 3, ed. Bridges, ii, 23.) Explaining the top diagram, he continued : 'Let *al* be the base of the pyramid, which is the visible object, whose species penetrates the cornea under the pyramidal form and enters the opening, and which tends naturally to the centre of the eye, and would go there if it were not met first by a denser body by which it is bent, namely, the vitreous humour, *chd*.' By 'centrum' he meant 'centre of curvature'. The 'centre of the eye' is the centre (*b*) of curvature of the anterior convex surface of the lens, which Bacon, following Avicenna (*Canon Medicinæ*, III, iii, i, 1), correctly held to be flattened. In the top diagram the refraction should be shown, not as drawn, but at the interface between the convex *posterior* surface of the lens and the concave anterior surface of the vitreous humour (*chd*; *h* is missing). The centre of curvature of this interface (*centrum vitrei*) is in front of *b*.

Plate 7. A white beam or 'column' of light coming from the left enters the hexagonal crystal at *k* and is refracted. Part of the light striking the farther surface (right) of the crystal is internally reflected, while the other part passes out and is immediately refracted again; both emerge coloured (cross-shaded). The coloured rays emerge in the order : red, yellow, green, blue; i.e. in the coloured beam on the right red is at the top (cf. Plate 8).

Plate 8. From the sun (*e*, top left) comes a stream of white light. Within the stream two separate beams (or 'columns') are drawn (each incorrectly shown diverging). Following the progress of one beam, we see it entering the transparent sphere (the large circle on the right) and being refracted By differential refraction, the beam now becomes differentiated into colours. Theodoric recognized four coloured rays, red (top), yellow, green, blue (bottom), but for simplicity only the red and blue rays are drawn in the diagram. These coloured rays are reflected at the internal surface of the sphere, intersect, and are refracted again on emerging into the air. The colours emerging from a raindrop are now reversed, with blue at the top and red at the bottom. The paths of the rays going to the eye of the observer (f, bottom left) are incorrectly shown here converging, although in another diagram in this MS they are correctly drawn.

Plate 9. This diagram is correctly drawn except that the incident rays going from the sun to the different drops should be parallel, which they would not be if they were all shown in the diagram; and the coloured rays emerging from the individual drops should be diverging instead of parallel. The paths of the individual rays inside each drop are not shown (see Plate 8). A particular drop sends only one colour to the observer at *c* (bottom centre). From the upper drop the red rays (emerging on the right) reach the observer, and from the other drops come yellow, green, and blue, respectively, thus giving the order of the colours seen in the rainbow.

Plate 10. The four lower circles represent the raindrops producing the primary bow (cf. Plate 9). The four upper circles represent the drops producing the secondary bow. Here the sunlight passes round inside each drop in the opposite direction to that in the drops producing the primary bow, and undergoes two internal reflections. The paths of the individual rays inside each drop are not shown. The colours go to the eye (*oculus*) in the reverse order to those seen in the primary bow, blue being uppermost in the secondary bow.

BIBLIOGRAPHY

IT is impossible to give a complete bibliography. Titles have been restricted to the most useful books on each topic, and to recent articles to which there is special indebtedness in the preparation of this volume. These may indicate lines of further exploration. The list has been limited so far as possible to works in English and French, but clearly for some essential topics it has been necessary to go outside those limits. For anyone wishing to explore the history of science further there is an extensive bibliographical apparatus available. The basic bibliographical work is G. Sarton, *Introduction to the History of Science*, Baltimore, 1927–47, 3 vols. in 5, which brings the subject down to the end of the 14th century and attempts to cover both the Eastern and the Western civilizations. This is supplemented by the Critical Bibliographies published at intervals since 1913 in *Isis* (Cambridge, Mass.). Besides these, two extremely useful recent bibliographical studies are: G. Sarton, *A Guide to the History of Science*, Waltham, Mass., 1952; and F. Russo, *Histoire des sciences et des techniques. Bibliographie* (*Actualités scientifiques et industrielles*, No. 1204), Paris, 1954. Also useful is H. Guerlac, *Science in Western Civilization. A Syllabus*, New York, 1952. An indispensable guide to the manuscript material of the medieval period is L. Thorndike and P. Kibre, *A Catalogue of Incipits of Mediaeval Scientific Writings in Latin*, Cambridge, Mass., 1937; continued by Thorndike in 'Additional incipits of mediaeval scientific writing in Latin,' *Speculum*, xiv (1939); 'More incipits of mediaeval scientific writings in Latin,' ibid., xvii (1942). Also relevant are the specialized bibliographical studies that are appearing, e.g.: I. M. Bochenski (editor), *Bibliografische Einführungen in das Studium der Philosophie*, xvii, *Philosophie des Mittelalters*, by F. van Steenberghen, Bern, 1950; W. Artelt, *Index zur Geschichte der Medizin, Naturwissenschaft und Technik*, Munich and Berlin, 1953– , i– ; A. C. Klebs et E. Droz, *Remèdes contre la peste. Facsimiles, notes et liste bibliographique des incunables sur la peste* (*Documents scientifiques du 15ᵉ siècle*, i) Paris, 1925; M. D. Knowles, 'Some recent advance in the history of medieval thought,' *The Cambridge Historical Journal*, ix (1947); G. E. Mohan, 'Incipits of logical writings

Bibliography

of the 13th–15th centuries,' *Franciscan Studies* (St Bonaventure, N.Y.), N.S. xii (1952).

The principal journals, in addition to *Isis*, are *Osiris* (Bruges), *Annals of Science* (London), *Archives internationales d'histoire des sciences* continuing *Archeion* (Paris), and *Revue d'histoire des sciences* (Paris). Several others specialize in the history of medicine, mathematics, technology, etc.: they are listed in Sarton's *Guide* and by Russo. Articles on the history of science also appear in the *Journal of the History of Ideas* (Lancaster, Pa. and New York), and for the philosophy of science there is *The British Journal for the Philosophy of Science* (Edinburgh and London). Of special importance also are the monographic series devoting particular attention to the publication of texts. Indispensable for the medieval period are *Beiträge zur Geschichte der Philosophie des Mittelalters* (Münster), *Études de philosophie médiévale* (Paris), and *Mediaeval Studies* (Toronto). *History of Science*, ed. by A. C. Crombie and M. A. Hoskin (Cambridge, 1962–   ), is an annual survey of publications and research.

The most useful general history of science is *Histoire générale des Sciences, publiée sous la direction de* René Taton, i, *La Science Antique et Médiévale* (to 1450), ii, *La Science Moderne* (1450–1800), iii, *La Science Contemporaine* (since 1800), Paris, 1957–   . M. Daumas (editor), *Histoire de la Science*, Paris, 1957 (Encyclopédie de la Pléiade) is also valuable. Numerous other general histories of science are listed in Sarton's *Guide* and by Russo. Valuable collections of papers are *Critical Problems in the History of Science*, ed. Marshall Clagett, Madison, Wisconsin, 1959, and the Oxford symposium, *Scientific Change: Historical Studies in the Intellectual, Technical and Social Conditions for Scientific Discovery and Technical Invention*, ed. A. C. Crombie, London, 1963. A useful collection of articles reprinted from the *Journal of the History of Ideas* is *Roots of Scientific Thought*, ed. P. P. Wiener and A. Noland, New York, 1957; another valuable collection of original material is J. R. Newman, *The World of Mathematics*, New York, 1956, 4 vols. Some of the older general histories are still useful, especially A. de Candolle, *Histoire des sciences et des savants depuis deux siècles*, Geneva, 1873; R. Caverni, *Storia del metodo sperimentale in Italia*, Florence, 1891–1900, 6 vols.; G. Cuvier, *Histoire des sciences naturelles*, complété par M. de Saint-Agy, Paris, 1831–45, 5 vols.; J. B. Delambre, *Histoire de l'astronomie ancienne*, Paris, 1817, *Histoire de l'astronomie au moyen âge*, Paris, 1819; G. Libri, *Histoire des sciences mathématiques en Italie, depuis la renaissance des lettres jusqu'à la fin du dix-septième siècle*, Paris, 1838–41, 2 vols.; J. É.

Bibliography

Montucla, *Histoire des mathématiques*, new ed. by J. de Lalande, Paris, 1799–1802, 4 vols.; W. Whewell, *History of the Inductive Sciences*, 2nd ed., London, 1847, *History of Scientific Ideas*, London, 1858, and his writings on the philosophy of science (see Vol. II under Chapter 2, Philosophy of Science, etc.). Studies devoted to the work of two of the greatest more recent historians of science are published in *Archeion*, xix (1937) on Pierre Duhem, and in *Revue d'histoire des sciences*, vii (1954) on Paul Tannery. There is no general history adequately covering all periods and all aspects of scientific thought and technology. General works relating to particular periods and civilizations are listed below under Chapters 1 and 2 and in the General sections of subsequent chapters. Special studies of Greek and Arabic as well as medieval Latin philosophy, science and technology are listed under the proper headings under Chapters 3 and 4 and in Vol. II under Chapter 1.

CHAPTER I

ANCIENT PHILOSOPHY AND SCIENCE: The indispensable background to the study of the scientific thought of the medieval West is a knowledge of both ancient Greek and medieval Arabic science and philosophy. The latter is discussed under Chapter 2. The character of Greek scientific thought is itself illuminated by the further background of thought in ancient Egypt and Mesopotamia, of which the following give some indication : J. H. Breasted, *The Edwin Smith Surgical Papyrus*, Chicago, 1930, *The Dawn of Conscience*, New York, 1933; H. and H. A. Frankfort, J. A. Wilson, and T. Jakobsen, *Before Philosophy: the Intellectual Adventure of Ancient Man; an essay on speculative thought in the Ancient Near East*, London (Pelican Books), 1949 (first published, Chicago, 1946); O. Neugebauer, *The Exact Sciences in Antiquity*, 2nd ed., Providence, R.I., 1957; H. J. J. Winter, *Eastern Science*, London, 1952. An excellent general study of the whole of ancient science, with French translations of texts and a useful bibliography, is P. Brunet et A. Mieli, *Histoire des Sciences: Antiquité*, Paris, 1935. Admirable short studies of Greek science are : M. Clagett, *Greek Science in Antiquity*, New York, 1956; J. L. Heiberg, *Mathematics and Physical Science in Classical Antiquity*, Oxford, 1922; A. Reymond, *Histoire des sciences exactes et naturelles dans l'antiquité greco-romane*, Paris, 1924 (English translation, London, 1927); cf. L. Bourgey, *Observation et expérience chez les medicins de la collection Hippocratique*, Paris, 1953; also W. A. Heidel, *The Heroic*

Bibliography

Age of Science: the conceptions, ideals, and methods of science among the ancient Greeks, Baltimore, 1933 – a perceptive analysis with emphasis on the biological sciences; S. Sambursky, *The Physical World of the Greeks*, London, 1956 – interesting for Stoic thought. An excellent selection of texts in translations is M. R. Cohen and I. E. Drabkin, *A Source Book in Greek Science*, New York, 1948. Further source material may be read in translation in the Loeb Classical Library (London and Cambridge, Mass.) and the Collection Budé (Paris), which between them include basic works of Plato, Aristotle, the Hippocratic corpus, Galen, the Greek mathematicians, Lucretius, etc. Useful for further source material and commentary are the special studies listed under Chapters 3 and 4 and the following: A. H. Armstrong, *An Introduction to Ancient Philosophy*, 2nd ed., London, 1949; E. Bréhier, *Histoire de la Philosophie*, i, Paris, 1943; J. D. Burnet, *Early Greek Philosophy*, 4th ed., London, 1930 – a basic work; R. G. Collingwood, *The Idea of Nature*, Oxford, 1945; F. M. Cornford, *The Unwritten Philosophy and other Essays*, Cambridge, 1950, *Principium Sapientiae: the origins of Greek philosophical thought*, Cambridge, 1952; B. Farrington, *Science in Antiquity*, London, 1936, *Greek Science*, London, 1944–9, 2 vols.; J. L. Heiberg, *Mathematics and Physical Science in Classical Antiquity*, London, 1922; H. I. Marrou, *Histoire de l'éducation dans l'antiquité*, Paris, 1950 – very useful; G. de Santillana, *The Origins of Scientific Thought*, London, 1961; P. M. Schuhl, *Essai sur la formation de la pensée grecque*, 2nd ed., Paris, 1949.

EARLY MEDIEVAL SCIENTIFIC THOUGHT: In addition to works listed under the General sections of Chapters 2 and 3, Adelardus von Bath, *Quaestiones Naturales*, ed. M. Müller (*Beitr. Ges. Philos. Mittelalt.*, xxxi. 2) Münster, 1923; R. Baron, 'Hvgonis de Sancto Victore Practica Geometriae,' *Osiris*, xii (1956); E. Brehaut, *An Encyclopaedist of the Dark Ages: Isidore of Seville*, New York, 1912; A. Clerval, 'L'enseignement des arts libéraux à Chartres et à Paris dans la première moitié du xiie siècle d'après l'*Heptateuchon* de Thierry de Chartres,' *Congrès scientifique international des catholiques, Paris, 1888*, Paris, 1889, ii., *Les Écoles de Chartres*, Paris, 1895; *Congrès international Augustinien, Paris, 1954*, tome iii, *Actes*, Paris, 1955; G. W. Coopland, *Nicole Oresme and the Astrologers*, Liverpool, 1952; O. G. Darlington, 'Gerbert the teacher,' *American Historical Review*, lii (1947); E. Gilson, *Introduction à l'étude de S. Augustin*, 2nd ed., Paris, 1943; R. M. Grant, *Miracle and Natural Law in Graeco-Roman and Early Christian*

Bibliography

Thought, Amsterdam, 1952; J. H. G. Grattan and C. Singer, *Anglo-Saxon Magic and Medicine. Illustrated specially from the semi-pagan text 'Lacnunga'*, London, 1952; C. W. Jones (editor), *Bedae Opera de Temporibus*, Cambridge, Mass., 1943 – for the early history of *computus* and the calendar; G. H. T. Kimble, *Geography in the Middle Ages*, London, 1938; H. Lattin, 'Astronomy : our views and theirs,' in *Symposium on the Tenth Century (Medievalia et Humanistica*, Fasc. ix), Boulder, Colorado, 1955; R. McKeon, *Selecting from Medieval Philosophers*, London, 1929, i; L. C. MacKinney, *Early Medieval Medicine*, Baltimore, 1937, 'Medical ethics and etiquette in the early middle ages,' *Bulletin of the History of Medicine*, xxvi (1952); H. I. Marrou, *St Augustin et la fin de la culture antique*, 2nd ed., Paris, 1938, 'Retractatio', 1949, *St Augustin et l'augustinisme*, Paris, 1955; E. C. Messenger, *Evolution and Theology*, London, 1931; *A Monument to St Augustin*, compiled by T. F. B[urns]., London, 1930; J. M. Parent, *La Doctrine de la création dans l'école de Chartres*, Paris and Ottawa, 1938; J. F. Payne, *English Medicine in the Anglo-Saxon Times*, Oxford, 1904; A. C. Pegis, 'The mind of St Augustine,' *Medieval Studies*, vi (1944); H. Pope, *Saint Augustine of Hippo*, London, 1937; F. Saxl and H. Meier, *Verzeichnis astrologischer und mythologischer illustrierter Handschriften des lateinischen Mittelalters*, vols. 1 and 2 (Hamburg, 1915, 1927) by Saxl, vol. 3 (London, 1953) by Saxl and Meier, edited by H. Bober; M. Schedler, *Die philosophie des Macrobius und ihr Einfluss auf die Wissenchaft des christlichen Mittelalters (Beitr. Ges. Philos. Mittelalt.*, xiii. 1), Münster, 1916; C. Singer, 'The scientific views and visions of Saint Hildegard of Bingen,' in *Studies on the History and Method of Science*, Oxford, 1917, i, *From Magic to Science*, London, 1928; C. and D. Singer, 'The origin of the medical school of Salerno, the first European university,' in *Essays on the History of Medicine presented to Karl Sudhoff*, ed. C. Singer and H. E. Sigerist, Oxford and Zurich, 1924; L. Spitzer, 'Classical and Christian ideas of world harmony,' *Traditio*, ii (1944), iv (1946); W. H. Stahl, *Macrobius. Commentary on the dream of Scipio*; translated with introduction and notes, New York, 1952, 'Dominant traditions in early medieval Latin Science,' *Isis* 1 (1959); C. Stephenson, 'In praise of medieval thinkers,' *Journal of Economic History*, viii (1948) – on Gerbert; A. Hamilton Thompson, *Bede: His Life, Times, and Writings*, Oxford, 1935 – very useful; C. C. J. Webb, *Studies in the History of Natural Theology*, Oxford, 1915; T. O. Wedel, *The Mediaeval Attitude toward Astrology*, New Haven and London, 1920; K. Werner, 'Die Kosmologie und Naturlehre des scholastischen Mittel-

Bibliography

alters mit spezieller Beziehung auf Wilhelm von Conches,' *Sitzungsberichte der kaiserlichen Akademie der Wissenschaften zu Wien,*
philos.-hist. Klasse, lxxv (1873); T. Whittaker, *Macrobius, or Philosophy, Science and Letters in the Year 400,* Cambridge, 1923; H. A.
Wolfson, *The Philosophy of the Church Fathers,* Cambridge, Mass.,
1956.

CHAPTER 2

There is no adequate short history of Arabic scientific thought. There
are sketches in A. Mieli, *Panorama general de historia de la ciencia
II: La época medieval, Mundo islámico y occidente christiana,* Buenos
Aires, 1946. *La science arabe et son rôle dans l'évolution scientifique
mondiale,* Leiden, 1938; H. J. J. Winter, *Eastern Science,* London, 1952.
Indispensable for bibliography and reference are Sarton, *Introduction,*
and C. Brockelmann, *Geschichte der arabischen Literatur,* Weimar
and Berlin, 1898–1902, 2 vols., Supplement, Leiden, 1937–42, 3 vols.;
2nd ed., Leiden, 1943– . For further details there are the special
studies listed under Chapters 3 and 4 and in Vol. II under Chapter 1,
and the following : S. M. Afnan, *Avicenna: his life and works,* London,
1958; A. J. Arberry (editor), *The Legacy of Persia,* Oxford, 1953; Sir
T. Arnold and A. Guillaume (editors), *The Legacy of Islam,* Oxford,
1931; *The Encyclopaedia of Islam,* ed. M. T. Houtsma *et alii,* Leiden
and London, 1908–38, 4 vols. and supplement, new ed. by J. H.
Kramers, H. A. R. Gibb, E. Lévi-Provençal and J. Schacht, 1954– ;
P. K. Hitti, *History of the Arabs,* 4th ed., London, 1949; M. Meyerhof,
'Von Alexandrien nach Bagdad,' *Sitzungsberichte der preussischen
Akademie der Wissenschaften zu Berlin,* philos.-hist. Klasse, 1930 –
for the translations from Greek into Arabic, 'A sketch of Arab science,'
Journal of the Egyptian Medical Association, xix (1936); De Lacy
O'Leary, *How Greek Science Passed to the Arabs,* London, 1948.

HEBREW PHILOSOPHY: E. R. Bevan and C. Singer (editors), *The
Legacy of Israel,* Oxford, 1927; Isaak Husik, *A History of Medieval
Jewish Philosophy,* Philadelphia, 1946; G. Vajda, *Introduction à la
pensée juive du moyen âge (Études de philos. médiévale,* xxxv) Paris,
1947; H. A. Wolfson, *Crescas' Critique of Aristotle,* English translation,
text, and commentary, Cambridge, Mass., 1929.

INDIAN SCIENTIFIC THOUGHT: S. R. Das, 'Scope and development of Indian astronomy,' *Osiris,* ii (1936); B. Datta and A. N. Singh,

Bibliography

History of Hindu Mathematics. A source book, Lahore, 1935–8, 2 vols.; G. T. Garratt (editor), *The Legacy of India*, Oxford, 1937; A. B. Keith, *Indian Logic and Atomism*, Oxford, 1921; P. Ray, *History of Chemistry in Ancient India*, Calcutta, 1956; Sir B. Seal, *The Positive Sciences of the Ancient Hindus*, London, 1915; D. E. Smith and L. C. Karpinski, *The Hindu-Arabic System of Numerals*, Boston, 1911; H. R. Zinner, *Hindu Medicine*, Baltimore, 1948.

CHINESE SCIENTIFIC THOUGHT AND TECHNOLOGY: See under Chapter 4, Building, etc., Industrial Chemistry, and Medicine, and J. T. Needham, *Science and Civilization in China*, Cambridge, 1954– , i– .

GENERAL HISTORY OF MEDIEVAL SCIENCE: Essential works of reference are Sarton, *Introduction*, and L. Thorndike, *A History of Magic and Experimental Science*, New York, 1923–58, 8 vols. Also indispensable is P. Duhem, *Le Système du Monde*, Paris, 1913–56, 7 vols., a classic work. A very useful survey is E. J. Dijksterhuis, *De Mechanisering van het Wereldbeeld*, Amsterdam, 1950 (English trans., Oxford, 1961). These studies include both ancient and medieval science. For the general history of medieval philosophy there are E. Bréhier, *Histoire de la philosophie*, Paris, 1943, i, et Fasc. supplémentaire ii, *La Philosophie Byzantine* par B. Tatakis, 1949; F. C. Copleston, *A History of Philosophy*, London, 1946–53, 3 vols.; E. Gilson, *La Philosophie au moyen âge*, 2nd ed., Paris, 1944, *History of Christian Philosophy in the Middle Ages*, London, 1955; F. Ueberweg and B. Geyer, *Grundriss der Geschichte der Philosophie*, ii, 11th ed., Berlin, 1928 – indispensable for reference; M. de Wulf, *Histoire de la philosophie médiévale*, 6th ed., vols. i, ii, Louvain and Paris, 1934–6 (English translation, London, 1938), vol. iii, Louvain and Paris, 1947 – with useful bibliographies. See also works listed below and under Chapter 3 and in Vol. II under Chapter 1.

Useful for background are M. Bloch, *La Société féodale*, Paris, 1939–40, 2 vols. – an excellent general survey; L. Bréhier, *Le monde byzantin*, Paris, 1947, 3 vols.; *The Cambridge Economic History of Europe:* vol. 1, *The Agrarian Life of the Middle Ages*, ed. J. H. Clapham and Eileen Power, Cambridge, 1941, vol. 2, *Trade and Industry in the Middle Ages*, ed. M. Postan and E. E. Rich, Cambridge, 1952; *The Cambridge Medieval History*, ed. C. W. Previté-Orton and Z. N. Brooke, Cambridge, 1911–36, 8 vols.; J. Coppens, *L'histoire critique de l'ancien Testament*, Tournai and Paris, 1938; C. Dawson, *Medieval*

Bibliography

Essays, 2nd ed., London, 1953; J. de Ghellinck, *Le Mouvement théologique du XIIe siècle*, 2nd ed., Bruges, 1948; C. H. Haskins, *The Renaissance of the Twelfth Century*, Cambridge, Mass., 1928; J. Huizinga, *The Waning of the Middle Ages*, London, 1924; J. M. Hussey, *Church and Learning in the Byzantine Empire, 867–1185*, Oxford, 1937; S. d'Irsay, *Histoire des universités*, Paris, 1933–5, 2 vols.; M. L. W. Laistner, *Thought and Letters in Western Europe, 500–900 A.D.*, 2nd ed., London, 1957; R. Latouche, *Les Origines de l'économie Occidentale*, Paris, 1956; E. Lesné, *Historie de la propriété ecclésiastique en France*, iv-v, Paris, 1938–40 – on schools, libraries, etc. to the end of the 12th century; F. Lot, *La Fin du monde antique et les débuts du moyen âge*, 2nd ed., Paris, 1956 (English trans. by P. and M. Leon, London, 1931); L. J. Paetow, *The Arts Course at Medieval Universities*, Urbana, Ill., 1910; G. Paré, A. Brunet et P. Tremblay, *La Renaissance du xiie siècle; les écoles et l'enseignement*, Paris and Ottawa, 1933; H. Pirenne, *Economic and Social History of Medieval Europe*, trans. by I. E. Clegg, London, 1936, *Histoire économique de l'Occident médiéval*, Paris, 1951; A. L. Poole (editor), *Mediaeval England*, Oxford, 1958; H. Rashdall, *The Universities of Europe in the Middle Ages*, 2nd ed. by F. M. Powicke and A. B. Emden, Oxford, 1936, 3 vols.; B. Smalley, *The Study of the Bible in the Middle Ages*, 2nd ed., Oxford, 1952; R. W. Southern, *The Making of the Middle Ages*, London, 1953 – an excellent and sensitive introduction; B. Spicq, *Esquisse d'une histoire de l'exégèse latine au moyen âge*, Paris, 1944; H. O. Taylor, *The Medieval Mind*, 4th ed., London, 1938, 2 vols.; J. W. Thompson, *The Literacy of the Laity in the Middle Ages*, Berkeley, 1939.

For the translations into Latin and their influence, useful guides for reference are Rashdall, *Universities*; Sarton, *Introduction*; Ueberweg-Geyer, *Grundriss*, ii; de Wulf, *Philosophie médiévale*. An indispensable work of reference is G. Lacombe, *Aristoteles Latinus*, Rome, 1939; cf. L. Minio-Paluello, 'Analytica posteriora ...,' *Aristoteles Latinus*, iv. 2, 3, Bruges and Paris, 1953–4. Useful for a general survey, although literary in main interest, are R. R. Bolgar, *The Classical Heritage and its Beneficiaries*, Cambridge, 1954; Sir J. E. Sandys, *A History of Classical Scholarship*, 3rd ed., Cambridge, 1904, i. For detailed studies there are, in addition to works mentioned under Chapter 3, General section, M. Alonso Alonso, articles in *Al-Andalus* (Madrid) 1943–9; H. Bédoret, articles in *Revue néoscolastique de philosophie*, 1938; D. J. Allan, 'Mediaeval versions of Aristotle, De Caelo, and of the Commentary of Simplicius,' *Medieval and Renais-*

sance Studies, ii (1950); A. Birkenmajer, 'Le Rôle joué par les médecins et les naturalistes dans la réception d'Aristotle au xiie et xiiie siècles,' *La Pologne au vi^e congrès international des sciences historiques, Oslo, 1928*, Warsaw, 1930; D. A. Callus, 'Introduction of Aristotelian Learning to Oxford,' *Proceedings of the British Academy*, xxix (1943); Marshall Clagett, 'Medieval mathematics and physics: a check-list of microfilm reproductions,' *Isis*, xliv (1953), and other articles in *Isis* (1952–5) and *Osiris* (1952–4), mainly on the translations of Euclid and Archimedes, M. B. Foster, 'The Christian doctrine of the creation and the rise of modern natural science,' *Mind*, N.S. xliii (1934), 'Christian theology and modern natural science of nature,' *Mind*, N.S. xliv (1935), xlv (1936); M. Grabmann, *Forschungen über die lateinischen Aristoteles-Ubersetzungen des xiii. Jahrhunderts (Beitr. Ges. Philos. Mittelalt., xvii. 5–6)* Münster, 1916; C. H. Haskins, *Studies in the History of Mediaeval Science*, 2nd ed., Cambridge, Mass., 1927; R. W. Hunt, 'English learning in the late twelfth century,' *Transactions of the Royal Historical Society*, 4th series, xix (1936), 'The Introductions to the "Artes" in the twelfth century,' in *Studia Mediaevalia in honorem admodum Reverendi Petro Raymondi Josephi Martin, O.P., S.T.M.*, Bruges, 1948; E. M. Jamison, *Admiral Eugenius of Sicily: His Life and Work*, Oxford, 1957; R. Klibansky, *The Continuity of the Platonic Tradition during the Middle Ages: Outlines of a Corpus Platonicorum Medii Aevi*, London, 1939; H. Liebeschütz, *Mediaeval Humanism in the Life and Writings of John of Salisbury (Studies of the Warburg Institute, ed. F. Saxl, xvii)*, London, 1950; J. C. Russell, 'Hereford and Arabic science in England about 1175–1200,' *Isis*, xviii (1932); T. Silverstein, 'Daniel of Morley, English cosmologist and student of Arabic science,' *Mediaeval Studies* (1948); H. O. Taylor, *The Classical Heritage of the Middle Ages*, New York, 1901; G. Théry, 'Notes indicatrices pour s'orienter dans l'étude des traductions médiévales' in *Mélanges Joseph Maréchal*, Brussels, 1950, ii; J. W. Thompson, 'The introduction of Arabic science into Lorraine in the 10th century,' *Isis*, xii (1929); F. van Steenberghen, *Aristote en Occident. Les Origines de l'Aristotélisme parisien*, Louvain, 1946 (English trans., Louvain, 1955); M. Steinschneider, *Die europäischen Ubersetzungen aus dem Arabischen, bis Mitte des 17. Jahrhunderts (Sitzungsberichte der kaiserlichen Akademie der Wissenschaften, Wein, philos.-hist. Klasse, cxlix. 4, cli. 1)*, Vienna, 1904–5; C. B. Vandewalle, *Roger Bacon dans l'histoire de la philologie*, Paris, 1929; R. de Vaux, 'La première entrée d'Averroës chez les Latins,' *Revue des sciences philosophiques et théologiques*, xii (1933), 'Notes et textes sur l'avicen-

nisme latin aux confins des xii-xiii siècles,' *Bibliothèque thomiste*, xx
(1934); A. van der Vyer, 'Les étapes du développement philosophique
du haut moyen âge,' *Revue belge de philologie et d'histoire*, viii (1929),
'Les premières traductions latines (xᵉ–xiᵉ siècles) de traités arabes sur
l'astrolabe,' *Ier congrès international de géographie historique*,
Brussels, 1931, ii, *Mémoires*, 'Les plus anciennes traductions latines
médiévales (xᵉ–xiᵉ siècles) de traités d'astronomie et d'astrologie,'
Osiris, (1936), 'L'évolution scientifique du haut moyen âge,' *Archeion*,
xix (1937); R. Walzer; 'Arabic transmission of Greek thought to
mediaeval Europe,' *Bulletin of the John Rylands Library*, Manchester,
xxix (1945); M. C. Welborn, 'Lotharingia as a center of Arabic and
scientific influence in the XI Century,' *Isis*, xvi (1931); S. D. Wingate,
The Mediaeval Latin Versions of the Aristotelian Scientific Corpus,
London, 1931; F. Wüstenfeld, *Die Übersetzungen arabischer Werke in
das Lateinische seit dem XI. Jahrhundert (Abhandlungen der
königlichen Gesellschaft der Wissenschaften zu Göttingen*, xxii. 3)
Göttingen, 1877.

CHAPTER 3

GENERAL: Excellent introductions to the general characteristics of
Aristotelianism are J. M. Le Blond, *Logique et méthode chez Aristote*,
Paris, 1939; A. Mansion, *Introduction à la physique aristotélicienne*,
2nd ed., Louvain, 1946, *Le judgement d'existence chez Aristote*, Paris,
1939; Sir W. D. Ross, *Aristotle*, 3rd ed., London, 1937; and J. de Ton-
quédec, *Questions de cosmologie et de physique chez Aristote et Saint
Thomas*, Paris, 1950. For the philosophy of science there is the detailed study by A. C. Crombie, *Robert Grosseteste and the Origins of
Experimental Science, 1100–1700*, Oxford, 1953 (3rd impression 1971;
bibliography). For details of 13th-century scientific thought in general
and its Greek and Arabic sources there are Albertus Magnus, *Opera
Omnia*, ed. P. Jammy, Lyons, 1651, 21 vols., revised by A. Borgnet,
Paris, 1890–9, 38 vols.; F. Alessio, *Mito e Scienza in Ruggero Bacone*,
Milan, 1957; Thomas Aquinas, *Opera Omnia*, Rome, 1882–1930, 15
vols.; Aristotle, *Complete Works*, trans. under the editorship of J. A.
Smith and W. D. Ross, Oxford, 1908–31, 11 vols.; Ibn Sīnā (Avicenne),
Livre des directives et remarques. Traduction avec introduction et
notes, par. A. M. Goichon, Paris, 1951; Roger Bacon, *Opera Quaedam
Hactenus Inedita*, ed. J. S. Brewer (Rolls Series), London, 1859, *Opus
Majus*, ed. J. H. Bridges, Oxford, 1897, i–ii, London, 1900, iii (with
De Multiplicatione Specierum) (English. trans. of the *Opus Majus*, by

R. B. Burke, Philadelphia, 1928, 2 vols.), *Opera Hactenus Inedita*, ed. R. Steele, Oxford, 1909–40, 16 Fasc. (contains most of the scientific writings not edited by Brewer and Bridges); L. Baur, *Die Philosophie des Robert Grosseteste* (*Beitr. Ges. Philos. Mittelalt.*, xviii. 4–6) Münster, 1917; L. Brunschwig, *Le Rôle du Pythagorisme dans l'évolution des idées* (*Actualités scientifiques et industrielles*, No. 446) Paris, 1937; D. A. Callus (editor), *Robert Grosseteste, Scholar and Bishop*, Oxford, 1955; M. H. Carré, *Realists and Nominalists*, Oxford, 1946; M. D. Chenu, *La théologie comme science au xiiie siècle*, 2e ed., Paris, 1943; F. M. Cornford, *Plato and Parmenides*, London, 1939; T. Crowley, *Roger Bacon: the problem of the soul in his philosophical commentaries*, Louvain and Dublin, 1950; H. C. Dales, 'Robert Grosseteste's *Commentarius in octo Libros physicorum Aristotelis*,' *Medievalia et Humanistica*, xi (1957); S. C. Easton, *Roger Bacon and his Search for a Universal Science*, Oxford, 1952; A. Forest, F. van Steenberghen, M. de Gandillac, *Le Mouvement doctrinal du xie au xive siècle* (*Histoire de l'Église*, fondée par A. Fliche et V. Martin; dirigée par A. Fliche et E. Jarry, xiii), Paris, 1951; A. Garreau, *Saint Albert le Grand*, Paris, 1932; L. Gauthier, *Ibn Rochd* (*Averroës*), Paris, 1948; A. M. Goichon, *La philosophie d'Avicenne et son influence en Europe médiévale*, Paris, 1944; M. Grabmann, *Die Geschichte der scholastischen Methode*, Freiberg-im-Breisgau, 1909–11, 2 vols., *Mittelalterliches Geistesleben*, Munich, 1926–36, 2 vols., *Der hl. Albert, der Grosse. Ein wissenschaftliches Characterbild*, Munich, 1932, *Bearbeitungen und Auslegungen der aristotelischen Logik* (*Abhandlungen der preussischen Akademie der Wissenschaften*, philos.-hist. Klasse, v), Berlin, 1937; Robert Grosseteste, *Die philosophischen Werke*, ed. L. Baur (*Beitr. Ges. Philos. Mittelalt.*, ix) Münster, 1912; G. von Hertling, *Albertus Magnus* (*Beitr. Ges. Philos. Mittelalt.*, xiv. 5–6), Münster, 1914; R. Hooykaas, 'Science and theology in the middle ages,' *Free University Quarterly* (Amsterdam), iii (1954); S. d'Irsay, 'Les sciences de la nature et les universités médiévales,' *Archeion*, xv (1933); K. H. Laurent et M. J. Congar, 'Essai de bibliographie Albertinienne,' *Revue thomiste*, N.S. xiv, 1931; A. G. Little (editor), *Roger Bacon Essays*, Oxford, 1914, 'The Franciscan school at Oxford in the thirteen century,' *Archivum Franciscanum Historicum*, xix (1926), 'Roger Bacon,' *Proceedings of the British Academy*, xiv (1928), *Franciscan Letters, Papers and Documents*, Manchester, 1943; A. O. Lovejoy, *The Great Chain of Being*, Cambridge, Mass., 1933; C. K. McKeon, *A Study of the Summa Philosophiæ of the Pseudo-Grosseteste*, New York, 1948; R. McKeon, 'The empiricist and experimentalist temper in the middle ages: a prolegomena to the study of medieval science,' *Essays in Honor of John Dewey*, New York, 1929, *Selections from Medieval Philosophers*, New York, 1929–30, 2 vols.; P. Mandonnet, *Siger de Brabant et l'averroism latin au xiiie siècle*, Fribourg, 1899; A. J. O. S. Mariétan, *Problème de la classification des sciences d'Aristote à S. Thomas*, Paris, 1901; H. Ostlander (editor), *Studia Albertina. Festschrift für Bernhard Geyer zum 70. Geburtstage (Beitr. Ges. Philos. Mittelalt., Supplementband iv)*, Münster, 1952; A. Gonzalez Palencia, *Alfarabi, Catálogo de las Ciencias*, Madrid, 1932; G. Quadri, *La philosophie arabe dans l'Europe médiévale des origines à Averroës*, Paris, 1947; R. Robinson, *Plato's Earlier Dialectic*, 2nd ed., Oxford, 1953; Sir W. D. Ross, *Plato's Theory of Ideas*, Oxford, 1951; J. C. Russell, *Dictionary of Writers of Thirteenth Century England (Bulletin of Institute of Historical Research, iii)* London, 1936; H. C. Scheeben, *Albert der Grosse: zur Chronologie seines Lebens (Quellen und Forschungen zur Geschichte des Dominikanerordens in Deutschland, xxvii)*, Vecht, 1931, 'Les Écrits d'Albert le Grand d'après les catalogues,' *Revue thomiste*, N.S. xiv (1931); L. Schütz, *Thomas-Lexikon*, Paderborn, 1895; D. E. Sharp, *Franciscan Philosophy at Oxford*, 1930, 'The *De ortu scientiarum* of Robert Kilwardby (d. 1279),' *The New Scholasticum*, viii (1934); F. Van Steenberghen, 'La littérature albertino-thomiste (1930–1937),' *Revue néoscholastique de philosophie*, xli (1938), *Siger de Brabant d'après ses œuvres inédits*, ii, 'Siger dans l'histoire de l'Aristotélisme' (*Les Philosophes Belges*, xiii), Louvain, 1942; J. Stenzel, *Plato's Method of Dialectic*, trans. and ed. by D. J. Allan, Oxford, 1940; A. E. Taylor, *Platonism and Its Influence*, London, 1925; P. A. Walz, A. Pelzer *et alii*, 'Serta Albertina,' *Angelicum* (Rome), xxi (1944); G. M. Wickens (editor), *Avicenna: Scientist and Philosopher*, London, 1952.

COSMOLOGY AND ASTRONOMY: Duhem, *Système du Monde*, is still the basic work. Other contributions are Roger Bacon, *Opera Hactenus Inedita*, ed. R. Steele, Oxford, 1926, vi – for work on the calendar; J. D. Bond, 'Richard Wallingford (1292?–1335),' *Isis*, iv (1922); F. J. Carmody, 'The planetary theory of Ibn Rushd,' *Osiris*, x (1952), *Al Bitruji de motibus celorum. Critical edition of the Latin translation of Michael Scot*, Berkeley, Calif., 1952; F. M. Cornford, *Plato's Cosmology. The Timæus of Plato translated with a running commentary*, London, 1937; J. B. J. Delambre, *Histoire de l'astronomie au moyen âge*, Paris, 1819 – including Arabic astronomy; J. Drecker. 'Hermanus Contractus. Über das Astrolab,' *Isis*, xvi (1931); J. L. E. Dreyer, *A History of Planetary Systems from Thales to Kepler*, Cam-

Bibliography

bridge, 1906 (reprinted as *A History of Astronomy* . . ., New York, 1953) – an excellent survey, 'Medieval astronomy,' in *Studies in the History and Method of Science*, ed. C. Singer, Oxford, 1921, ii; P. Duhem, 'Essai sur la notion de théorie physique de Platon à Galilée,' *Annales de philosophie chrétienne*, vi (1908) (reprinted Paris, 1908); R. T. Gunther, *Early Science in Oxford*, Oxford, 1923, ii, 1929, v, *The Astrolabes of the World*, Oxford, 1932; W. Hartner, 'The principle and use of the astrolabe' in *A Survey of Persian Art*, ed. A. U. Pope, London and New York, 1939, iii – the best description in English of the use of the astrolabe. 'The Mercury horoscope of Marcantonio Michiel of Venice,' in *Vistas in Astronomy*, ed. A. Beer, London and New York, 1955; Sir Thomas Heath, *Aristarchus of Samos, the Ancient Copernicus*, Oxford, 1913, a history of Greek astronomy to Aristarchus, *Greek Astronomy*, London, 1932; F. Kaltenbrunner, *Die Vorgeschichte der gregorianischen Kalenderreform* (*Sitzungsberichte der kaiserlichen Akademie der Wissenschaften*, philos.-hist. Klasse, lxxxii), Vienna, 1876; L. O. Kattsoff, 'Ptolemy and scientific method,' *Isis*, xxxviii (1947); H. Michel, 'Le Rectangulus de Wallingford précédé d'une note sur le Torquetum,' *Ciel et Terre*, Brussels, Nos. 11–12 (1944), *Traité de l'astrolabe*, Paris, 1947 – fundamental; J. M. Millás-Vallicrosa, *Etudios sobre Azarquiel*, Madrid and Granada, 1943–50 – on the astrolabe; O. Neugebauer, 'The origin of the Egyptian calendar,' *Journal of Near Eastern Studies*, i (1942), 'The history of ancient astronomy: problems and methods,' *Journal of Near Eastern Studies*, iv (1945), (reprinted, with some amplification, in *Publication of the Astronomical Society of the Pacific*, xlviii, 1946), 'The early history of the astrolabe. Studies in ancient astronomy, ix,' *Isis*, xl (1949) – an important study, *The Transmission of Planetary Theories in Ancient and Medieval Astronomy*, Scripta Mathematica, New York, 1955; M. A. Orr, *Dante and the Early Astronomers*, 2nd ed., London, 1956; A. Pannekoek, *A History of Astronomy*, London, 1961; D. J. Price and R. M. Wilson, *The Equatorie of the Planetis*, Cambridge, 1955; Claude Ptolémée, *Composition mathématique*, traduite . . . par M. Halma, Paris, 1813–16, 2 vols. (reprinted Paris, 1927), Ptolemy, *The Almagest*, trans. R. C. Taliaferro (*Great Books of the Western World*, xvi) Chicago, 1952; G. V. Schiaparelli, *Scritti sulla storia della astronomia antica*, Bologna, 1925 – pioneer studies; E. L. Stevenson, *Terrestrial and Celestial Globes*, New Haven, Conn., 1921, 2 vols.; H. Suter, *Die Mathematiker und Astronomen der Araber und ihre Werke* (*Abhandlungen zur Geschichte der mathematischen Wissenschaften*, x), Leipzig, 1900, 'Nachträge und Berichtigungen . . .' (ibid. xiv, 1902) –

Bibliography

fundamental; F. Sherwood Taylor, 'Mediaeval scientific instruments,' *Discovery*, xi (1950); L. Thorndike, *The Sphere of Sacrobosco*, Chicago, 1949, *Latin Treatises on Comets. Between 1238 and 1368 A.D.*, Chicago, 1950; M. C. Welborn, *Calendar Reform in the Thirteenth Century*, University of Chicago, unpublished dissertation, 1932; P. W. Wilson, *The Romance of the Calendar*, New York, 1937; J. K. Wright, 'Notes on the knowledge of latitudes and longitudes in the middle ages,' *Isis*, v (1923); E. Zinner, 'Die Tafeln von Toledo,' *Osiris*, i (1936).

METEOROLOGY AND OPTICS: For optics see Crombie, *Robert Grosseteste* and 'The mechanistic hypothesis and the scientific study of vision,' *Proc. Roy. Microscopical Soc.*, ii (1967) 3–110 (bibliography); G. F. Vescovini, *Studi sulla prospettiva medievale*, Turin, 1965; also under Ch. 4, Medicine, and Vol. II under Ch. 2, Scientific Instruments. For further details there are C. Baeumker, *Witelo, ein Philosoph und Naturforscher des XIII. Jahrhunderts (Beitr. Ges. Philos. Mittelalt.*, iii 2) Münster, 1908; H. Bauer, *Die Psychologie Alhazeus (Beitr. Ges. Philos. Mittelalt.*, x. 5), Münster, 1911; A. Birkenmajer, 'Études sur Witelo, i–iv,' *Bulletin international de l'Académie Polonaise des Sciences et des Lettres* (Cracow), Classe d'hist. et de philos., Années 1918, 1920, 1922; C. B. Boyer, 'Aristotelian references to the law of reflection,' *Isis*, xxxvi (1946), 'The theory of the rainbow: medieval triumph and failure,' *Isis*, xlix (1958); Euclid, 'The Optics of Euclid,' trans. H. E. Burton, *J. Optical Soc. of America*, xxxv (1945); G. Hellmann, *Neudrucke von Schriften und Karten über Meteorologie und Erdmagnetismus*, Nos. xii–xv, Berlin, 1899–1904 – on weather prediction and optics, *Die Wettervorhersage im ausgehenden Mittelalter (XII. bis XV. Jahrhundert), (Beiträge zur Geschichte der Meteorologie*, viii), Berlin, 1917; D. Kaufmann, *Die Sinne. Beiträge zur Geschichte der Physiologie und Psychologie im Mittelalter aus hebräischen und arabischen Quellen*, Leipzig, 1884; E. Krebs, *Meister Dietrich (Theodoricus Teutonicus de Vriberg). Sein Leben, seine Werke, seine Wissenschaft (Beitr. Ges. Philos. Mittelalt.*, v. 5–6), Münster, 1906; A. Lejeune, *Euclide et Ptolémée*, Louvain, 1948; *Recherches sur la Catoptrique grecque d'après les sources antiques et médiévales*, Brussels, 1957, *L'Optique de Claude Ptolémée*, ed. A. Lejeune, Louvain, 1956; G. Sarton, 'The tradition of the optics of Ibn al-Haitham,' *Isis*, xxix (1938), xxxiv (1942–3); A. Sayili, 'The Aristotelian explanation of the rainbow,' *Isis*, xxx (1939); F. M. Shuja, *Cause of Refraction as explained by the Moslem Scientists*, Delhi, 1936; Theodoricus Teutonicus de Vriberg, *De Iride*, ed. J. Würschmidt (*Beitr. Ges. Philos. Mittelalt.*,

261

Bibliography

xii. 5–6) Münster, 1914; C. M. Turbayne, 'Grosseteste and an ancient optical principle,' *Isis*, i (1959); W. A. Wallace, *The Scientific Methodology of Theodoric of Freiberg*, Freibourg, 1959; E. Wiedemann, an important series of articles on Arabic optics published mainly in *Annalen der Physik und Chemie, Sitzungsberichte der preussischen Akademie der Wissenschaften zu Berlin*, philos.-hist. Klasse, and *Archiv für die Geschichte der Naturwissenschaften und der Technik*, 1890–1930 : see Sarton, *Introduction*, i. 722–23 and H. J. Seemann, 'Eilhard Wiedemann,' *Isis*, xiv (1930); H. J. J. Winter, 'The optical researches of Ibn al-Haitham,' *Centaurus*, iii (1954).

MECHANICS: H. Carteron, *La notion de force dans le système d'Aristote*, Paris, 1923; M. Clagett, *The Science of Mechanics in the Middle Ages*, Madison, Wisconsin, 1959 – an indispensable study, with texts and commentary; F. M. Cornford, *The Laws of Motion in the Ancient World*, Cambridge, 1931; I. E. Drabkin, 'Notes on the laws of motion in Aristotle,' *American Journal of Philology*, lix (1938); R. Dugas, *Histoire de la mécanique*, Neuchâtel, 1950 (English trans. New York, 1955) – medieval mechanics largely based on Duhem; P. Duhem, *Les Origines de la Statique*, Paris, 1905–6, 2 vols. – indispensable; B. Ginzburg, 'Duhem and Jordanus Nemorarius,' *Isis*, xxv (1936); E. A. Moody and M. Clagett, *The Medieval Science of Weights*, Madison, 1952 – a critical source book; P. Tannery, *Mémoires scientifiques*, publiée par J. L. Heiberg, v, 'Sciences exactes au moyen âge (1877–1921),' Toulouse and Paris, 1922; H. J. J. Winter, articles on Arabic physics in *Endeavour*, ix–x (1950–1).

MAGNETISM: H. D. Harradon, 'Some early contributions to the history of geomagnetism – I,' *Terrestrial Magnetism and Atmospheric Electricity*, xlviii (1943), with an English translation of the *Epistola* of Petrus Peregrinus; E. O. von Lippmann, *Geschichte der Magnetnadel bis zur Erfindung des Kompasses* [gegen 1300] (*Quellen und Studien zur Geschichte der Naturwissenschaften und der Medizin*, iii. 1) Berlin, 1932; A. C. Mitchell, 'Chapters in the history of terrestrial magnetism,' *Terrestrial Magnetism and Atmospheric Electricity*, xxxvii (1932), xlii (1937), xliv (1939); P. F. Mottelay, *Bibliographical History of Electricity and Magnetism*, London, 1922; Petrus Peregrinus Maricurtensis, *De Magnete*, ed. G. Hellmann, *Neudrucke von Schriften und Karten über Meteorologie und Erdmagnetismus*, x, Berlin, 1898; Petrus Peregrinus, *The Epistle, Concerning the Magnet*, done into English by S. P. Thompson, London, 1902; E. Schlund, 'Petrus Peregrinus von Mari-

Bibliography

court: sein Leben und seine Schriften,' *Archivum Franciscanum Historicum*, iv (1911), v (1912) – an exhaustive study; Li Shu-hua, 'Origine de la boussole,' *Isis*, xlv (1954); S. P. Thompson, 'Petrus Peregrinus de Maricourt and his Epistola de Magnete,' *Proceedings of the British Academy*, ii (1905–6).

GEOLOGY: F. D. Adams, *The Birth and Development of the Geological Sciences*, Baltimore, 1938 (reprinted, New York, 1954); Avicenna, *De Congelatione et Conglutinatione Lapidum*, ed. E. J. Holmyard and D. C. Mandeville, Paris, 1927; P. Duhem, *Études sur Léonard de Vinci*, ii, Paris, 1909; K. Klauck, 'Albertus Magnus und die Erdkunde,' in *Studia Albertina*, ed. H. Ostlender (*Beitr. Ges, Philos. Mittelalt., Supplementband* iv), 1952.

CHEMISTRY: K. C. Bailey, *The Elder Pliny's Chapters on Chemical Subjects*, edited, with translation and notes, London, 1929–32, 2 vols.; P. E. M. Berthelot, *Les origines de l'alchimie*, Paris, 1885, *Collections des anciens alchimistes grecs*, texte et traduction, 3 vols., Paris, 1888 – basic sources, *La chimie au moyen âge*, Paris, 1893, 3 vols.; H. H. Dubs, 'The beginnings of alchemy,' *Isis*, xxxviii (1947); D. I. Duveen, *Bibliotheca alchemica et chemica* – an annotated catalogue of printed books on alchemy, chemistry and related subjects, London, 1949; M. Eliade, *Forgerons et Alchimistes*, Paris, 1956; R. J. Forbes, *Bitumen and Petroleum in Antiquity*, Leiden, 1936 – for 'Greek fire' etc., *A Short History of the Art of Distillation*, Leiden, 1948; W. Ganzen-müller, *L'Alchimie au moyen âge*, traduit de l'allemand par G. Petit-Dutaillis, Paris, no date (German ed., Paderborn, 1938); E. J. Holmyard, *Makers of Chemistry*, Oxford, 1931, *Alchemy*, London (Pelican Books), 1957 – an excellent survey; P. Kraus, 'Djabir,' *Encyclopaedia of Islam*, Leiden and London, 1938, Supplement, *Jabir ibn Hayyan*, Cairo, Impr. de l'Institut français d'archéologie orientale, 1942–3, 2 vols.; P. Kraus and S. Pines, 'al-Razi,' *Encyclopaedia of Islam*, Leiden and London, 1936, iii; E. O. von Lippmann, *Entstehung und Ausbreitung der Alchemie*, 2 vols., Berlin, 1919–31; Robert P. Multhauf, 'John of Rupescissa and the origin of medical chemistry,' *Isis*, xlv (1954), 'The significance of distillation in Renaissance medical chemistry,' *Bulletin of the History of Medicine*, xxx (1956); J. R. Partington, 'Albertus Magnus on alchemy,' *Ambix*, i (1937); M. Plessner, 'The place of the Turba Philosophorum in the development of alchemy,' *Isis*, xlv (1954) – a very useful critical discussion of recent work on the history of alchemy; J. F. Ruska, *Tabula Smaragdina; ein Beitrag zur Geschichte der hermetischen Literatur*, Heidelberg, 1926, *Turba Philosophorum, ein Beitrag zur Geschichte der Alchemie* (*Quellen und Studien zur Geschichte der Naturwissenschaften und der Medizin*, i) Berlin, 1931; J. A. Stillman, *The Story of Early Chemistry*, New York, 1924; F. Strunz, *Geschichte der Naturwissenschaften in Mittelalter*, Stuttgart, 1910; F. Sherwood Taylor, 'A survey of Greek alchemy,' *Journal of Hellenic Studies*, i (1930), 'The Origin of Greek alchemy,' *Ambix*, i (1937), 'The evolution of the still,' *Annals of Science*, v (1945), *The Alchemists*, New York, 1949 – with a useful short bibliography; F. A. Yates, 'The art of Ramón Lull (1232–c. 1316). An approach to it through Lull's theory of the elements,' *Journal of the Warburg and Courtauld Institutes*, xvii (1954).

BIOLOGY: Botany, zoology, anatomy, physiology: besides the following, cf. under Chapter 4, Agriculture and Medicine: P. Aiken, 'The animal history of Albertus Magnus and Thomas of Cantimpré,' *Speculum*, xxii (1947); Albertus Magnus, *De Vegetabilibus*, ed. C. Jessen, Berlin, 1867, *De Animalibus*, ed. H. Stadler (*Beitr. Ges. Philos. Mittelalt.* xv–xvi), 1916–20, *Quaestiones super de Animalibus*, ed. E. Filthaut (*Opera Omnia*, ed. Institutum Alberti Magni Coloniense, B. Geyer Praeside, xii) Münster, 1955; Anonymus Londinensis, *Medical Writings*, ed. W. H. S. Jones, Cambridge, 1947; A. Arber, *Herbals*, Cambridge, 1938, *The Natural Philosophy of Plant Form*, Cambridge, 1950; H. Balss, *Albertus Magnus als Zoologie*, Stuttgart, 1947; H. S. Bennett, 'Science and information in English writings of the 15th century,' *Modern Language Review*, xxxix (1944); A. Biese, *The Development of the Feeling for Nature in the Middle Ages and Modern Times*, London, 1905; M. De Bouard, 'Encyclopédies médiévales,' *Revue des questions historiques*, cxii (1930); G. S. Brett, *A History of Psychology*, London, 1912–21, 3 vols.; A. J. Brock, *Greek Medicine*, London, 1929; J. V. Carus, *Geschichte der Zoologie*, Munich, 1872; A. C. Crombie, 'Cybo d'Hyères: a 14th century zoological artist,' *Endeavour*, xi (1952); A. Delorme, 'La morphogenèse d'Albert le Grand dans l'embryologie scolastique,' *Revue thomiste*, N.S. xiv (1931); A. Fellner, *Albertus Magnus als Botaniker*, Vienna, 1881; D. Fleming, 'Galen on the motions of the blood in the heart and lungs,' *Isis*, xlvi (1955); H. W. K. Fischer, *Mittelalterliche Pflanzenkunde*, Munich, 1929; A. Fonahn, *Arabic and Latin Anatomical Terminology* (Norwegian Acad., hist.-philos. Klasse, 1921, No. 7), Christiana, 1922; Emperor Frederick II, *De Arte Venandi Cum Avibus*, ed. C. A. Willemsen, Leipzig, 1942; Galen, *Opera Omnia*, ed. C. G. Kühn, Leipzig, 1821–

Bibliography

33. 20 vols., *On the Natural Faculties*, translated by A. J. Brock (Loeb Classical Library) London and New York, 1916, *On Anatomical Procedures*, translation ... with introduction and notes by C. Singer, London, 1956; R. W. T. Gunther, *The Herbal of Apuleius Barbarus*, Oxford, 1925, *The Greek Herbal of Dioscorides*, Oxford, 1934; W. A. Heidel, *Hippocratic Medicine, Its Spirit and Method*, New York, 1941; D. Jalabert, 'La flore gothique : ses origines, son évolution du XIIe au XVe siècles,' *Bulletin monumental*, xci (1932); K. F. W. Jessen, *Botanik der Gegenwart und Vorzeit in kulturhistorischer Entwicklung*, Leipzig, 1864, Waltham, Mass., 1948; W. H. S. Jones, *Philosophy and Method in Ancient Greece (Bull. of the History of Medicine*, Suppl. viii), Baltimore, 1946; S. Killermann, *Die Vogelkunde des Albertus Magnus*, 1270–80, Regensburg, 1910, 'Das Tierbuch des Petrus Candidus, 1460,' *Zoologische Annalen*, vi (1914); E. O. von Lippmann, *Urzeugung und Lebenskraft*, Berlin, 1933; G. Loisel, *Histoire des ménageries de l'antiquité à nos jours*, Paris, 1913, i; T. E. Lones, *Aristotle's Researches into Natural Science*, London, 1912; E. Mâle, *L'Art religieux du 13e siècle en France*, 3e ed., Paris, 1910 (English trans. by D. Nussy, London, 1913); E. H. F. Meyer, *Geschichte der Botanik*, Königsberg, 1857, iv; L. L. F. Moncourier, *L'École médicale d'Alexandrie*, Bordeaux, 1931; Claus Nissen, *Die botanische Buchillustration. Ihre Geschichte und Bibliographie*, Stuttgart, 1951–2, 2 vols. *Die Illustrierten Vogelbücher*, Stuttgart, 1953; H. Ostlender (editor), *Studia Albertina (Beitr. Ges. Philos. Mittelalt.*, Supplementband iv); N. Pevsner, *The Leaves of Southwell*, London, 1945; A. Platt, 'Aristotle on the heart,' in *Studies in the History and Method of Science*, ed. Singer, Oxford, 1921, ii; E. S. Russell, *Form and Function*, London, 1916; G. Senn, *Die Entwicklung der biologischen Forschungsmethode in der Antike und ihre grundsätzliche Förderung durch Theophrast von Eresos*, Aarau, 1933 – very important; C. Singer, *Greek Biology and Greek Medicine*, Oxford, 1922, 'Greek biology and its relation to the rise of modern biology,' in *Studies in the History and Method of Science*, ii, *The Evolution of Anatomy*, London, 1925 (reprinted as *A Short History of Anatomy and Physiology from the Greeks to Harvey*, New York, 1957); F. Strunz, *Albertus Magnus*, Vienna and Leipzig, 1926; K. Sudhoff, *Ein Beitrag zur Geschichte der Anatomie im Mittelalter, speziell der anatomischen Graphik, nach Handschriften des 9. bis 15. Jahrhundert (Studien zur Geschichte der Medizin*, iv), Leipzig, 1908, illustrated articles on medieval anatomy and embryology in *Archiv für Geschichte der Medizin*, iv (1910), vii (1913); W. Sudhoff, 'Die Lehre von den Hirnventrikeln in textlicher und graphischer

Bibliography

Tradition des Altertums und Mittelalter,' ibid, vii (1913); H. O. Taylor, *Greek Biology and Medicine*, London, 1923; Sir D'Arcy W. Thompson, *On Aristotle as a Biologist*, Oxford, 1913; L. Thorndike and F. S. Benjamin (editors), *The Herbal of Rufinus*, Chicago, 1945; G. Verbeke, *L'Evolution de la doctrine du pneuma du stoïcisme à St Augustin*, Paris, 1945; J. Walsh, 'Galen's writings and influences inspiring them,' *Annals of Medical History*, vi (1934), vii (1935), viii (1936), ix (1937); Lynn White, jr., 'Natural science and naturalistic art in the middle ages,' *American Historical Review*, lii (1947); T. H. White, *The Book of Beasts*, New York, 1954 – English translation of a 12th-century bestiary; E. Wickersheimer, 'L'"Anatomie" de Guido da Vigevano,' *Archiv für Geschichte de Medizin*, vii (1913), *Anatomies de Mondino dei Luzzi et de Guido de Vigevano*, Paris, 1926, with illustrations; J. Wimmer, *Deutsche Pflanzenkunde nach Albertus Magnus*, Halle a/S., 1908; C. A. Wood and M. F. Fyfe, *The Art of Falconry ... of Frederick II*, Stanford, 1943; Conway Zirkle, 'The inheritance of acquired characters and the provisional hypothesis of pangenesis,' *American Naturalist*, lxix (1935), lxx (1936), 'The early history of the idea of the inheritance of acquired characters of pangenesis,' *Transactions of the American Philosophical Society*, xxxv (1946).

CHAPTER 4

GENERAL: A. E. Berriman, *Historical Metrology, A new analysis of the archaeological and historical evidence relating to weights and measures*, New York, 1953; M. Bloch, 'Les "inventions" médiévales,' *Annales d'histoire économique et sociale*, vii (1935); P. Boissonade, *Le Travail dans l'Europe chrétienne au moyen âge (5e–15e siècles)*, Paris, 1921, *Life and Work in Medieval Europe*, trans. by Eileen Power, London, 1927; J. Delevsky, 'L'évolution des sciences et les techniques industrielles,' *Revue d'histoire économique et sociale*, XXV (1939); F. M. Feldhaus, *Die Technik der Vorzeit, der geschichtlichen Zeit und der Naturvölker*, Leipzig and Berlin, 1914, *Die Technik der Antike und des Mittelalters*, Potsdam, 1931; R. J. Forbes, *Man the Maker*, New York, 1950; A. T. Geoghegan, *The Attitude towards Labor in Early Christianity and Ancient Culture (Catholic University of American Studies in Christian Antiquity*, No. 6), Washington, D.C., 1945; Bertrand Gille, 'Les développements technologiques en Europe de 1100 à 1400,' *Cahiers d'histoire mondiale*, iii (1956); W. Hallock and H. T. Wade, *Outlines of the Evolution of Weights and Measures and the Metric System*, New York, 1906; Lefebvre des Noettes, 'La "nuit"

Bibliography

du moyen âge et son inventaire,' *Mercure de France*, ccxxxv (1932); L. Mumford, *Technics and Civilization*, London, 1934; J. U. Nef, *War and Human Progress. An essay on the rise of industrial civilization*, London, 1950; A. Neuburger, *The Technical Arts and Sciences of the Ancients*, London, 1930; L. F. Salzman, *English Life in the Middle Ages*, Oxford, 1926; C. Singer, E. J. Holmyard, A. R. Hall, and T. I. Williams (editors), *A History of Technology*, Oxford, 1954–8, 5 vols. – the basic work; A. Uccelli *et alii*, 'La Storia della Tecnica,' in *Enciclopedia Storica delle Scienze e delle loro Applicazione*, Milan, 1944, ii; A. P. Usher, *A History of Mechanical Inventions*, 2nd ed., Cambridge, Mass., 1954; James C. Webster, *The Labors of the Months in Antique and Mediaeval Art to the End of the Twelfth Century*, Evanston, 1938; Lynn White, jr., 'Technology and invention in the Middle Ages,' *Speculum*, xv (1940) – with an excellent bibliography, *Medieval Technology and Social Change*, Oxford, 1962.

EDUCATION AND TECHNOLOGY: Cf. the works of Clerval, Crombie, Grabmann (1909–11), Hunt, d'Irsay, Paré *et alii*, Gonzaléz Palencia, Rashdall, Sharp (1934), listed under Chapters 1, 2, 3; also R. Baron, 'Sur l'introduction en Occident des termes "geometria theorica et practica," ' *Revue d'histoire des sciences*, viii (1955); G. Beaujouan, *L'interdépendence entre la science scolastique et les techniques utilitaires (xiie, xiiie, et xive siècles)* (Conférence du Palais de la Découverte) Paris, 1957; B. Gille, *Esprit et civilisation technique au moyen âge* (Conférence du Palais de la Découverte) Paris, 1952; Theophilus the Presbyter, *De Diversis Artibus*, Latin text and English trans. by C. R. Dodwell, Edinburgh, 1960.

MUSIC: Willi Apel, 'Early history of the organ,' *Speculum*, xxiii (1948); R. d'Erlanger, *La musique arabe*, Paris, 1930–9, 4 vols.; H. G. Farmer, *The Influence of Music: From Arabic Sources*, London, 1926, *History of Arabian Music to the Thirteenth Century*, London, 1929, *Historical Facts for the Arabian Musical Influence*, London, 1930, *Al-Farabi's Arabic-Latin Writings on Music (A Collection of Oriental Writers on Music, ii)*, Glasgow, 1934; G. Reese, *Music in the Middle Ages*, London, 1941; K. Schlesinger, *Oxford History of Music*, Oxford, 1929.

AGRICULTURE AND STOCK BREEDING: D. Bois, *Les plantes alimentaires chez tous les peuples et à travers les âges*, Paris, 1927–8, 2 vols.; Sir F. Crisp, *Medieval Gardens*, London, 1924; H. C. Darby, *The Medieval Fenland*, Cambridge, 1940; Lord Ernle, *English Farming*,

Bibliography

Past and Present, 5th edition, edited by Sir A. D. Hall, London, 1936; M. L. Gothein, *A History of Garden Art,* trans. by Mrs Archer-Hind, London, 1928; N. B. S. Gras, *A History of Agriculture in Europe and America,* New York, 1925; Lefebvre des Noettes, *L'Attelage, le cheval de selle à travers les âges,* Paris, 1931, 2 vols.; L. Moulé, *Histoire de la médecine vétérinaire,* Paris, 1891–1911, 4 parts; Eileen Power, *The Wool Trade in English Medieval History,* Oxford, 1941; Sir F. Smith, *The Early History of Veterinary Literature,* London, 1919, i.

BUILDING, PRINTING, MACHINES AND INSTRUMENTS: Most informative is Usher, *History of Mechanical Inventions;* cf. Vol. II under Chapter 2; in addition there are A. S. Blum, *La route du papier,* Grenoble, 1946; Pierce Butler, *The Invention of Printing in Europe,* Chicago, 1940; T. F. Carter, *The Invention of Printing in China and its Spread Westwards,* 3rd ed., revised by L. Carrington Goodrich, New York, 1955; E. M. Carus-Wilson, 'An industrial revolution in the 13th century,' *Economic History Review,* xii (1941); M. Destrez, *La Pecia,* Paris, 1936; B. Gille, 'La machinisme au moyen âge,' *Actes du VIᵉ Congrès international d'Histoire des Sciences, Amsterdam, 1950,* Paris, 1953; D. Hunter, *Papermaking,* 2nd ed., London, 1947; D. Knoop and G. P. Jones, *The Mediaeval Mason,* Manchester, 1933; V. Mortet et P. Deschamps, *Recueil de textes relatifs à l'histoire de l'architecture,* Paris, 1911–29, 2 vols.; Douglas C. McMurtrie, *The Book. The Story of Printing and Bookmaking,* 3rd ed., New York, 1938; E. Panofsky, *Gothic Architecture and Scholasticism,* Latrobe, Pa., 1951; P. Pelliot, *Les débuts de l'imprimerie en Chine,* Paris, 1953; A. Ruppel, *Johannes Gutenburg. Sein Leben und sein Werk,* 2nd ed., Berlin, 1947; C. L. Sagui, 'La meunerie de Barbegal (France) et les roues hydrauliques chez les anciens et au moyen âge,' *Isis,* xxxviii (1948); E. A. Thompson (ed. and trans.), *A Roman Reformer and Inventor. Being a new text of the Treatise De rebus bellicis,* Oxford, 1952; Villard de Honnecourt, *Kritische Gesamtausgabe des Bauhüttenbuches, MS fr. 19093 der Pariser Nationalbibliothek,* ed. H. R. Hahnloser, Vienna, 1935; E. E. Viollet-Le-Duc, *Dictionnaire raisonné de l'architecture française du XIᵉ au XVIᵉ siècle,* Paris, 1854–68, 10 vols.; G. H. West, *Gothic Architecture in England and France,* London, 1927; E. Zinner, 'Aus der Frühzeit der Räderuhr. Von der Gewichtsuhr zur Federzugsuhr,' *Abhandlungen deutsche Museum,* xxii (1954).

MAPS AND GEOGRAPHY: R. Almagia, 'Quelques questions au sujet des cartes nautiques et des portulans d'après les recherches récentes,' *Actes du Ve Congrès international d'Histoire des Sciences, Lausanne, 1947,* Paris, 1948; L. Bagrow, 'The origin of Ptolemy's Geographia,' *Geografiska Annaler,* Stockholm, xxvii (1945), *Geschichte der Kartographie,* Berlin, 1951; C. R. Beazley, *The Dawn of Modern Geography,* London, 1897–1906, 3 vols.; Lloyd A. Brown, *The Story of Maps,* London, 1951; A. Cortesão, *The Nautical Chart of 1424 and the Early Discovery and Cartographical Representation of America,* Coimbra, 1954; M. Destombes, *Cartes catalanes du XIVᵉ siècle* (Rapport de la commission pour la bibliographie des cartes anciennes, Fascicule i), Paris, 1952; D. B. Durand, 'The earliest modern maps of Germany and Central Europe,' *Isis,* xix (1933), *The Vienna-Klosterneuburg map corpus of the fifteenth century. A study in the transition from medieval to modern science,* Leiden, 1952; *Four Maps of Great Britain by Matthew Paris,* London, 1928, K. Kretschmer, *Die italienischen Portolane des Mittelalters,* Berlin, 1909; D. J. Price, 'Medieval land surveying and topographical maps,' *Geographical Journal,* cxxi (1955); E. L. Stevenson, *Portolan Charts, their origin and characteristics,* New York, 1911; R. V. Tooley, *Maps and Map-Makers,* London, 1949; R. Vaughan, *Matthew Paris,* Cambridge, 1958; J. K. Wright, *Geographical Lore at the Time of the Crusades,* New York, 1925.

INDUSTRIAL CHEMISTRY, MINING, METALLURGY, FIRE-ARMS: Cf. above under Chapter 3, and G. Agricola, *De Re Metallica,* English trans. by H. C. and L. H. Hoover, New York, 1950; *Bergwerk- und Probierbüchlein,* trans. A. E. Sisco and C. S. Smith, New York, 1949 – 16th-century works on mining, geology and assaying; Vanoccio Biringuccio, *Pirotechnia,* trans. C. S. Smith and M. Gnudi, New York, 1943; Lazarus Erker's *Treatise on Ores and Assaying,* trans. from the German ed. of 1580 by A. E. Sisco and C. S. Smith, Chicago, 1951; R. J. Forbes, *Metallurgy in Antiquity,* Leiden, 1950, 'Metallurgy and technology in the middle ages,' *Centaurus,* iii (1953); L. C. Goodrich and Feng Chia-Sheng, 'The early development of firearms in China,' *Isis,* xxxvi (1946); E. B. Haynes, *Glass,* London, 1948; H. W. L. Hime, *The Origin of Artillery,* London, 1915; J. B. Hurry, *The Wood Plant and its Dye,* London, 1930; R. P. Johnson, 'Compositiones variae,' in *Illinois Studies in Language and Literature,* xxiii (1939); J. U. Nef, 'Mining and metallurgy in medieval civilization,' in *The Cambridge Economic History,* ii; J. R. Partington, *Origins and Development of Applied Chemistry,* London, 1935; B. Rathgen, *Das Geschütz im Mittelalter,* Berlin, 1928; T. A. Rickard, *Man and Metals,* New York, 1932, 2 vols.; E. Salin et A. France-Lanord, *Le Feu à l'époque mérovingienne,* Paris,

Bibliography

1943; L. F. Salzmann, *English Industries in the Middle Ages*, Oxford, 1923; C. Singer, *The Earliest Chemical Industry*, London, 1949; D. V. Thompson, jr., *The Materials of Medieval Painting*, London, 1936; E. Turrière, 'Le développement de l'industrie verrière d'art depuis l'époque vénitienne jusqu'à la fondation des verreries d'optique,' *Isis*, vii (1925); Wang Ling, 'On the invention and use of gunpowder and firearms in China,' *Isis*, xxxvii (1947).

MEDICINE: In addition to works listed under Chapter 3, Sir T. C. Allbutt, *The Historical Relations of Medicine and Surgery to the End of the Sixteenth Century*, London, 1905; W. R. Bett (ed.), *A Short History of Some Common Diseases*, Oxford, 1934; E. Bock, *Die Brille und ihre Geschichte*, Vienna, 1903; E. G. Browne, *Arabian Medicine*, Cambridge, 1821; A. M. Campbell, *The Black Death and Men of Learning*, New York, 1931; D. Campbell, *Arabian Medicine and its Influence on the Middle Ages*, London, 1926, 2 vols.; A. Castiglioni, *History of Medicine*, trans. by E. B. Krumbhaar, 2nd ed., New York, 1947 – very useful; K. Chiu, 'The introduction of spectacles into China,' *Harvard Journal of Asiatic Studies*, i (1936); H. P. Cholmeley, *John of Gaddesden and the Rosa Medicinae*, Oxford, 1912; C. Creighton, *History of Epidemics in Great Britain*, Cambridge, 1891–4, 2 vols.; P. Diepgen, 'Die Bedeutung des Mittelalters für den Fortschritt in der Medizin,' in *Essays Presented to Karl Sudhoff*, ed. Singer and Sigerist, Oxford and Zürich, 1924, *Geschichte der Medizin*, ... I. Band: *Von den Anfängen der Medizin bis zur Mitte des 18. Jahrhunderts*, Berlin, 1949; Cyril Elgood, *A Medical History of Persia and the Eastern Caliphate*, New York, 1951; P. L. Entralgo, *Mind and Body. Psychosomatic pathology: a short history of the evolution of medical thought*, London, 1955; Fielding H. Garrison, *An Introduction to the History of Medicine*, 4th ed., Philadelphia, 1929 – with much bibliographical material; J. Grier, *A History of Pharmacy*, London, 1937; O. Cameron Gruner, *A Treatise on the Canon of Medicine of Avicenna*, incorporating a translation of the first book, London, 1930; D. Guthrie, *A History of Medicine*, Edinburgh, 1945 – with a useful bibliography; J. F. K. Hecker, *The Epidemics of the Middle Ages*, trans. by Babington, London, 1859; L. F. Hirst, *The Conquest of Plague*, Oxford, 1953; T. Husemann, 'Die schlafschwämme und andere Methoden der allgemeinen und örtlichen Anäthesie im Mittelalter,' *Deutsche Zeitschrift für Chirurgie*, xlii (1896), 'Weitere Beiträge ... ,' ibid., liv (1900); S. d'Irsay, 'The Black Death and the mediaeval universities,' *Annals of Medical History*, vii (1925); E. Kremers and G. Udang,

Bibliography

History of Pharmacy, Philadelphia, 1940; M. Laignel-Lavastine, *Histoire générale de la médecine, de la pharmacie, de l'art dentaire et de l'art vétérinaire*, Paris, 1934–6, 2 vols.; R. A. Leonardo, *A History of Surgery*, New York, 1942; D. P. Lockwood, *Ugo Benzi, medieval philosopher and physician, 1376–1439*, Chicago, 1951; E. R. Long, *History of Pathology*, Baltimore, 1928; C. A. Mercier, *Leper Houses and Mediaeval Hospitals*, London, 1915; Maître Henri de Mondeville, *Chirurgie*, traduction française avec des notes, une introduction et une biographie par E. Nicaise, Paris, 1893; M. Neuburger, *History of Medicine*, trans. by E. Playfair, London, 1910–25, 2 vols.; Johannes Noll, *The Black Death. A chronicle of the plague*, trans. by C. H. Clarke, London, 1926 (German ed., Potsdam, 1924); G. H. Oliver, *History of the Invention and Discovery of Spectacles*, London, 1913; Petrus Hispanus, *Die Ophthalmologie*, ed. A. M. Berger, Munich, 1899; W. A. Pussey, *The History and Epidemiology of Syphilis*, Baltimore, 1933; Rhazes, *A Treatise on the Smallpox and Measles*, trans. by W. A. Greenhill, London, 1848 (ed. E. C. Kelly, New York, 1939); E. Rieseman, *The Story of Medicine in the Middle Ages*, New York, 1935; M. von Rohr, 'Aus der Geschichte der Brille,' *Beiträge zur Geschichte der Technik und Industrie*, xvii (1927), xviii (1928), 'Gedanken zur Geschichte der Brillenherstellung,' *Forschungen zur Geschichte der Optik (Beilagehefte zur Zeitschrift für Instrumentenkunde*, Berlin), ii (1937); E. Rosen, 'Did Roger Bacon invent eyeglasses?' *Archives internationales d'histoire des sciences*, xxxiii (1954), 'The invention of eyeglasses,' *Journal of the History of Medicine*, xi (1956) – an important critical study; E. Sachs, *The History and Development of Neurological Surgery*, New York, 1952; H. E. Sigerist, 'Die Geburt der abendländischen Medizin,' in *Essays Presented to Karl Sudhoff*, ed. Singer and Sigerist, Oxford and Zürich, 1924, *The Great Doctors*, New York, 1933, 'On Hippocrates,' *Bull. Inst. Hist. Medicine*, ii (1934), 190–214, *Civilization and Disease*, Cornell, 1943. *A History of Medicine*, New York, 1951– , i– ; C. Singer, 'Steps leading to the invention of the first optical apparatus,' in *Studies in the History and Method of Science*, ii, *A Short History of Medicine*, Oxford, 1928; K. Sudhoff, *Tradition und Naturbeobachtung in den Illustrationen medizinischer Handschriften und Frühdrucke vornehmlich 15. Jahrhunderts*, Leipzig, 1907, on the *Tractatus pestilentiae, Archiv für Geschichte und Medizin*, v (1912), *Beiträge zur Geschichte der Chirurgie im Mittelalter; graphische und textliche Untersuchungen in mittelalterlichen Handschriften (Studien zur Geschichte der Medizin*, x–xii) Leipzig, 1914–18, 'Pestschriften aus der ersten 150 Jahren nach der Epidemie des

Bibliography

"schwarzen Todes" 1348,' *Archiv für Geschichte der Medizin*, ix (1916), xvii (1925); O. Temkin, *The Falling Sickness. A history of epilepsy from the Greeks to the beginnings of modern neurology*, Baltimore, 1971; C. J. S. Thompson, *The History of Evolution of Surgical Instruments*, New York, 1942; E. A. Underwood (editor), *Science, Medicine and History, Essays ... in honour of Charles Singer*, Oxford, 1953; R. Verrier, *Études sur Arnald de Villeneuve*, Leiden, 1947; J. J. Walsh, *Medieval Medicine*, London, 1920; C. E. A. Winslow, *Man and Epidemics*, Princeton, 1952.

INDEX

Index

al-Shirazi, Qutb al-din (1236–1311), 124
Aluminibus et Salibus, De, 57
al-Zarqali (d.c. 1087), 64, 107, 109–10
Amalfi, 52
Amiens, 212
Amodio, Andrea, 155
Anæsthesia, 231, 238
Analytics, 54
Anatomia, 177–8
Anatomia Porci, 176, 229
Anatomia Ricardi, 176–7, 229
Anatomia Vivorum, 171
Anatomy, 164, 240; avian, 152; of eye, 114, 116; human, 171–80, 188
Anaxagoras (5th century B.C.), 47n., 48, 136
Anaximander, 158
Angels, 39
Anima, De, 113
Animalibus, De, 54, 151, 160, 163, 170, 189
Animals, 35, 36, 42, 48; classification of, 165–6, 167; illustrations, 152, 153–5. See also Husbandry
Anthropology, 30
Antidotarium Nicolai, 132, 231
Antiseptics, 237–9
Ants, 164
Antwerp, 209
Apennines, 138
Aphorisms, 59
Apollodorus, 183
Apollonius, 113
Apparatus, 184; chemical, 140, 143, 144–8; scientific, 122–3, 227. See also Instruments
Apuleius Barbarus, 43
Apuleius (pseudo-), 54, 154, 156
Aquinas, St Thomas (1225–74), 54, 76, 77–8, 102, 103, 160, 187
Arabs and Arabic science, 25–6, 29, 51, 184, 196, 205; agriculture, 199; alchemy, 140–41, 142–4; astronomy, 64, 104; biology, 150; charts, 214–15; chemistry, 140–41; clocks, 217; dentistry, 239–40; glass making, 226; horse breeding, 202; hospitals, 241; invasion by, 32; medicine, 64–5, 230; ophthalmology, 230, 236; optics, 227; spectacles, 237; surgery, 237
Arber, Agnes, 158
Arches, 210–12
Archimedes, 127, 183, 195

Architectura, De, 55
Architecture, 153, 209–13
Architettura, 212
Arderne, John (1307–77), 166, 180, 239
Aretino, Spinello (c. 1333–1410), 155
Aristarchus of Samos, 24, 96, 101n., 103
Aristotle and Aristotelianism, 31, 54, 70, 94, 99–100, 102, 117, 123, 124, 149–50, 156, 157, 169–70, 182, 189; anatomy, 171; astronomy, 93–5, 99–100, 102–3; biology, 149–50, 169–70; causation, 83–7, 133; chemistry, 140–42; Christian theology, 71–9; classifications, 165–6, 186–7; contradictions, 80; cosmology, 71–3, 89–92; criticism, 26, 88, 95–6, 103, 126; embryology, 162–3; generation, 160–64, 167; geology, 133–4, 138; induction, 82–3; infinity, 87; influence, 63; logic, 229; mathematics, 83–4, 113; mechanics, 125–7; medicine, 232; meteorology, 110–11; mineralogy, 143; optics, 110, 112–13, 114; physics and metaphysics, 82–9; physiology, 171–3, 178–9; schools of thought, 76–9
Aristotle (pseudo-), 133, 135, 150, 156
Arithmetic, 189
Arithmetic, 31, 43, 65, 185, 187, 194
Armati, Salvino degl' (d. 1317), 185
Armillary sphere, 108
Arms, 186. See also Weapons
Arnald of Villanova (c. 1235–1311), 146, 148, 231, 232, 236
Arras, Bishop of, 198
Ars Cantus Mensurabilis, 192, 194
Art of the Black Land, 140–41
Art of Falconry, 151–2
Art of Medicine, 229
Arteries, 172–5
Arte Venandi cum Avibus, De, 151
Artillery, 223. See also Cannons
Arts, classification of, 186–9; fine, 30; liberal, 31, 33, 88; mechanical, 187; seven liberal, 186, 189
Aryabhata (b. 476), 65
Asia, 51, 151, 211, 215
Aspectibus, De, 57
Assyria, 24
Astrolabe, 43, 117, 122, 184, 189, 215, 222; Arabic, 167–8; construction of, 106–7; in Middle Ages, 104–5
Astrolabum catholicum, 107

274

Index

Book production, 208–9
Borgognoni, Hugh (d. 1252–8), 231, 238
Borgognoni, Theodoric (1205–98), 231, 238
Boring, 225
Boring machines, 216
Botanical garden, 166
Botanical iconography, 153–6
Botany, 30, 150, 155–6; medieval, 54
Bow-drill, 216
Brace, 216
Bradwardine, Thomas (d. 1349), 109
Brahmagupta (b. 598), 65
Brain, 171, 175
Breeding, 201–2
Brewing, 227
Britain, 40, 41
Brothers of the Holy Ghost, 241
Bruges (city), 136
Bruno, Giordano, 160
Bruno of Longoburgo (13th century), 180, 238
Brunschwig, Hieronymus, 147
Buch der Nature, Das, 139, 168
Buckwheat, 200
Bugia, 66
Buildings, 210–13
Burgos, 225
Burgundio of Pisa, 52, 168
Burgundy, 210
Buridan, Jean (d. after 1358), 103, 138, 170
Burley, Walter, 170
Bury St Edmunds, 154
Buttresses, 211–12
Byzantine empire, 51
Byzantine learning, 196
Byzantines, 224, 239
Byzantium, 53, 201, 208, 214, 237, 241; capture in 1204, 53; glass, 226; numerals, 66–7

Cælo, De, 54, 100, 189
Cairo, 241
Calabria, 200
Calais, 223
Calendar, 38, 39–40, 41, 43, 67, 71, 190; Christian, 39, 41; Gregorian, 41, 103; Hebrew, 39; Julian, 39; Omar Khayyam, 103; reform of, 108–9
Caliphs, 51
Callippus (4th century B.C.), 89, 93
Cambrai, 223

Cambrensis, Giraldus (*c*. 1147–1223), 36
Cambridge University, 211
Camera obscura, 124
Camerarius, R. J., 161
Campanus of Novara, John (d. after 1292), 108
Campbell, A. M., 234
Canals, 213
Cancer, tropic of, 105, 107
Canistris, Opicinus de (d. 1352), 215
Cannons, 223, 224–5
Canon, 160
Canones Azarchelis (Toledan Tables), 64
Canon of Medicine, 171, 230
Canterbury, 218
Canterbury, Archbishop of, 78
Canterbury Tales, 37n., 64n.
Canute, 44
Capella, Martianus (*c*. 600 B.C.), 33, 34, 103
Capricorn, tropic of, 105, 107
Cardano, Hieronymo, 139
Carolingian period, 197, 229
Carolingian renaissance, 41
Carpathians, 221
Carte Pisane, 214
Carthusians, 197
Cartography, 190, 214–16; Majorcan school, 215
Case-histories, 231
Cassiodorus (*c*. 490–580?), 31, 33, 38, 229
Cast iron, 222
Catalania, 214
Catapults, 203
Categories, 31
Catelan Mappemonde, 215
Cathedrals, 210–12, 219
Cathedral schools, 41, 43
Cato the Elder, 168, 183, 196
Causality, 85–6, 112
Causis Proprietatum Elementorum, De, 136
Celsus, Aurelius, 38, 239
Cennini, Cennino, 184
Cereals, 196, 197, 199
Cesalpino, Andrea, 157
Chalcidius (4th century), 34
Changes, 46n.–47n., 83–5
Chants, 191–4
Chaos, 46
Chapel of King's College, 211
Charcoal, 200, 226
Charlemagne, 32, 41, 196, 197, 201

Index

Index

Index

Fibonacci of Pisa, Leonardo (d. after 1240), 52, 66, 185
Fire, 39, 46, 46–7 n., 47, 90, 91, 92, 110–11
Firearms, 217, 223
Fishing, 151, 199, 201
Fitzherbert, Sir Anthony, 199
Flanders, 155, 224
Flax, 197
Flight, 213–14
Florence, 222, 235
Flowers, 199, 200
Flying buttresses, 211–12
Foliage, 153
Folk-medicine, 229. *See also* Medicine
Fontana, Giovanni de', 185
Forge hammers, 205
Formatione Corporis Humani in Utero, De, 169
Forms, 33, 48 n., 82, 83, 85, 141. *See also* Substance
Fossils, 133, 134, 135, 138, 139
Four elements, 141–2
Four humours, 170, 171–2, 230
France, 67, 76, 77, 108–9, 179, 225, 226, 234; northern, 154, 199, 205; southern, 208
Franciscans, 77
Franco of Cologne, 192
Frederick II, Emperor, 110, 151–2, 169, 188, 202, 236
Free will, 36, 72, 78, 181
Frescobaldi, 225
Frisius, Gemma, 107
Froissart, 223
Frontinus, 183
Fruit trees, 199, 200
Fuels, 220
Fuggers, house of, 225
Furnaces, 220–21
Furness, 220

Gaddesden, John of (d. 1361), 239–40
Galen (129–200), 38, 52, 54, 64, 71, 78, 80, 160, 178, 183, 229; anatomy, 171–7; biology, 150, 171; botany, 161; medicine, 229, 230, 235, 238; physiology, 171–5
Galileo, 23–4, 26, 37, 74, 75, 127, 130; architecture, 213; medicine, 229
Gariopontus, 229
Gaston de Foix, 168
Gears, 203, 204
Geber, 147. *See also* Jabir ibn Hayyan

Geminus, 100, 102
Generatione Animalium, De, 160, 161
Genesi ad Litteram, De, 74, 75
Genesis, 25, 46, 48
Genoa, 52, 232
Gentile da Foligno (14th century), 169, 179, 232, 233, 235, 239
Geographia, 216
Geography, 30. *See also* Cartography
Geology, 133–9
Geometry, 25, 65, 83, 88–9, 112–13, 188, 194, 210. *See also* Mathematics
Geometry of Boethius, 31
Geoponica, 168
Gerard of Cremona, 52, 66, 104
Gerbert, 43, 104, 184, 217
Germany, 138–9, 151, 164, 181, 184, 189, 196, 197, 202, 224, 229
Gersdorff, Hans von, 239
Ghent, 224
Ghiberti, Lorenzo, 222
Giacomo d'Angelo, 215
Gilbert, William, 131, 215
Gilbert the Englishman, 234, 238
Giles of Rome (*c.* 1247–1316), 102, 169, 185, 187
Giotto (*c.* 1276–1336), 155
Giovanni da San Gimignano (d. 1323), 185
Giovanni de' Dondi (b. 1318), 138
Giraffes, 152
Giraldus Cambrensis, 137
Glass, 140, 219, 226–7, 236
Glossary, Latin-Arabic, 52
Gnostics, Alexandrian, 140
Gold, 141, 144, 221
Gordon, Bernard of (d.c. 1320), 179, 237
Gothic architecture, 210–12; iconography, 153
Gough map, 215
Grafting, 159, 199
Grain, 197
Granada, 235
Grassus, Benvenutus, 236
Gravity and specific gravity, 48–9, 54, 128–9, 130, 139
Greek fire, 224
Greek grammar, 53
Greeks and Greek science, 24–6, 30–32, 34, 42, 46 n.–48 n., 51–4, 64, 80, 104, 140–43, 183, 191, 193, 202, 204, 217, 229–31, 234, 240
Gregory IX, Pope, 76

Index

Index

Illustrations, 153–4, 168–9; of dissections, 179, 180; in herbals, 166, 168
Immortality, 72
Incidence, angle of, 113, 114, 119–20
India, 24, 231
Induction, 82, 117, 229
Industry, 197; mechanization, 203–19
Infections, 234–6
Infinity, 87
Infirmaries, 38, 241
Ink, 228
Innocent III, Pope, 53, 241
Inquiry into Plants, 158
Insect pests, 199
Institutio Divinarum Litterarum, 38
Instruments: astronomical, 107–8, 189–90 (*see also* Astrolabe); dentistry, 240; musical, 193; optical, 227; precision, 219, 222; surgical, 238, 239. *See also* Apparatus
Interpretatione, De, 31
Inventions, prehistoric, 194
Ireland, 32, 40, 204
Iride, De, 114
Iride et Radialibus Impressionibus, De, 122
Iron working, 219–21
Irrigation, 198, 203
Isaac Israeli, 229
Isidore of Seville (560–636), 31–2, 33, 36, 38, 39, 229
Islam, 51, 53, 54, 192, 208, 217, 236
Isolation, 235
Italy, 43, 50, 53, 67, 104, 111, 137–8, 155, 166, 181, 190, 198, 200, 202, 206, 208, 221, 240; northern, 115, 215; southern, 52, 199

Jabir ibn Hayyan (8th century), 142–3, 144
Jacob ben Makir, 109
James of Venice, 52
Jarrow, 40
Jean de Jandum (d. 1328), 170
Jean de Linières, 109, 189
Jean de Murs, 193
Jehan de Brie, 169
Jerusalem, 241
John of Burgundy, 234
John of Garland, 192
John of Milano, 166
John of Rupescissa (d. after 1356), 146
John of St Amand, 133

John of Salisbury (c. 1115–80), 46, 54
John of Seville, 52
Jordanus Nemorarius, 126, 127–30, 210
Jordanus Saxo, 126
Josquin des Prés, 193
Journal of the History of Medicine, 237
Jumièges, 206
Jundishapur, 51
Jung, J., 157
Jupiter, 49

Kant, Immanuel, 26
Kegs, 201
Kent, 41, 220
Kepler, John, 102
Khayyam, Omar, 103
Khem, 141
Kilwardby, Robert (d. 1279), 78, 89, 187–8
Kinematics, 126, 203
Kitab al-Shifa, 143
Konungs Skuggsja, 137, 151
Korea, 209
Kyeser, Konrad, 185, 213

Laboratories, 140
Lanfranchi (d. before 1306), 177, 238
Langres, 210
Languages, 51–2; glossary, 52; new words, 53; vernacular, 54
Laon, 33; cathedral of, 212; school of, 33
Lapidary, 36
Lathes, 216
Latin Averroïsts, 73, 75–6
Latitudes, 104, 215
Lead, 221, 222
Leech Books, 42, 229
Legumes, 199
Leibniz, Gottfried Wilhelm von, 160
Le Mans cathedral, 242
Lenses, 114–24, 236–7
Leprosy, 234–5, 241
Letters to Posterity, 237
Leucippus, 47 n.
Lever, 54
Levi ben Gerson (1288–1344), 109–10, 124
Liber Abaci, 59, 66
Liber Almansoris, 57
Liber Astronomiæ, 58, 93
Liber Charastonis, 57
Liber de Coloribus Faciendis, 184

281

Index

Mediterranean, 32, 52, 131, 204, 232
Megenburg, Conrad von (1309–74), 139
Melancholia, 242
Menageries, 152
Mental disorders, 241–2
Mercuric sulphide, 143
Mercury, 49, 90, 103, 143
Meridians, 215–16
Merlee, William, 111
Mesopotamia, 53, 194
Messina, 53
Metallica, De Re, 185
Metallurgy, 140, 185, 221–6
Metals, 134, 216–17, 219–22, 225, 226
Metaphysics, 83
Metaphysics, 54
Meteorologica (Aristotle), 110, 113, 117, 133, 134, 135, 189
Meteorologica (Posidonius), 100
Meteorology, 29, 110–11
Métier d'historien, 23
Meyronnes, François de, 103
Miasma, 234
Microcosm, 37
Micrologus de Disciplina Artis Musicæ, 191
Midwives, 229
Milan, 218, 235
Military technology, 185. *See also* Warfare; Weapons
Mills, 203–6, 207–8
Millstones, 204, 205, 206
Mineral acids, 222
Mineralibus, De, 134, 135, 143, 144
Mineralibus et Rebus Metallicis, De, 135
Mineralogy, 30, 136
Minerals, 36, 133, 134, 225, 226, 229
Mining, 140, 225
Mirfield, John (d. 1407), 239
Miroir de Phœbus, Le, 168
Mist, 111
Mittelalterliches Hausbuch, 185
Mock suns, 111, 118
Mohammedan invasions, 32
Mohammedanism, 72. *See also* Islam
Monasteries, 32, 184, 190, 209, 217, 219, 241; Benedictine, 154, 229; Monte Cassino, 32, 38, 52
Monch, Philip, 225
Mondeville, Henry of (d.c. 1325), 67, 176, 238–9; on dissection, 179; medicine, 240
Mondino of Luzzi (c. 1275–1326),

239; on dissection, 177–8; physiology, 178–9
Mongols, 53, 224
'Mons Meg', 224
Monte Cassino, 32, 38 n., 52
Montpellier, 108, 109, 179, 180, 241; university of, 37, 150, 230, 233
Moon, 35, 39–40, 47, 49, 72, 90, 91, 110, 111; eclipses, 104; tides, 41
Morality, 34–5
Morphology, 157
Mortar, 210
Mosaics, 154
Motion, 82–4, 90–92, 112, 125–6, 127–8
Motor car, 70
Motors, 92
Motu Animalium, De, 170
Motu Cordis, De, 170
Motu et Tempore, De, 62
Mountains, 133, 134, 135, 136, 138, 139
Movement. *See* Motion
Multiplication of species, 112, 133
Mumford, Lewis, 190, 214, 219, 227
Mural quadrant, 108
Music, 31, 83, 191–4; scales, 191–4
Mutation, 159–60

Nantwich, 228
Naples, 52, 156, 232
Natural causes, 45
Natural Faculties, On The, 174
Natural History, 30, 38, 56, 150
Naturalibus Facultatibus, De, 174
Natural philosophy, 54
Natural science, 25, 68, 78, 81–3
Natura Rerum, De, 151, 164
Nature, 111 n.
Naturis Rerum, De, 35, 131, 150
Navigation, 131, 186, 214
Naviglio Grande, 198
Nazareth church, 242
Neckam, Alexander (1157–1217), 35, 131, 150–51, 184, 220
Nemesius, 229
Neolithic period, 216
Neoplatonism, 33, 34, 68, 72, 88, 92, 112, 140
Neptunism, 134–5
Nervous system, 171, 178–80
Nestorian Christians, 51
Netherlands, 199, 201, 205, 209
Newton, Isaac, 23, 118
Nicholas of Cusa (1401–64), 215

283

Index

Petrus Hispanus, 228, 236
Pfolspeundt, Heinrich von, 239
Pharmacy, 155
Philip Augustus, 53
Philippe de Vitri (1291–1361), 193
Philippe le Bel, 225
Philosopher's Stone, 67, 142
Physica, 189
Physica Elementa, 63
Physics, 54, 100–101, 125
Physiologus, 35
Physiology, 30, 171–6, 178–9, 230–31
Physique divisé en trois tomes, La, 189
Picardy, 131
Pier Candido Decembrio (1399–1477), 169
Pierre d'Ailly (15th century), 36, 109, 215
Piers Plowman, 207
Pietro d'Abano, 103, 228
Pillars, 210, 212
Pirotechnia, 225
Pisa, 52, 236
Pisano, Andrea, 222
Pistoia, 235
Plagues, 232–4
Plain Chants, 192
Planetaria, 110, 190, 217
Planets, 32, 39, 47, 49, 90, 91, 93–4, 97, 98–9
Plantis, De, 150, 156
Plastic surgery, 239
Platearius, Matthæus, 155–6
Plato and Platonism, 25, 33, 34, 45, 48 n., 49–50, 73, 82, 84, 89, 91, 93, 113, 136
Platonicus, 43
Plato of Tivoli, 52
Pleurisy, 42
Pliny (23–79), 30–31, 33, 35–6, 38, 39, 41, 150, 154, 161, 168, 204
Plotinus (*c*. 203–70), 33
Ploughs, 197–8, 203
Plummer, H. C., 111 n.
Plutonic theory, 135
Pneumatica, 54
Poetry, 54
Poissy, 212
Poland, 200, 225
Pole-lathes, 216
Pole Star, 106, 107
Polo, Marco, 151
Ponderoso et Levi, De, 54
Popes, 152. *See also under names*
Po River, 198

Porrée, Gilbert de la (*c*. 1076–1154), 45
Portolan maps, 131, 214, 215
Portugal, 198
Posidonius (b.c. 135 B.C.), 100, 102, 137
Posterior Analytics, 115
Post mortem examinations, 177, 240
Pottery, 228
Poultry, 201
Power : animal, 198, 201–2, 203, 204; mechanical, 203; steam, 226; transmission of, 203, 204; wind, 213
Practica, 43, 229
Practica Chirurgica, 237
Prague, 166
Premnon Fisicon, 229
Presses, 203, 209, 216
Primum movens, 90, 92, 94, 95, 99
Printing, 208–9
Probierbüchlein, 185
Proclus (410–85), 63
Properties of Things, On the, 153
Proprietatibus Elementorum, De, 133
Provence, 110
Prussia, 200
Psychiatry, 242
Ptolemy and Ptolemaic system, 31, 31 n., 54, 71, 80, 99–100, 114, 118, 189; astrolabe, 104; astronomy, 64, 93, 95–100, 103; cartography, 215–16; optics, 113, 114; star catalogue, 109
Public health, 235–6
Pumps, 203, 226
Puy-Guillaume, 223
Pyrotechnics, 140
Pythagoras, 188
Pythagoreanism, 33, 47 n.–48 n.

Qibla, 107
Quadrant, 108, 189
Quadrivium, 46, 186, 191
Quæstiones de Cælo et Mundo, 138
Quæstiones Naturales, 29, 35, 44–5
Qualea, Leonardo (*c*. 1470), 138
Quarantine, 234, 235

Rabbi ben Ezra, 66
Radolf of Liège, 104
Ragusa, 235
Rain, 111
Rainbows, 111, 114, 115, 117, 121; colours of, 122–4
Raindrops, 122–3

285

Index

Rakes, 198
Ramelli, A., 213
Rationalism, 23–4, 25–6, 73, 75–6, 181–2
'Rational soul', 149
Ratione Ponderis, De, 129, 130
Ravenna, 154, 209
Rays, 113
Reality, 37; aspects of, 83; search for, 81
Reason, 26, 73, 74, 76, 77–8
Rectangulus, 110
Reflection, law of, 113
Refraction, 113–14
Régime du Corps, 236
Regimen Sanitatis Salernitanum, 230
Regimine Principum, De, 185
Regiomontanus, J. M., 124
Renaissance, Carolingian, 41
Reproduction, 160–64
Rerum Natura, De, 38, 168
Reservoirs, 198
Rete, 107
Retort, 147
Retrogradations, 93
Revelation, 73, 74, 78
Rhazes (d.c. 924), 64, 64 n., 114; chemistry, 143, 144; medicine, 230
Rhine River, 198, 221
Rhinocerus indomitus, 44
Rhône River, 198
Rhumb-lines, 214
Richard I, 191
Richard of Wallingford (*c.* 1292–1335), 109–10, 190
Richard of Wendover, 171
Rinio, Benedetto, 155, 168
Ristoro d'Arezzo, 137
Rivers, 133, 134, 136, 139
Road building, 213
Robert of Chester, 52, 140
Robert of Cricklade (*c.* 1141–71), 150
Robert Grosseteste, 122 n.
Roger II, King of Sicily, 188
Roger of Salerno, 230, 237
Roland of Parma (13th century), 230, 237
Romans and Roman science, 26, 29, 40, 154, 195–6, 208, 219, 223, 229; agriculture, 195–6, 198–9; architecture, 209; contributions to science, 30; glass, 226; hospitals, 240; overshot wheels, 204; salt making, 227–8; ships, 213; technology, 195

Roman wall, 40, 215
Roofs, 210–12
Roscelinus (11th century), 44
Rotary lathe, 216
Rotation, 199, 200; scientific, 200
Royal Bethlehem hospital, 242
Royal Society, 68
Rudders, 213, 214
Rufinus, 69, 155, 156
Rufus of Ephesus, 114
Ruralia Commoda, 199
Russia, 164, 205, 232
Rye, 197

Sacrobosco (John Holywood), 67
Saffron, 200
Sagres, 205
Sails, 206, 213
St Albans, 218; monks of, 206
St Ambrose, 35, 38
St Augustine and Augustinianism (354–430), 25, 33, 34, 35, 36, 38, 44, 46, 48, 73–5, 77, 78, 88, 112, 159, 169
St Bartholomew's hospital, 241
St Basil the Great, 38
St Benedict, 32, 38n.
St Clement of Alexandria, 34–5
St Cloud, Guillaume de, 108
Sainte Chapelle, La, 212
St Francis of Assisi, 181
St Germain des Prés, 212
St Gregory the Great, 38
St Jerome, 74
St John the Baptist, 192
St Louis of France, 214
St Malo, 223
St Thomas's hospital, 241
St Vitus's Dance, 234
Salamanca University, 191
Salerno, 37, 77, 166, 231; Archbishop of, 229; medical school, 43, 150, 176–7, 188, 229, 230
Salt making, 227–8
Sanatio, 143
Sanitation, 236
Saphœa Arzachelis, 107
'Saracen corn', 200
Saturn, 49, 90
Savonarola, Michael (1384–1464), 146
Sawmills, 205
Saws, 204
Saxony, 219
Scales, musical, 191–4
Scandinavia, 214
Scholasticism, 85, 87, 90, 144

286

Index

Schools: cathedral, 34, 41, 43; at Chartres, 34, 76; at Laon, 32–3. *See also* Monasteries; Universities
Science, 32, 34–5, 54, 69, 76, 80, 82, 88–9, 182; ancient sources of, 55–63; applied, 69; classification, 186–8; development, 183; empirical, 26; history, 21–2, 25–7; hydraulic, 198; and sports, 169; theory of, 81
Scientific Revolution, 26, 27, 81, 126
Scientific rotation, of crops, 200
Scot, Michael, 52, 93, 137, 160, 231
Scotland, 220
Screw, 183, 203
Screw-cutting lathe, 216
Scripture, 74, 75
Sea coal, 220
Seafood, 201
Seeds of disease, 235
Seine River, 53
Semi-heliocentric system, 103
'Seminal causes', 48, 159
Seneca (1st century), 35
Sens, 212
Sense organs, 33
Senses, 81–3
'Sensitive soul', 149
Septem Diebus et Sex Operum Distinctionibus, De, 46
Shakespeare, William, 153
Sheep, 199, 200
Shipbuilding, 213
Ships, 216
Shuttle, 206, 208
Sicily, 52, 68, 197, 200, 229
Siena, 125
Siger of Brabant, 77
Signatures, doctrine of, 36
Silesia, 166
Silk, 197, 200, 207, 208
Simon of Genoa, 155
Simplicius (6th century), 63, 100
Singing, 191–4
Slavery, 203
Soap making, 227
Soils, 198–200
Solar year, 39–40
Soranus (2nd century), 229
Souls, 86, 92, 149
Sound, 115
Space, 39, 46, 49, 84
Spain, 40, 52, 53, 54, 64, 104, 111, 192, 197, 198, 205, 208, 221, 225, 226
Spectacles, 115, 185, 227, 236–7
Spectrum, 120, 123

Speculum, 242
Speculum Doctrinale, 151
Speculum Majus, 181
Speculum Regale, 137
Spina, Alessandro della (d. 1313), 185
Spinning wheels, 216
Spirits, 146
Sponge iron process. 220
Sports, 169
Spring crop, 197–8
Stadler, H., 163 n.
Stained-glass windows, 227
Standardization, 222, 225
Star-maps, 217
Stars, 32, 39, 47, 91, 103, 104, 105, 217, 219; catalogue of, 109; fixed, 89–90, 93, 95; North Star, 132; Pole Star, 106, 107; shooting, 111
Statics, 126
Steam, 203
Steam engine, 202
Steel making, 220
Steel wire, 222
Stellar Rays, On, 68
Stills, 144–7
Stock breeding. *See* Breeding
Stoicism, 34, 37, 48, 113, 160
Strato (c. 288 B.C.), 130
Strip land system, 197
Styria, 225
Submarines, 70, 213
Substance, conception of, 82–7, 90, 92, 141–2. *See also* Forms
Substantial form. *See* Substance
Sudhoff, Karl, 229
Sugar cane, 200
Sulphur, 143, 147
Sumeria, 203
Summa Perfectionis, 147–8
Summa Philosophiæ, 121, 137
Summa Theologica, 102
Sun, 39, 49, 90, 103, 104; mock suns, 111, 118
Sundial, 108
Surgery, 41–2, 176–80, 184, 188, 230, 238–40
Sussex, 219, 220, 221
Sweden, 201, 221
Sydenham, Thomas, 242
Sylvaticus, Matthæus, 155, 166
Symbolism, 44
Symbols, 35–6, 68, 140
Synesius, 140
Synod of Whitby, 40
Syracuse, 53
Syria, 52

287

Index

Taccola, Jacopo Mariano, ships, 213
Talisman, 67
Tanning, 227
Tartaglia, 130, 185
Tears, cause of, 29
Teasdale, John (Dimsdale), 170
Technology, 80; in ancient civilizations, 24; development of, 182; fields of, 80; Greek, 194–6; Latin writers on, 184–6; military, 185; historic, 194–5
Teleology, 150
Tempier, Étienne, 78
Temporibus, De, 40
Temporum Ratione, De, 41
Textile processes, 206–8; raw materials for, 200
Thabit ibn Qurra, 99
Thales, 131
Thames River, 213
Theatre, 186, 217
Themon Judæi, 124
Theodoric, optics, 122–4. *See also* Dietrich of Freiberg
Theodoric the Great, 196
Theophilus the Priest (7th century), 178, 184, 230; bells, 222; glass, 226; paints, 228
Theophilus (of Byzantium, 7th century), 170, 224
Theophrastus (c. 372–288 B.C.), 36, 134, 134n., 150, 157; botany, 158, 161; cosmogony, 159; mutation, 159
Theory of the Magic Art, The, 68
Therapeutics, 38, 229
Thierry d'Hireçon (d. 1328), 198
Thierry of Chartres (d.c. 1150), 45–6, 47–8, 189
Thomas of Cantimpré (c. 1228–44), 151, 156, 160, 164, 168
Thomas of Sarepta, 166
Thorndike, L., 133n., 155
Three-field system, 197–8
Three spirits, 171–2
Thrust, 210–12
Thunder, 111
Thunderbolts, 111
Tidal tables, 41
Tides, 41, 137
Timæus, 33–4, 37, 45–6, 48n., 49, 50, 140
Timber, 221
Time: calculation of, 106; continuum, 84; measurement, 190, 218–19

Tin, 221
Toledan Tables (Canon Azarchelis), 64
Toledo, 50, 52, 68, 104, 107, 220; meridian of, 64
Toledo steel, 220
Toledo Tables, 109
Tools, agricultural, 197–8
Torrigiano, Pietro, 179
Toulouse University, 76
Trade, 51–2
Trajectory, 128–9; oblique, 127
Transformation, 92, 141–2
Translations, 51–3, 54, 64, 184; collections of, 80
Transmutations, 68, 144, 147
Transport, 213–14
Transylvania, 225
Treadle hammer, 205
Treatise on the Astrolabe, 106
Treatise on the Epidemic Sickness, 234
Treatments: mental disorders, 242; wounds, 238
Treatyse of Fysshynge with an Angle, 169
Trebuchets, 223
Trentina, 235
Trigonometry, 64, 108–9
Trip hammers, 205
Trutfetter of Eisenach, Jodocus, 124
Tunis, 214
Tuscany, 225
Two-field system, 197–8
Tyler, Wat, 206
Type, 209
Tyrol, 225
Tzetzes, John (12th century), 25

Universals, 44, 48n., 82
Universe, 37, 38, 41, 47, 49, 72, 88, 89–90, 140; age, 31; Arabic system, 72; Aristotelian system, 71; atomic, 47n.; centre of, 90–91; duration, 87–8; in 8th century, 38–9; elements of, 46; essence of, 33; movement in, 49; Neoplatonic, 34; Platonic system, 34, 48–50; size, 31; and void, 49. *See also* Cosmos; Macrocosm
Universities: curriculum, 188–90; of Erfurt, 124; medicine in, 228–9; of Perugia, 233; surgery, 240; of Toulouse, 76
Utrecht Psalter, 203

288

Index

Alistair Cameron Crombie teaches at the University of Oxford, where he has been responsible for introducing the history of scientific and medical thought into normal courses and research in history, philosophy and the sciences. He is a Fellow of Trinity College. He is a graduate in natural science of the Universities of Melbourne and Cambridge, where he taught and carried out research in biology: this work was published in the *Proceedings of the Royal Society* and other scientific journals. He took up the history of science professionally in 1946 and taught the subject first at University College, London, before moving to Oxford in 1953. He has been a visiting professor at American, Australian, Japanese and Indian Universities and has lectured in many European countries. He has been President of the British Society for the History of Science and of the International Academy of the History of Science; and he is a Fellow of the Royal Historical Society and a member of the British Society for the Philosophy of Science, the International Academy of the History of Medicine and the Academia Leopoldina. In 1969 he was awarded the Galileo Prize for an original work on Galileo in relation to contemporary intellectual culture. His main historical work has been an extensive comparative study of methods and styles of inquiry and explanation, especially in the biomedical sciences, within the intellectual, social and technical history of medieval and early modern Europe. His publications, besides *Augustine to Galileo* which is now available in several languages, include *Robert Grosseteste and the Origins of Experimental Science 1100–1700*, *Scientific Change* and *The Mechanistic Hypothesis and the Scientific Study of Vision*, as well as numerous articles and chapters. He was the original editor of the *British Journal for the Philosophy of Science* and joint founding editor of the annual review *History of Science*. He is married with four children. Among his interests are literature, travel and landscape gardening.

AUGUSTINE TO GALILEO

AUGUSTINE
TO
GALILEO

VOLUME II

SCIENCE IN THE LATER MIDDLE AGES
AND EARLY MODERN TIMES

13TH TO 17TH CENTURIES

CONTENTS

7

Contents

8

Contents

PLATES

1 Nicole Oresme with an armillary sphere. From *Le Livre du ciel et du monde*, Bibliothèque Nationale, Paris, MS français 565 (14th century).

2 The earliest known graph; showing the changes in latitude (vertical divisions) of the planets relative to longitude (horizontal divisions). From MS Munich 14436 (11th century).

3 A page from Descartes, *La Géométrie* (1637), in which he discusses the algebraic equation of a parabola.

4 The mathematical disciplines and philosophy. From N. Tartaglia, *Nova Scientia*, Venice, 1537.

5 Diagram of vortices. From Descartes, *Principia Philosophiæ*, Amsterdam, 1644.

6 The Copernican system. From Copernicus, *De Revolutionibus Orbium Cœlestium*, Nuremburg, 1543.

7 Kepler's demonstration of the elliptical orbit of Mars. From *Astronomia Nova*, Prague, 1609.

8 Page from Thomas Harriot's papers at Petworth House, describing his observations on Jupiter's satellites made at Syon House, on the Thames near Isleworth, and from the roof of a house in London.

9 Telescope and other instruments in use; and an apparatus for showing sun-spots by projection on to a screen. From C. Scheiner, *Rosa Ursina*, Bracciani, 1630.

10 The earth as a magnet, and magnetic dip. From Gilbert, *De Magnete*, London, 1600.

11 The heart and its valves. From Vesalius, *De Humani Corporis Fabrica*, Basel, 1543.

12 Leonardo's drawing of the heart and associated blood vessels. From *Quaderni d'Anatomia*,iv,Royal Library, Windsor, MS; by gracious permission of H.M. the Queen.

13 Harvey's experiments showing the swelling of nodes in veins at the valves. From *De Motu Cordis*, London, 1639 (1st ed. 1628).

14 The *sensus communis* and the localized functions of the brain. From G. Reisch, *Margarita Philosophica*, Heidelberg, 1504.

15 Descartes' theory of perception showing the transmission of the nervous impulse from the eye to the pineal gland and thence to the muscles. From *De Homine*, Amsterdam, 1677 (1st ed.Leiden, 1662).

16 A cross-staff in use for surveying. From Petrus Apianus, *Cosmographia*, Antwerp, 1539.

17 A water-driven suction pump in use at a mine. From Agricola, *De Re Metallica*, Basel, 1561 (1st ed. 1556).

18 Diagram from Descartes, *Principia Philosophiæ* (1644), illustrating his explanation of magnetism.

19 Botanist drawing plants. From Fuchs, *De Historia Stirpium*, Basel, 1542.

20 Leonardo's drawing of the head and eye in section. From *Quaderni d'Anatomia*,v, Royal Library, Windsor MS; by gracious permission of H.M. the Queen.

21 A dissection of the muscles. From Vesalius, *De Humani Corporis Fabrica* (1543).

22 Diagrams illustrating the comparison between the skeletons of a man and a bird, from Belon, *Histoire de la nature des oyseaux*, Paris, 1555.

23a Embryology of the chick. From Fabrizio, *De Formatione Ovi et Pulli*, Padua, 1621.

23b Embryology of the chick, showing the use of the microscope. From Malpighi, *De Formatione Pulli in Ovo* (first published 1673), in *Opera Omnia*, London, 1686.

24 The comparative anatomy of the ear ossicles from Casserio, *De Vocis Auditisque Organis*, Ferrara, 1601.

ACKNOWLEDGEMENTS

VOLUME II

ACKNOWLEDGEMENTS are made to the following for supplying photographs for illustrations: the Librarian of Cambridge University (Fig. 5 and Plates 4, 9, 11, 12, 13, 14, 16, 17, 20, 21, 23A and B); the Director of the British Museum, London, (Fig. 3); Bodley's Librarian, Oxford, (Fig. 4 and Plates 2, 3, 5, 6, 7, 8, 10, 15, 18, 22); the Director of the Bibliothèque Nationale, Paris, (Plate 1). The following lent blocks for illustrations: Messrs William Heinemann, Ltd. (Plate 19); Messrs Macmillan & Co., Ltd (Plate 24). Plates 12 and 20 are reproduced by gracious permission of H.M. the Queen.

AUGUSTINE TO GALILEO

VOLUME II

I

SCIENTIFIC METHOD
AND DEVELOPMENTS
IN PHYSICS IN THE LATER
MIDDLE AGES

I. THE SCIENTIFIC METHOD OF
THE LATER SCHOLASTICS

THE activity of mind and hand that showed itself in the additions of scientific fact and in the development of technology made in the 13th and 14th centuries is to be seen also in the purely theoretical criticism of Aristotle's theory of science and fundamental principles that took place at the same time. This criticism was to lead later to the overthrow of the whole system of Aristotelian physics. Much of it developed from within Aristotle's scientific thought itself. Indeed Aristotle can be seen as a sort of tragic hero striding through medieval science. From Grosseteste to Galileo he occupied the centre of the stage, seducing men's minds by the magical promise of his concepts, exciting their passions and dividing their allegiances. In the end he forced them to turn against him as the real consequences of his undertaking gradually became clear; and yet, from the depths of his own system, he provided many of the weapons with which he was attacked.

The most important of these weapons were made by new ideas on scientific method, especially by new ideas on induction and experiment and on the role of mathematics in explaining physical phenomena. These gradually led to an entirely different conception of the kind of question that should be asked in natural science, the kind of question, in fact, to which the experimental and mathematical methods could give an answer. The field in which the new kind of question was to produce its greatest effects from the middle of the 16th century was in dynamics, and it was precisely Aristotle's ideas on space and motion that came in for the most radical criticism during the later Middle Ages. The effect of this scholastic criticism was to undermine the foundations of his whole system of physics (with the exception of biology) and so to clear the way for the new system constructed by the experimental and mathematical methods. At the end of the medieval period a fresh impetus was given to mathematics and mathematical physics by the translation into Latin and printing of some previously unknown or little known Greek texts.

It must always be remembered when reading medieval scientific writings that these were composed, just as a modern scientific paper is composed, within the context of an accepted manner of discussion and of a given nexus of problems. The academic context of discussions of logic and method and of mathematics and natural science was primarily the arts course, and further scope in certain branches of science was provided for those who went on to study medicine. The normal manner of discussion was in the form of the commentary, which by the fourteenth century had developed into the method of proposing and discussing specific problems or *quæstiones* (see Vol. I, pp. 33, 156, 188–90, 229). A modern reader may be puzzled by a commentary or treatise that takes up the discussion of a problem in the middle and assumes not only a knowledge of the background but also the appropriateness of the manner and methods of proposing a solution. Certainly medieval scientific writings are not always self-explanatory or easy to read. Many of them almost seem to be specially designed to mislead the 20th-century reader. We will be certainly misled if we fail to realize that the commentary was not simply an exposition of the text of Aristotle or some other 'authority' but that it, and even more the *quæstiones*, were the manner of offering criticisms and proposing original results and solutions. And we will be equally misled if we translate the more modern-sounding of those original solutions into 20th-century terms, and overlook the context of assumptions and conceptions in which they were proposed and the actual questions to which they were offered as answers. The fact that so many questions in medieval (and ancient) science overlap with similar questions in the context of modern science may be the greatest obstacle to historical understanding.

The great idea recovered during the 12th century, which made possible the immediate expansion of science from that time, was the idea of rational explanation as in formal or geometrical demonstration; that is, the idea that a particular fact was explained when it could be deduced from a more general principle. This had come through the gradual recovery of Aristotle's logic and of Greek and Arabic mathematics. The idea of mathematical demonstration had, in fact, been the great discovery of the Greeks in the history of science, and it was the basis not only of their considerable contributions to mathematics itself and to physical sciences like astronomy and geometrical optics, but also of much of their biology and medicine. Their bent of mind was to conceive of science, where possible, as a matter of deductions from indemonstrable first principles.

In the 12th century, this notion of rational explanation developed first among logicians and philosophers not primarily concerned with natural science at all but engaged in grasping and expounding the principles, first, of the *logica vetus* or 'old logic' based on Boethius and, later in the century, of Aristotle's *Posterior Analytics* and various works of Galen. What these logicians did was to make use of the distinction, ultimately deriving from Aristotle, between experiential knowledge of a fact and rational knowledge of the reason for, or cause of, the fact; they meant by the latter knowledge of some prior and more general principle from which they could deduce the fact. The development of this form of rationalism was, in fact, part of a general intellectual movement in the 12th century, and not only scientific writers such as Adelard of Bath and Hugh of St Victor, but also theologians such as Anselm, Richard of St Victor and Abelard tried to arrange their subject-matter according to this mathematical-deductive method. Mathematics was for these 12th-century philosophers the model rational science and, like good disciples of St Augustine and Plato, they held that the senses were deceitful and reason alone could give truth.

Though mathematics was regarded in the 12th century as the model science, it was not until the beginning of the 13th century that Western mathematics became worthy of this reputation. The practical mathematics kept alive in Benedictine monasteries during the early Middle Ages, and taught in the cathedral and monastery schools founded by Charlemagne at the end of the 8th century, was very elementary and limited to what was necessary to keep accounts, calculate the date of Easter, and measure land for the purposes of surveying. At the end of the 10th century Gerbert had initiated a revival of interest in mathematics, as he did also in logic, by collecting Boethius's treatises on those subjects. Although Boethius's treatise on arithmetic contained an elementary idea of the treatment of theoretical problems based on the properties of numbers, the so-called 'Geometry of Boethius' was, in fact, a later compilation from which most of his own contributions had dropped out. It contained certain of Euclid's axioms, definitions and conclusions but consisted mainly of a description of the abacus, the device generally used for calculations, and of practical surveying methods and the like. The writings of Cassiodorus and Isidore of Seville, the other sources of the mathematical knowledge of the time, contained nothing fresh (Vol. I, pp. 31–3).

Gerbert himself wrote a treatise on the abacus and even improved the current type by introducing apices, and a few other additions were made to practical mathematics during the 11th and 12th centuries, but until the end of the 12th century Western mathematics remained almost entirely a practical science Eleventh- and 12th-century mathematicians were able to use the conclusions of the Greek geometers for practical purposes, but were unable to demonstrate those conclusions, even though the theorems of the first book of Euclid's *Elements* became known during the 11th century and the whole of that work was translated by Adelard of Bath early in the 12th. Examples of 11th-century geometry are Francon of Liège's attempt to square the circle by cutting up pieces of parchment, and the correspondence between Raimbaud of Cologne and Radolf of Liège in which each vainly tried to outdo the other in an unsuccessful attempt to demonstrate that the sum of the angles of a triangle equals two right angles. Little better work was done till the end of the 12th century.

In arithmetic, the situation was somewhat better owing to the preservation of Boethius' treatise on the subject. For instance, Francon himself was able to show that it was impossible to express rationally the square root of a number not a perfect square. The marked improvements that took place in Western mathematics early in the 13th century occurred first in the fields of arithmetic and algebra, and this was due largely to the development of this earlier tradition by two scholars of originality. The first was Leonardo Fibonacci of Pisa, who had given the earliest complete Latin account of the Arabic, or Hindu, system of numerals in his *Liber Abaci* in 1202 (see Vol. I, p. 66). In later works he made some highly original contributions to theoretical algebra and geometry, his basic knowledge being derived primarily from Arabic sources, but also from Euclid, Archimedes, Hero of Alexandria and the 3rd-century A.D. Diophantus, the greatest of the Greek algebraists. Fibonacci on some occasions replaced numbers by letters in order to generalize his proof. He developed indeterminate analysis and the sequence of numbers such that each is equal to the sum of the two preceding (now called 'Fibonacci sequences'), gave an interpretation of a negative solution as a debt, used algebra to solve geometrical problems (a striking innovation), and gave solutions of various problems involving quartic equations.

The second mathematician of originality in the 13th century was Jordanus Nemorarius, who shows no trace of Arabic influence but developed the Greco-Roman arithmetical tradition of Nicomachus and Boethius, in particular the theory of numbers. Jordanus habitually made use of letters for the sake of generality in arithmetical problems and he developed certain algebraic problems leading to linear and quadratic equations. He was also an original geometer. His treatises contained discussions of old problems, such as the determination of the centre of gravity of a triangle, and also the first general demonstration of the fundamental property of stereographic projection, that circles are projected as circles (cf. Vol. I, pp. 126–31).

After Jordanus there was a gradual improvement in Western geometry as well as in other parts of mathematics. A number of important original ideas were added. In an edition of Euclid's *Elements*, which he produced in about 1254 and which remained a standard text-book until the 16th century, Campanus of Novara included a study of 'continuous quantities', to which he was led by considering the angle of contingence between a curve and its tangent smaller than any angle between two straight lines. By using a mathematical induction ending in a *reductio ad absurdum*, he also proved the irrationality of the 'golden section' or 'golden number', that is the division of a straight line so that the proportion of the smaller section to the larger equals that of the larger to the whole. He also calculated the sum of the angles of a stellated pentagon. In the 14th century the grasp of the principle of geometrical proof made possible the improvements introduced into trigonometry by John Maudith, Richard of Wallingford and Levi ben Gerson (see Vol. I, pp. 109–10), and into the theory of proportions by Thomas Bradwardine and his followers in Merton College, Oxford, and by Albert of Saxony and others in Paris and Vienna. This work on proportions, like the striking work of Nicole Oresme on the use of coordinates and the use of graphs to represent the form of a function, was developed chiefly in connexion with certain physical problems; it will be considered on a later page. Of considerable importance also were the improvements introduced into the methods of calculation in the Hindu system of numerals during the 13th and 14th centuries. The methods of multiplying and dividing used by the Hindus and Moslems had been very uncertain. The modern method of multiplication was introduced from Florence, and the modern technique of division was also invented during the later Middle Ages. This made division into an ordinary matter for the counting house, whereas it had formerly been a formidably difficult operation even for skilled mathematicians. The Italians also invented the system of double-entry book-keeping, and the commercial nature of their interests is shown by their arithmetic books in which problems are concerned with such practical questions as partnership, exchange, simple and compound interest, and discount.

The recovery of the idea of a demonstrative science in which a fact was explained when it could be deduced from a prior and more general principle, and the great improvements in mathematical technique that took place in Western Christendom during the 13th century, were the chief intellectual achievements that made 13th-century science possible at all. But the medieval natural philosophers did not stop there in their thinking on scientific method. The new knowledge in fact raised important methodological problems, as general problems of scientific thinking. Specially important were the problems how, in natural science, to arrive at the prior principles or general theory from which the demonstration or explanation of particular facts was to proceed; and how, among several possible theories, to distinguish between the false and the true, the defective and the complete, the unacceptable and the acceptable. In their study of these problems the medieval philosophers investigated the logical relationship between facts and theories, or data and explanations, the processes of the acquisition of scientific knowledge, the use of inductive and experimental analysis to break down a complex phenomenon into its component elements, the character of the verification and falsification of hypotheses, and the nature of causation. They began to form the conception of natural science as in principle inductive and experimental as well as mathematical, and they began to develop the logical procedures of experimental inquiry, which chiefly characterize the difference between modern and ancient science.

In classical antiquity several quite different conceptions of scientific method had been formed within the general scheme of demonstrative science. The postulational method expounded by Euclid became most effective in application to the highly abstract subjects of pure mathematics and of mathematical astronomy, statics and optics. At its purest it was not experimental: long chains of deduction followed from premisses accepted as self-evident. For example, most of the problems investigated by Archimedes, its greatest Greek exponent, required, even in mathematical physics, no actual experiments: in formulating the law of balance and lever Archimedes appealed not to experiment but to symmetry. But in more complicated subjects, especially in astronomy, the postulated hypotheses had to be tested by checking quantitative conclusions deduced from them against observation.

Related to this form of argument was the dialectical method of Plato, in which the argument was conducted by provisionally accepting a proposition and then proceeding to show either that it led to a self-contradiction or a contradiction with something accepted as true, or that it did not. This gave grounds for rejecting or accepting it. The mathematical equivalent of this form of argument is the *reductio ad absurdum* widely used by Greek mathematicians.

When attempting to deal, not simply with abstract mathematical subjects, but with the more difficult problems of matter (living and dead), many Greek physicists again adopted a form of the postulational method, proposing theoretical, unobservable, particles out of which to construct a theoretical world to match the world that is observed. The outstanding example of this is Democritus' theory of atoms and the void; another is the physics of Plato's *Timæus* (see Vol. I, pp. 48–50, and below, pp. 50–51).

In contrast with this abstractly theoretical approach stands the strongly empirical method of Aristotle. Instead of explicitly postulating unobservable entities to explain the observed world, his basic procedure was to analyse observable things immediately into their parts and principles and then to reconstruct the world rationally from the discovered constituents (see Vol. I, pp. 82–3). This method involved no long chains of deduction such as are found in Euclid, but kept its conclusions as close as possible to things as they were observed.

The history of Greek thought about scientific method can be dramatized by seeing it as an attempt by the mathematicians to impose a clearly postulational scheme, which provoked the resistance of those, especially in medicine, with greater experience of the enigmas of matter. The drama can be followed within the Hippocratic medical writings themselves and continued among the physicists and physiologists of Alexandria. It gave rise at one extreme to an excessive dogmatism about the possibility of discovering causes, and at the other to the sceptical views of the Sophists and the Empirical school of medicine. It continued in the Middle Ages, with the added complication that the translations available did not always allow the actual views of classical writers to be clearly appreciated or respected. Grosseteste notoriously interpreted Aristotle in a Platonic sense and introduced into his logic postulational examples taken from Euclid.

Among ancient Greek writers known in the early 13th century, only Aristotle and certain medical writers, especially Galen, had seriously discussed the inductive and experimental side of science; Aristotle himself was, of course, a doctor. Certain of Aristotle's followers in the Lyceum and in Alexandria, in particular Theophrastus and Strato, had had a very clear understanding of some of the general principles of the experimental method, and experiment seems to have been practised fairly generally by members of the medical school at Alexandria. But the writings of these authors were almost unknown in the Middle Ages. Even in their own time their methods did not have the transforming effect on Greek science which the methods begun in the Middle Ages were to have in the modern world.

Among the Arabs, experiments had been carried out by a number of scientific writers: for example by Alkindi and Alhazen, al-Shirazi and al-Farisi in optics, and by Rhazes, Avicenna and others in chemistry; and certain Arab medical men, especially Ali ibn Ridwan and Avicenna, had made contributions to the theory of induction. But for one reason or another Arabic science failed to become thoroughly experimental in outlook, though it was certainly the example of Arab work that stimulated some of the experiments made by Christian writers, for instance Roger Bacon and Theodoric of Freiberg and possibly Petrus Peregrinus, discussed on earlier pages.

Before the Greek conception of science had been fully

recovered, some Western scholars in the 12th century had shown both that they were aware of the need for proofs in mathematics, even though they could not give them, and that they held at least in principle that nature must be investigated by observation. The saying, *nihil est in intellectu quod non prius fuerit in sensu*, became a commonplace, and such a natural philosopher as Adelard of Bath described simple experiments and may actually have carried out some of them. At the same time scholars gave an increasing value to the practical applications of science and to the accuracy and manual skill found in the practical arts (see Vol. I, pp. 183 *et seq.*). By the 13th century the knowledge of the Greek conceptions of theoretical explanation and mathematical proof, gained from the translations of classical and Arabic works, had put philosophers in a position to convert the naïve theoretical empiricism of their predecessors into a conception of science that was both experimental and demonstrative. Characteristically, in receiving ancient and Arabic science into the Western world, they made an attempt not only to master its technical content but also to understand and to prescribe its methods, and so found themselves embarking on a new scientific enterprise of their own.

It must not be supposed that this philosophical conception of experimental science, developed largely in commentaries on Aristotle's *Posterior Analytics* and the problems found in it, was accompanied by a single-minded reliance on the experimental method such as is found in the 17th century. Medieval science remained in general within the framework of Aristotle's theory of nature, and deductions from that theory were by no means always rejected even when contradicted by the results of the new mathematical, logical, and experimental procedures. Even in the midst of otherwise excellent work, medieval scientists sometimes showed a strange indifference to precise measurements, and could be guilty of misstatements of fact, often based on purely imaginary experiments copied from earlier writers, which the simplest observation would have corrected. Nor must it be supposed that when the new experimental and mathematical methods were applied to scientific problems, this was always the *result* of the theoretical discussions of method. In fact the examples of scientific investigations undertaken in application of a conscious conception of method were often of little scientific interest, whereas some of the most interesting scientific treatises, especially those written in the 13th century, for example those of Jordanus on statics, of Gerard of Brussels on kinematics, of Petrus Peregrinus on magnetism, contain little or no discussion of problems of method. This does not mean that their authors were necessarily uninfluenced by discussions of method; certainly the work of Gerard of Brussels illustrates the influence, not of the ideas of the philosophers, but of the model of Archimedes, the greatest of the Greek mathematical physicists, the role of whose writings in the development of scientific thought in the Middle Ages is still under historical investigation.* In the 14th century the influence of philosophical discussions of method on the inquiry into problems is both evident and important. But the examples given do show that in the Middle Ages, as in other periods, discussions of method and actual scientific investigations belonged to two separate streams, even though their waters were so often and so profoundly mingled, as certainly they were in the entire period covered by the pages that follow.

Among the first to understand and use the new theory of experimental science was Robert Grosseteste, who was the real founder of the tradition of scientific thought in medieval Oxford and, in some ways, of the modern English intellectual tradition. Grosseteste united in his own work the experimental and the rational traditions of the 12th century and he set forth a systematic theory of experimental science. He seems to have studied medicine as well as mathematics and philosophy, so he was well equipped. He based his theory of science in the first place on Aristotle's distinction between knowledge of a fact (*demonstratio quia*) and knowledge of the reason for the fact (*demonstratio propter quid*). His theory had three essentially different aspects which, in fact, characterize all the discussions of methodology

* See Marshall Clagett, *Archimedes in the Middle Ages*, i–, Madison, Wisc., 1964–.

down to the 17th century and indeed down to the present day: the inductive, the experimental, and the mathematical.

The problem of induction, Grosseteste held, was to discover the cause from knowledge of the effect. Knowledge of particular physical facts, he said, following Aristotle, was had through the senses, and what the senses perceived were composite objects. Induction involved the breaking up of these objects into the principles or elements that produced them or caused their behaviour, and he conceived of induction as an upward process of abstraction going from what Aristotle had said was 'more knowable to us', that is, the composite objects perceived through the senses, to abstract principles prior in the order of nature but at first less knowable to us. We must proceed inductively from effects to causes before we can proceed deductively from cause to effect. What had to be done in trying to explain a particular set of observed facts was, therefore, to arrive at a statement or definition of the principle or 'substantial form' that caused them. As Grosseteste wrote in his commentary on Aristotle's *Physics*:

Since we search for knowledge and understanding by means of principles, in order that we may know and understand natural things we must first determine the principles pertaining to all things. Now the natural way for us to reach knowledge of principles is to start from universal applications and go to these principles, to start from wholes corresponding to these very principles. ... Then as, speaking generally, the procedure for acquiring knowledge is to go from universal compound wholes to more determined species, so, from complete wholes which we know confusedly ... we can go back to those very parts by means of which it is possible to define the whole and, from this definition, to reach a determinate knowledge of the whole. ... Every agent has that which is to be produced, in some way already described and formed within it, and so the 'nature' as an agent has the natural things which are to be produced in some way described and formed within itself. This description and form (*descriptio et formatio*), existing in the nature itself of the things to be produced, before they are produced, is therefore called knowledge of this nature.*

* See A.C.Crombie, *Robert Grosseteste and the Origins of Experimental Science*, 1100–1700, 3rd impression, revised, Oxford, 1971, p. 55.

The Scientific Method of the Later Scholastics

All discussions of scientific method must presuppose a philosophy of nature, a conception of the kinds of causes and principles the method will discover. In spite of the Platonic influence shown in the fundamental significance he gave to mathematics in the study of physics, the framework of Grosseteste's philosophy of nature was essentially Aristotelian. He saw the definition of the principles explaining a phenomenon, in effect a definition of the conditions necessary and sufficient to produce it, entirely within the categories of the four Aristotelian causes. As he wrote in *De Natura Causarum* (published by L. Baur in his edition of Grosseteste's philosophical works in *Beiträge zur Geschichte der Philosophie des Mittelalters*, Münster, 1912, vol. 9, p. 121):

Thus we have four genera of causes and from these, when they exist, there must be a caused thing in its complete being. For a caused thing cannot follow upon the being of any other cause except those four, and that alone is a cause from whose being something else follows. Therefore there is no other cause beyond these, and so there is in these genera a number of causes that is sufficient.

To arrive at such a definition Grosseteste described, first, a dual process which he called 'resolution and composition'. These names came from the Greek geometers and from Galen and other later classical writers, and were of course simply the Latin translations of the Greek words for 'analysis and synthesis'.* The central principle of his method Grosseteste derived, in fact, from Aristotle, but he developed it more fully than Aristotle had done. The method followed a definite order. By the first procedure,

* For the history of these terms, and of the 'resolutive-compositive' method, see Crombie, *Robert Grosseteste and the Origins of Experimental Science 1100–1700*, especially pp. 27–29, 52–90, 193–94, 297–318. For the method in Plato's dialectic, e.g. in *Republic*, book 6, see L. Brunschvicg, *Les Étapes de la philosophie mathématique*, 3e éd., Paris, 1947, pp. 49 *et seq.* Other important Greek discussions of the method are by Galen, *Techne* or *Ars Medica*, ed. C. G. Kühn (*Medicorum Graecorum Opera*), Leipzig, 1821, vol. 1; and by Pappus of Alexandria, *Collectio Mathematica*, 7. 1–3, English trans. by T. L. Heath, *History of Greek Mathematics*, Cambridge, 1921, vol. ii, pp. 400–1. Cf. also Hippocrates, *Techne* (*The Art*), English trans. by W. H. S. Jones (Loeb Classical Library), London and Cambridge, Mass., 1923; and Archimedes, *Method*, English trans. by T. L. Heath, Cambridge, 1912.

resolution, he showed how to sort out and classify, by likeness and difference, the component principles or elements constituting a phenomenon. This gave him what he called the nominal definition. He began by collecting instances of the phenomenon under examination and noting the attributes all had in common, till he arrived at the 'common formula' which stated the empirical connexion observed, a causal connexion being suspected when attributes were found frequently associated together. Then, by the opposite process of composition, by rearranging the propositions so that the more particular were seen to follow deductively from the more general, he showed that the relation of general to particular was one of cause and effect. That is, he arranged the propositions in causal order. He illustrated his method by showing how to arrive at the common principle causing animals to have horns which, he said in his commentary on the *Posterior Analytics*, book 3, chapter 4, 'is due to the lack of teeth in the upper mandible in those animals to which nature does not give other means of preservation in place of horns,' as she does to the deer with its rapid flight and to the camel with its large body. In horned animals the earthy matter that would have gone to form the upper teeth went instead to form the horns. He added: 'Not having teeth in both jaws is also the cause of having several stomachs,' a correlation which he traced to the poor mastication of food by animals with only one row of teeth.

Besides this orderly process by which the causal principle was reached by resolution and composition, Grosseteste also envisaged the possibility, as Aristotle had, of a theory or principle explaining repeatedly observed facts being reached by a sudden leap of intuition or scientific imagination. In either case, the further problem then presented itself, namely, how to distinguish between false and true theories. This introduced the use of specially arranged experiments or, where it was not possible to interfere with natural conditions, for example in the study of comets or heavenly bodies, the making of observations that would give the answer to specific questions.

Grosseteste held that it was never possible in natural science to arrive at a complete definition or an absolutely certain knowledge of the cause or form from which effects followed, as it was, for example, with the abstract subjects of geometry like triangles. A triangle could be completely defined by certain of its attributes, for instance by defining it as a figure bounded by three straight lines; from this definition all its other properties could be analytically deduced, so that cause and effect were reciprocal. This was not possible with material subjects because the same effect might follow from more than one cause and it was never possible to know all the possibilities. 'Can the cause be reached from knowledge of the effect in the same way as the effect can be shown to follow from its cause?' he wrote in book 2, chapter 5 of his commentary on the *Posterior Analytics*. 'Can one effect have many causes? For, if one determinate cause cannot be reached from the effect, since there is no effect that has not some cause, it follows that an effect, just as it has one cause, so it may have another, and so there may be several causes of it.' Grosseteste's point seems to be that there may be an ostensible plurality of causes, which our available methods and knowledge may not enable us to reduce to one actual cause in which the effect is univocally prefigured. In natural science, as he wrote in book 1, chapter 11, owing to the remoteness of causes from immediate observation and to the mutability of natural things, there is thus 'minor certitudo'. Natural science offered its explanations 'probably rather than scientifically ... Only in mathematics is there science and demonstration in the strictest sense.' It was precisely because they were in the nature of things hidden from our direct inspection that a scientific method was necessary to bring these causes 'more knowable in nature but not to us' as certainly as possible to light. By making deductions from the various theories advanced and by eliminating theories whose consequences were contradicted by experience, it was possible, Grosseteste held, to approach closer to a true knowledge of the causal principles or forms really responsible for events in the world of our observation.

As he said in his commentary on the *Posterior Analytics*, book 1, chapter 14:

This therefore is the way by which the abstracted universal is reached from singulars through the help of the senses ... For when the senses several times observe two singular occurrences, of which one is the cause of the other or is related to it in some other way, and they do not see the connexion between them, as, for example, when someone frequently notices that the eating of scammony happens to be accompanied by the discharge of red bile, and does not see that it is the scammony that attracts and withdraws the red bile; then, from the constant observation of these two observable things, he begins to form a third unobservable thing, namely, that scammony is the *cause* that withdraws red bile. And from this perception repeated again and again and stored in the memory, and from the sensory knowledge from which the perception is built up, the functioning of the reasoning begins. The functioning reason therefore begins to wonder and to consider whether things really are as the sensible recollection says, and these two lead the reason to the experiment, namely, that he should administer scammony after all other causes purging red bile have been isolated and excluded. When he has administered scammony many times with the sure exclusion of all other things that withdraw red bile, then there is formed in the reason this universal, namely, that all scammony of its nature withdraws red bile, and this is the way in which it comes from sensation to a universal experimental principle.

His method of elimination or falsification Grosseteste based on two assumptions about the nature of reality. The first was the principle of the uniformity of nature, meaning that forms are always uniform in the effect they produce. 'Things of the same nature are productive of the same operations according to their nature,' he said in his tract *De Generatione Stellarum* (published by Baur in his edition of Grosseteste's philosophical works), Aristotle had stated the same principle. Grosseteste's second assumption was the principle of economy, which he generalized from various statements of Aristotle. This principle Grosseteste used both as describing an objective characteristic of nature and as a pragmatic principle. 'Nature operates in the shortest way possible,' he said in his *De Lineis, Angulis et Figuris*, and he used this as an argument to support the law of reflection of light and his own 'law' of refraction. He said also, in his commentary on the *Posterior Analytics*, book I, chapter 17:

that demonstration is better, other circumstances being equal, which necessitates the answering of a smaller number of questions for a perfect demonstration, or requires a smaller number of suppositions and premises from which the demonstration proceeds ... because it gives us knowledge more quickly.

In the same chapter and elsewhere Grosseteste spoke explicitly of applying the method of *reductio ad absurdum* to the investigation of nature. His method of falsification is an application of this method in an empirical situation. He used it explicitly in several of his scientific opuscula where it was appropriate, for instance in his studies on the nature of the stars, on comets, on the sphere, on heat, and on the rainbow. A good example is in the tract *De Cometis*, in which he considered in turn four different theories put forward by earlier writers to account for the appearance of comets. The first was that put forward by observers who thought that comets were produced by the reflection of the sun's rays falling on a heavenly body. The hypothesis was falsified, he said, by two considerations: first, in terms of another physical theory, because the reflected rays would not be visible unless they were associated with a transparent medium of a terrestrial and not a celestial nature; and secondly, because it was observed that

the tail of the comet is not always extended in the opposite direction to the sun, whereas all reflected rays would go in the opposite direction to the incident rays at equal angles.*

He considered the other hypotheses in the same way in terms of 'reason and experience', rejecting those contrary either to what he regarded as an established theory verified by experience or to the facts of experience (*ista opinio falsificatur*, as he said), till he came to his final definition which he held survived these tests, that 'a comet is sublimated fire assimilated to the nature of one of the seven planets'. This theory he then used to explain various

---

* In fact the tails of comets are repelled by the sun, though the angles would differ from those made by reflected light. Good examples of the same kind of empirical analysis are Aristotle's discussions of comets in the *Meteorology* (book 1, chapter 6) and his refutation of pangenesis in *De Generatione Animalium* (book 1, chapters 17, 18).

further phenomena, including the astrological influence of comets.

Of even greater interest is the method Grosseteste used in his attempt to explain the shape of the rainbow (see Vol. I, pp. 114–15), when he seized upon simpler phenomena which could be investigated experimentally, the reflection and refraction of light, and tried to deduce the appearance of the rainbow from the results of a study of these. Grosseteste's own work on the rainbow is somewhat elementary, but the experimental investigation of the subject which Theodoric of Freiberg undertook is truly remarkable both for its precision and for the conscious understanding he shows of the possibilities of the experimental method (see Vol. I, p. 122 *et seq.*). The same characteristics are to be found in the work of other experimental scientists who came after Grosseteste, for instance in that of Albertus Magnus, Roger Bacon, Petrus Peregrinus, Witelo and Themon Judæi, even though nearly all these writers could also be guilty of elementary mistakes. The influence of Grosseteste is especially noticeable on those who studied the rainbow. For example the initial inquiries of Roger Bacon and Witelo were aimed at discovering the conditions necessary and sufficient to produce this phenomenon. The 'resolutive' part of their investigations gave them a partial answer by defining the species to which the rainbow belonged and by distinguishing it from the species to which it did not belong. It belonged to a species of spectral colours produced by the differential refraction of sunlight passing through drops of water; as Bacon pointed out, this was different, for example, from the species including the colours seen in iridescent feathers. Moreover a further defining attribute of the rainbow was that it was produced by a large number of discontinuous drops. 'For,' as Themon wrote in his *Quæstiones super Quatuor Libros Meteorum*, book 3, question 14, 'where such drops are absent there no rainbow or part of it appears, although all the other requisite conditions are sufficient.' This, he said, could be tested by experiments with rainbows in artificial sprays. Roger Bacon had made such experiments. Postulating the requisite conditions – the sun at a definite position in relation to raindrops and to the observer – a rainbow would result.

Having defined these conditions, the purpose of the next stage of the inquiry was to discover how they would in fact produce a rainbow; that is, to construct a theory incorporating them in such a manner that a statement describing the phenomena could be deduced from it. The two essential problems were to explain, first, how the colours were formed by the raindrops, and secondly, how they were sent back to the observer in the shape and order seen. Particularly significant features of the whole inquiry were the use of model raindrops in the shape of spherical flasks of water, and the procedures of verification and falsification to which each theory was submitted, especially by the authors of rival theories. For example, the discovery of the differential refraction of the colours having pointed the way to the solution of the first problem, Witelo tried to solve the second by supposing that the sunlight was refracted right through one raindrop and the resulting colours then reflected back to the observer from the convex external surfaces of other drops behind. Theodoric of Freiberg showed that this theory would not give the observed effects, but that these would follow from the theory he based on his own discovery of the internal reflection of light within each drop. Thus by theory and experiment he solved the problem he set himself. For, as he wrote in the preface to *De Iride*, 'it is the function of optics to determine what the rainbow is, because, in doing so, it shows the reason for it, in so far as there is added to the description of the rainbow the manner in which this sort of concentration may be produced in the light going from any luminous heavenly body to a determined place in a cloud, and then by particular refractions and reflections of rays is directed from that determined place to the eye.'

Different from, though in many cases (as indeed in the case of Galileo himself) scarcely to be separated from the experimental method and the making of special observations to verify or falsify theories, was the use of mathematics in natural science. Grosseteste himself, because of his 'cosmology of light' (see Vol. I, pp. 88–9, 112 *et seq.*), said in his little work, *De Natura Locorum*, that from the 'rules and principles and fundamentals . . given by the power of geometry, the careful observer of natural things can

35

give the cause of all natural effects.' And elaborating this idea in his *De Lineis* he said :

The usefulness of considering lines, angles and figures is the greatest because it is impossible to understand natural philosophy without these ... For all causes of natural effects have to be expressed by means of lines, angles and figures, for otherwise it would be impossible to have knowledge of the reason for those effects.

Grosseteste, in fact, regarded the physical sciences as being subordinate to the mathematical sciences in the sense that mathematics could provide the reason for observed physical facts, though, at the same time, he maintained the Aristotelian distinction between the mathematical and the physical propositions in a given theory, and asserted the necessity of both for a complete explanation. Essentially the same attitude was taken by many leading scientists throughout the Middle Ages, and, indeed, in a different form by most writers in the 17th century. Mathematics could describe what happened, could correlate the concomitant variations in the observed events, but it could not say anything about the efficient and other causes *producing* the movement because it was, in fact, explicitly an abstraction from such causes (see Vol. I, pp. 88–9). This is the attitude seen in both optics and astronomy in the 13th century (see Vol. I, pp. 112, 93 *et seq.*).

As time went on, the retention of causal, 'physical' explanations, which usually meant explanations taken from Aristotle's qualitative physics, became more and more of an embarrassment. The great advantage of mathematical theories was just that they could be used to correlate concomitant variations in a series of observations made with measuring instruments so that the truth or falsehood of these theories, and the precise occasions where they failed, could easily be determined experimentally. It was just this consideration which brought about the triumph of Ptolemaic over the Aristotelian astronomy at the end of the 13th century (see Vol. I, p. 102). In contrast with this clearly understood role of mathematics in a scientific investigation, it was difficult to see what to do with a theory of 'physical' causes, however necessary they might seem to be theoretically for a complete explanation of the observed occurrences. Moreover, many aspects of Aristotle's physical philosophy were a positive hindrance to the use of mathematics. From the beginning of the 14th century attempts were made to circumvent these difficulties by devising new systems of physics, partly under the influence of a revived Neoplatonism and partly under the influence of the 'nominalism' revived by William of Ockham.

Improvements were made in the theory of induction by several writers after Grosseteste and the enormous and sustained interest taken in this purely theoretical and logical question is a good indication of the intellectual climate in which natural science was conducted before the middle of the 17th century. Perhaps it does something to explain why the brilliant beginnings of experimental science seen in the 13th and early 14th centuries did not at once go on to bring about what, in fact, only happened in the 17th century. For some four centuries from the beginning of the 13th century, the question guiding scientific inquiry was to discover the real, the enduring, the intelligible behind the changing world of sensible experience, whether this reality was something qualitative, as it was conceived of at the beginning of that period, or something mathematical, as Galileo and Kepler were to conceive of it at the end. Some aspects of this reality might be revealed by physics or natural science, others by mathematics, others again by metaphysics; yet though these different aspects were all aspects of a single reality, they could not all be investigated in the same way or known with the same certainty. For this reason it was essential to be clear as to the methods of investigation and explanation legitimate in each case, and what each could reveal of the underlying reality. In most scientific writings down to the time of Galileo a discussion of methodology is carried on *pari passu* with the account of a concrete investigation, and this was a necessary part of the endeavour of which modern science is the result. But from the beginning of the 14th century to the beginning of the 16th there was a tendency for the best minds to become increasingly interested in problems of pure logic divorced from experimental practice, just as in another field they became more interested in making purely theoretical, though also necessary, criticisms of Aristotle's physics without bothering to make observations (see below, p. 49 *et seq.*).

Perhaps the first writer after Grosseteste seriously to discuss the problem of induction was Albertus Magnus. He had a good grasp of the general principles as they were then understood, but of greater interest is the work done by Roger Bacon. In chapter 2 of the sixth part of his *Opus Majus*, 'On Experimental Science', Bacon said:

This experimental science has three great prerogatives with respect to the other sciences. The first is that it investigates by experiment the noble conclusions of all the sciences. For the other sciences know how to discover their principles by experiments, but their conclusions are reached by arguments based on the discovered principles. But if they must have particular and complete experience of their conclusions, then it is necessary that they have it by the aid of this noble science. It is true indeed, that mathematics has universal experiences concerning its conclusions in figuring and numbering, which are applied likewise to all sciences and to this experimental science, because no science can be known without mathematics. But if we turn our attention to the experiences which are particular and complete and certified wholly in their own discipline, it is necessary to go by way of the considerations of this science which is called experimental.

The first prerogative of Roger Bacon's experimental science was thus to confirm the conclusions of mathematical reasoning; the second was to add to deductive science knowledge at which it could not itself arrive, as, for instance, in alchemy; and the third was to discover departments of knowledge still unborn. His experimental science was, he admitted, as much a separate applied science, in which results of the natural and speculative sciences were put to the test of practical utility, as an inductive method. His attempt to discover the cause of the rainbow (see Vol. I, pp. 117–18), with which he illustrated the first prerogative of experimental science, shows that he had grasped the essential principles of induction by which the investigator passed from observed effects to the discovery of the cause and isolated the true cause by eliminating theories contradicted by facts.

With Roger Bacon the programme for mathematicizing physics and a shift in the object of scientific inquiry from the Aristotelian 'nature' or 'form', to laws of nature in a recognizably modern sense, becomes explicit (cf. below, p. 97 *et seq.*). Echoing Grosseteste, he wrote for example in the *Opus Majus*, part 4, distinction 4, chapter 8 : 'In the things of this world, as regards their efficient and generating causes, nothing can be known without the power of geometry.' The language he used in discussing the 'multiplication of species' seems to relate this general programme unequivocally to the inquiry for predictive laws. In *Un fragment inédit de l'Opus Tertium* edited by Duhem (p. 90), he wrote : 'That the laws (*leges*) of reflection and refraction are common to all natural actions I have shown in the treatise on geometry.' He claimed to have demonstrated the formation of the image in the eye 'by the law of refraction', remarking that the 'species of the thing seen' must so propagate itself in the eye 'that it does not transgress the laws which nature keeps in the bodies of the world'. Normally the 'species' of light were propagated in straight lines, but in the twisting nerves 'the power of the soul makes the species relinquish the common laws of nature (*leges communes naturæ*) and behave in a way that suits its operations' (Ibid., p. 78).

For some three hundred years from the middle of the 13th century a most interesting series of discussions of induction was made by members of the various medical schools, and in this the tendency towards pure logic becomes very marked. Galen himself had recognized the need for some method of discovering the causes which explained the observed effects, when he drew a distinction between the 'method of experience' and the 'rational method'. He referred to effects or symptoms as 'signs', and he said that the 'method of experience' was to go inductively from these signs to the causes which produced them, and that this method necessarily preceded the 'rational method' which demonstrated syllogistically * from causes to effects. Galen's

---

* The syllogism is a form of reasoning in which, from two given propositions, the premisses, with a common or middle term, is deduced a third proposition, the conclusion, in which the non-common terms are united. For example, from the major premiss, 'whatever has an opaque body inter-

ideas had been developed by Avicenna in his *Canon of Medicine* and this work contained an interesting discussion of the conditions which must be observed in inducing the properties of medicines from their effects. The subject was taken up in the 13th century by the Portuguese doctor Petrus Hispanus, who died in 1277 as Pope John XXI, in his *Commentaries on Isaac*, a work on diets and medicines. First, he said, the medicine administered should be free from all foreign substances. Secondly, the patient taking it should have the disease for which it was especially intended. Thirdly, it should be given alone without admixture of other medicine. Fourthly, it should be of the opposite degree to the disease.* Fifthly, the test should be made not once only but many times. Sixthly, the experiments should be with the proper body, on the body of a man and not of an ass. On the fifth of these conditions a contemporary, John of St Amand, repeated the warning that a medicine which had had a heating effect on five men would not necessarily always have a heating effect, for the men in question might all have been of a cold and temperate constitution, whereas a man of hot nature would not find the medicine heating.

From the beginning of the 14th century the subject of induction was taken up in the medical school of Padua where, owing to the influence of the Averroïsts who had come to dominate the university, the philosophical climate was thoroughly Aristotelian. From the time of Pietro d'Abano in his famous *Conciliator* in 1310, down to Zabarella in the early 16th century, these medical logicians developed the methods of 'resolution and composition' into a theory of experimental science very different from the method simply of observing ordinary, everyday occurrences with

---

posed between it and its source of light loses its light,' and the minor premiss, 'the moon has an opaque body interposed between it and its source of light,' the conclusion follows, 'therefore the moon loses its light,' that is, suffers an eclipse. In this way an eclipse of the moon is explained as an instance of a more general principle.

* I.e., if the disease causes an excess of one quality such as heat the medicine should cause a decrease in that quality, that is, have a cooling effect (cf. Vol. I, p. 171 *et seq.*).

40

which Aristotle and some of the earlier scholastics had been content to verify their scientific theories. Starting from observations, the complex fact was 'resolved' into its component parts:

the fever into its causes, since any fever comes either from the heating of the humour or of the spirits or of the members; and again the heating of the humour is either of the blood or of the phlegm, etc.; until you arrive at the specific and distinct cause and knowledge of that fever,

as Jacopo da Forlì (d. 1413) said in his commentary *Super Tegni Galeni*, comm. text 1. A hypothesis was then excogitated from which the observations could again be deduced, and these deduced consequences suggested an experiment by which the hypothesis could be verified. This method was followed by doctors of the period in the autopsies performed to discover the origin of a disease or the causes of death, and in the clinical study of medical and surgical cases recorded in their *consilia*. It has been shown that Galileo himself derived much of the logical structure of his science from his Paduan predecessors, whose technical terms he used (see below, p. 146 *et seq.*), though he did not go so far as to accept the conclusion of a late member of this school, Agostino Nifo, who said (1506), that since the hypotheses of natural science rested simply on the facts they served to explain, therefore all natural science was merely conjectural and hypothetical. The double procedure of resolution and composition was given in Padua the Averroïst name *regressus*. Discussing this 'regress', beginning with the inquiry for the cause of an observed effect, Nifo wrote in his *Expositio super Octo Aristotelis Libros de Physico*, published at Venice in 1552, book 1, commentary 4:

When I more diligently consider the words of Aristotle, and the commentaries of Alexander and Themistius, of Philoponus and Simplicius, it seems to me that in the regress made in demonstrations in natural science the first process, by which the discovery of the cause is put into syllogistic form, is a mere hypothetical (*coniecturalis*) syllogism. ... But the second process, by which is syllogized the reason why the effect is so through the discovered cause, is demonstration *propter quid* – not that it makes us know *simpliciter*, but conditionally (*ex conditione*), provided that that really is the cause, or provided that the propositions are true that represent it to be the cause, and that nothing else can be the cause. ... Alexander ... asserts that the discovery of the circles of epicycles and eccentrics from the appearances which we see is conjectural ... The opposite process he says to be a demonstration, not because it makes us know *simpliciter*, but conditionally, provided that those really are the cause and that nothing else can be the cause: for if those exist, then so do the appearances, but whether anything else can be the cause is not known to us *simpliciter*. ... But you object that in that case the science of nature is not a science at all. To that it can be replied that the science of nature is not a science *simpliciter*, like mathematics. Yet it is a science *propter quid*, because the discovered cause, gained through a conjectural syllogism, is the reason why the effect is so ... That something is a cause can never be so certain as that an effect exists (*quia est*); for the existence of an effect is known to the senses. That it is the cause remains conjectural ...

The whole of the pre-Galilean tradition of scientific method at Padua was finally summed up by Jacopo Zabarella (1533–89) in a series of treatises on the subject. Sharing the conception that had been developing since the 13th century that natural scientific explanations were hypothetical, he wrote in chapter 2 of *De Regressu*: 'demonstrations are made by us and for us ourselves, not for nature.' And he went on in chapter 5:

There are, I judge, two things that help us to know the cause distinctly. One is the knowledge *that* it is, which prepares us to discover *what* it is. For when we form some hypothesis about the matter we are able to search out and discover something else in it; where we form no hypothesis at all, we shall never discover anything ... Hence when we find that cause to be suggested, we are in a position to seek out and discover what it is. The other help, without which this first would not suffice, is the comparison of the cause discovered with the effect through which it was discovered, not indeed with the full knowledge that this is the cause and that the effect, but just comparing this thing with that. Thus it comes about that we are led gradually to the knowledge of the conditions of that thing; and when one of the conditions has been discovered we are helped to the discovery of another, until we finally know this to be the cause of that effect. ... The regress thus consists necessarily of three parts. The first in a 'demonstration that', by which we are led from a confused knowledge of the effect to a confused knowledge of the cause. The second is this 'mental consideration' by which, from a confused knowledge of the cause, we acquire a distinct knowledge of it. The third is demonstration in the strictest sense, by which we are at length led from the cause distinctly known to the distinct knowledge of the effect. ... From what we have said it can be clear that it is impossible to know fully that this is the cause of this effect, unless we know the nature and conditions of this cause, by which it is capable of producing such an effect.

Of great importance for the whole of natural science were the discussions of induction made by two Franciscan friars of Oxford living at the end of the 13th and the beginning of the 14th centuries. With them, and particularly with the second of them, began the most radical attack on Aristotle's system from a theoretical point of view. Both were preoccupied with the natural grounds of certainty in knowledge and the first, John Duns Scotus (c. 1266–1308), may be considered as summing up the tradition of Oxford thought on 'theory of science' which began with Grosseteste, before that tradition was projected violently in new directions by his successor, William of Ockham (c. 1284–1349). Each of them set out his essential point of view early in life in a theological work, their commentaries on the *Sentences* of Peter Lombard.

The principal contribution made by Scotus to the problem of induction was the very clear distinction he drew between causal laws and empirical generalizations. Scotus said that the certainty of the causal laws discovered in investigating the physical world was guaranteed by the principle of the uniformity of nature, which he regarded as a self-evident assumption of inductive science. Even though it was possible to have experience of only a sample of the correlated events under investigation, the certainty of the causal connexion underlying the observed correlation was known to the investigator, he said (in his *Oxford Commentary*, book 1, distinction 3, question 4, article 2), 'by the following

43

proposition reposing in the soul : *Whatever occurs as in a great many cases from some cause which is not free* [*i.e.*, not free-will] *is the natural effect of that cause.*' The most satisfactory scientific knowledge was that in which the cause was known, as, for instance, in the case of an eclipse of the moon deducible from the proposition : 'an opaque object interposed between a luminous object and an illuminated object impedes the transmission of light to such an illuminated object'. Even when the cause was not known and 'one must stop at some truth which holds as in many cases, of which the extreme terms [of the proposition] are frequently experienced united, as, for example, that a herb of such and such a species is hot' – even, that is, when it was impossible to get beyond an empirical generalization – the certainty that there was a causal connexion was guaranteed by the uniformity of nature.

William of Ockham, on the other hand, was sceptical about the possibility of ever knowing particular causal connexions or ever being able to define particular substances, though he did not deny the existence of causes or of substance as the identity persisting through change. He believed, in fact, that empirically established connexions had a universal validity by reason of the uniformity of nature, which he held, like Scotus, to be a self-evident assumption of inductive science. His importance in the history of science comes partly from some improvements he introduced into the theory of induction, but much more from the attack he made on contemporary physics and metaphysics as a result of the methodological principles which he adopted.

His treatment of induction Ockham based on two principles. First, he held that the only certain knowledge about the world of experience was what he called 'intuitive knowledge' gained by the perception of individual things through the senses. Thus, as he said in the *Summa Totius Logicæ*, part 3, part 2, chapter 10, 'when some sensible thing has been apprehended by the senses ... the intellect also can apprehend it,' and only propositions about individual things so apprehended were included in what he called 'real science'. All the rest, all the theories constructed to explain the observed facts, comprised 'rational science', in which names stood merely for concepts and not for anything real.

Ockham's second principle was that of economy, the so-called 'Ockham's razor'. This had already been stated by Grosseteste, and Duns Scotus and some other Oxford Franciscans had said that it was 'futile to work with more entities when it was possible to work with fewer'. Ockham expressed this principle in various ways throughout his works, a common form being one that was used in his *Quodlibeta Septem*, quodlibet 5, question 5: 'A plurality must not be asserted without necessity.' The well-known phrase *Entia non sunt multiplicanda præter necessitatem* was introduced only in the 17th century by a certain John Ponce of Cork, who was a follower of Duns Scotus.

The improvements Ockham made in the logic of induction were based principally on his recognition of the fact that 'the same species of effect can exist through many different causes,' as he said in the same chapter of the *Summa Totius Logicæ* as quoted from above. To establish causal connexions in particular cases he formulated rules, as in the following passage from his *Super Libros Quatuor Sententiarum*, book 1, distinction 45, question 1, D:

Although I do not intend to say universally what an immediate cause is, nevertheless I say that this is sufficient for something being an immediate cause, namely that when it is present the effect follows, and when it is not present, all the other conditions and dispositions being the same, the effect does not follow. Whence everything that has such a relation to something else is an immediate cause of it, although perhaps not *vice versa*. That this is sufficient for anything being an immediate cause for anything else is clear, because if not there is no other way of knowing that something is an immediate cause of something else. ... It follows that if, when either the universal or the particular cause is removed, the effect does not occur, then neither of them is the total cause but rather each a partial cause, because neither of those things from which by itself alone the effect cannot be produced is the efficient cause, and consequently neither is the total cause. It follows also that every cause properly so called is an immediate cause, because a so-called cause that can be absent or present without having any influence on the effect, and which when present in other circumstances does not produce the effect, cannot be considered a cause, but this is how it is with every other cause except the immediate cause, as is clear inductively.

This amounts to something like J. S. Mill's Method of Agreement and Difference. Since the same effect might have different causes, it was necessary to eliminate rival hypotheses. 'So,' said Ockham in the same work, prologue, question 2, G :

let this be posited as a first principle: all herbs of such and such species cure a fevered person. This cannot be demonstrated by syllogism from a better-known proposition, but it is known by intuitive knowledge and perhaps of many instances. For since he observed that after eating such herbs the fevered person was cured and he removed all other causes of his recovery, he knew evidently that his herb was the cause of recovery, and then he has experimental knowledge of a particular connexion.

Ockham denied that it could be proved either from first principles or from experience that any given effect had a final cause. 'The special characteristic of a final cause,' he said in his *Quodlibeta Septem*, quodlibet 4, question 1, 'is that it is able to cause when it does not exist'; 'from which it follows that this movement towards an end is not real but metaphorical,' he concluded in his *Super Quatuor Libros Sententiarum*, book 2, question 3, G. This phrase was in fact a commonplace and was used for example by Albertus Magnus and Roger Bacon. But for Ockham only immediate or proximate causes were real, and the 'total cause' of an event was the aggregate of all the antecedents which sufficed to bring about the event.

The effect of Ockham's attack on contemporary physics and metaphysics was to destroy belief in most of the principles on which the 13th-century system of physics was based. In particular, he attacked the Aristotelian categories of 'relation' and 'substance' and the notion of causation. He held that relations, such as that of one thing being above another in space, had no objective reality apart from the individual perceptible things between which the relation was found. Relations, according to him, were simply concepts formed by the mind. This view was incompatible with the Aristotelian idea of the cosmos having an objective principle of order according to which its constituent substances were arranged, and it opened the way for the notion that all motion was relative in an indifferent geometrical space without qualitative differences.

In discussing 'substance', Ockham said that experience was had only of attributes and that it could not be demonstrated that any given observed attributes were caused by a particular 'substantial form'. He held that the regular sequences of events were simply sequences of fact, and that the primary function of science was to establish these sequences by observation. It was impossible to be certain about any particular causal connexions, for experience gave evident knowledge only of individual objects or events and never of the relation between them as cause and effect. For example, the presence of fire and the sensation of burning were found associated together, but it could not be demonstrated that there was any causal connexion between them. It could not be proved that any particular man was a man and not a corpse manipulated by an angel. In the natural course of things a sensation was had only from an existing object, but God could give us a sensation without an object. This attack on causation was to lead Ockham to make revolutionary statements on the subject of motion (see below, pp. 75–9).

An even greater degree of philosophical empiricism, and one not to be attained again until the writings of David Hume in the 18th century, was reached by a French contemporary of Ockham, Nicholas of Autrecourt (d. after 1350). He doubted the possibility of knowing the existence of substance or causal relations at all. As with Ockham, from a limitation of evidential certitude to what was known through 'intuitive experience' and logically necessary implications, he concluded, in a passage published by J. Lappe in *Beiträge zur Geschichte der Philosophie des Mittelalters* (1908, vol. 6, part 2, p. 9 *): 'from the fact that one thing is known to exist, it cannot be evidently inferred that another thing exists,' or does not exist; from which it followed that from a knowledge of attributes it was impossible to infer the existence of substances. And, he said in the *Exigit ordo Executionis*, edited by J. R. O'Donnell in *Mediæval Studies* (1939, vol. 1, p. 237):

concerning things known by experience in the manner in which it is said to be known that rhubarb cures cholera or that the magnet attracts iron, we have only a conjecturative habit (*solum habitus conjecturativus*) but not certitude. When it is said that we have certitude concerning such things in virtue of a proposition reposing in the soul that *that which occurs as in many cases from an unfree course is the natural effect of it,* I ask what you call a natural cause, *i.e.,* do you say that that which produced in the past as in many cases and up to the present will produce in the future if it remains and is applied? Then the minor [premiss] is not known, for allowing that something was produced as in many cases, it is nevertheless not known that it ought to be thus produced in the future.

And so, he said, in a passage published by Hastings Rashdall in the *Proceedings of the Aristotelian Society,* N.S. vol. 7:

Whatever conditions we take which may be the cause of any effect, we do not evidently know that, when those conditions are posited, the effect posited will follow.

The effect on philosophy in general of this search for evident knowledge was to divert interest within the discussions of the schools away from the traditional problems of metaphysics to the world of experience. Ockhamite nominalism or, as it may more properly be called, 'terminism', went to show that in the natural world all was contingent and therefore that observations were necessary to discover anything about it.

The relation of faith to reason remained a central problem in medieval speculation, and a diversity of attitudes was taken to it by Augustinians, Thomists, Averroïsts and Ockhamites. 'The spirit and the enterprise' of early medieval philosophy was, as R. McKeon put it in his *Selections from Medieval Philosophers* (vol. 2, pp. ix–x), 'of faith engaged in understanding itself.' Between Augustine and Aquinas philosophy had passed from the consideration of truth as a reflection of God to truth in the relation of things to each other and to man, leaving their relation to God for theology. Ockham himself firmly divorced theology from philosophy, the former deriving its knowledge from revelation and the latter from sensory experience, from which alone it took its origin. And whereas the Averroïsts were driven to entertain the possibility of 'double truth' (see Vol. I, p. 78), the Ockhamites, for instance Nicholas of Autrecourt, sought a solution to the problem in their doctrine of 'probabilism'. By this they meant that natural philosophy could offer a probable but not a necessary system of explanations, and that where this probable system contradicted the necessary propositions of revelation, it was wrong. In his own attempt to reach the most probable system of physics Nicholas made a thorough-going attack on the Aristotelian system and arrived at the conclusion that the most probable system was one based on atomism. After this time, no further attempts were made to construct systems rationally synthesizing the contents of both faith and reason. Instead, there began a period of increasing reliance on the literal word of the Bible instead of the teaching of a divinely instituted Church, a period of speculative mysticism seen in Eckhart (*c.* 1260–1327) and Henry Suso (*c.* 1295–1365), and of empiricism and scepticism seen in Nicholas of Cusa (1401–64) and Montaigne (1533–92). Nicholas of Cusa, for example, held that though it was possible to approach closer and closer to truth, it was never possible to grasp it finally, just as it was possible to draw figures approximating more and more closely to a perfect circle, yet no figure we drew would be so perfect that a more perfect circle could not be drawn. Montaigne was even more sceptical. Indeed, since the 14th century the stream of sceptical empiricism has flowed strongly in European philosophy, and it has done its work of directing attention to the conditions of human knowledge which has produced some of the most important clarifications of scientific methodology.

2. MATTER AND SPACE IN LATE MEDIEVAL PHYSICS

The most radical attacks made in the 14th century on Aristotle's whole system of physics concerned his doctrines about matter and space, and about motion. Aristotle had denied the possibility of atoms, void, infinity and plural worlds, but when his strict determinism had been condemned by the theologians in 1277 this opened the way to speculation on these subjects. With the assertion of God's omnipotence, philosophers argued that God could create a body moving in empty space or an infinite universe and proceeded to work out what the consequences would be if He did. This seems a strange way to approach science, but there is no doubt that it was science they were approaching. They discussed the possibility of plural worlds, the two infinities, and centre of gravity; and they discussed also the acceleration of freely falling bodies, the flight of projectiles, and the possibility of the earth's having motion. Not only did the criticisms of Aristotle remove many of the metaphysical and 'physical' restrictions his system had placed on the use of mathematics, but also many of the new concepts reached were either incorporated directly into 17th-century mechanics or were the germs of theories to be expressed in the new language created by mathematical and experimental techniques.

Central to the whole discussion of matter, space and gravitation in the 13th and 14th centuries were the two conceptions of dimensionality coming respectively from the atomists and Plato, and from Aristotle (cf. Vol. I, pp. 48–50, 88–92). In the *Timæus* Plato had put forward a clearly mathematical conception of space, which he conceived as dimensions independent of bodies but in which bodies could exist and could move; space was in fact the receptacle of all things, as real as the eternal ideas and more real than the bodies occupying it. The part of space occupied by the dimensions of a body was the body's 'place'; the part not so occupied was a vacuum. This was essentially the atomists' view.

To this view Aristotle objected in his *Physics* (book 4) that dimensions could not exist apart from bodies with dimensions; he conceived dimensions as quantitative attributes of bodies, and no attribute could exist apart from the substance in which it inherited (cf. Vol. I, pp. 83–4). Moreover, Aristotle maintained that the conception of space held by Plato and the atomists was useless in explaining the actual movements of bodies: for example, why should a given body go up rather than down, or *vice versa*? His own explanation of the different movements actually observed in bodies was in terms of his conception of 'place'.

This had two essential characteristics. Primarily it was the physical environment of the body, the 'innermost boundary' of whatever contained the body. Aristotle maintained that the bodies making up the universe were all contiguous with each other, thus composing a *plenum*. The innate preference of a body for a particular physical environment within this *plenum* was the cause of the natural motions all bodies were observed to have (cf. Vol. I, pp. 84–6, 125–6). To this notion of place as a physical ambience moving each body according to its nature by final causality, Aristotle added also a geometrical, spatial characteristic. He held that each place in the universe was itself motionless; and in his *De Cælo* he gave to each of the places making up the universe as a whole a position in absolute space relative to the centre of the earth fixed at the centre of the universe. This gave him his conception of 'up' and 'down' as absolute directions from the centre to the circumference of the outermost sphere.

Aristotle's conceptions of dimensionality and of place are good examples of the empirical concreteness so noticeable in all his thought. Much of the character of 14th-century physics is a result of a renewed application of the more abstract thinking of Plato and the atomists.

The form of atomism found in Plato's *Timæus* and Lucretius' *De Rerum Natura* (see below, p. 116), and in the works of several other ancient Greek writers,* had been developed by

---

* The development of the atomic theory in the Ancient World after the time of Plato and Aristotle (for development down to Plato see note in Vol. I. p. 46) was largely the work of Epicurus (340–270 B.C.), Strato of Lampsacus (*fl.c.* 288 B.C.), Philo of Byzantium (2nd century B.C.), and Hero of Alexandria (1st century B.C.). The theory of Epicurus was expounded by Lucretius (*c.* 95–55 B.C.) in his poem, *De Rerum Natura*. Epicurus made two changes in Democritus's theory. He held, first, that the atoms fell perpendicularly in empty space owing to their weight and secondly, that interactions between them which resulted in the formation of bodies took place as a result of 'swerves' which occurred by chance and led to collisions. He assumed a limited number of shapes but an infinite number of atoms of each shape. Different kinds of atoms had different weights, but all fell with the same velocity. Epicurus also stated a principle which had been held by certain previous atomists, namely, that all bodies of any

some 13th-century philosophers. Grosseteste, for example, had said that the finite space of the world was produced by the infinite 'multiplication' of points of light, and he also regarded heat as due to a scattering of molecular parts consequent on

---

weight whatever would fall in a void with the same velocity. Differences in velocity of given bodies in a given medium, e.g., air, were due to differences in the proportion of resistance to weight. On collision, atoms became interlocked by little branches or antlers; only the atoms of the soul were spherical. To meet Aristotle's objection based on the change of properties in compounds, he assumed that a 'compound body' formed by the association of atoms could acquire new powers not possessed by individual atoms. The infinite number of atoms produced an infinite number of universes in infinite space. It seems that Strato's treatise *On the Void* was the basis of the introduction to Hero's *Pneumatica*. Strato combined atomist with Aristotelian conceptions and took an empirical view of the existence of void, which he used to explain the differences in density between different bodies. In this he was followed by Philo in his *De ingeniis Spiritualibus* (which was not widely known in the Middle Ages) and by Hero, who denied the existence of a continuous extended vacuum but made use of interstitial vacua between the particles of bodies to explain the compressibility of air, the diffusion of wine into water, and similar phenomena. These writers also carried out experiments to demonstrate the impossibility of an extended void. Aristotle had proved that air had body by showing that a vessel must be emptied of air before it could be filled with water. Philo and Hero both performed the experiment, also described by Simplicus, showing that in a water clock or clepsydra, water could not *leave* a vessel unless there was a means for air to enter it. Philo also described two other experiments proving the same conclusion. He fixed a tube to a globe containing air and dipped the end of the tube under water, and showed that when the globe was heated air was expelled and when it cooled the contracting air drew water up the tube after it. The air and water remained in contact, preventing a vacuum. He also showed that when a candle was burnt in a vessel inverted over water, the water rose as the air was used up. Apart from these and some other Alexandrian writers, such as the doctor Erasistratus and members of the Methodical sect, atomism was not favourably regarded in antiquity. It was opposed by the Stoics, although they believed in the possibility of void within the universe and in an infinite void beyond its boundaries; and it was opposed also by a number of other writers such as Cicero, Seneca, Galen and St Augustine. But atomism was briefly discussed by Isidore of Seville, Bede, William of Conches and several Arab and Jewish writers such as Rhazes (d. *c.* 924) and Maimonides (1135–1204).

movement. Even Roger Bacon, though he followed Aristotle and tried to show that atomism led to consequences which contradicted the teachings of mathematics, for instance the incommensurability of the diagonal and side of a square (see Vol. I, p. 46, note), agreed with Grosseteste in regarding heat as a form of violent motion. Towards the end of the 13th century several writers adopted atomist propositions, though these were refuted by Scotus while discussing the question whether angels could move from place to place with continuous movement. Similar propositions were refuted again early in the 14th century by Thomas Bradwardine (*c.* 1295–1349). The propositions refuted were that continuous matter consisted either of *indivisibilia*, that is, discontinuous atoms separated from each other, or of *minima*, that is, atoms joined to each other continuously, or of an infinite number of actually existing points.

At the turn of the 13th century, a complete form of atomism was put forward by Giles of Rome (1247–1316), who derived the basis of it from Avicebron's theory of matter as extension successively specified by a hierarchy of forms (see Vol. I, p. 88). Giles held that magnitude might be considered in three ways : as a mathematical abstraction, and as realized in an unspecified and in a specified material substance. An abstract cubic foot and a cubic foot of unspecified matter were then potentially divisible to infinity, but in the division of a cubic foot of water a point was reached at which it ceased to be water and became something else. The geometrical arguments against the existence of natural *minima* were therefore irrelevant. Nicholas of Autrecourt was led, by the impossibility of demonstrating that there was in a piece of bread anything beyond its sensible accidents, to abandon altogether the explanation of phenomena in terms of substantial forms and to adopt a completely Epicurean physics. He came to the probable conclusion that a material *continuum* was composed of minimal, infra-sensible indivisible points, and time of discrete instants, and he asserted that all change in natural things was due to local motion, that is, to the aggregation and dispersal of particles. He also believed that light was a movement of particles with a finite velocity. That some of these conclusions were proposed with reference to a discussion of the theological doctrine of transubstantiation shows how closely all cosmological questions were linked together, and was one reason why he was obliged to retract some of his theses. These discussions survived in nominalist teaching in the 15th and 16th centuries, in writings of Nicholas of Cusa and Giordano Bruno (1548–1600), and eventually led to the atomic theory being used to explain chemical phenomena in the 17th century.

Concerning the problem of void, which arose partly out of the discussion of whether there were plural worlds – for if there were what lay between them? – such writers at the end of the 13th and beginning of the 14th century as Richard of Middleton (or Mediavilla, *fl.c.* 1294) and Walter Burley (1275–1344) went so far as to say that it was a contradiction of God's infinite power to say that He could not maintain an actual void. Nicholas of Autrecourt went further and affirmed the probable existence of a vacuum : 'There is something in which no body exists, but in which some body can exist,' he said in a passage published by J. R. O'Donnell in *Mediæval Studies* (1939, vol. 1, p. 218). Most writers accepted Aristotle's arguments and rejected an actually existing void (see Vol. I, p. 84), though they might accept Roger Bacon's description of void as a mathematical abstraction. 'In a vacuum nature does not exist,' he said in the *Opus Majus*, part 5, part 1, distinction 9, chapter 2.

For vacuum rightly conceived of is merely a mathematical quantity extended in the three dimensions, existing *per se* without heat and cold, soft and hard, rare and dense, and without any natural quality, merely occupying space, as the philosophers maintained before Aristotle, not only within the heavens, but beyond.

Some of the physical arguments against the existence of void were taken from such ancient Greeks as Hero and Philo, whose experiments with the candle and the water clock or clepsydra were known to several writers, in particular Albertus Magnus, Pierre d'Auvergne (d. 1304), Jean Buridan (d. probably in 1358) and Marsilius of Inghen (d. 1396). Some of these writers also mentioned another experiment in which water was shown to mount in a J-tube when air was sucked out of the long arm with the short arm under water. Another experiment was made with a water clock, with which it was shown that water would not run out of the holes in the bottom when the hole at the top was closed with the finger. This was contrary to the natural motion of water downwards and Albertus Magnus explained this as due to the impossibility of void, which meant that water could not run out unless air could enter and maintain contact with it. Roger Bacon was not satisfied with such a negative explanation. He held that the final cause of the phenomenon was the order of nature, which did not admit void, but the efficient cause was a positive 'force of universal nature', and adaptation of the 'common corporeity' of Avicebron (see Vol. I, p. 88), which pressed on the water and held it up. This was similar to the explanation already given by Adelard of Bath. Giles of Rome later substituted another positive force, *tractatus a vacuo* or suction by a vacuum, a universal attraction which kept bodies in contact and prevented discontinuity. The same force, he held, caused the magnet to attract iron. Another 14th-century writer, John of Dumbleton (*fl.c.* 1331–49), said that to maintain contact celestial bodies would, if necessary, abandon their natural circular motions as particular bodies and follow their universal nature or 'corporeity', even though this involved an unnatural rectilinear movement. In the 15th and 16th centuries, Roger Bacon's full theory was forgotten in Paris and condensed into the 'nature abhors a vacuum' that provoked the sarcasms of Torricelli and Pascal.

The possibility of both infinite addition and infinite division of magnitude led to interesting discussions on the logical basis of mathematics. It was asserted by Richard of Middleton and later by Ockham that no limit could be assigned to the size of the universe and that it was potentially infinite (see Vol. I, p. 88). It was not actually infinite, for no sensible body could be actually infinite. Richard of Middleton tried to show also that this last conclusion was incompatible with Aristotle's doctrine of the eternity of the universe, which Albertus Magnus and Thomas Aquinas had said could be neither proved nor disproved by reason but must be denied from revelation. Richard said that as indestructible human souls were continually being generated, if the universe had existed from eternity there would now be an infinite multitude of such beings. An actually infinite multitude could not exist, therefore the universe had not existed from eternity. The whole discussion led to an examination of the meaning of infinity. The development of the geometrical paradoxes that would arise from the categorical assertion of an actually existing infinity, such as in Albert of Saxony's discussion of whether there could be an infinite spiral line on a finite body, led Gregory of Rimini (1344) to try to give precise signification to the words 'whole', 'part', 'greater', 'less'. He pointed out that they had a different meaning when referring to finite and infinite magnitudes, and that 'infinity' had a different signification according to whether it was taken in a distributive or collective sense. This problem was discussed in the *Centiloquium Theologicum* formerly attributed to Ockham but of uncertain authorship. Conclusion 17, C shows that the author had achieved a logical subtlety which was to be recovered only in the 19th and 20th centuries in the mathematical logic of Cantor, Dedekind and Russell.

There is no objection to the part being equal to its whole, or not being less, because this is found, not ... only intensively but also extensively ... for in the whole universe there are no more parts than in one bean, because in a bean there is an infinite number of parts.

These discussions of infinity and other problems, such as the maximum resistance a force could, and the minimum it could not overcome, laid the logical basis of the infinitesimal calculus. Medieval mathematics was limited in range and it was only when humanists had drawn attention to Greek mathematics, and especially to Archimedes, that the mathematical developments which actually took place in the 17th century became a possibility.

Associated with the problem of infinite magnitude was that of plural worlds. In 1277 the Bishop of Paris, Etienne Tempier, condemned the proposition that it was impossible for God to create more than one universe. The problem was usually discussed in connexion with gravity and the natural place of the elements (see Vol. I, pp. 90–91, 139).

In his *De Cælo* (book 1, chapter 8) Aristotle had briefly considered the possibility of a mechanical explanation of gravitation by external forces either pulling or pushing bodies, but he rejected this on the grounds that it was made unnecessary by the whole conception that the movements of gravity and levity were the spontaneous movements of a 'nature' towards its natural place (cf. Vol. I, pp. 84–5, below, p. 61 *et seq.*). It was to this view that Averroës lent his authority, making gravity an *intrinsic* tendency belonging to the 'nature' or 'form' of a body and thus causing its movement. This conception of gravity and levity as intrinsic properties causing natural movement became the normal one in the 13th century, accepted for example by Albertus Magnus and Thomas Aquinas, although opinion differed as to the precise manner in which the 'form' caused a body to move.

But already in the 13th century there were natural philosophers who held that, over and above the natural spontaneity of the form and the final causality of the natural place, it was necessary to look for some further efficient causality of gravitation. Some writers conceived this as an *external* cause. Bonaventura and Richard of Middleton, for example, suggested that an attracting force (*virtus loci attrahentis*) should be attributed to natural place and an expelling force to unnatural place. Roger Bacon developed a complete 'field' theory to account for gravitation (cf. Vol. I, pp. 87–8, 112–18, below, p. 75). He proposed that the natural place exercised not only final causality but also efficient causality through a *virtus immaterialis*, an immaterial power coming from the heavenly bodies and filling all space. Gravity and levity were diffused immaterial forces which, although derived from 'celestial virtue', produced their effects by being concentrated more intensely in various natural places. This explanation is to be found also in the *Summa Philosophiæ* of pseudo-Grosseteste.

An even more extreme form of this explanation by external forces seems to have been put forward by some 14th-century writers who conceived natural place as a total efficient cause of gravitation. For example Buridan in his *Quæstiones de Cælo et Mundo* (book 2, question 12) mentions the opinion of 'certain people' (*aliqui*) who 'say that place is the motive cause of the heavy body by means of attraction, just as a magnet attracts iron'. He attacked this opinion on the grounds of experience. Since heavy bodies accelerate as they fall, he said, there must be an increase in the motive force commensurate with the increase in velocity (cf. Vol. I, pp. 90, 125–6, below, p. 80 *et seq.*). Those who hold that the motive force is attraction by the natural place must therefore suppose that this is greater near the natural place than farther off, as is the case with the magnet. But if two stones are dropped from a tower, one from the top and the other from lower down, the first has a much greater velocity than the second when both have reached, for example, a point a foot from the ground. Hence it is not simply nearness to the natural place that determines velocity, but, whatever the cause is, velocity depends on the length of the fall. 'Nor is it similar to the magnet and the iron,' he concluded, 'because if iron is near a magnet, it immediately starts moving more quickly than if it were farther removed; but this is not the case of heavy bodies with respect to their natural place.'*

A further objection to the natural place exerting any kind of force, any *vis trahens* on the body moving towards it, was made by Albert of Saxony (*c.* 1316–90). He pointed out that to such a force a heavier body would offer a greater resistance than a lighter body and so should fall more slowly than a lighter body, which was contrary to experience.

These arguments are a good example of the extreme difficulty which the dynamical problems whose solutions we now take for granted presented to those who first attacked them.

All these writers accepted the principle that action at a distance simply speaking was impossible, and those who proposed

* In fact both magnetism and gravity give bodies an acceleration inversely proportional to the square of the distance.

the analogy of the magnet usually had in mind the explanation of its action given by Averroës (see Vol. I, p. 133). According to this theory the force that moved the iron was a quality induced in it by the *species magnetica* that went out from the magnet through the medium and altered the iron, thus giving it the power *to move itself*. Thus was preserved the essential principle of Aristotelian dynamics, that the motive power must accompany the moving body.

An exception was William of Ockham. Arguing that intermediate 'species' and agents postulated simply to avoid having to accept action at a distance were not necessary to 'save the appearances', he boldly declared that there was no objection to action at a distance as such. The sun in illuminating the earth acted at a distance immediately. The magnet, he asserted in his *Commentary on the Sentences* (book 2, question 18), 'pulls [the iron] immediately and not by means of a power existing in some way in the medium or in the iron; therefore the lodestone acts at a distance immediately and not through a medium'. As for the general principle that the motive power must accompany the moving body, Ockham's attack on the whole contemporary conception of motion altogether denied this as a premiss for dynamical explanations (see below, pp. 75–9).

At least one other 14th-century writer, John Baconthorpe, followed Ockham in accepting the possibility of action at a distance, asserting, as Dr Maier quotes him in her book, *An der Grenze von Scholastik und Naturwissenschaft* (p. 176, note), that the magnet 'attracts the iron effectively'. But the common opinion on gravitation in the 14th century, as in the 13th, rejected both action at a distance and external forces of any kind and took Aristotle's and Averroës' view of it as an intrinsic tendency. This was the view, for example, of Jean de Jandun, Walter Burley, Buridan, Albert of Saxony, and Marsilius of Inghen. The attempt by Buridan and others to give quantitative precision to this intrinsic cause of motion led to the most interesting dynamical theorizing before Galileo (see below, pp. 80 *et seq.*, 160 *et seq.*).

The question then arose, what was the natural place of an element, for example earth, at which it came to rest? In discussing this problem, Albert of Saxony (*c.* 1316–90) distinguished between the centre of volume and the centre of gravity. The weight of each piece of matter was concentrated at its centre of gravity and earth was in its natural place when its centre of gravity was at the centre of the universe. The natural place of water was in a sphere round the earth, so that it exerted no pressure on the surface of the earth which it covered.

Although Aristotelians like Buridan and Albert of Saxony rejected the explanation of gravity by external forces, the Aristotelian explanation did not remain alone in the field. With the revival of Platonism, especially in the 15th century, an argument for the existence of plural worlds was found in the conception of gravity of the Pythagoreans and Plato.

Heraclides of Pontus and the Pythagoreans maintain that each of the stars constitutes a world, that it consists of an earth surrounded by air and that the whole is swimming in illimitable ether,

the 5th-century A.D. Greek writer Joannes Stobæus had said in his *Eclogarum Physicarum*, chapter 24. The theory of gravity derived from the *Timæus* was that the natural movement of a body was to rejoin the element to which it belonged, in whichever world it was, while violent movement had the opposite effect (see Vol. I, pp. 48–9). This explanation of gravity as the tendency of all similar bodies to congregate, as *inclinatio ad simile*, was generally adopted by those who rejected Aristotle's conception of absolute space. The Aristotelian objection that if there were plural worlds there would be no natural place thus lost its point. Matter would simply tend to move towards the world nearest it. This theory was mentioned by Jean Buridan, himself a critic of Aristotle's absolute space although not of course of his natural place. It was adopted by Nicole Oresme (see below, pp. 78, 86–97) and later by the leading 15th-century Platonist, Nicholas of Cusa, who said that gravitation was a local phenomenon and each star a centre of attraction capable of keeping together its parts. Nicholas of Cusa also believed that each star had its inhabitants, as the earth did. Albert of Saxony had retained the essential structure of the Aristotelian universe; Ockham, though he held, like Avicebron, that the matter of elementary and celestial bodies was the same, said that only God could corrupt the celestial substance. Nicholas of Cusa said that there was absolutely no distinction between celestial and sublunary matter and that since the universe, while not actually infinite, had no boundaries, neither the earth nor any other body could be its centre. It had no centre. Each star, of which our earth was one, consisted of the four elements arranged concentrically round a central earth and each was separately suspended in illimitable space by the exact balance of its light and heavy elements.

3. DYNAMICS—TERRESTRIAL AND CELESTIAL

Aristotle's dynamics involved several propositions all of which came to be criticized in the later Middle Ages. In the first place, there was Aristotle's conception of local motion, like all kinds of change, as a process by which the potentialities of any body to movement were made actual by a motive agent (see Vol. I, pp. 84–5, 90, 125–6). In natural motion this agent was an intrinsic principle, acting either as an efficient cause, for example the 'soul' in living things (cf. Vol. I, p. 149), or as a principle producing characteristic spontaneous motion in a particular environment, as in the motion of bodies towards their 'natural place'. Each of the celestial spheres was also moved by a 'soul', which became with later writers an 'Intelligence' that pushed the sphere round. In unnatural or forced 'violent' motion, the agent was always an external mover which accompanied the moving body and imposed its alien form of movement on it. But whether the motion was produced by the natural activity of the 'nature' or 'form' or was imposed by an external agent, the essential principle was preserved: 'Everything that is moved must be moved by something.' If the cause ceased, so did the effect. Basic to the whole conception of natural motion was that it proceeded towards an end, a goal, for example the earth as the goal of a naturally falling stone. Unnatural motion was the imposition of a motion alien to the natural goal, and such motion continued only so long as the external agent remained in contact with the body moved. Aristotle held further that the velocity of a moving body was directly proportional to the motive power and inversely proportional to the resistance of the medium in which movement took place. This gave the law,

$$\text{velocity } (v) \propto \frac{\text{motive power } (p)}{\text{resistance } (r)}$$

It was an important limitation, coming from the Greek conception of proportion and from Aristotle's vague formulation, that Aristotle himself did not in fact express his 'law' in the manner in which, for convenience, it is written in the preceding line. According to the Greek conception, a magnitude could result only from a 'true' proportion, that is from a ratio between 'like' quantities, for example between two distances or two times. A ratio, between two 'unlike' quantities such as distance (s) and time (t) would thus not have been considered as a magnitude, so that the Greeks did not in fact give a metric definition of velocity as a magnitude representing a ratio between space and

time, i.e. $v = k\dfrac{s}{t}$. Such a metric definition was one of the achieve-

ments of the 14th-century scholastic mathematicians. Aristotle himself could express the relation of velocity to power and resistance only by taking the problem in separate stages. Thus

$\dfrac{s_1}{s_2} = \dfrac{t_1}{t_2}$, i.e. speed is uniform, when $p_1 = p_2$ and $r_1 = r_2$; $\dfrac{s_1}{s_2} = \dfrac{p_1}{p_2}$ when

$t_1 = t_2$ and $r_1 = r_2$; and $\dfrac{s_1}{s_2} = \dfrac{r_2}{r_1}$ when $t_1 = t_2$ and $p_1 = p_2$.

Aristotle's 'law' expressed his belief that any increase in velocity in a given medium could be produced only by an increase in motive power. It also followed from the 'law' that in a void bodies would fall with instantaneous velocity; as he regarded this conclusion as absurd he used it as an argument against the possibility of a void. He held that in a given medium bodies of various materials but of the same shape and size fell with velocities proportional to their various weights.

This conception and classification of motion was based on

direct observation and it was confirmed by many everyday phenomena. But three phenomena presented difficulties that were
ultimately to prove fatal to the mathematical formulation drawn
from Aristotle's account. First, according to Aristotle's 'law',
there should be a finite velocity (v) with any finite values of
power (p) and resistance (r), yet in fact if the power were smaller
than the resistance it might fail to move the body at all. Aristotle
himself recognized this and made reservations for his law, for
example in the case of a man trying to move a heavy weight
and not succeeding.

Secondly, what was the source of the increase in motive power
required to produce the acceleration of freely falling bodies? He
had seen that bodies falling vertically in air accelerated steadily,
and he thought that this was because the body moved more
quickly as it got nearer to its natural place in the universe as
the goal and fulfilment of its natural motion.

Thirdly, what was the motive power that kept a projectile in
motion after it had left the agent of projection? If the upward
movement of a stone was not due to the stone itself but to the
hand that threw it, what was responsible for its continued movement after it ceased to be in contact with the hand? What kept
an arrow in flight after it had left the bowstring? Aristotle himself in the *Physics* (book 8) proposed this problem and discussed
two solutions, Plato's and his own. In the *Timæus* Plato had
given to bodies only one proper motion, that towards their proper
place in space forming the receptacle of all things, and this
motion he explained by the geometrical shape of the elementary
bodies and the shaking of the receptacle by the World Soul. All
other movements he attributed to collision and mutual replacement, *antiperistasis*: a projectile, for example, at the moment of
discharge compressed the air in front of it, which then circulated
to the rear of the projectile and pushed it forward, and so on in
a vortex. Aristotle's objection to this explanation was that unless
the original mover gave, to what it moved, not only motion but
also the power to be a mover itself, the motion would cease. He
therefore proposed that the bowstring or hand communicated a
certain quality or 'power of being a movent' (as he said in book 8,

chapter 10 of the *Physics*, 267a 4) to the air in contact with it, that this transmitted the impulse to the next layer of air, and so on, thus keeping the arrow in motion until the power gradually died away. This power, he said, came from the fact that air (and water), being intermediate elements, were heavy or light, depending on their actual environment. The air could thus move a projectile upwards from its own natural motion. If actual space were a void, he argued in book 4 of the *Physics*, not even forced motion would be possible; a projectile would not be able to move in void space.

As seen in the light of the classical mechanics completed in the 17th century, the notorious defect of Aristotle's mechanics was its failure to deal adequately with *acceleration* as distinct from velocity. From the point of view of these later conceptions, his fundamental difficulty arose from the fact that by analysing motion entirely in terms of velocities continuing over a period of time, he was unable to deal with *initial* velocity, or with the force required to *start* a body moving. His idea of force or power is restricted to that causing motions continuing over a period of time. All the difficulties found in his treatment were finally overcome when motion was analysed in terms of velocity *at an instant*. Using this conception, Newton was able to show that the same initial force that started a body moving must, if it continued to act, produce not just continued velocity but the same constant change in velocity, that is, constant acceleration. The moves towards clarity in these problems that were made before Newton will be seen in the sequel.

Parts of Aristotle's dynamics had already been criticized in the Ancient World by members of other schools of thought. The Greek atomists had considered it an axiom that all bodies of whatever weight would fall in a void with the same velocity, and that differences in the velocity of given bodies in a given medium, for instance air, were due to differences in the proportion of resistance to weight (see above, p. 52, note). The Alexandrian mechanicians and the Stoics had also admitted the possibility of void, but Philo had said that differences in velocity of fall were due to different 'weight-forces' (corresponding to different 'masses') and from this Hero drew the corollary that if two bodies of a given weight were fused, the speed of fall of the united body would be greater than that of each singly. The Christian Neo-platonist, John Philoponus of Alexandria, writing in the 6th century A.D., had also rejected both Aristotle's and the atomists' laws regarding falling bodies and maintained that in a void a body would fall with a finite velocity characteristic of its gravity, while in air this finite velocity was decreased in proportion to the resistance of the medium. The rotation of celestial spheres provided an example of a finite velocity that took place in the absence of resistance. Philoponus also pointed out that the velocities of bodies falling in air were not simply proportional to their weights, for when a heavy and a less heavy body were dropped from the same height, the difference between their times of fall was much smaller than that between their weights. Philoponus did accept Aristotle's theory for explaining the continuous acceleration of falling bodies, though this was not accepted by other late Greek physicists. Some of these put forward an adaptation of the Platonic conception of *antiperistasis*, according to which the falling body forced down the air which then drew the body after it and so on, natural gravity both receiving continuously increasing assistance from the traction of the air and continuously causing an increase in that assistance.

Philoponus seems to have been the first to show that the medium cannot be the cause of projectile motion. If it is really the air that carries the stone or the arrow along, why, he asked, must the hand touch the stone at all or the arrow be fitted to the bow? Why does not violent beating of the air move the stone? Why can a heavy stone be thrown farther than a very light one? Why do two bodies have to collide to be deflected and not simply pass close to each other through the air? These every-day observations, which were to form the staple of criticism of Aristotle's dynamics down to the time of Galileo himself, led Philoponus to propose an alternative explanation of the 'forced' motion of projectiles. Obviously the air did not produce the motion but resisted it. He put forward the original idea that the instrument of projection imparted motive power not to the air but to the projectile itself: 'a certain incoporeal motive power must be given to the projectile through the act of throwing,' he said in his commentary on Aristotle's *Physics* (book 4, chapter 8). But this motive power, or 'energy' (*energeia*), was only borrowed and was decreased by the natural tendencies of the body and by the resistance of the medium, so that the projectile's unnatural motion eventually came to an end.

Philoponus' theory has been claimed by some scholars, notably by Duhem, as the origin of certain medieval conceptions that have been supposed in turn to have given rise to the modern conception of inertia, which was to be the basis of the revolution in dynamics in the 17th century (see below, p. 78, note). We will see on a later page that this view of complete continuity may be questioned on the grounds both of the actual historical derivation and of the character of the conception of motion concerned. But the theory that unnatural motion could be maintained by a motive power imparted to the unnaturally moving body itself was an important innovation and it was mentioned by several writers before it reappeared as the theory of *impetus* in the 14th century. Philoponus himself was attacked by Simplicius (d. 549) in the 'Digressions against John the Grammarian' which he appended to his own commentary on the *Physics*. He objected specifically to Philoponus' denial of the fundamental principle that whatever is moved unnaturally must be moved by an external agent in contact with it. His own explanation of projectile motion was a development of the *antiperistasis* theory he held that the projectile and the medium alternately acted on each other until eventually the motive power became exhausted. At the same time he put forward an explanation of the acceleration of freely falling bodies by supposing that their weight increased as they approached the centre of the world.

The first Arabic writer known to have taken up Philoponus' theory was Avicenna, who defined the power imparted to the projectile, as S. Pines translates him in an important article in *Archeion* (1938, vol. 21, p. 301), as 'a quality by which the body pushes that which prevents it moving itself in any direction'. He called this also a 'borrowed power', a quality given to the projectile by the projector as heat was given to water by a fire. Avicenna made two important modifications of the theory. First, whereas Philoponus had held that even in a void, if this were possible, the borrowed power would gradually disappear and the projectile's 'forced' motion cease, Avicenna argued that in the absence of any obstacle this power, and the 'forced' motion it produced, would persist indefinitely. Secondly, he tried to express the motive power quantitatively, saying in effect that bodies moved by a given power would travel with velocities inversely proportional to their weights, and that bodies moving with a given velocity would travel (against the resistance of the air) distances directly proportional to their weights. A further development of the theory was made by Avicenna's 12th-century follower Abu'l Barakat al-Baghdadi, who proposed an explanation of the acceleration of falling bodies by the accumulation of successive increments of power with successive increments of velocity.

The main points at issue between the Aristotelian conception of motions and this ultimately Neoplatonic conception, first expounded by Philoponus, were taken up by Averroës in a discussion that was to determine the main lines of the debate that began in the West in the 13th century. Philoponus had maintained that in all cases, in falling bodies and in projectiles, velocity was proportional only to motive power, and that the resistance of the medium merely reduced it from a definite finite velocity. This 'law of motion' was advocated by the 12th-century Spanish Arab Ibn Badga, or Avempace as he was called in Latin, as an alternative to Aristotle's. It meant substituting for Aristotle's 'law of motion' the formula: velocity $(v)=$power $(p)-$resistance (r). Avempace argued that even in a void a body would move with finite velocity because, although there was no resistance, the body would still have to traverse *distance*. Like Philoponus he cited the motion of the celestial spheres as an example of finite velocity without resistance. In his commentary on Aristotle's *Physics* Averroës attacked not only Avempace's account of motion (which he thought was original) but the whole conception of 'natures' on which it was based. Avempace's mistake, he maintained, was to treat the 'nature' of a heavy body as if it were an entity distinct from the matter of the body, and as if the matter were moved by the 'form' acting as an efficient cause in the same way as an immaterial Intelligence moved its celestial sphere or the 'soul' caused the movements of a living organism. Averroës specifically objected to Avempace's assumption that the medium was an impediment to natural motion, for this would mean that all actual bodies moved unnaturally, since all do in fact move through corporeal media.

The natural point of departure for the scholastic commentators on Aristotle's *Physics* and *De Cælo* were the commentaries by Averroës that accompanied the most popular early Latin versions. Averroës' exposition and criticism of Avempace thus became the source of a major divergence in the attempts to formulate a law relating the velocities of natural motions. But it marked more than that. It has been claimed that it reflected a major cleavage in the conception of nature which runs through the whole history of philosophy.* Philoponus and Avempace had followed Plato in looking for the real natures and causes of phenomena not in immediate experience but in factors abstracted by reason from experience. It might be that all observed bodies do in fact move through a medium; the *law* of their motion was nevertheless to be sought, not in immediate experience, but by abstract analysis which discovered the intelligible real world as an idealization of which the multifarious diversity of the world of experience was the composite product and in a sense the 'appearance'. Against this Averroës identified the real world with the directly observable and the concrete, and looked for the law of motion close to the data of experience in all their immediate diversity.

The conclusion of Averroës' line of argument would be to attribute the abstract factors into which we analyse immediate experience to our ways of thinking rather than to the things thought of, to regard these factors as mere concepts or even

* See E. A. Moody, 'Galileo and Avempace,' *Journal of the History of Ideas*, 1951, vol. 12.

names, not as discoveries of something real. This was the issue between the 'nominalists' and 'realists' in the Middle Ages and between the 'empiricists' and 'rationalists' in the 17th and 18th centuries. It represents a major difference not only in philosophy of nature but in scientific method. Certainly Averroës and his Western followers saw their close empiricism as a true expression of Aristotelian methods, whereas Avemplace was described by Albertus Magnus and Aquinas as a Platonist, and Galileo was to claim his method of mathematical idealization as a triumph for Plato over Aristotle. The methods applied on the different sides of the debate in the 13th and 14th centuries can be seen from these two points of view, although the positive contributions to the problem of motion by no means all came from one side.

In the 13th century it was mainly the philosophical issues that determined the terms of the discussion of motion, but this gave way in the 14th century to a greater attention to the mathematical and quantitative formulation of laws of motion. Attention began to turn from the 'why' to the 'how'. Practically without exception – the most significant was William of Ockham – the natural philosophers of this period based their discussions on the accepted Aristotelian principle that being in motion meant being moved by something. Differences of opinion concerned the nature of the moving power in the different cases and the quantitative relations between the different determinants of velocity.

The first scholastic philosopher to take up the debate between Averroës and Avemplace was Albertus Magnus. He stood firm for Averroës, and in this he was followed by Giles of Rome and others, until in the 14th century Thomas Bradwardine produced an original version of the Aristotelian 'law' expressing the proportionality between velocity and power and resistance. Averroës had taken up Aristotle's own reservations about the law $v \alpha \, p/r$, in the case where power failed to overcome resistance and produce any movement at all (see above, p. 62). He had tried to overcome this difficulty by saying that velocity followed the *excess* of power over resistance, and some 13th-century Latin writers supposed movement to arise only when p/r was greater than 1. Thomas Bradwardine, in his *Tractatus Proportionum* (1328), limited comparisons of the proportion of power to resistance to cases when this was so. He tried, in what seems to be one of the earliest attempts to use algebraic functions to describe motion, to show how the dependent variable v was related to the two independent variables p and r.

The formulation of the Aristotelian 'law of motion' metrically as a function, so that it became quantitatively refutable, was an achievement of the greatest importance, even though neither Bradwardine nor any of his contemporaries discovered an expression that fitted the facts or indeed applied any empirical quantitative tests. The first requirement was to give a metric definition of velocity as a magnitude representing a ratio between space and time. Aristotle had not only failed to do this, but his method of expression had not clearly distinguished the static analysis of the relationship between power (p), resistance (r) and distance (s), where time (t) is not considered, for example in dealing with the lifting of weights, from the kinematic-dynamic analysis where time is considered (cf. Vol. I, pp. 126–7). The first writer, at least in the West, to attempt a purely kinematic analysis of motion seems to have been Gerard of Brussels, whose important treatise *De Motu* was composed, according to Clagett, possibly between 1187 and 1260. It appears to have been associated in some way with the activities of Jordanus, and it shows the strong influence of Euclid and Archimedes, making use of the latter's characteristic type of proof by *reductio ad absurdum* (or proof *per impossibile*) and method of exhaustion. Dealing with movements of rotation, Gerard took an approach that has become characteristic of modern kinematics, seeing as the basic objective of analysis the representation of non-uniform velocities by uniform velocities. Although he fell short of defining velocity as a ratio of unlike quantities, his analysis inevitably involved the concept of velocity, and he seems to have assumed that the speed of a motion can be assigned some number or quantity making it a magnitude like space or time. Bradwardine specifically discussed some of Gerard's propositions, and it seems probable that *De Motu* directed the attention of the Oxford mathematicians of the 14th century to the kinematic description of variable movements and to the metric definition of velocity required for their treatment (cf. below, p. 105 *et seq.*).

Using his metric formulation, Bradwardine was able to show that Aristotle's analysis and various other current formulae, including Avempace's, did not fit the facts of *moving* bodies, as he understood them. He rejected them all because they did not satisfy his physical presuppositions or hold for all values. In their place he proposed an interpretation of Aristotle's law based on the theorem given in Campanus of Novara's commentary on Euclids' fifth book in which it was proved that if $a/b = b/c$, then $a/c = (b/c)^2$. Bradwardine argued that Aristotle's law meant that if a given ratio p/r produced a velocity v, then the ratio that would double this velocity was not $2p/r$ but $(p/r)^2$, and the ratio that would halve it was $\sqrt{p/r}$. The exponential function by which he related these variables may be written, in modern terminology, $v = \log (p/r)$. Since the logarithm of $1/1$ is zero, the condition is satisfied that when force and resistance are equal, no motion results, and the formula gives a continuous gradual change as v as p/r approaches 1. Although Bradwardine's treatment of dynamics suffered from the serious defect (by no means unique in the period) that he did not test his 'law' by making measurements, his formulation of the problem in terms of an equation in which the complexity of the relations involved were recognized was an important contribution to the methods of mathematical physics. His shifting of the ground of the discussion of motion from 'why' to 'how' had an immediate and lasting influence. His equation was accepted by the Oxford mathematicians Heytesbury, Dumbleton and Richard Swineshead (see below, p. 105) and by Buridan, Albert of Saxony and Nicole Oresme, and down to the 16th century it was almost universally held to be the true Aristotelian 'law of motion'.

The earliest and most important critic of Aristotle's 'law of motion' from the point of view of Avempace was Thomas Aquinas. The main point at issue was whether a body would move with finite velocity in a void. In his commentary on the

71

Physics Aquinas supported Avempace's argument that even without any resistance, all motion must take time because it traverses extended distance. Hence he accepted Avempace's 'law', $v = p - r$. He was even prepared to accept Averroës' assertion that this would imply an 'element of violence' in all actual natural motions, for these *all started* from an unnatural place. Roger Bacon, Peter Olivi (1245/49–98), Duns Scotus and other 13th-century writers followed Aquinas in defending Avempace. In the 14th century his 'law' was generally rejected under the influence of Averroës and Bradwardine, but it found a supporter towards the end of the century in a certain Magister Claius. He held that heavy bodies would fall in a void faster than light bodies, but that none would have an infinite velocity. It was an expression for motion identical with Avempace's that Galileo was to use in his early work on dynamics at Pisa.

Associated with Avempace's quantitative analysis of motion, there were new attempts in the 13th century to explain the cause of the acceleration of freely falling bodies and of the continued velocity of projectiles. Clearly the medium could be of no assistance if these were considered *in vacuo*. It is a disputed point whether Aquinas himself accepted the theory that the original agent impressed on the projectile some kind of power, some *virtus impressa*, which acted as the *instrument* of its continued motion. Certainly he discussed this theory, but he also distinguished clearly between natural motive powers such as the intrinsic power of growth given by the father to the seed in reproduction, and the unnatural extrinsic power moving a projectile. The latter he seems in fact to have attributed to the medium. Olivi did propose an explanation of projectile motion by what he called, in his *Quæstiones in secundum librum Sententiarum*, 'violent impulses or inclinations given by the projector', comparable with the natural impulses of heaviness and lightness. The context of Olivi's explanation was the problem of action at a distance in a discussion of causality in general. He cited projectile motion as an example of action caused not by direct contact, or by the medium, but by 'species' or 'similitudes' or 'impressions' impressed by the agent of projection on the projectile and moving it after separation from the thrower. In fact Olivi's explanation was an adaptation of the theory of the 'multiplication of species' of Grosseteste and Roger Bacon (cf. Vol. I, pp. 88, 112–13, above, p. 57 *et seq.*). It was basically a Neoplatonic emanation, and essential to it was that it moved towards a goal.

The first scholastic natural philosopher to put forward a theory of 'impressed force' as an Aristotelian motive power, a *vis motrix* determined not by the goal but by the projecting agent, seems to have been an Italian follower of Duns Scotus, Franciscus de Marchia. In his commentary on the *Sentences*, written about 1320 in Paris, Marchia followed Aquinas in discussing the problem of instrumental causality. The context of the problem, moving by analogy with ease from theology to physics, is characteristic of much scholastic natural· philosophy. In inquiring whether any power to produce grace resided in the sacraments themselves or came only direct from God, Marchia raised the question of projectile motion in order to show that both in the sacraments and in projectiles there was a certain residual power that was capable of producing effects. Rejecting Aristotle's theory that projectile motion was caused by the air, he concluded that it must be explained, as translated from the passage quoted by Dr Maier in her *Zwei Grundprobleme der Scholastischen Naturphilosophie* (p. 174), 'by the motion or impulse of a power left behind (*virtus derelicta*) in the stone by the primary mover,' that is, by the hand or the bowstring. Marchia was careful to point out that this power was not innate or permanent. It was an accidental quality, which was extrinsic and violent, and being opposed to the natural inclinations of the body it endured only for a certain time. The motive power of a projectile was, he said, a 'form' that was neither wholly permanent, like whiteness or the heat of fire, nor wholly transient (*fluens, successiva*) like the process of heating or of moving, but something intermediate which endured for a limited time.

The existence of a similar 'law of motion' and similar conceptions of motive power in the writings of Philoponus and Avempace and of the scholastics of the 13th and 14th centuries has naturally led historians to look for a possible historical connexion between them. Certainly nearly all these writers belong to the Neoplatonic tradition, but no actual documentary derivation has been traced. So far as is known, Philoponus' own writings were not known in the Middle Ages. Direct medieval knowledge of his views seems to have been largely limited to the incomplete and not very clear presentation of his position by Simplicius, whose commentary on the *Physics* was translated into Latin in the 13th century. Avicenna's discussion of projectile motion and 'impressed power' does not occur in the part of his commentary that was translated into Latin under the name *Sufficientia Physicorum*, which contains only the first four books (cf. Vol. I, p. 58). Alpetragius is known to have been strongly influenced by a disciple of Avempace, Ibn Tofail, and the Latin translation of Alpetragius' work made in 1528 and published in Venice in 1531 as *Theorica Planetarum* gave a clear account of Philoponus' theory, though not under his name. But in the medieval translation, made by Michael Scot in 1217 with the name *Liber Astronomiæ*, the theory is abridged out of existence in the passage concerned. So far as the evidence goes, Dr Maier has concluded that the theory of 'impressed power', and that of *impetus* which succeeded it in the 14th century, were developed independently by the scholastics, mainly through their discussions of instrumental causality in reproduction and in the sacraments.

Not all natural philosophers in the 13th and 14th centuries accepted this view of the cause of projectile motion, and there were many, for example Giles of Rome, Richard of Middleton, Walter Burley and Jean de Jandun, who continued to accept Aristotle's explanation, however unsatisfactory, because they were even more dissatisfied with the alternatives. They objected both to mediated action at a distance by the 'multiplication of species' and to 'impressed power' as being equally impossible. The author of *De Ratione Ponderis*, of the school of Jordanus Nemorarius (see Vol. I, pp. 129–30), held that the air caused both the continued velocity and a supposed initial acceleration of projectiles; in the 16th century this theory was still partly accepted even by such physicists as Leonardo da Vinci, Cardano and Tartaglia.

To explain the acceleration of freely falling bodies, many natural philosophers continued to follow either Aristotle or the theory using the air and *antiperistasis*. An original account of falling bodies was put forward by Roger Bacon. He supposed that each particle in a heavy body naturally tended to fall by the shortest route towards the centre of the universe, but that each tended to be displaced from this straight path by the particles lateral to it. The resulting mutual interference by the different particles acted as an internal resistance, which would make movement take time even in a void, where there was no external resistance, and so Aristotle's argument that it would be instantaneous did not hold.

As to the nature of the 'form' that was the physical cause of movement, that is, the nature of the motive power that all these theories presupposed as necessary for the state of being in motion, at least two different views were hotly argued in the 14th century. The first view was that usually associated with Duns Scotus, namely, the theory that motion was a 'fluent form' or *forma fluens*. According to this theory, motion was an incessant flow in which it was impossible to divide or isolate a state, and a moving body was successively determined by a form distinct at once from the moving body itself and from the place or space through which it moved. This theory was held by Jean Buridan and Albert of Saxony. The second view was that motion was a 'flux of form' of *fluxus formæ*, according to which motion was a continuous series of distinguishable states. One form of this theory was held by Gregory of Rimini, who identified motion with the space acquired during the movement, and said that during motion the moving body acquired from instant to instant a series of distinct attributes of place.

A third conception of motion, starting from a radically different point of view, was put forward by Ockham. One of the principal objects of Ockham's logical inquiries was to define the criteria by which a thing could be said to exist (cf. above, pp.

44–7). Nothing really existed, he held, except what he called *res absolutæ* or *res permanentes*, individual things, substances determined by observable qualities. 'Apart from *res absolutæ*, that is substances and qualities,' he said in the *Summa Totius Logicæ*, part 1, chapter 49, 'no thing is imaginable either in actuality or in potentiality.' Words like 'time' and 'motion' did not designate *res absolutæ* but relations between *res absolutæ*. They designated what Ockham called *res respectivæ*, without real existence. It is this careful analysis of the references of terms that is so striking a feature of Ockham's work, and it was through this that he and the other 'terminists' did so much to clarify many issues in 14th-century philosophy. As he said in his *Summulæ in Libros Physicorum*, book 3, chapter 7 : 'If we sought precision by using words like "mover", "moved", "movable", "to be moved", and the like, instead of words like "motion", "mobility" and others of the same kind, which according to the form of language and to the opinion of many do not seem to stand for permanent things, many difficulties and doubts would be excluded. But now, because of these, it seems as if motion were some independent thing quite distinct from the permanent things.

Applying these distinctions to the problems of dynamics, Ockham rejected altogether Aristotle's basic principle that local motion was a realized potentiality. He defined motion as the successive existence, without intermediate rest, of a continuous identity existing in different places, and for him movement itself was a concept having no reality apart from the moving bodies that could be perceived. It was unnecessary to postulate any inhering form to cause the movement, any real entity distinct from the moving body, any flux or flow. All that need be said was that from instant to instant a moving body had a different spatial relationship with some other body. Every new effect required a cause, but motion was not a new effect, since it was nothing except that the body existed successively in different places. Ockham therefore rejected all three current explanations of the cause of projectile motion, the impulse of the air, action at a distance mediated by 'species', and 'impressed power' given

to the projectile itself (cf. above, p. 57). 'I say therefore,' he said
in his *Commentary on the Sentences*, book 2, question 26, M,

that that which moves (*ipsum movens*) in motion of this kind, after
the separation of the moving body from the original projector, is the
body moved by itself (*ipsum motum secundum se*) and not by any
power in it or relative to it (*virtus absoluta in eo vel respectiva*), for it
is impossible to distinguish between that which does the moving and
that which is moved (*movens et motum est penitus indistinctum*). If
you say that a new effect has some cause and that local motion is a
new effect, I say that local motion is not a new effect in the sense of
a real effect ..., because it is nothing else but the fact that the moving
body is in different parts of space in such a manner that it is not in
any one part, since two contradictories cannot both be true ...
Though any particular part of space which the moving body traverses is new with respect to the moving body, seeing that the body
now moves through those parts and previously was not doing so, yet
that part is not new really speaking. ... It would indeed be astonishing if my hand were to cause some power in the stone by the mere
fact that through local motion it came into contact with the stone.*

This conception he amplified with an application of the principle of economy in the so-called *Tractatus de Successivis* edited
by Boehner, asserting in part 1 (p. 45):

Motion is not such a thing wholly distinct in itself from the permanent body, because it is futile to use more entities when it is possible
to use fewer ... That without such an additional thing we can save
motion, and everything that is said about motion, is made clear by
considering the separate parts of motion. For it is clear that local
motion is to be conceived as follows: positing that the body is in one
place and later in another place, thus proceeding without any rest or
any intermediate thing other than the body itself and the agent itself
which moves, we have local motion truly. Therefore it is futile to
postulate such other things.

The same applied, he said, to change in quality and to growth
and decrease (cf. Vol. I, pp. 77–8). He continued in part 3 (pp.
121–2):

* Translated from the Latin text published by Anneliese Maier, *Zwei
Grundprobleme der Scholastischen Naturphilosophie*, Rome, 1951, pp. 157–
8.

It is clear how 'now before' and 'now after' are to be assigned, treating 'now' first : this part of the moving body is now in this position, and later it is true to say that now it is in another position, and so on. And so it is clear that 'now' does not signify anything distinct but always signifies the moving body itself which remains the same in itself, so that it neither acquires anything new nor loses anything existing in it. But the moving body does not remain always the same with respect to its surroundings, and so it is possible to assign 'before and after', that is, to say : 'this body is now at A and not at B', and later it will be true to say : 'this body is now at B and not at A', so that contradictories are successively made true.

It has been claimed by some historians that by rejecting the basic Aristotelian principle expressed by the phrase *Omne quod movetur ab alio movetur*, Ockham took the first step towards the principle of inertia* which was to revolutionize physics in the 17th century. Certainly by asserting the possibility of motion under the action of *no* motive power, a possibility formally excluded by the Aristotelian principle, Ockham opened the way to the principle of inertia and to the 17th-century definition of force as that which *alters* the state of rest or of uniform velocity, in other words, that which produces acceleration. The relevance of Ockham's conception of motion to 17th-century ideas becomes even more suggestive when taken in conjunction with the ideas of some other 14th-century writers. Nicholas Autrecourt, for example, related it to his conception of the atomic nature of a continuum and of time. Marsilius of Inghen, though himself rejecting Ockham's conception of motion, discussed it in connexion with the conception of infinite space, an idea closely related to the 'geometrization of space' in the 17th century. Nicole Oresme (d.

---

* According to the principle of inertia a body will remain in a state of rest or of motion with uniform velocity in a straight line unless acted on by a force. This conception was the basis of Newton's mechanics. For Newton uniform rectilinear motion was a condition or state of the body equivalent to rest and no force was required to maintain such a state. The principle of inertia was thus directly contrary to Aristotle's principle according to which motion was not a state but a process and a moving body would cease to move unless continually acted on by a moving force.

78

1382), though he retained the *forma fluens* to explain motion, put forward the idea that absolute motion could be defined only by reference to an immovable infinite space, placed beyond the fixed stars and identified with the infinity of God. From such passages Newton does not seem so far away, both as a physicist and as a natural theologian.

But the relation, both logical and historical, of Ockham's conception of motion to the principle of inertia is by no means straightforward. If we are tempted to read his statements in the light of Descartes' similar assertion that he made no distinction between motion and body in motion, we must also remember that for Descartes and for Newton the change in spatial relationships in passing from a state of rest to a state of motion *was* a new effect. It was an effect that required for its production not only a cause, but a precisely determined one. From Ockham's conception of motion it is not in fact possible to deduce some of the essential properties of the conservation of speed and direction implied by the modern principle of inertia. Yet Ockham had not overlooked the dynamical aspects of motion. In his *Expositio super Libros Physicorum*, when discussing the controversy between the supporters of Averroës and of Avempace, he defended Aquinas for asserting that where there was no resistance motion would take time, the length of time depending on the distance. But where there was a material resistance, he said that the time would depend on the proportion of the motive power to the resistance. In this way he distinguished what we would now call the kinematic measure of velocity from the dynamical measure of the motive power or force in terms of the work done. The confusion of these measures is another example of the difficulty with which the (to us) apparently most elementary mechanical concepts were grasped, a difficulty which even the entire 17th century did not wholly overcome. When Bradwardine rejected Avempace's 'law of motion', he made use of arguments similar to Ockham's, and it is difficult not to see a connexion in the common shift of the problem from the 'why' to the 'how' which Ockham made as a logician and Bradwardine as a mathematical physicist.

In the event it was not Ockham who produced the most significant and influential new dynamical theory in the 14th century, but a physicist whose outlook was profoundly opposed to that of the 'terminists', Jean Buridan, twice Rector of the University of Paris between 1328 and 1340. Buridan discussed the classical problems of motion in his *Quæstiones super Octo Libros Physicorum* and in his *Quæstiones de Cælo et Mundo*. To the existing criticisms of the Platonic and the Aristotelian theories of projectile motion, he added that the air could not account for the rotational motion of a grindstone or a disc, for the motion continued even when a covering was placed close to the bodies, thus cutting off the air. He rejected likewise the explanation of the acceleration of freely falling bodies by their attraction to the natural place, because he maintained that the mover must accompany the body moved (cf. above, p. 58 *et seq.*). The theory of *impetus* by means of which he explained the various phenomena of persistent and accelerated motion was based, like the earlier theory of *virtus impressa*, on Aristotle's principles that all motion requires a motive power and that the cause must be commensurate with the effect. In this sense the theory of *impetus* was the historical conclusion of a line of development within Aristotelian physics, rather than the beginning of a new dynamics of inertia, of which, since it lay in the future, Buridan himself naturally knew nothing. But, under the influence of Bradwardine, Buridan formulated his theory with much greater quantitative precision than any of its predecessors. It is this aspect of some of his essential definitions that looks to the future.

Since other explanations of the persistence of motion of a body after separation from the original mover failed, Buridan concluded that the mover must impress on the body itself a certain *impetus*, a motive power by which it continued to move until affected by the action of independent forces. In projectiles this *impetus* was gradually reduced by air resistance and natural gravity downwards; in freely falling bodies it was gradually increased by natural gravity acting as an accelerating force which added successive increments of *impetus*, or 'accidental gravity', to that already acquired. The measure of the *impetus* of a body was the quantity of matter in it multiplied by its velocity.

'Therefore it seems to me,' wrote Buridan in his *Quæstiones super Octo Libros Physicorum*, book 8, question 12,

that we must conclude that a mover, in moving a body, impresses on it a certain *impetus*, a certain power capable of moving this body in the direction in which the mover set it going, whether upwards, downwards, sideways or in a circle. By the same amount that the mover moves the same body swiftly, by that amount is the *impetus* that is impressed on it more powerful. It is by this *impetus* that the stone is moved after the thrower ceases to move it; but because of the resistance of the air and also because of the gravity of the stone, which inclines it to move in a direction opposite to that towards which the *impetus* tends to move it, this *impetus* is continually weakened. Therefore the movement of the stone will become continually slower, and at length the *impetus* is so diminished or destroyed that the gravity of the stone prevails over it and moves the stone down towards its natural place.

One can, I think, accept this explanation because the other explanations do not appear to be true, whereas all the phenomena accord with this one.

For if it is asked why I can throw a stone farther than a feather and a piece of iron or lead suited to the hand farther than a piece of wood of the same size, I say that the cause of this is that the reception of all forms and natural dispositions is in matter and by reason of matter. Hence, the greater quantity of matter a body contains, the more *impetus* it can receive and the greater the intensity with which it can receive it. Now in a dense, heavy body there is, other things being equal, more *materia prima* than in a rare, light body.* Therefore a dense, heavy body receives more of this *impetus* and receives it with more intensity [than a rare, light body]. In the same way a certain

* Buridan's *materia prima* was, like that in the *Timæus*, already extended with dimensions. Quantity of matter was then proportional to volume and density. Duhem (*Études sur Léonard de Vinci*, 3ᵉ série, 1913, pp. 46–49) suggests that he approached the notion of density through that of specific weight, to which it was proportional. The Greek pseudo-Archimedean *Liber Archimedis de ponderibus* defined specific weight and showed how to compare the specific weights of different bodies by the hydrostatic balance or areometer. This work was well known in the 13th and 14th centuries.

quantity of iron can receive more heat than an equal quantity of wood or water. A feather receives so feeble an *impetus* that it is soon destroyed by the resistance of the air and, similarly, if one projects with equal velocity a light piece of wood and a heavy piece of iron of the same size and shape, the piece of iron will go farther because the *impetus* impressed on it is more intense, and this does not decay as fast as the weaker *impetus*. It is for the same cause that it is more difficult to stop a big mill wheel, moved rapidly, than a smaller wheel: there is in the big wheel, other things being equal, more *impetus* than in the small. In virtue of the same cause you can throw a stone of one pound or half a pound farther than the thousandth part of this stone: in this thousandth part the *impetus* is so small that it is all soon overcome by the resistance of the air.

This seems to me to be also the cause on account of which the natural fall of heavy bodies goes on continually accelerating. At the beginning of this fall, gravity alone moved the body: it fell then more slowly; but, in moving, this gravity impressed on the heavy body an *impetus*, which *impetus* moves the body at the same time as gravity. The movement therefore becomes more rapid, and by the amount that it is made more rapid, so the more intense the *impetus* becomes. It is thus evident that the movement will go on accelerating continually.

Anyone who wants to jump far draws back a long way so that he can run faster and so acquire an *impetus* which, during the jump, carries him a long distance. Moreover, while he runs and jumps he does not feel that the air moves him, but he feels the air in front of him resist with force.

One does not find in the Bible that there are Intelligences charged to communicate to the celestial spheres their proper motions; it is permissible then to show that it is not necessary to suppose the existence of such Intelligences. One could say, in fact, that God, when he created the universe, set each of the celestial spheres in motion as it pleased him, impressing on each of them an *impetus* which has moved it ever since. God has therefore no longer to move these spheres, except in exerting a general influence similar to that by which he gives his concurrence to all phenomena. Thus he could rest on the seventh day from the work he had achieved, confiding to created things their mutual causes and effects. These *impetūs* which God impressed on the celestial bodies have not been reduced or destroyed by the passage of time, because there was not, in celestial bodies, any inclination towards other movements, and there was no resistance which could corrupt and restrain these *impetūs*. All this I do not give as certain; I would merely ask theologians to teach me how all these things could come about . . .*

He went on to define the relation of his theory of *impetus* to other contemporary theories of motion. First he insisted that while the *impetus* of a projectile was an intrinsic principle of motion that inhered in the body it moved, it was a violent and unnatural principle impressed on the body by an extrinsic agent, and was opposed to the body's natural gravity. But what was *impetus*? It could not be identified with the motion itself, he argued evidently with an eye to Ockham, for the purpose of the theory was to propose a cause of the motion. So it was something distinct from the moving body. Nor could it be something purely transient, like motion itself, for this required a continuous agent to produce it. So, he concluded,

this *impetus* is an enduring thing (*res naturæ permanentis*), distinct from local motion, by which the projectile is moved ... And it is probable that this *impetus* is a quality designed by nature to move the body on which it is impressed, just as it is said that a quality impressed by a magnet on a piece of iron moves the iron to the magnet. And it is probable that just as this quality is impressed by the mover on the moving body together with the motion, so also it is decreased, corrupted and hindered, just as the motion is, by resistance [of the medium] or the contrary [natural] tendency.

It has been claimed that by making *impetus* a *res permanens*, an enduring motive power that would maintain the body in motion unchanged so long as it was not acted on by forces that either diminished or increased it, Buridan took a strategic step towards the principle of inertia. Certainly his *impetus* was from this point of view an improvement on Marchia's *virtus*, which endured only *ad modicum tempus*. Certainly also there are striking resemblances between some of the basic definitions found in Buridan's and in 17th-century dynamics. Buridan's measure of

* Translated from the Latin published by Anneliese Maier, *Zwei Grundprobleme der Scholastischen Naturphilosophie*, Rome, 1951, pp. 211–12; the passages below are translated from pp. 213–14, 223.

the *impetus* of a body as proportional to the quantity of matter and the velocity suggests Galileo's definition of *impeto* or *momento*, Descartes' *quantité de mouvement*, and even Newton's *momentum* as the product of mass multiplied by velocity. It is true that, in the absence of independent forces, Buridan's *impetus* would endure in a circle in celestial bodies, as well as in a straight line in terrestrial bodies, whereas Newton's momentum would persist only in a straight line in all bodies and required a force to bend it in a circle. But in this Galileo was not with Newton but stood somewhere between him and Buridan.

There is a certain resemblance also between Buridan's *impetus* and Leibniz's '*force vive*', or kinetic energy. In explaining the acceleration of freely falling bodies, Buridan said in his *Quæstiones de Cælo et Mundo*, book 2, question 12 :

It must be imagined that a heavy body not merely acquires motion from its primary mover, namely from its gravity, but that it also acquires in itself a certain *impetus* together with that motion, which has the power of moving that same body, together with the constant natural gravity. And because this *impetus* is acquired commensurately with the motion, therefore the faster the motion, the greater and stronger is the *impetus*. So, therefore, the heavy body is initially moved only by its natural gravity, and hence slowly, but afterwards it is moved by the same natural gravity and simultaneously by the *impetus* that has been acquired, and thus it is moved more rapidly; ... and so again it is moved more rapidly, and thus it is continuously accelerated always, to the end.

Some people, he concluded, call this *impetus* 'accidental gravity'.

It is interesting to look for analogies between terms appearing in systems of dynamics so widely separated in time, but these can also hide from us the gap that may separate their content. Can it really be said that Buridan's formulation of the theory of *impetus* implied the 17th-century definition of force as that which did not simply maintain velocity but altered it? Everything Buridan wrote about *impetus* indicates that he was proposing it as an Aristotelian cause of motion that should be commensurate with the effect; therefore if the velocity increased, as in falling bodies, so must the *impetus*. It is true that, as a result of his attempt at quantitative formulations, Buridan's *impetus* can be seen as something more than an Aristotelian cause, as a force or power possessed by a body, by reason of being in motion, of altering the state of rest or motion of other bodies in its path. It is true also that there is too much similarity between this and the definition of *impeto* or *momento* given by Galileo in his *Two New Sciences* for it to be supposed that he owed nothing to Buridan (cf. below, p. 160 *et seq.*). But considering it in its own period, and not as a precursor of something in the future, it is clear that Buridan himself saw his theory as a solution of the classical problems that arose within the context of Aristotelian dynamics, from which he never escaped.

This is illustrated by the most suggestive question 9 of book 12 of his *Quæstiones in Libros Metaphysicæ*.

Many people posit that the projectile, after leaving the projector, is moved by an *impetus* given by the projector, and that it is moved as long as the *impetus* remains stronger than the resistance. The *impetus* would last indefinitely (*in infinitum duraret impetus*) if it were not diminished by a resisting contrary, or by an inclination to a contrary motion; and in celestial motion there is no resisting contrary, so that when, at the creation of the world, God moved any sphere with whatever velocity he wished, he ceased from moving and that motion endured forever afterwards because of the *impetus* impressed on that sphere. Hence it is said that God rested on the seventh day from all the works that he had performed.

Did this mean that *impetus* would in fact always endure forever in all bodies in the absence of opposing forces? Buridan asserts this only for the celestial bodies, whose continuing motion was naturally circular. But in terrestrial bodies the *impetus* impressed violently, for example on a projectile, would always be opposed by the intrinsic natural tendency of the body towards its natural place, there to come to rest. Moreover, according to the basic dynamical law, which Buridan accepted in Bradwardine's formulation, that velocity was proportional to power and resistance, if there were no resistance velocity would be infinite. Sharing the empiricism common to all Aristotelians, Buridan did not consider abstracting the effects of *impetus* alone from those of its interaction with natural tendencies and with resistance. He stayed close to the actual world as he saw it. He did not conceive the principle of inertial motion in empty space.

But in a profound sense Buridan and his contemporaries did anticipate the great cosmological reform of the 16th and 17th centuries. Buridan's theory of *impetus* was an attempt to include both terrestrial and celestial movements in a single system of mechanics. In this he was followed by Albert of Saxony, Marsilius of Inghen and Nicole Oresme; although Oresme, holding that in the terrestrial region there were only accelerated and retarded motions, adapted the theory of *impetus* to this assumption and seems to have regarded it not as a *res naturæ permanentis* but as something that 'lasted only for a certain time'. In one form or another the theory became widely accepted in the 14th, 15th and 16th centuries in France, England, Germany and Italy.

As to questions of terrestrial dynamics, Buridan himself explained the bouncing of a tennis ball by analogy with the reflection of light, by saying that the initial *impetus* compressed the ball by violence when it struck the ground, and when it sprang back this imparted a new *impetus* which caused the ball to bounce up.* He gave a similar explanation for the vibration of plucked strings and the oscillation of a pendulum.

Albert of Saxony used Buridan's theory in his explanation of the trajectory of a projectile by compound *impetus*, an idea which itself went back to the 2nd-century B.C. Greek astronomer, Hipparchus, whose account was preserved in Simplicius' commentary on *De Cælo*. According to Aristotelian principles an elementary body could have only one simple motion at any time, for a substance could not have two contradictory attributes simultaneously. If it did, one would destroy the other. Albert of Saxony held that the trajectory of a projectile was divided into three periods: (1) an initial period of purely violent motion during which the impressed *impetus* annihilated natural gravity; (2) an intermediate period of compound *impetus* during which

---

* By contrast Descartes in *La Dioptrique* explained the reflection and refraction of light by analogy with the mechanics of a tennis ball. Cf. below, pp. 130, 258.

movement was both violent and natural; and (3) a final period of purely natural movement vertically downwards after natural gravity and air resistance had overcome the impressed *impetus* (Fig. 1). He considered air resistance as having a definite frictional value even when the projectile was at rest. In a horizontally fired projectile, motion during the first period was in a

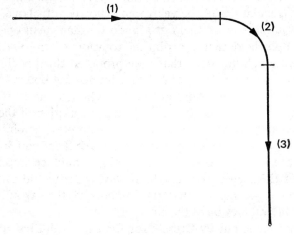

Fig. 1

horizontal straight line until it suddenly curved during the second period to fall vertically in the third. When fired vertically upwards the projectile came to rest during the second period (or *quies media*) and then descended when natural gravity overcame air resistance. This theory was accepted by Blasius of Parma (d. 1416), Nicholas of Cusa, Leonardo da Vinci and other followers of Albert of Saxony, until it was modified in accordance with mathematical principles by Tartaglia in the 16th century and finally replaced by Galileo in the 17th.

The most significant developments of the new dynamics in the celestial region took place in application to the possibility of the daily rotation of the earth on its axis (cf. Vol. I, p. 103). This had been discussed and rejected in the 13th century by two Persian astronomers, al-Katibi and al-Shirazi, though no connexion has been established between them and the Latin writers of the 14th century. For the latter the question involved not only the dynamical explanation of the persistence of motion, but also the conceptions of space and of gravitation. The most important writers to discuss the possibility of the motion of the earth and to relate it to these cognate problems were Buridan and Oresme. The frequency with which they referred to the Parisian condemnations of 1277 is a further illustration of the significance of these in the scientific speculations of the years that followed (cf. above, p. 49).

In his *Quæstiones de Cælo et Mundo* Buridan mentioned that many people held that the diurnal rotation of the earth was probable, though he added that they proposed this possibility as a scholastic exercise. He realized that immediate observation of the bodies themselves could not decide whether the heavens or the earth were in motion, but he rejected the motion of the earth on the grounds of the observations. For example, he pointed out that an arrow shot vertically fell to the place from which it was shot. If the earth revolved, he said, this would be impossible; and as for the suggestion that the revolving air would carry the arrow round, he replied that the *impetus* of the arrow would resist the lateral motion of the air.

The case made out by Oresme for the earth's diurnal rotation was far more elaborate. He discussed the problem in his *Livre du Ciel et du Monde*, a French commentary on Aristotle's *De Cælo* written in 1377 by command of Charles V of France, who commissioned him also to translate from the Latin into French Aristotle's *Ethics, Politics*, and *Economics*.* A lover of learning and of his own language, Charles' *cabinet de livres* at the Louvre contained a large number of books translated into the vernacular at his own command, and these he encouraged the members of his entourage to read for their education and enjoyment. Although he concluded his *Livre du Ciel* by deciding in favour of the geostatic system, Oresme's analysis of the whole question was the most detailed and acute made between the Greek astronomers and Copernicus. In its treatment of the mixture of scientific,

---

* Like Copernicus later, Oresme also wrote an extremely penetrating treatise on money : see *The De Moneta of Nicholas Oresme and English Mint Documents*, trans. C. Johnson, London and Edinburgh, 1956.

philosophical and theological issues involved it foreshadowed the controversial writings of Galileo.

An important question discussed by Oresme in expounding the geostatic system was that of the constant motion of the spheres. Since his version of the *impetus* theory could not account for constant motion, he fell back on a vague theory of a balance between 'motive qualities and powers' which God gave to the spheres at the creation to correspond to the gravity (*pesanteur*) of terrestrial bodies, and commensurate 'resistance' which opposed these powers (*vertus*). In fact he said that at the creation these powers and resistances had been bestowed by God upon the 'Intelligences' that moved the heavenly bodies; the Intelligences moved with the bodies whose movement they caused and were related to them as the human soul was to the body. Comparing the celestial machine to a clock, he concluded, in book 2, chapter 2 of *Le Livre du Ciel:*

And these powers are so controlled, tempered and harmonized with the resistances that the movements are made without violence; and apart from violence, it is not in the least like a man making a clock and letting it go and be moved by itself. Thus God left the heavens to be moved continually in accordance with the proportions which their motive powers have to their resistances, and with the established order.

But was it possible to accept the assumptions on which the geostatic system, and the traditional objections to the earth's movement, were based? One of the essential assumptions of Aristotle's cosmology was that there must be at the centre of the universe a fixed body about which the celestial spheres revolved and in relation to which the natural movements of terrestrial bodies took place. Against this Oresme argued that the directions of space, motion, and natural gravitation and levitation must, in so far as they were observable, all be regarded as relative.

Oresme agreed with those who argued that God, by his infinite power, could create an infinite space and as many universes as he chose. 'And so,' he wrote in book 1, chapter 24 of *Le Livre du Ciel*, 'beyond the sky is an empty, incorporeal space quite different from ordinary full and corporeal space, just exactly as the duration known as eternity is quite different from temporal duration, even if it were perpetual . . . Further, this space mentioned above is infinite and indivisible and is the immensity of God and is even God, just as the duration of God known as eternity is infinite and indivisible and even God . . .'

So far as directions were distinguished within our universe, Oresme showed that, considering right and left, before and behind, 'these 4 differences in the sky are not absolutely and really distinct, but only relatively, as it is said' (book 2, chapter 6). Only up and down could be said to be absolutely and really distinct, but then only relative to a particular universe. We could, for example, distinguish up and down according to the motion of light and heavy bodies. 'I say then that high and low in this . . . way are nothing else but the natural order of heavy and light things, which is such that all the heavy things, so far as is possible, are in the middle of the light things, without determining any other immovable place for them' (book 1, chapter 24). By combining this Pythagorean or Platonic theory of gravity with the conception of infinite space, Oresme was thus able to dispense with a fixed centre of the universe to which all natural gravitational movements were related. Gravity was simply the tendency of heavier bodies to go to the centre of spherical masses of matter. Movements were produced by gravity only relative to a particular universe; there was no absolute direction of gravity applying to all space.

There was then no ground for arguing that, supposing that the skies revolved, the earth must necessarily be fixed in the centre. On the analogy of a revolving wheel, Oresme showed that it was only necessary in circular motion that an imaginary mathematical point in the centre be at rest, as was in fact assumed in the theory of epicycles. Moreover he said that it was not part of the definition of local motion that it should be referred to some fixed point or body. For example, 'beyond the universe is a space conceived as infinite and immobile, and it is possible without contradiction for the whole universe to be moved in this space in a straight line. And to say the contrary is an article condemned in Paris. This postulated, there is no other body to which

the universe is related in any other way according to place ...
Further, imagining that the earth was moved through space for
one day of daily motion and that the heavens were at rest, and
after this time that things were again as they were' (book 2,
chapter 8) : then everything would again be as it was before.

In chapter 25 of book 2 of *Le Livre du Ciel* Oresme said that,
'subject to correction,' it seemed to him possible to maintain the
opinion 'that the earth is moving with daily motion and the
heavens not. And first, I will declare that it is impossible to show
the contrary by any observation (*expérience*); secondly, from
reason (*par raisons*); and thirdly, I will give reasons in favour of
the opinion.' The objections which Oresme quoted against the
earth's motion had all been mentioned by Ptolemy and were to
be used against Copernicus; he met them with arguments that
were to be used again by Copernicus and by Bruno.

The first objection from experience was that the skies were
actually observed to revolve about their polar axis. To this
Oresme replied, citing the fourth book of Witelo's *Perspective*,
that the only motion that could be observed was relative motion.

I assume that local motion cannot be observed except in so far as a
body can be seen to change its position in respect to another body.
Thus, if a man is in a boat A, moving very smoothly, either fast or
slowly, and he can see nothing outside except another boat B, moving
in exactly the same way as the boat A in which he is, I say that it
will seem to this man that neither of the boats is moving. If A is at
rest and B is moving, it will seem to him that B is moving; and if A is
moving and B is at rest, it will seem to him just as before that B is
moving. And so if A was at rest for an hour and B was moving, and
then in the next hour, *e converso*, A was moving and B remained at
rest, this man would not be able to perceive this change or variation,
but it would seem to him all the time that B was moving; and this is
evident from experience ... It would seem to us all the time that the
place where we are was at rest and that the other always moved, just
as it seems to a man in a moving boat that the trees outside are
moving. Similarly, if a man was on the sky, supposing that he was
moving with daily motion, ... it would seem to him that the earth
was moving with daily motion, just as the sky seems to be, to us on the
earth. Similarly, if the earth were moving with daily motion and the

sky was not, it would seem to us that the earth was at rest and that the sky was moving. Any intelligent person can easily imagine this.

The second objection from experience was if the earth were turning through the air from west to east, there should be a continuous strong wind blowing from the east. To this Oresme replied that the air and the water would share the earth's rotation, so that there would be no such wind. The third objection was that which had convinced Buridan : that if the earth were rotating an arrow or stone sent vertically upwards should be left behind to the west when it fell, whereas in fact it fell to the place whence it was sent up. Oresme's answer to this was profoundly significant. He said that the arrow 'is moved very rapidly eastwards with the air through which it goes and with the whole mass of the inferior part of the universe indicated before which is moved with daily motion, and thus the arrow returns to the place on the earth from which it was sent.' The arrow would in fact have not one movement but two, a vertical movement from the bow, and a circular movement from being on the rotating globe. The actual trajectory of the arrow, he said, would be comparable with that of a particle of fire (*a*) which rose from one position to a higher one nearer the celestial spheres. This he illustrated with a diagram, showing that the particle of fire would not simply rise to a position *b* directly above *a*, but as it rose would be carried laterally by the circular motion to the position *c* to one side of *b*.

I say that just as in the case of the arrow discussed above, so in this case it can be said that the motion of *a* is composed (*composé*) partly of a rectilinear motion and partly of a circular motion, for the region of the air and the sphere of fire through which *a* passes are moving, according to Aristotle, with circular motion. If they were not so moved, *a* would rise straight up on the line *ab*; but since *b* is meanwhile translated, by daily circular motion, to the point *c*, it is clear that as it rises *a* describes the line *ac*, and that the motion of *a* is composed of rectilinear and circular motion. The motion of the arrow will be of the same kind, as has been said; it will be a composition or mixture of motions (*composition ou mixcion de movemens*) . . .*

* This would seem incompatible with acceptance of the three-fold division of the trajectory of a projectile : see above, pp. 86–7.

Thus, just as to a person on a moving ship any movement rectilinear with respect to the ship appears rectilinear, so to a person on the earth the arrow would appear to fall vertically to the point from which it was fired. The movement would appear the same to an observer on the earth whether the earth rotated or was at rest. 'I conclude then that it is impossible to show by any observation that the heavens are moving with daily motion and that the earth is not moving in this way.' This conception of the composition of movements was to become one of the most fruitful in Galileo's dynamics.

The objections 'from reason' against the earth's motion came mainly from the Aristotelian principle, used later by Tycho Brahe against Copernicus, that an elementary body could have only one simple movement which, for earth, was rectilinearly downwards. Oresme asserted that each of the elements except the skies might well have two natural movements, one being rotation in a circle when it was in its natural place, and the other being rectilinear motion by which it returned to its natural place when displaced from it. The *'vertu'* that moved the earth in rotation was its 'nature' or 'form', just as was that which moved it rectilinearly back to its natural place. As for the objection that the earth's rotation would ruin astrology, Oresme replied that all the calculations and tables would be just as before.

The main positive arguments that Oresme brought in favour of the earth's rotation all turned on this being simpler and more perfect than the alternative, once more a striking anticipation of the arguments, Platonic in inspiration, that were to be used by Copernicus and Galileo. If the earth rotated, he said, all the apparent celestial motions would take place in the same sense, from east to west; the habitable part of the globe would be on its right or nobler side; the heavens would enjoy the nobler state of rest and the base earth would move; the more distant celestial bodies would make their revolutions proportionately more slowly than those nearer the east, instead of more rapidly as in the geocentric system. Moreover,

all philosophers say that anything done by many or by large operations that could be done by less or smaller operations would be done in

vain. And Aristotle says ... that God and Nature do nothing in vain.
... And so, since all the effects which we see can be produced and all
appearances saved by one small operation, namely the daily motion
of the earth, which is very small compared with the heavens, without
so multiplying operations which are so diverse and outrageously
large, it follows that God and Nature would have made and ordered
such operations for nothing, and that is not fitting, as the saying goes.

Among the advantages of simplicity was that the ninth sphere
would no longer be necessary.

Throughout his discussions Oresme, the Bishop, after all, of
Lisieux, had taken into account the support apparently given to
the geostatic system by many passages of Scripture, but these he
had turned by remarking for example: 'One can say that it (*scil.*
Scripture) conforms in this part to the manner of common
human speech, just as it does in several places, as where it is
written that God repented and that he became angry and calm
again, and things of the same kind, which are not in fact at all as
the letter puts it.' Again we are reminded of Galileo, and in the
same spirit Oresme dealt with the celebrated problem of Joshua's
miracle and asserted that no arguments could be found against
the earth's motion.

When God performs any miracle, it must be supposed and held that
he does this without disturbing the common course of nature more
than the least that is necessary for the miracle. And so, if one can say
that God lengthened the day in the time of Joshua by stopping only
the motion of the earth or the inferior region, which is so small,
indeed a mere point compared with the heavens, without bringing it
about that the whole universe outside this little point has been put out
of its common course and order, and likewise the heavenly bodies,
then this is much more reasonable ... and one can say the same thing
about the sun going back on its course in the time of Hezekiah.

After finally reviewing all the arguments he has brought
against the accepted cosmology, it is somewhat surprising to find
Oresme concluding his chapter by returning to it once more.

Nevertheless everyone holds and I think that it (*scil.* the heavens) is
moving and not the earth: for God fixed the earth, so that it does not

move (*Deus enim firmavit orbem terre, qui non commovebitur**), notwithstanding the reasons to the contrary, for these are persuasive arguments that do not prove evidently. But considering everything that has been said, one could believe from this that the earth is moving and not the heavens, and there is nothing evident to the contrary. In any case this seems *prima facie* as much contrary to natural reason as the articles of our faith, or more so, all or several. And so what I have said for amusement (*par esbatement*) can in this way acquire a value for confuting and regaining those who want to use reason to call our faith in question.

Was this last remark related to the purpose for which Oresme in his concluding chapter said he composed *Le Livre du Ciel*: 'to stimulate, excite and move the hearts of young men of fine and noble intelligence and with a desire for knowledge, so that they will study to contradict and correct me, for love and affection for the truth'? On the issue, so delicate, so fundamental and so passionate in Western thought from the arrival of the new Aristotle in the 13th century down to the controversies of Galileo, of the relation of reason to revelation, of the cosmology of natural science to the cosmology of Scripture, Oresme seems to have taken a position not uncommon among contemporaries who were at once Christian believers and philosophical sceptics. He was prepared to submit reason unconditionally to revelation, and at the same time to use reason to confound reason. 'And all this I say and put forward without insistence, from great humility and fearfulness of heart, saluting always the majesty of the Catholic faith, and in order to hold in check the curiosity or presumption of any of those who, perhaps, might want to slander or attack it or to inquire too boldly, to their confusion.'

But whatever the reasons why Oresme finally rejected the cosmology of the earth's motion in support of which he gave so many arguments, he leaves no doubt about his final opinion. 'But in fact there never was and never will be but a single corporeal universe,' he declared in chapter 24 of book 1 of *Le Livre du Ciel*; that universe was the accepted geostatic one of Aristotle

* *Vulgate*, Psalm 92. 'The world also is stablished, that it cannot be moved'. (Authorized Version, Psalm 93.)

and Ptolemy. And indeed, as Oresme well understood, none of his arguments positively proved the motion of the earth; he declared simply, as Galileo was to declare three centuries later, that he had shown that it was impossible to prove the contrary. But Oresme's conception of motion did not contain the dynamical potentialities that Galileo was to exploit, however unsuccessfully, in the cosmological debate. His conception of relative motion in fact resembled that of Descartes in ignoring what came to be called the inertial properties of matter. It provided him with no criteria for deciding between dynamically possible and impossible astronomical systems.

Albert of Saxony claimed in his *Quæstiones in Libros de Cælo et Mundo*, book 2, question 26 :

we cannot in any manner, by the movement of the earth and the repose of the sky, save the conjunctions and oppositions of the planets, any more than the eclipses of the sun and moon.

But in fact, as Oresme said in book 2, chapter 25 of his commentary, in pointing out that astrology would not be affected by the earth's rotation, 'all conjunctions, oppositions, constellations, figures and influences of the sky would be just as they are, in every way, ... and the tables of movements and all other books would be as true as they are now, except only that one would say that the daily movement is apparent in the heavens and real in the earth.' It was for philosophical and physical reasons that astronomers continued to use the geostatic hypothesis, and natural philosophers did no more than toy with alternatives. For example, Nicholas of Cusa (1401–64) in the next century threw out the suggestion that in every twenty-four hours the eighth sphere revolved twice about its poles while the earth revolved once. Oresme's treatise was never printed and it is not known whether Copernicus ever saw it. The question of plural worlds on which, for instance, Leonardo da Vinci sided with Nicholas of Cusa against Albert of Saxony, continued to excite passionate debates at the end of the 15th century and long afterwards, and these authors were read in northern Italy when Copernicus was at Bologna and Padua. Cusa had given Buridan's dynamics a

Platonic twist by attributing the permanence of celestial rotation to the perfect spherical form of the spheres. The circular movement of a sphere on its centre would continue indefinitely, he said in his *De Ludo Globi*, and just as the movement given to the ball in a game of billiards would continue indefinitely if the ball were a perfect sphere, so God had only to give the celestial sphere its original *impetus* and it has continued to rotate ever since and kept the other spheres in motion. This explanation was adapted by Copernicus for his system. By giving the earth and planets an annual motion round the sun Copernicus offered a mathematical as well as physical alternative to Ptolemy. When he came to consider gravitation and the other physical problems involved, his work appears as a direct development of that of his predecessors.

4. MATHEMATICAL PHYSICS IN THE LATER MIDDLE AGES

One of the most important changes facilitating the increasing use of mathematics in physics was that introduced by the theory that all real differences could be reduced to differences in the category of quantity; that, for example, the intensity of a quality, such as heat, could be measured in exactly the same way as could the magnitude of a quantity. This change was what chiefly distinguished the mathematical physics of the 17th century from the qualitative physics of Aristotle. It was begun by the scholastics of the later Middle Ages.

As with so many scientific concepts in the Middle Ages, the problem was first discussed in a theological context and the principles worked out there were later applied to physics. It was Peter Lombard who opened the question by asserting that the theological virtue of charity could increase and decrease in a man and be more or less intense at different times. How was this to be understood? Two schools of thought developed, one supporting Aristotle's view of the relations of quality to quantity and the other opposing it.

For Aristotle, quantity and quality belonged to absolutely

different categories. A change in quantity, for instance growth, was brought about by the addition of either continuous (length) or discontinuous (number) homogeneous parts. The larger contained the smaller actually and really and there was no change of species. Although a quality, for instance heat, might exist in different degrees of intensity, a change of quality was not brought about by the addition or subtraction of parts. If one hot body was added to another the whole did not become hotter. A change of intensity in a quality therefore involved the loss of one attribute, that is, one species of heat, and the acquisition of another. This was the view, for example, of Aquinas.

Those who, in the 14th century, took the opposite side to Aristotle in this discussion of the relation of quality to quantity, or, as it was called, the 'intension and remission of qualities or forms' *(intensio et remissio qualitatum seu formarum)*, maintained that when two hot bodies were brought into contact, not only the heats but also the bodies were added together. If it were possible to abstract the heat from one body and add it alone to another body, the latter would become hotter. In the same way if it were possible to abstract the gravity from one body and add it to the mass of another body, the latter would become heavier. It was thus asserted, and supported by the authority of Scotus and Ockham, that the intensity of a quality such as heat was susceptible to measurement in numerical degrees, in the same way as the magnitude of a quantity.

Aristotle had analysed physical phenomena into irreducibly, qualitatively, different species, but mathematical physics reduces the qualitative differences of species to differences of geometrical structure, number and movement, in other words, to differences of quantity, and for mathematics one quantity is the same as another. 'I hold that there exists nothing in external bodies for exciting in us tastes, odours and sounds except sizes, shapes, numbers and slow or swift motions,' Galileo was to declare famously in *Il Saggiatore* (question 48) (cf. below, pp. 303–4), matching the equally famous exclamation of Descartes: *'Qu'on me donne l'étendue et le mouvement, et je vais refaire le monde. ... l'univers entier est une machine où tout se fait par figure et mouvement.'* The origin of this idea is to be found in Pythagoras and in Plato's *Timæus,* well known throughout the Middle Ages, and it was the Platonists who were mainly responsible for developing it in the Middle Ages, as later in the 17th century.

Grosseteste, for example, in developing his theory of the 'multiplication of species' (cf. Vol. I, pp. 88–9, 112–13, above, p. 36), distinguished between the physical activity by which the species or *virtus* were propagated through the medium and the sensations of light or heat which they produced when they acted on the appropriate sense organs of a sentient being. The physical activity was independent, as he put it in *De Lineis,* of 'whatever it may meet, whether something with sense perception or something without it, whether something animate or inanimate; but the effect varies with the recipient.' * For, he went on, 'when received by the senses this power produces an operation in some way more spiritual and more noble; on the other hand when received by matter, it produces a material operation, as the sun by the same power produces diverse effects in different subjects, for it cakes mud and melts ice.' In this passage Grosseteste was in effect implying a distinction between primary and secondary qualities in the same sophisticated manner as this was made in the 17th century; the distinction became methodologically and metaphysically significant in physics when the primary qualities were attributed to a physical activity that need not be directly observable (cf. below, pp. 151, 305 *et seq.*).

The physical mode of operation of the fundamental material substance and power, which he held to be light, he conceived to be by means of a succession of pulses or waves on the analogy of sound, and he attempted to express this activity and its diversified effects in mathematical form (cf. Vol. I, p. 115). A similar distinction between light as sensation and light as an external physical activity to be expressed geometrically was made by Roger Bacon, Witelo, and Theodoric of Freiberg. Though no

---

* 'Uno modo agit, quicquid occurrat, sive sit sensus, sive sit aliud, sive animatum, sive inanimatum. Sed propter diversitatem patientis diversificantur effectus.' (ed. L. Baur, *Beiträge zur Geschichte der Philosophie des Mittelalters,* 1912, vol. 9, p. 60.)

medieval writer seems to have conceived the fundamental idea that different colours as perceived were correlated with anything corresponding to the 'wave-length' of light, the optical writers did propose that the differences in the qualitative effects of light were produced by quantitative differences in the light itself. For example, Witelo and Theodoric of Freiberg said that the colours of the spectrum – each a different species of colour according to a strict Aristotelian view – were produced by the progressive weakening of white light by refraction (cf. Vol. I, pp. 122–3). Grosseteste correlated the intensity of illumination and of heat with the angle at which the rays were received and their concentration. John of Dumbleton was to attempt to formulate a quantitative law relating intensity of illumination to distance.

As Roger Bacon expressed the point in his *Opus Majus* (part 4, distinction 1, chapter 2), 'all categories depend on a knowledge of quantity, concerning which mathematics treats, and therefore the whole excellence of logic depends on mathematics.' In medical writings also it became a commonplace to discuss Galen's suggestion that heat and cold should be represented in numerical degrees. There was a general move in many different fields towards finding means of representing qualitative differences by concepts that could be expressed quantitatively and manipulated by mathematics. The interest of the scholastics was seldom directed purely towards solving actual scientific problems. They were nearly always primarily interested in some question of principle in natural philosophy or method, and if particular scientific problems were tackled, it was nearly always so to speak accidentally by way of illustration of a more general quasi-philosophical point. But it is nevertheless possible to see in the 14th-century discussions the origins of some of the most powerful procedures of mathematical physics that became fully effective only in the 17th century. At the same time motion, where the statically conceived Greek geometry had been impotent, was first treated mathematically, thus leading to the foundation of the science of kinematics, that is, the analysis of movement in terms of distance and time.

The new methods of mathematical physics were developed in

the first place in connexion with the idea of functional relationships. This is the natural complement of a systematic conception of concomitant variations between cause and effect; by expressing the phenomenon to be explained (the dependent variable as we now call it) as an algebraic function of the conditions necessary and sufficient to produce it (the independent variables), it can be shown precisely how changes in the former are related to changes in the latter. To be effective in practice the method depends on making systematic measurements, and these were few and far between before the 17th century, although some were made, for example in astronomy, and in Witelo's account of the systematic variation of angles of refraction with angles of incidence of light (see Vol. I, pp. 118–21). In the 14th century the idea of functional relationships was developed without actual measurements and only in principle; that represented the extent of contemporary interest in this as in most other aspects of scientific method.

Two main methods of expressing functional relationships were developed. The first was the 'word-algebra' used in mechanics by Bradwardine at Oxford, in which generality was achieved by the use of letters of the alphabet instead of numbers for the variable quantities, while the operations of addition, division, multiplication, etc. performed on these quantities were described in words instead of being represented by symbols as in modern algebra (cf. above, p. 70 *et seq.*; below, pp. 138–9). Bradwardine was followed in this method at Oxford by numerous writers of treatises on 'proportions', and by a group at Merton College during the 1330s and 1340s known as the *calculatores*, especially William of Heytesbury (c. 1313–72), Richard Swineshead (fl.c. 1344–54), the author of the *Liber Calculationum* who was specifically known as *Calculator,** and John of Dumbleton (fl.c. 1331–49). None of these Oxford writers seems to have been interested in the dyn-

---

* I am indebted to Dr J. A. Weisheipl for the following note distinguishing this Richard Swineshead from two contemporaries, John and Roger, who also bear the place-name of Swineshead. It would seem that John, also a Fellow of Merton College (c. 1343–55), became a lawyer, but no writings of his are known. Roger wrote the treatise *De Motibus Naturalibus*, 'datus

amical aspects of motion; indeed, apparently under the influence of Ockham and Bradwardine, Heytesbury and Dumbleton specifically rejected the theory of *virtus impressa*, without adopting Buridan's alternative theory of *impetus*. It was in Paris that Bradwardine's methods were developed in the context of a physical dynamical theory, and all the principal writers on *impetus* show his direct influence and used his dynamical function: Buridan himself, Oresme, Albert of Saxony, Marsilius of Inghen.

Applied to the problem of giving quantitative expression to changes of quality, the problem of *intensio et remissio qualitatum seu formarum* or 'the latitude of forms' (*latitudo formarum*) as it was called, the purpose of the methods developed at Oxford was to express the amounts by which a quality or 'form' increased or decreased numerically in relation to some fixed scale. A 'form' was any variable quantity or quality in nature, for example local motion, growth and decrease, qualities of all kinds, or light and heat. The intensity (*intensio*) or 'latitude' of a form was the numerical value that was to be assigned to it, and thus it was possible to speak of the rate at which the *intensio*, for example of velocity or of heat, changed in relation to another invariable form known as the 'extension' (*extensio*) or 'longitude' (*longitudo*), for example distance or time or quantity of matter. A change was said to be 'uniform' when, as in uniform local motion, equal distances were covered in equal successive intervals of time, and 'difform' when, as in accelerated or retarded motion, unequal distances were covered in equal intervals of time. Such a 'difform' change was said to be 'uniformly difform' when the acceleration or retardation was uniform; otherwise it was 'difformly difform'.

Oxonie ad utilitatem studencium' (Erfurt MS Amplon, F.135, f.47), and probably the well-known logical text-book *De Insolubilibus et Obligationibus* before 1340; nothing is known about him, but he may have become a Benedictine monk of Glastonbury and Master in Sacred Theology, the *subtilis Swynshed, proles Glastoniæ*, of Richard Tryvytlam's poem in *Collectanea* (vol. 3, ed. M. Burrows). The date of his death is given as 1365 in British Museum MS Arundel 12, f.80.

It was this conception of the relationship between the *intensio* and *extensio* of forms that gave rise to the second method of expressing functional relationships in the 14th century, a geometrical method by means of graphs. The Greeks and Arabs had sometimes used algebra in connexion with geometry, and the idea of plotting the position of a point in relation to rectangular co-ordinates had been familiar to geographers and astronomers since classical times (cf. Plate 2). The graphical representation of the degrees of *intensio* of a quality against *extensio* by means of rectilinear co-ordinates had become fairly common in both Oxford and Paris by the early years of the 14th century. Representing *extensio* by a horizontal straight line (*longitude*), each degree of *intensio* corresponding to a given *extensio* was represented by a perpendicular vertical line (*latitudo vel altitudo*) of specified height. The line connecting the summits of these 'latitudes' could then assume different shapes. For example, if velocity ('intensity or latitude of motion') were plotted against time ('longitude'), uniform velocity would be represented by a horizontal straight line at a height corresponding to the velocity; uniformly difform velocity (i.e. uniform acceleration or retardation) by a straight line making an angle with the horizontal; difformly difform velocity (i.e. changing acceleration or retardation) by a curve.

One of the first to use this geometrical method was Dumbleton, who discussed the subject in his *Summa Logicæ et Philosophiæ Naturalis*, a vast critical discussion of most of the major topics of contemporary physics. In the second part of this important work * Dumbleton made an interesting distinction between a change in quality 'in reality and in name', asserting that in fact no species of quality really changed, but that each degree of intensity was a different species; the mathematical methods gave merely a quantitative and 'nominal' representation of such differences. In the fifth part of the *Summa* he applied the method to the problem of the variation of the intensity or strength of action of light with distance from the source. There can be few writers in any period whose argument is more difficult to follow

* Cambridge MS Peterhouse 272; Oxford MS Merton 306; both 14th cent.

than Dumbleton's, but in the course of a succession of propositions, objections, objections to objections, going on almost endlessly, he did begin the analysis of some basic questions of optics that were only answered in the 17th century. He said that the intensity of illumination at a given point was directly proportional to the strength of the luminous source and inversely proportional to the 'density' of the medium. With a given source and medium he said that the intensity of illumination decreased with distance but not 'uniformly difformly', that is, not in simple proportion. It was Kepler who in his *Ad Vitellionem Paralipomena* (1604) first formulated the photometric law according to which the intensity of illumination is proportional to the inverse square of the distance from the source (see below, pp. 200–201).

The graphical method of representing the 'latitude of forms' was used in Paris in connexion with kinematic problems by Albert of Saxony and Marsilius of Inghen, but the most striking advances were made by Oresme. There are many examples of Oresme's originality as a mathematician: he conceived the notion of fractional powers, afterwards developed by Stevin (cf. below, p. 138), and gave rules for operating them. It has been claimed that he anticipated Descartes in the invention of analytical geometry. Leaving aside the obscure question whether Descartes had any actual direct or indirect knowledge of Oresme's work, it is clear from the latter itself that Oresme had other ends in view than those of the 17th-century mathematician.

Following the common practice, Oresme represented *extensio* by a horizontal straight line and made the height of perpendiculars proportional to *intensio*. His object was to represent the 'quantity of a quality' by means of a geometrical figure of an equivalent shape and area. He held that properties of the representing figure could represent properties intrinsic to the quality itself, though only when these remained invariable characteristics of the figure during all geometrical transformations. He even suggested the extension of these methods to figures in three dimensions. Thus Oresme's horizontal *longitudo* was not strictly equivalent to the abscissa of Cartesian analytical geometry; he was not interested in plotting the positions of points in relation to the rectilinear co-ordinates, but in the figure itself. There is in his work no systematic association of an algebraic relationship with a graphical representation, in which an equation in two variables is shown to determine a specific curve formed by simultaneous variable values of *longitudo* and *latitudo*, and *vice versa*. Nevertheless, his work was a step towards the invention of analytical geometry and towards the introduction into geometry of the idea of motion which Greek geometry had lacked. He used his method to represent linear change in velocity correctly.

According to the definitions given above, the velocity of a body moving with uniform acceleration would be uniformly difform with respect to time. Taking acceleration as 'the velocity of a velocity', Heytesbury in his *Regulæ Solvendi Sophismata* defined uniform acceleration and uniform retardation very clearly as a movement in which equal increments of velocity were acquired, or lost, in any equal periods of time. He also gave an analysis and definition of instantaneous velocity, the measure of which he made (as Galileo was to later) the space that *would* be described by a point if it were allowed to move for some given time at the velocity it had at the given instant. Using these and similar definitions Heytesbury and his contemporaries at Merton College gave kinematic descriptions of various forms of movement, but one was to prove of special significance. Some time before 1335 (the date of Heytesbury's *Regulæ*) it was discovered at Oxford that a uniformly accelerated or retarded movement is equivalent, so far as the space traversed in a given time is concerned, to a uniform movement of which the velocity is equal throughout to the instantaneous velocity possessed by the uniformly accelerated or retarded movement at the middle instant of time. This was proved arithmetically by Heytesbury,* by Richard Swineshead, and by Dumbleton, and it may be called the

---

* The proof is given in *De Probationibus Conclusionum* (Venice, 1494) attributed to Heytesbury, but the authenticity of this work is not beyond dispute. Swineshead's proof occurs in the *Liber Calculationum* and Dumbleton's in the *Summa*, both of which were written certainly after Heytesbury's *Regulæ*.

Mean Speed Rule of Merton College. Oresme, in his *De Configurationibus Intensionum*, or *De Configuratione Qualitatum*, part 3, chapter 7, afterwards gave the following geometrical proof of this rule. He said:

> Any uniformly difform quality has the same quantity as if it uniformly informed the same subject according to the degree of the mid-point. By 'according to the degree of the mid-point' I understand: if the quality be linear. For a quality of a surface it would be necessary to say: 'according to the degree of the middle line ...'

We will demonstrate this proposition for a linear quality.

Let there be a quality which can be represented by a triangle, ABC (Fig. 2). It is a uniformly difform quality which, at point B, terminates at zero. Let D be the mid-point of the line representing the subject; the degree of intensity that affects this point is represented by the line DE. The quality that will have everywhere the degree thus designated can then be represented by the quadrangle AFGB ... But by the 26th

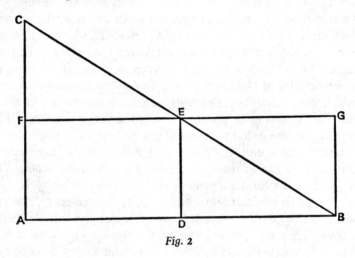

Fig. 2

proposition of Euclid, Book I, the two triangles EFC and EGB are equal. Triangle ABC, which represents the uniformly difform quality, and the quadrangle AFGB, which represents the uniform quality, according to the degree of the mid-point, are then equal. The two qualities which can be represented, the one by the triangle and the other by the quadrangle, are then also equal to one another, and it is that which was proposed for demonstration.

The reasoning is exactly the same for a uniformly difform quality which terminates in a certain degree ...

On the subject of velocity, one can say exactly the same thing as for a linear quality, only, instead of saying: 'mid-point', it would be necessary to say: 'middle instant of the time of duration of the velocity'.

It is then evident that any uniformly difform quality or velocity whatever is equalled by a uniform quality or velocity.*

The treatment of kinematic problems in the 14th century remained almost entirely in the realm of the theoretical. Especially in Oxford, problems were posed *secundum imaginationem*, as imaginary possibilities for theoretical analysis and without empirical application. In Paris the physical and dynamical context of the discussion did direct interest to the kinematics of actual natural motion, but this was treated largely without reference to observation or experiment. A good example is the treatment of the kinematics of freely falling bodies given by Albert of Saxony in his *Quæstiones in Libros de Cælo* (book 2, question 14). After discussing various possible ways in which the natural velocity of a freely falling body might increase with time and with the space traversed, he concluded that the velocity of fall increased in direct proportion to the distance of the fall.† This erroneous opinion was also to seduce Galileo before he decided on the correct solution, namely that velocity increased in direct proportion to the time of the fall, or in other words that a freely falling body moved according to Heytesbury's definition of uniformly accelerated velocity (see below, pp. 154–7). This correct solution was indeed *implied* elsewhere by Albert of Saxony, when like Buridan he said that the longer a movement took the more *impetus* was required and thus the more velocity was acquired. But he did not say this when discussing the kinematic

* Translated from the Latin published by H. Wieleitner, *Bibliotheca Mathematica*, 3rd series, 1914, vol. 14, pp. 230–31.

† Some writers have supposed that Albert of Saxony proposed the correct law of fall as an alternative possibility, but his technical language does not permit this interpretation. See M. Clagett, *Isis*, 1953, vol. 44, p. 401.

problem and there is no evidence that he was himself aware of the kinematic implications of his dynamics. The correct law of acceleration in free fall was given, with considerable confusion, by Leonardo da Vinci and later unequivocally by the Spanish scholastic Domingo de Soto, and finally with quantitative deductions by Galileo.

Certainly the first two of these writers based their work either directly or indirectly on that of their 14th-century predecessors in Oxford and Paris, and Galileo also had a knowledge of the 14th-century kinematics and dynamics. The *calculatores* of Merton College in fact enjoyed a long period of considerable popularity, first at Paris and in Germany, then in Italy and especially at Padua in the 15th and 16th centuries, and again at Paris in the 16th. Between about 1480 and 1520 the new printing presses, especially of Venice and Paris, published editions of the most important writings of Heytesbury, Richard Swineshead and Bradwardine, and of Buridan and Albert of Saxony. Oresme's own principal writings escaped publication, but indirect knowledge of his kinematical theorems was available. Galileo in his *Juvenilia* apparently notes on lectures by his master Francesco Bonamico at Pisa, mentioned, among many other medieval writers on physics, Burley, Heytesbury, Calculator, Albert of Saxony and Marliani. This does not, of course, imply that he read their books. He also mentioned Ockham and Soto, and Philoponus and Avempace; but the names of Buridan and Oresme do not occur.

Resolving Albert of Saxony's hesitations, Soto in 1545 took the velocity of free fall as proportional to time, and declared it to be 'uniformly difform', that is, uniformly accelerated. The violent movement of a projectile fired vertically upwards he also declared to be 'uniformly difform', but in this case uniformly retarded. To both he applied the Mean Speed Rule relating distance and time, thus transcending the qualitative difference between natural and violent motion by means of mathematics.* When Galileo finally stated the correct law of free fall and clearly

---

* Another fundamental aspect of falling bodies, that the acceleration is the same for all bodies of any substance, was first fully appreciated, though only slowly, by Galileo.

elucidated 'the intimate relationship between time and motion,' as he said in the Third Day of his *Two New Sciences* (1638), he used Oresme's theorem in setting out his proof (see below, p. 160).

But there is a world of difference between Galileo's discussion of free fall and that of his scholastic predecessors, and the main direction of the interests of the latter could not be better illustrated than by the contrast. Where the 14th-century scholastics had discussed possible kinds of motion with only the most casual references to empirical actuality, Galileo turned his attention firmly towards the motions actually found in nature as the real object whose elucidation was the main if not the only purpose of theoretical kinematic analysis. Between the 14th century and the 17th, scientific thinkers had transferred their main attention from questions of principle and possibility to questions of actual fact. 'For anyone may invent an arbitrary type of motion and discuss its properties,' Galileo wrote in a famous passage in the Third Day of the *Two New Sciences*; and the properties which these motions and curves possessed in virtue of their definitions might be interesting, even though not met with in nature. 'But we have decided to consider the phenomena of bodies falling with an acceleration such as actually occurs in nature and to make this definition of accelerated motion exhibit the essential features of observed accelerated motions.' And this, he concluded, he had eventually succeeded in doing, and was confirmed in this belief by the exact agreement of his theoretical definition with the results of experiments with a ball rolling down an inclined plane (see below, p. 154 *et seq.*).

The 14th-century attempt to express the quantitative equivalent of qualitative differences led to genuine discoveries concerning both mathematics and physical fact. The latter were extended by the encouragement given to physical measurement, although here ideas were ahead of practical possibilities determined by the scope and accuracy of the available instruments. For example, Ockham said that time could be considered objectively only in the sense that by enumerating the successive positions of a body moving with uniform motion, this motion could be used to measure the duration of the motion or rest of other things. The sun's motion could be used to measure terrestrial movements, but the ultimate reference of all movement was the sphere of the fixed stars, which was the fastest and most nearly uniform motion there was. Other writers elaborated systems for measuring time in fractions (*minutæ*) and the division of the hour into minutes and seconds was in use early in the 14th century. Although mechanical clocks had come in during the 13th century, they were too inaccurate for measuring small intervals of time, and the water-clock and sand-glass continued to be used. The accurate measurement of long intervals was not possible before the invention of the pendulum clock by Huygens in 1657.

The idea of representing heat and cold in numerical degrees was also familiar to physicians. As a zero point Galen had suggested a 'neutral heat' which was neither hot nor cold. Since the only means of determining the degree of heat was by direct sense-perception and a person of hotter temperature would perceive this 'neutral temperature' as cold, and *vice versa*, he had suggested, as a standard neutral degree of heat, a mixture of equal quantities of what he regarded as the hottest (boiling water) and coldest (ice) possible substances. From these ideas Arab and Latin physicians developed the idea of scales of degrees, a popular scale being one ranging from 0 to 4 degrees of heat or of cold. Drugs also were supposed to have something analogous to a heating or cooling effect and were given their place on a scale. Natural philosophers adopted a scale of 8 degrees for each of the four primary qualities. Though in these attempts to estimate degrees of heat, it was known that heat caused expansion, the only thermometer was still the senses. Moreover, a fundamental conceptual difficulty can be seen in the attempt to measure both heat and cold. It was only when the classical conception of pairs of opposites – hot, cold; up, down; and all the rest – had been replaced by the concept of homogeneous linear measures that a workable system of measurement became possible for physics as a whole. The change was made first in mechanics, and modern thermometry followed that example (cf. below, p. 162, note).

Besides the water-clock and sand-glass, the mechanical clock, the astronomical instruments already described, and such 'mathematical instruments' as the straight-edge, square, compass and dividers, the only other scientific measuring instruments available in the 14th and 15th centuries were, in fact, the rules, measures, balances and weights for employing the standards of length, capacity and weight recognized in trade. Balances of both the equal-arm and steelyard type date from antiquity and were used by alchemists and by assayers in metallurgy.

Further attempts to make use of measurement and experiment in science were made during the 15th century, when the scientific leadership of Europe passed from the Anglo-French universities to Germany and Italy. Attempts had been made in the 14th century to express the relationship between the elements graphically on a chart and to state the proportions of the elements and the degrees of the primary qualities for each of the metals, spirits (quicksilver, sulphur, arsenic, sal ammoniac), etc. In the fourth book of his *Idiota*, entitled *De Staticis Experimentis*, Nicholas of Cusa suggested that such problems should be solved by weighing. His conclusions imply the idea of the conservation of matter.

Idiot. ... For weighing a piece of Wood, and then burning it thoroughly, and then weighing the ashes, it is knowne how much water there was in the wood, for there is nothing that hath a heavie weight but water and earth. It is knowne moreover by the divers weight of wood in aire, water and oyle, how much the water that is in the wood, is heavier or lighter than clean spring water, and so how much aire there is in it. So by the diversity of the weight of ashes, how much fire there is in them : and of the Elements may bee gotten by a nearer conjecture, though precision be always inattingible. And as I have said, of Wood, so may be done with Herbs, flesh and other things.

Orator. There is a saying, that no pure element is to be given, how is this to be prov'd by the Ballance?

Idiot. If a man should put a hundred weight of earth into a great earthen pot, and then should take some Herbs, and Seeds, and weigh them, and then plant or sow them in that pot, and then should let them grow there so long, untill hee had successively by little and little, gotten an hundred weight of them, he would finde the earth but very little diminished, when he came to weigh it againe: by which he might gather that all the aforesaid herbs, had their weight from the water. Therefore the waters being ingrossed (or impregnated) in the earth, attracted a terrestreity, and by the opperation of the Sunne upon the Herb were condensed (or were condensed into an Herb). If those Herbs bee then burn't to ashes, mayst thou not guesse by the diversity of the weights of all, how much earth thou foundest more than the hundred weight, and then conclude that the water brought all that? For the Elements are convertible one into another by parts, as wee finde by a glass put into the snow, where wee shall see the aire condensed into water, and flowing in the glass.*

The *Statick Experiments* contained several other suggested applications of the balance. One of these, the comparison of the weights of herbs with those of blood or urine, was directed towards understanding the action of medicines. This was investigated in a different way in the *Liber Distillandi* published by Hieronymus Brunschwig in Strassburg in 1500, in which it was recognized that the action of drugs depended on pure principles, 'spirits' or 'quintessences' which could be extracted by steam distillation and other chemical methods. Cusa also suggested that the time a given weight of water took to run through a given hole might be used as the standard of comparison for pulse rates. The purity of samples of gold and other metals, he said, could be discovered by determining their specific weights, using Archimedes' principle. The balance might be used also to measure the 'virtue' of a lodestone attracting a piece of iron and, in the form of a hygrometer consisting of a piece of wool balancing a weight, to determine the 'weight' of the air. The same device was described by Leon Battista Alberti (1404–72) and by Leonardo da Vinci (1452–1519). The air might also be 'weighed', Cusa said, by determining the effect of air resistance on falling weights while time was measured by the weight of water running through a small hole.

Whether might not a man, by letting a stone fall from a high tower, and letting water run out of a narrow hole, into a Bason in the meane time; and then weighing the water that is runne out, and doing the

* Cusanus, *The Idiot in Four Books*, London, 1650.

same with a piece of wood of equall bignesse, by the diversity of the weights of the water, wood, and stone, attain to know the weight of the aire?

Cusa's suggestions were sometimes a little vague and it is rather tantalizing that the last experiment should have been described without reference to the dynamics of falling bodies. This problem was taken up, suggestively but inadequately, by the Italian doctor Giovanni Marliani (d. 1483). Marliani had made some observations on heat regulation in discussing the intensity of heat in the human body. He developed Bradwardine's modification of Aristotle's law of motion. In criticizing the Aristotelian law he mentioned experiments based on dynamical deductions from the statics of Jordanus Nemorarius, which had been kept alive at Oxford, and had been made known to the Italians by the *Tractatus de Ponderibus* of Blasius of Parma (d. 1416). Marliani noted in his *De Proportione Motuum in Velocitate* that the period of a pendulum decreased with decreasing length and that the rate at which balls rolled down inclined planes increased with the angle of inclination. But he did not determine the precise quantitative relations involved. His main criticisms of the laws of motion of Aristotle and of Bradwardine were directed to pointing out their internal inconsistency, and the experiments he described were no doubt for the most part 'thought experiments'.

Better work was done in astronomy by Georg Peurbach (1423–61) and Johannes Müller or Regiomontanus (1436–76). Peurbach, who held a professorship at Vienna, assisted in a revision of the *Alfonsine Tables*. Perceiving, as some 14th-century writers had done, the advantage of using sines instead of chords, he computed a table of sines for every 10′. Regiomontanus, who knew the work of Levi ben Gerson (see Vol. I, p. 109), wrote a systematic treatise on trigonometry which was to have a great influence, computing a table of sines for every minute and a table of tangents for every degree. He completed a text-book begun by Peurbach and based on Greek sources, the *Epitome in Ptolemœi Almagestum*, which was printed at Venice in 1496. Another work by Peurbach, his *Theoricœ Novœ Planetarum*, published in

Nuremberg in 1472 or 1473, is interesting for its diagrams of the systems of solid spheres. Regiomontanus' pupil, Bernard Walther (1430–1504), with whom he collaborated in the observatory built at Nuremberg, was the first to employ for purposes of scientific measurement a clock driven by a hanging weight. In this the hour wheel was fitted with 56 teeth so that each tooth represented a fraction more than a minute.

The precise manner in which, granting the overriding importance of the conceptual revolution that accompanied the dynamics of inertia, there is continuity of historical development from the mathematical physics of the 14th century to that of the 16th and 17th, presents a delicate problem on which much scholarship has been spent. Of the basic differences in philosophical aims and methods associated with the new dynamics, changes whose establishment was the achievement of Galileo, there can be no question, as will be discussed more fully on a later page. But in comparison with 17th-century physics, that of the 14th century was also limited in both experimental and mathematical technique. The failure to put into general practice the experimental method so brilliantly initiated in the 13th century and the excessive passion for logic, which affected science as a whole, meant that the factual basis of the theoretical discussions was sometimes very slight. The mathematical expression of qualitative intensity in the 'art of latitudes', as it was called, thus gave rise to the same naïve excesses as the analogous attempts, to which this was the father, at omnicompetent mechanism in the 17th and 18th centuries. For instance, Oresme extended the *impetus* theory to psychology. One of his followers, Henry of Hesse (1325–97), while doubting whether the proportions and intensions of the elements in a given substance were knowable in detail, seriously considered the possibility of the generation of a plant or animal from the corpse of another species, for example of a fox from a dead dog. For although the number of permutations and combinations was enormous, during the corruption of a corpse the primary qualities might be altered to the proportions in which they occurred in some other living thing. Dumbleton and other writers had discussed at length latitudes of moral qualities like truth, faith, and perfection. Gentile da Foligno (d. 1348) applied the method to Galen's physiology and this was elaborated in the 15th century by Jacopo da Forli and others who treated health as a quality like heat and expressed it in numerical degrees. Such elaborately subtle and in practice sterile applications of a method called down the ridicule of humanists like Luis Vives (1492–1540) and Pico della Mirandola (1463–94), and made Erasmus (1467–1536) groan when he remembered the lectures he had had to endure at the university. The same geometrical ideal was expressed again in 1540 by Rheticus when he said that medicine could achieve the perfection to which Copernicus had brought astronomy, and again by Descartes.

5. THE CONTINUITY OF MEDIEVAL AND 17TH-CENTURY SCIENCE

Many scholars now agree that 15th-century humanism which arose in Italy and spread northwards, was an interruption in the development of science. The 'revival of letters' deflected interest from matter to literary style and, in turning back to classical antiquity, its devotees affected to ignore the scientific progress of the previous three centuries. The same absurd conceit that led the humanists to abuse and misrepresent their immediate predecessors for using Latin constructions unknown to Cicero and to put out the propaganda which, in varying degrees, has captivated historical opinion until quite recently, also allowed them to borrow from the scholastics without acknowledgement. This habit affected almost all the great scientists of the 16th and 17th centuries, whether Catholic or Protestant, and it has required the labours of a Duhem or a Thorndike or a Maier to show that their statements on matters of history cannot be accepted at their face value.

This literary movement performed some important services for science. Ultimately perhaps the greatest of these was the simplification and clarification of language, although this occurred mainly in the 17th century when it applied particularly to French, but also, under the influence of the Royal Society, to

English. The most immediate service was to supply the means of developing mathematical technique. The development and physical application of the many problems discussed in Oxford, Paris, Heidelberg or Padua in terms of logic and simple geometry were sharply limited by lack of mathematics. It was unusual for medieval university students to progress beyond the first book of Euclid, and although the Hindu system was known, Roman numerals continued in use, although not among mathematicians, into the 17th century. Competent mathematicians, such as Fibonacci, Jordanus Nemorarius, Bradwardine, Oresme, Richard of Wallingford and Regiomontanus were, of course, better equipped and made original contributions to geometry, algebra and trigonometry, but there was no continuous mathematical tradition comparable with that in logic. The new translations by the humanists, presented to the public through the newly invented printing press, placed the wealth of Greek mathematics within easy grasp. Some of these Greek authors, such as Euclid and Ptolemy, had been studied in the preceding centuries; others, such as Archimedes, Apollonius and Diophantus, were available in earlier translations but not generally studied. Among works on applied mathematics Ptolemy's *Cosmographia* and *Geographia* were both printed several times, but the *Almagest* was not printed, except as epitomized by Regiomontanus, until early in the 16th century. Few Arabic astronomical writings were printed. By far the most editions of any author were those of Aristotle's writings, often accompanied by the glosses of Averroës and other commentators.

The whole conception of nature was affected by the systematic atomism found in the full text of Lucretius' *De Rerum Natura* discovered in a monastery in 1417 by a humanist scholar, Poggio Bracciolini. Certainly Lucretius' ideas were not unknown before this date. They appear, for example, in the writings of Hrabanus Maurus, William of Conches, and Nicholas of Autrecourt. But Lucretius' poem seems to have been known only in part, in quotations in the books of grammarians. It was printed later in the 15th century and thereafter many times.

Not only mathematics and physical science, but also biology

benefited from the texts and translations published by the humanists. The humanist press made readily available the works of authors who had been either, like Celsus (*fl.* 14–37 A.D.), previously unknown or, like Theophrastus, known only through secondary sources, and new translations of Aristotle and Galen and of Hippocrates. The last came to replace Galen as the chief medical guide, greatly to the advantage of empirical practice. Pliny's *Natural History* was printed many times and Dioscorides' *De Materia Medica* twice, and there were many editions of Arabic medical writers in Latin translation: Avicenna, Rhazes, Mesue, Serapion. The new texts acted as a stimulus to the study of biology in what was at first a very curious way, for not the least important motive was the desire of humanist scholars, with their excessive adulation of antiquity, to identify animals, plants and minerals mentioned by classical authors. The limitations of this motive were eventually made evident by the very biological studies which it inspired, for these revealed the limitations of classical knowledge, and this was shown still further by the new fauna and flora discovered as a result of geographical exploration, by the increasing practical knowledge of anatomy being acquired by the surgeons, and by the brilliant advances in biological illustration stimulated by naturalistic art. But the original humanist motive draws attention to a feature of 16th- and early 17th-century science in nearly all its branches which historians of science of an earlier generation than the present would have been inclined to associate rather with the preceding centuries; for it was just this extravagant reverence for the ancients, just this devotion to the texts of Aristotle or Galen, that provoked the sarcastic hostility of the contemporary scientists who were trying to use their eyes to look at the world in a new way. And the beginning of this new science dates from the 13th century.

The principal original contributions made during the Middle Ages to the development of natural science in Europe may be summarized as follows:

1. In the field of scientific method, the recovery of the Greek idea of theoretical explanation in science, and especially of the 'Euclidean' form of such explanation and its use in mathematical

physics, raised the problems of how to construct and to verify or falsify theories. The basic conception of scientific explanation held by the medieval natural scientists came from the Greeks and was essentially the same as that of modern science. When a phenomenon had been accurately described so that its characteristics were adequately known, it was explained by relating it to a set of general principles or theories connecting all similar phenomena. The problem of the relation between theory and experiment presented by this form of scientific explanation was analysed by the scholastics in developing their methods of 'resolution and composition'. Examples of the use of the scholastic methods of induction and experiment are seen in optics and magnetics in the thirteenth and fourteenth centuries. The methods involved everyday observations as well as specially devised experiments, simple idealizations, and 'thought experiments', but also mention of imaginary and impossible experiments.

2. Another important contribution to scientific method was the extension of mathematics to the whole of physical science, at least in principle. Aristotle had restricted the use of mathematics, in his theory of the subordination of one science to another, by sharply distinguishing the explicative roles of mathematics and 'physics'. The effect of this change was not so much to destroy this distinction as to change the kind of question scientists asked. One principal reason for the change was the influence of the Neoplatonic conception of nature as ultimately mathematical, a conception exploited in the notion that the key to the physical world was to be found in the study of light. Certainly the medieval scientists did not press this conception to the limit, but they did begin to show less interest in the 'physical' or metaphysical question of cause and to ask the kind of question that could be answered by a mathematical theory within reach of experimental verification. Examples of this method are seen in mechanics, optics and astronomy in the 13th and 14th centuries. It was through the mathematicization of nature and of physics that the inconvenient classical concept of pairs of opposites was replaced by the modern concept of homogeneous linear measures.

3. Besides these ideas on method, though often closely con-

nected with them, a radically new approach to the question of space and motion began at the end of the 13th century. Greek mathematicians had constructed a mathematics of rest, and important advances in statics had been made during the 13th century, progress assisted by Archimedean methods of manipulating ideal quantities such as the length of the weightless arm of a balance. The 14th century saw the first attempts to construct a mathematics of change and motion. Of the various elements contributing to this new dynamics and kinematics, the ideas that space might be infinite and void, and the universe without a centre, undermined Aristotle's cosmos with its qualitatively different directions and led to the idea of relative motion. Concerning motion, the chief new idea was that of *impetus*, and the most significant characteristic of this concept was that a measure was given of the quantity of *impetus* in which this was proportional to the quantity of matter in the body and the velocity imparted to it. Also important was the discussion of the persistence of *impetus* in the absence of resistance from the medium and of the action of gravity. *Impetus* was still a 'physical' cause in the Aristotelian sense; in considering motion as a state requiring no continuous efficient causation, Ockham made another contribution perhaps related to the 17th-century idea of inertial motion. The theory of *impetus* was used to explain many different phenomena, for instance the motion of projectiles and falling bodies, bouncing balls, pendulums and the rotation of the heavens or of the earth. The possibility of the last was suggested by the concept of relative motion, and objections to it from the argument from detached bodies were met by the idea of 'compound motion' advanced by Oresme. The kinematic study of accelerated motion began also in the 14th century, and the solution of one particular problem, that of a body moving with uniform acceleration, was to be applied later to falling bodies. Discussions of the nature of a continuum and of maxima and minima began also in the 14th century.

4. In the field of technology, the Middle Ages saw some remarkable progress. Beginning with new methods of exploiting animal-, water- and wind-power, new machines were developed

for a variety of purposes, often requiring considerable precision. Some technical inventions, for instance the mechanical clock and magnifying lenses, were to be used as scientific instruments. Measuring instruments such as the astrolabe and quadrant were greatly improved as a result of the demand for accurate measurement. In chemistry, the balance came into general use. Empirical advances were made and the experimental habit led to the development of special apparatus.

5. In the biological sciences, some technical advances were made. Important works were written on medicine and surgery, on the symptoms of diseases, and descriptions were given of the flora and fauna of different regions. A beginning was made with classification, and the possibility of having accurate illustrations was introduced by naturalistic art. Perhaps the most important medieval contribution to theoretical biology was the elaboration of the idea of a scale of animated nature. In geology observations were made and the true nature of fossils understood by some writers.

6. Concerning the question of the purpose and nature of science, two medieval contributions may be singled out. The first is the idea, first explicitly expressed in the 13th century, that the purpose of science was to gain power over nature useful to man. The second is the idea insisted on by the theologians, that neither God's action nor man's speculation could be constrained within any particular system of scientific or philosophical thought. Whatever may have been its effects in other branches of thought, the effect of this idea on natural science was to bring out the relativity of all scientific theories and the fact that they might be replaced by others more successful in fulfilling the requirements of the rational and experimental methods.

Thus the experimental and mathematical methods were a growth, developing within the medieval system of scientific thought, which was to destroy from within and eventually to burst out from Aristotelian cosmology and physics. Though resistance to the destruction of the old system became strong among certain of the late scholastics, and especially among those whose humanism had given them too great a devotion to the ancient texts and those by whom the old system had been too closely linked with theological doctrines, there can be little doubt that it was the development of these experimental and mathematical methods of the 13th and 14th centuries that at least initiated the historical movement of the Scientific Revolution culminating in the 17th century.

But when all is considered, the science of Galileo, Harvey and Newton was not the same as that of Grosseteste, Albertus Magnus and Buridan. Not only were their aims sometimes subtly and sometimes obviously different and the achievements of the later science infinitely the greater; they were not in fact connected by an unbroken continuity of historical development. Towards the end of the 14th century, the brilliant period of scholastic originality came to an end. For the next century and a half all that Paris and Oxford produced on astronomy, physics, medicine or logic were dreary epitomes of the earlier writings. One or two original thinkers like Nicholas of Cusa and Regiomontanus appeared in Germany in the 15th century. Italy fared better but rather with the new group of 'artist-engineers' like Leonardo da Vinci than in the universities. Interest and intellectual originality were directed towards literature and the plastic arts rather than towards natural science.

Apart from anything else, the enormously greater achievements and confidence of the 17th-century scientists make it obvious that they were not *simply* carrying on the earlier methods though using them better. But if there is no need to insist on the historical fact of a Scientific Revolution in the 17th century, neither can there be any doubt about the existence of an original scientific movement in the 13th and 14th centuries. The problem concerns the relations between them. Whatever may have happened earlier, must the new science of the 17th century after all be considered a completely new beginning, as some historians of the past have claimed? Did the 'new philosophy', the 'Physico-mathematical Experimental Learning' of the early Royal Society, spring unheralded from the heads of Galileo and Harvey and Francis Bacon and Descartes? Granting the great and fundamental differences between medieval and

17th-century science, the equally striking underlying similarities, apart from other evidence, indicate that a more accurate view of 17th-century science is to regard it as the second phase of an intellectual movement in the West that began when the philosophers of the 13th century read and digested in Latin translation the great scientific authors of classical Greece and Islam.

It may be asked then what the scientists of the 16th and 17th centuries in fact knew of the medieval work, and how the similarities and differences of their aims may be characterized?

As to the first question, the products of the early printing presses show that the principal medieval scientific writings were certainly made readily available, and this in turn indicates that there was an academic demand for them. The available data indicates, as would be expected, that the early presses of the late 15th and early 16th centuries, for example at Venice and Padua and Basel and Paris, continued to reproduce by the new process of printing the same kinds of writings that had formerly been reproduced by hand. A large proportion of these printed works were scientific, and consisted of editions of the writings of the standard classical, Arabic (in Latin translation), and medieval authors. A considerable improvement over the old manuscript copies was the publication of critical *opera omnia* in collected editions.

Although there were some notable exceptions, most of the most important medieval scientific writings were made available in print. Without going into elaborate details, these included, among the more philosophical authors, the principal writings on scientific method and philosophy of science by Grosseteste, Albertus Magnus, Aquinas, Roger Bacon, Duns Scotus, Burley, Ockham, Cusa, and the Italian Averroïsts from Pietro d'Abano down to Nifo and Zabarella in the early 16th century. The dynamical and kinematical writings of Bradwardine, Heytesbury, Richard Swineshead, Buridan, Albert of Saxony, and Marliani were all printed more than once, and so were some of the mathematical writings of Oresme, although not the important *De Configurationibus Intensionum* and *Livre du Ciel*. Dumbleton's writings also remained in manuscript. On statics the *Liber Jordani de Ponderibus* was published in 1533, and the *De Ratione Ponderis* of the 'school' of Jordanus Nemorarius was published by Tartaglia, in 1565. On optics the writings of Grosseteste, Roger Bacon, Witelo (together with Alhazen's treatise), Pecham, and Themo Judæi all found publishers. The most notable exception was the *De Iride* of Theodoric of Freiberg, but an account of his theory of the rainbow with the essential diagrams was published in Erfurt in 1514. Petrus Peregrinus' *Epistola de Magnete* was printed twice in the 16th century, in 1558 and 1562; it was known to and acknowledged by Gilbert. The most popular astronomical text was Sacrobosco's *Sphere*, but astronomical tables and related mathematical writings like those of Jean de Linières, Jean de Murs, Peurbach and Regiomontanus were also printed in representative quantity. Chaucer's *Treatise on the Astrolabe* was printed, but Richard of Wallingford's manuscripts were not. Another very important mathematician whose writings escaped publication was Leonardo Fibonacci.

The most important medieval biologist was Albertus Magnus; his *De Animalibus* was printed and so were his geological and chemical writings. Among other printed biological works were *The Art of Falconry* of the Emperor Frederick II and the writings of Thomas of Cantimpré, Peter of Crescenzi and Conrad von Megenburg. The herbals of Rufinus and Rinio remained unprinted, but other works in this field were printed, notably Matthæus Sylvaticus' *Pandectæ*, and new herbals in Latin and in the vernacular were also issued by the presses (see below, p. 267 *et seq.*). The most popular work on natural history was Bartholomew the Englishman's *On the Properties of Things*. On anatomy, surgery and medicine the treatises, for example, of Mondino, Guy de Chauliac, Arnald of Villanova, Gentile da Foligno, and John of Gaddesden were printed many times, in some cases in several languages. Other excellent writings in this field, like those of Henri de Mondeville and Thomas of Sarepta, remained un-published. On chemistry and alchemy the writings of Arnald of Villanova and those attributed to Raymond Lull were printed. So also were a number of practical treatises on various subjects, those of Brunschwig, Agricola and Biringuccio including much of earlier chemical practice.

The extent to which the scientists of the period showed an interest in these medieval treatises varied with different individuals. In the 16th century the strong classical leanings of men like Copernicus and Vesalius perhaps prevented them from paying much attention in print to medieval authors, but other leading scientists certainly did so. For example the Italian anatomists Achillini and Berengario da Carpi wrote commentaries on Mondino's anatomy (see below, p. 276). The theory of *impetus* and other aspects of medieval dynamics, kinematics and statics were studied and taught by mathematicians and philosophers such as Tartaglia, Cardano, Benedetti, Bonamico and the young Galileo himself. In England Dr John Dee collected manuscripts especially of the mathematical and physical writings of Grosseteste, Roger Bacon, Pecham, Bradwardine and Richard of Wallingford, while Robert Recorde recommended the writings of Grosseteste and other Oxford writers to students of astronomy. Dee and Recorde and Leonard and Thomas Digges were early supporters of the Copernican theory, and all saw their work as a revival of the great days of Oxford in the 13th and 14th centuries. Thomas Digges, in describing his father's pioneering work on telescopes, acknowledged Roger Bacon as an authority in optics. Leonardo da Vinci, Maurolico, Marc' Antonio de Dominis, Giambattista della Porta, Johann Marcus Marci and Christopher Scheiner all referred in their optical writings to Roger Bacon, Witelo and Pecham. Kepler wrote a commentary on Witelo, correcting his tables of angles of refraction; Snell's work on the law of refraction seems to have been stimulated by the edition of Witelo and Alhazen by Frederick Risner in 1572; and many other 17th-century optical writers, for example Descartes himself, Fermat, James Gregory, Emanuel Maignan and Grimaldi used the same source. As for Descartes, he seldom mentioned those to whom he was indebted, but his *Météores* follows the exact order of the subjects of Aristotle's *Meteorology* and is in more ways than one the last of the medieval commentaries on that much glossed work (cf. below, pp. 256–9).

Enough has been said to show that leading scientists of the 16th and early 17th centuries both knew and used the writings of their medieval predecessors. The story is the same in biology as in physics, where Albertus Magnus was the principal medieval writer. In the conceptions of scientific method and explanation the medieval part of the ancestry is equally visible, especially for example in Galileo's use of the methods of 'resolution and composition' to elucidate the relation between theory and experiment and to develop the 'Euclidean' form of scientific explanations. So it is also in the Neoplatonic conception of nature as ultimately mathematical, first exploited in the Middle Ages in Grosseteste's 'cosmology of light' and apparent in different ways in the thought of Galileo, Kepler and Descartes. But did the scientists, especially of the 17th century, simply accept and continue the aims and methods of the scholastics? It will appear in greater detail in the chapter that follows that clearly they did much more. One characteristic may be singled out as indicating an essential difference.

The central doctrines of medieval science developed almost entirely within the context of academic discussions based at some stage, near or far, on the books used in university teaching. The commentaries and *quæstiones* on the subjects treated in these books may have travelled far from the originals of Aristotle or Ptolemy or Euclid or Alhazen of Galen; they never escaped from them altogether. It is true that the applications of academic sciences, such as of astronomy in determining the calendar and making proposals for its reform, or of arithmetic in the work of the exchequer and of commercial houses, or of anatomy and physiology and chemistry in surgery and medicine, were put into practice outside the universities. It is true also that in other fields outside the university system altogether, for example in technology of different kinds and in art and architecture with their increasing tendency to naturalism, developments took place that were to be of profound importance for science. Certainly the reasons for the development of science within the universities, and for the growth and spread of the university system itself, must be related to the reasons for the development of national political states based on an expanding commercial capitalism that could give employment to the men responsible for these technological and artistic activities outside. The latter, becoming the 'artist-engineers' of the 15th and 16th centuries and the *virtuosi* and independent scientific gentlemen of the 17th, were to take over the leadership of science, making it more an activity of the Accademia dei Lincei or the Royal Society or the Académie Royale des Sciences than of the universities. This was true even though in these scientific societies there was a predominance of university men, who were in fact to bring the new science back into the universities themselves.

But in the 13th and 14th centuries it was within the framework of the university faculty of arts, its curriculum expanded to include the new translations from Greek and Arabic and some technical treatises on applied mathematics, and of the higher faculties of medicine and of theology, that the central conceptions of science were cultivated. The men who cultivated them were clerics and academic teachers. The academic exercise was never far away in the background of the treatises they have left behind, those unliterary writings that form the great collections of manuscripts and early printed books that show us their ways of thought. Certainly many of them were original and ingenious thinkers. But the great scientific and cosmological problems with which they dealt were seldom seen by them as purely scientific. The greatest problem of all was the relation of the cosmology of Christian theology based on revelation to the cosmology of rational science dominated by Aristotle's philosophy. Although some of the best medieval scientific work was done on particular problems studied without any reference to theology or philosophy or even methodology, it was within a general framework of philosophy closely bearing on theology, and specifically within the system of university studies run by clerics, that the central development of medieval science took place.

The result of this was that science in the Middle Ages was nearly always at the same time philosophy of science. No doubt the same characteristics will appear in any age that is still determining the direction and objectives of its inquiries, as they did eminently in the 17th century, for example in the scientific thought and controversies of Galileo, Descartes and Newton. In contrast with both medieval and 17th-century scientists, those of the 20th century know in general how they are going to deal with problems, the kinds of questions they are going to put to nature and the methods they will employ to get their answers. It is only in the profoundest and most general problems, when a line of explanation seems to meet with an *impasse*, that philosophy need nowadays disturb the even course of the bulk of the scientific work actually being done.

But there is one basic difference between the aims of medieval philosophy of science and of all the philosophy of science since Galileo. The latter is *primarily* concerned with clarifying and facilitating the processes and further advances of science itself. The main interest of scientists since Galileo has been in the ever-increasing range of concrete problems that science can solve, and if philosophical investigations are undertaken by scientists, it is usually because certain concrete and specific scientific problems can be satisfactorily solved only by a thorough reform of fundamental principles. The essays in philosophy by Galileo and Newton had essentially this purpose. But medieval natural philosophers were *primarily* interested less in the concrete problems of the world of experience than in the *kind of knowledge* natural science was, how it fitted into the general structure of their metaphysics, and, if it extended so far, how it bore on theology. Many scientific problems were discovered as analogies that could illuminate a theological problem, as was the case with instrumental causality and the theory of *impetus*. Being taken up in the interests of something else, this was no doubt one reason why in the course of development they were so often so peremptorily dropped.

The contrast is one of general emphasis and is certainly not exclusive. In the 18th century Berkeley and Kant, for example, were primarily concerned not with science but with the bearing of Newtonian cosmology on metaphysics, while in the 13th century Jordanus, Gerard of Brussels and Petrus Peregrinus seem to have been innocent of any philosophical interests and purely concerned with the immediate scientific problems in hand. But if what has been said does truly characterize the general intellectual ambience of medieval science, it explains much that is puzzling and seemingly downright perverse in otherwise excellent work. It helps to explain, for example, the gap between the repeated insistence on the principle of empirical verification and the many general assertions never tested by observation; worse, the satisfaction with imaginary experiments either incorrect or impossible; even worse, the false figures given, for example, by scientists of the calibre of Witelo or Theodoric of Freiberg allegedly as the results of measurements plainly never made. There are of course examples of medieval science not marred by such defects, but it was a peculiarity of the period that they could occur in the course of even the best-conceived investigations. The impression is left that the investigator was not strongly interested in mere details of fact and measurement. Certainly the strong interest in the theory and logic of experimental science and in related philosophical conceptions of nature, sustained from Grosseteste down to the threshold of Galileo's activities, stands in striking contrast with the comparative scarcity of actual experimental investigations. This becomes intelligible if we see the medieval natural philosophers not as modern scientists *manqués* but as primarily philosophers. They gave an account of experimental inquiries often as an exercise in what could be done in one branch of philosophy in distinction from others. Certainly this had the desirable effect of clarifying the problems of natural science and helping to extricate them from alien contexts of metaphysics and theology. In what was actually found out by experiment they were less interested.

It was a direction of interest that could have been fatal to Western science. Excellent as may have been much of their general characterization of the methodology of experimental science, it meant that the methodologists seldom really put their methods to the practical test. So they rarely made them really precise or really adequate. Undirected experiments and simple everyday observations abound in the work of medieval scientists. Certainly there was no general movement to conceive of experimental inquiry as a sustained testing of a series of precisely and quantitatively formulated hypotheses, pressing on to the reformulation of a whole area of theory. The examples of experimental inquiries, even the best of them, remained isolated without general effect on the accepted doctrines of light or of cosmology. They were thought sufficient to illustrate the method, and methodology was an end in itself. It would have become a dead end had not Galileo and his contemporaries, with a new direction of interest, pursued the subjects of the examples for their own sakes. It was through taking these seriously, through paying attention to the detailed facts of experiment and measurement and mathematical functions actually exemplified in nature, that the 17th-century scientists were led to their radical revolution in the whole theoretical framework of physics and cosmology, where the medieval natural philosophers had only revised some limited sections.

If it is true that a fundamental change in the interests of scientists and in the conception of science can be charted about the time of Galileo, a further point would indicate another detail of the general line of change. Perhaps the most powerful feature of the medieval philosophy of science that remained strongly influential in the early 17th century was the Neoplatonic conception that nature was ultimately to be explained by mathematics. In the Middle Ages this belief was exploited mainly in the field of optics. Within the ambience of Platonism, and encouraged by the story in *Genesis* of the first day of creation, leading thinkers of the 13th and 14th centuries focused their attention on the study of light as the key to the mysteries of the physical world, and in optics they did some of their best scientific work. But, as in the Aristotelian classification, optics remained, together with astronomy and music, one of the *mathematica media*, mathematical sciences applied to the physical world as distinct on the one hand from pure mathematics, and on the other from physics as the science of 'natures' and causes. Medieval scientists seemed to feel no overwhelming desire or need to dispense with these philosophical distinctions. Mathematical physics never really became a universal science rendering Aristotelian physics unnecessary.

Perhaps it was pointed of Descartes, the most medieval of the great 17th-century scientists in the sense of being the most dominated by a philosophy of nature, to call his reforming work on cosmology *Le Monde, ou Traité de la Lumière*. But Descartes' physics were not based on a theory of light; rather his theory of light was based on his conception of motion. It was in the study of motion and not of light that the 17th-century scientists looked for the key to physics. It was there too that to their satisfaction they found it.

Certainly in giving special weight to the study of motion as distinct from other aspects of nature the 17th-century physicists made a fortunate choice. But Aristotle and the medieval Aristotelians had already made the study of motion the basis of their physics. The choice made by the 17th-century scientists was not fortuitous, nor was the success with which it was exploited. By taking the empirical phenomena of motion seriously as a problem and seeing the solution through to the end, they had no alternative but to reform the whole of cosmology, to invent new mathematical techniques in the process, and to provide the eminent example for the methods of science as a whole. This, it may be suggested, was the advance made by the secular *virtuosi* of the 17th century over the clerics of the medieval universities to whom in other ways they owed so much.

2

THE REVOLUTION IN
SCIENTIFIC THOUGHT IN
THE SIXTEENTH AND
SEVENTEENTH CENTURIES

I. THE APPLICATION OF MATHEMATICAL
METHODS TO MECHANICS

How the scientific revolution of the 16th and 17th centuries came about is easier to understand than the reason why it should have taken place at all. So far as the internal history of science is concerned, it came about by men asking questions within the range of an experimental answer, by limiting their inquiries to physical rather than metaphysical problems, concentrating their attention on accurate observation of the kinds of things there are in the natural world and the correlation of the behaviour of one with another rather than on their intrinsic natures, on proximate causes rather than substantial forms, and in particular on those aspects of the physical world which could be expressed in terms of mathematics. Those characteristics that could be weighed and measured could be compared, could be expressed as a length or number and thus represented in a ready-made system of geometry, arithmetic or algebra, in which consequences could be deduced revealing new relations between events which could then be verified by observation. The other aspects of matter were ignored.

The systematic use of the experimental method by which phenomena could be studied under simplified and controlled conditions, and of mathematical abstraction which made possible new classifications of experience and the discovery of new causal laws, enormously speeded up the tempo of scientific progress. One outstanding fact about the Scientific Revolution is that its initial and in a sense most important stages were carried through before the invention of the new measuring instruments, the telescope and microscope, thermometer and accurate clock, which were later to become indispensable for getting the accurate and satisfactory answers to the questions that were to come to the forefront of science. In its initial stages, in fact, the Scientific Revolution came about rather by a systematic change in intellectual outlook, in the type of question asked, than by an increase in technical equipment. Why such a revolution in methods of thought should have taken place is obscure. It was not simply a continuation of the increasing attention to observation and to the experimental and mathematical methods that had been going on since the 13th century, because the change took on an altogether new speed and a quality that made it dominate European thinking. It is not an adequate explanation to say that the new approach was simply the result of the work done on inductive logic and mathematical philosophy by the scholastic philosophers down to the 16th century or the result of the revival of Platonism in the 15th century. It cannot be attributed simply to the effect of the renewed interest in some hitherto poorly known Greek scientific texts, such as the work of Archimedes, though these certainly stimulated mathematical thought.

Various aspects of the social and economic conditions of the 16th and 17th centuries certainly provided motives and opportunities that might stimulate science. At the beginning of the 16th century some outstanding scholars showed a vigorous interest in the study of the technical processes of manufacture, and this helped to unite the mind of the philosopher with the manual skill of the craftsman. Luis Vives wrote in 1531 in his *De Tradendis Disciplinis* advocating the serious study of the arts of cooking, building, navigation, agriculture and clothmaking, and specifically urged that scholars should not look down on manual workers or be ashamed of asking them to explain the mysteries of their crafts. Rabelais, writing two years later, suggested that a proper branch of study for a young prince was to learn how the objects he used in ordinary life were made. Rabelais described how Gargantua and his tutor visited goldsmiths and jewellers, watchmakers, alchemists, coiners and many other craftsmen. In 1568 a Latin reader published in Frankfort for the use of school children seems to have been inspired by the same respect for skilled craftsmanship, for it took the form of a series of Latin verses each describing the work of a different craftsman, for example a printer, a papermaker, a pewterer or a turner. A marked advance was made during the 16th century also in the publication of treatises written by the educated on various technical processes. Of these, *De Re Metallica* (1556) by Georg Bauer (1490–1555), or Agricola, as he called himself, on mining and metallurgy and the treatises by Besson, Biringuccio, Ramelli and, in the early 17th century, by Zonca are the most outstanding examples (cf. Vol. I, pp. 184–6). This interest in the technical achievements of the various crafts was expressed most clearly by Francis Bacon (1561–1626), first in 1605 in *The Advancement of Learning* and later more fully in the *Novum Organum*. Bacon was of the opinion that technics or, as he called it, the mechanical arts, had flourished just because they were firmly founded on fact and modified in the light of experience. Scientific thought, on the other hand, had failed to advance just because it was divorced from nature and kept remote from practical experiment. In his view the learning of the schoolmen had been 'cobwebs of learning ... of no substance or profit' and the new humanistic learning must be directed to the benefit of man. Descartes took precisely the same view of the matter. In the 16th century several mathematicians such as Thomas Hood (*fl.* 1582–98) and Simon Stevin (1548–1620) were specially employed by governments to solve problems of navigation or fortification. In the latter part of the 17th century the Royal Society interested itself in the technical processes of various trades in the hope that the information collected would not only provide a solid foundation for the speculations of scholars, but also would be of practical value to mechanics and artificers themselves. Several treatises were collected on special subjects: Evelyn wrote a *Discourse of Forest-*

Trees and the Propagation of Timber, Petty on dyeing, and Boyle a general essay entitled *That the Goods of Mankind may be much increased by the Naturalist's Insight into Trades*. The English History of Trades did not get written, but the idea was attractive and almost a century later twenty volumes on arts and crafts were published by the Paris Academy of Sciences.

There are also, certainly, examples of this active interest of the learned in technical questions leading scientists to make contributions to fundamental problems. The attempt to calculate the angle at which a gun must be fired to give the maximum range led Tartaglia (*c.* 1500–57) to criticize the whole Aristotelian conception of motion and attempt new mathematical formulations, though the problem was solved only by Galileo. The experience of engineers who built water pumps is said to have influenced Galileo and Torricelli in their experiments leading to the barometer, and the rumour that some Dutch lens-grinders had invented a telescope is known to have stimulated Galileo to study the laws of refraction with the object of constructing one himself. Descartes wrote his *Dioptrique* (1637) explicitly to give a scientific basis for constructing lenses for telescopes and spectacles. When they did their fundamental work on the pendulum, both Galileo and Huygens had in mind the need for an accurate clock for determining longitude, which became increasingly pressing with the extension of ocean voyages.

The existence of motives and opportunities, even when they brought fundamental scientific problems into prominence, does not explain the intellectual revolution that made it possible for scientists to solve these problems, and the history of the interaction between motives, opportunities, skills and intellectual changes that brought about the Scientific Revolution has, in fact, yet to be written.

The internal revolution in scientific thought that took place during the 16th and 17th centuries had, then, two essential aspects, the experimental and the mathematical, and it was precisely those branches of science which were most amenable to measurement that showed the most spectacular developments. In antiquity mathematics had been used most successfully in astronomy, optics and statics, and to these the medieval schoolmen, less successfully, added dynamics. These were also the branches of science which showed the greatest advances in the 16th and 17th centuries and, in particular, it was the successful application of mathematics to mechanics that changed men's whole conception of nature and brought about the destruction of the whole Aristotelian system of cosmology. It was only after they had, following the Greek example, successfully applied their new methods to these abstract and comparatively manageable problems, that scientists found themselves in a position to attack the more difficult mysteries of dead and living matter. Chemistry, physiology, and the sciences of electricity and magnetism cannot until the 19th century compare in performance with Newtonian mechanics (cf. above, p. 24, below, p. 325).

One of the first to try to express nature in terms of the new mathematics was Leonardo da Vinci (1452–1519). Leonardo received his early education in the Platonic city of Florence and later worked in Milan and the other northern Italian towns where the scientific ideal was Aristotelian. Nearly all his physical conceptions were suggested by scholastic writers, such as Jordanus Nemorarius, Albert of Saxony and Marliani, but he was able to develop their mechanical ideas through his new knowledge of Greek mathematicians like Archimedes, to whose *On the Equilibrium of Planes* he had access in manuscript.

Among ancient mathematicians Archimedes had been the most successful in combining mathematics with experimental inquiry; because of this he became the ideal of the 16th century. His method was to select definite and limited problems, and it would be truer to say that he proceeded rather by the mathematical manipulation of ideal quantities than by actual measurement. He formulated hypotheses which he either regarded, in the Euclidean manner, as self-evident axioms or could verify by simple experiments. The consequences of these he then deduced and, in principle, experimentally verified. Thus, in the work just mentioned, he began with the axioms that equal weights suspended at equal distances are in equilibrium, that equal weights suspended at unequal distances are not in equilibrium but that which hangs at the greater distance descends, and so on. These axioms contained the principle of the lever, or, what is equivalent, of the centre of gravity, and from them Archimedes deduced numerous consequences.

Leonardo's mechanics, like those of his predecessors, was based on Aristotle's axiom that motive power is proportional to the weight of the body moved and the velocity impressed on it. Jordanus Nemorarius and his school had developed this axiom to express the principle of virtual velocity or work, and applied it, with the notion of the statical moment, to the lever and the inclined plane. Leonardo used the conclusions of this school and made various advances on them. He recognized that the effective (or potential) arm of a balance was the line which, passing through the fulcrum, was at right angles to the perpendicular passing through the suspended weight. He recognized that a sphere on an inclined plane moved until it reached a point where its centre of gravity was vertically above its point of contact, though he rejected Jordanus' correct treatment of equilibrium on an inclined plane for an incorrect solution given by Pappus. He did recognize that the velocity of a ball rolling down an inclined plane was uniformly accelerated, and showed that the velocity of a falling body increased by the same amount for a given vertical fall whether it descended vertically or down an incline. He also recognized that only the vertical component need be considered in estimating motive power, and that the principle of work was incompatible with perpetual motion: he said that if a wheel were moved for a time by a given quantity of water and if this water were neither added to nor allowed a greater fall, then its function was finished. The principle of work, with that of the lever, he used also to develop the theory of pulleys and other mechanical appliances. In hydrostatics he recognized the fundamental principles that liquids transmit pressure and that the work done by the mover equals that done by the resistance. In hydrodynamics he developed the principle which the school of Jordanus had derived from Strato, that, with a given fall, the smaller the cross-section of the passage the greater the velocity of a flowing liquid.

Leonardo's dynamics was based on the theory of *impetus*, which, he held, carried the moving body in a straight line. But he adhered (like Cardano, Tartaglia and other later 16th-century Italian mechanicians) to the Aristotelian view that the supposed acceleration of a projectile after leaving the projector was due to the air. He accepted also Albert of Saxony's division of the trajectory of a projectile into three periods, but he recognized that the actual motion of a body might be the resultant of two or more different forces or velocities. He applied the principle of compound *impetus*, together with that of a centre of gravity which he derived from Albert of Saxony and developed for solid figures, to a number of problems including percussion and the flight of birds.

In addition to his studies in mechanics Leonardo also used Greek geometry in an attempt to improve the theory of lenses and the eye which he derived from an edition of Pecham's *Perspectiva Communis*, printed in 1482. He made certain advances but suffered, like his predecessors, from the belief that the visual function resides in the lens instead of the retina, and from the inability to understand that an inverted image on the latter was compatible with seeing the world in the way we do. His devotion to the ideal of measurement is shown by the scientific instruments which he tried to improve or devise, such as a clock, a hygrometer similar to Cusa's to measure moisture in the atmosphere, a hodometer similar to Hero's to measure distance travelled, and an anemometer to measure the force of the wind. Though he wrote no book and his illegible mirror-written notes covered with sketches were not deciphered and published until much later, many of them not until the 19th century, his work was not lost to his immediate posterity. His manuscripts were copied in the 16th century and his mechanical ideas were pillaged by Hieronymo Cardano (1501–76), and may have passed to Stevin and through Bernardino Baldi to Galileo, Roberval and Descartes. The Spaniard Juan Batiste Villalpando (1552–1608) made use of his ideas on the centre of gravity, and from him they were transmitted, through the wide scientific correspondence of the learned Minim friar, Marin Mersenne, to the 17th century.

The natural philosophers who succeeded Leonardo developed still further the powerful mathematical technique that was becoming possible with the recovery and printing of some hitherto unknown or little studied Greek texts. The earliest printed Latin edition of Euclid appeared in Venice in 1482, and Latin editions of Archimedes, Apollonius and Diophantus were made by Francesco Maurolico (1494–1575) and of Euclid, Apollonius, Pappus, Hero, Archimedes and Aristarchus by Federigo Commandino (1509–75).

The first advances in mathematical technique were in algebra. The first comprehensive printed *Algebra*, that of Luca Pacioli (1494), contained the problem of cubic equations (those involving cubes of numbers, e.g., x^3), which were first solved by Tartaglia (whose real name was Nicolo Fontana of Brescia). His work was pirated by Hieronymo Cardano, who anticipated him in publication (1545). Cardano's former servant and pupil, Lodovico Ferrari (1522–65), first solved quartic equations (involving x^4). Limitations in the general theory of numbers prevented the understanding of quintics (involving x^5) until the 19th century, but François Viète (1540–1603) gave a method of obtaining numerical values of the roots of polynomials and introduced the principle of reduction. The theory of equations was also developed by the English mathematician, Thomas Harriot (1560–1621). To the earlier algebraists negative roots had seemed unintelligible. These were first understood by Albert Girard (1595–1632), who also extended the idea of number to include 'imaginary' quantities like $\sqrt{-1}$, which had no place in the ordinary numerical scale extending from zero to infinity in both the negative and positive directions. At the same time improvements were made in algebraic symbolism. Viète used letters for unknowns and constants as an essential part of algebra. Stevin invented the present mode of designating powers and introduced fractional exponents. His symbolism was later generalized by Descartes in the form x^2, x^3, etc. Other symbols such as $+$, $-$, $=$, $>$, $<$, $\sqrt{}$, etc. to represent operations which had previously been written out in words, had been gradually introduced from the end of the 15th century, so that by the first decades of the 17th century algebra and arithmetic had been standardized into something like their present form.

About the same time two important advances were also made in geometry. The first was the introduction of analytical geometry, the second the emergence of infinitesimal calculus. A step towards analytical geometry had been made by Nicole Oresme and there are reasons for believing that Descartes, who was not in the habit of mentioning those to whom he was indebted, knew his work. The man to whom Descartes was probably most indebted here was Pierre de Fermat (1601–65), who fully grasped the equivalence of different algebraic expressions and geometrical figures traced by loci moving with reference to co-ordinates. If his predecessors invented the method, it was Descartes who, in his *Géométrie* (1637), first developed its full power. He rejected the dimensional limitation on algebra and by letting, for instance, squares or cubes of terms (x^2, y^3) represent lines, he was able to put geometrical problems into algebraic form and to use algebra to solve them. Problems of motion thus received fruitful development when a curve could be represented as an equation (see Plate 3). Descartes also showed that the entire conic sections of Apollonius were contained in some equations of the second degree.

Descartes' analytical geometry depended on the assumption that a length was equivalent to a number; this no Greek would have accepted. The second mathematical advance made during the early years of the 17th century depended on a similar pragmatic illogicality. To compare rectilinear and curvilinear figures, Archimedes had used the 'method of exhaustion'. In this the area of a curvilinear figure could be determined from that of inscribed and circumscribed rectilinear figures, by making them approach the curve by increasing the number of their sides. When determining elliptical areas Kepler had introduced the idea of the infinitely small into geometry and Francesco Bonaventura Cavalieri (1598–1647) made use of this idea to develop Archimedes' method into the 'method of indivisibles'. This depended on considering lines as composed of an infinite number of points, surfaces of lines and volumes of surfaces. The relative magnitudes of two surfaces or solids could then be found simply by summation of series for points or lines. In contrast with Descartes' analytical geometry, which was not generally used in physics until the end of the 17th century, the 'method of indivisibles' arose directly out of physical problems. It was later developed by Newton and Leibniz into the infinitesimal calculus.

Aristotle had maintained, as against the Pythagorean theory of Plato, that mathematics, though useful in defining the relations between certain events, could not express the 'essential nature' of physical things and processes, for it was an abstraction excluding from consideration irreducible qualitative differences which, nevertheless, existed. According to Aristotle, the study of physical bodies and events was the proper object not of mathematics but of physics. In studying them, he arrived at such essential distinctions not only as those between irreducibly different qualities perceived through the senses but also, in the consideration of observed motions, as those between natural and violent movements, gravity and levity, and terrestrial and celestial substance. This point of view had been shared by Euclid and was accepted by Tartaglia in his commentary on the *Elements*. Tartaglia said that the subject-matter of physics, which was gained through sensory experience, was distinct from the subject-matter of geometrical demonstration. A physical speck, for instance, was divisible to infinity, but a geometrical point, being without dimensions, was by definition indivisible. The subject-matter of geometry, he said, was continuous quantity-point, line, volume, and its definitions were purely operational. Geometry was not concerned with what exists; it could deal with physical properties like weight or time only when these had been translated into lengths by measuring instruments. Since its principles were known by abstraction from material things, the conclusions it demonstrated were applicable to them. Thus physics might use mathematics, but was left with an independent non-mathematical field of its own.

With the increasing success of mathematics in solving concrete physical problems during the 16th century, the area of this purely physical preserve was reduced. The practical geometers of the 16th century developed the idea of using measurements, for which instruments of increasing accuracy were required, to determine whether what held true in mathematical demonstrations also held true in physical things. For instance, Tartaglia accepted the Aristotelian principle, which had led to a three-fold division of the trajectory of a projectile (cf. Fig. 1), that an elementary body could have only a single movement at any time (since if it had two one would eliminate the other). When he came to make a mathematical study of the flight of a projectile, he realized that when fired out of the vertical it began its descent under the action of gravity *immediately* after leaving the gun (cf. Plate 4). He had to maintain, therefore, that natural gravity was not entirely eliminated by *impetus*. Cardano (who also developed Leonardo's ideas on the balance and virtual velocities) went a step further. He drew a distinction in mechanics between mathematical relations and moving powers or principles, the proper subject of 'metaphysics', and accepted the old forms of such powers. He objected altogether to the arbitrary separation of mathematical subject-matter into irreducibly different classes, such as in the different periods in the trajectory of a projectile. Viète took the same view.

The old problem of projectiles had, in fact, gained a new importance in the 16th century when improved types of bronze cannon with accurately bored barrels began to replace the 14th- and 15th-century cast-iron monsters, and when a more powerful kind of gunpowder was produced in Germany. At the same time improvements were made in small arms, particularly in methods of firing, and from the end of the 15th century the old method of touching off the powder by applying a burning match to the touch hole was replaced by a number of improved devices. First came the match-lock which enabled the burning match to be brought down by pressing a trigger. This was applied to the arquebus, the common infantryman's weapon after the battle of Pavia, in 1525. Then came the wheel-lock using pyrites instead of a burning match, though this was too dangerous to be much used. Finally, by 1635, came a device using flint which became the flint-lock used by the soldiers of Marlborough and Wellington. Problems of theoretical ballistics did not arise in the use of small arms, but with the heavy guns, as the range increased with more powerful gunpowder, problems of sighting became serious. Tartaglia devoted much time to these problems and the invention of the gunner's quadrant has been attributed to him. Later, Galileo, Newton and Euler made further contributions, though it was not until the second half of the 19th century that accurate ballistic tables were constructed on the basis of experiments.

Another 16th-century mathematician and physicist who made a critical scrutiny of Aristotelian theories and exposed some of their contradictions, even as a system of physics, was Giovanni Battista Benedetti (1530–90). He knew of the criticisms that had been made in late Greek times of Aristotle's ideas on falling bodies (see above, p. 64 *et seq*.). He imagined a group of bodies of the same material and weight falling beside each other, first connected and then separately, and he concluded that their being in connexion could not alter their velocity. A body the size of the whole group would, therefore, fall with the same velocity as each of its components. He therefore concluded that all bodies of the same material (or 'nature'), whatever their size, would fall with the same velocity, though he made the mistake of believing that the velocities of bodies of the same volume but different material would be proportional to their weights. Inspired by Archimedes, he thought of weight as proportional to the relative density in a given medium.* He then used the same argument as Philoponus to prove that velocity would not be infinite in a void (see above, pp. 65, 73). Benedetti also held that in a projectile natural gravity was not entirely eliminated by the *impetus* of flight, and he followed Leonardo in maintaining that *impetus* engendered movement only in a straight line, from which it might be deflected by a force, such as the 'centripetal' force exerted by a string which prevented a stone swung in a circle flying off at a tangent.

* Archimedes' principle asserts that when a body floats its weight is equal to the weight of the liquid displaced and when it sinks its weight decreases by that amount.

The Application of Mathematical Methods to Mechanics

Sixteenth-century physicists turned increasingly from Aristotle's qualitative 'physical' explanations to the mathematical formulations of Archimedes and to the experimental method. Although their enunciations were not always rigorous, their instincts were usually sound. Like Archimedes, they tried to conceive of a clear hypothesis and put it to the test of experience. Thus, beginning with the assumption that perpetual motion was impossible, Simon Stevin was led to a clear appreciation of the basic principles of both hydrostatics and statics. In the former science he concluded (1586) that any given mass of water was in equilibrium in all its parts, for if it were not it would be in continuous movement, and he then used this theory to show that the pressure of a liquid on the base of the containing vessel depended only on depth and was independent of shape and volume. Equipotential points were those on the same horizontal surface.

With the same assumption of the impossibility of perpetual motion he showed also why a loop of cord, on which weights were attached at equal distances apart, would not move when hung over a triangular prism (Fig. 3). He showed that as long as the bottom of the prism was horizontal no movement occurred in the upper section of the cord when the suspended section was removed, and from this he arrived at the conclusion that weights on inclined planes were in equilibrium when proportional to the lengths of their supporting planes cut by the horizontal. The same conclusion had, in fact, been reached in the 13th century in *De Ratione Ponderis*, which had been published in 1565 (see Vol. I, pp. 129–30). This conclusion implied the idea of the triangle or parallelogram of forces, which Stevin applied to more complicated machines.

An important statical principle arising out of this work of Stevin, though the germ of it came from Albert of Saxony, seems to have been taught by Galileo Galilei (1564–1642). This was that a set of connected bodies, such as those on Stevin's inclined plane, could not set themselves in motion unless this resulted in the approach of their common centre of gravity towards the centre of the earth. The work done was then equal to the product of the weight moved multiplied by the vertical distance. The

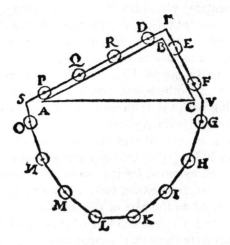

Fig. 3 Stevin's demonstration of the equilibrium of the inclined plane. From Beghinselen des Waterwichts, *Leiden, 1586.*

precise enunciation of this principle and its fruitful application to mathematical physics was made by Galileo's pupil, Torricelli.

Stevin performed the experiment, also attributed to Galileo, of dropping simultaneously two leaden balls, one ten times heavier than the other, from a 30-foot height on to a plank. They struck the plank at the same instant and he asserted that the same held for bodies of equal size but different weight, that is, of different material. Similar experiments had, in fact, been mentioned in the writings of critics of Aristotle since Philoponus, though the result was not always the same because of the appreciable effect of air resistance on the lighter bodies. Stevin and his predecessors recognized that their observations were incompatible with the Aristotelian law of motion, according to which velocity should be directly proportional to the moving cause, with falling bodies their weight, and inversely proportional to air resistance. But Stevin did not develop the dynamical consequences of his observations.

It was, in fact, Galileo who was chiefly responsible for carrying the experimental and mathematical methods into the whole field of physics and for bringing about the intellectual revolution by which first dynamics, and then all science, were established in the direction from which there was no return. The revolution in dynamics in the 17th century was brought about by the substitution of the concept of inertia, that is, that uniform motion in a straight line is simply a state of a body and is equivalent to the rest, for the Aristotelian conception of motion as a process of becoming which required for its maintenance continuous efficient causation. The problem of the persistence of motion was brought to the fore because it was this Aristotelian conception that lay behind some of the most important objections to Copernicus' theory of the earth's rotation, for instance that based on the argument from detached bodies (see above, p. 89, below, p. 182), and the question of the truth of the Copernican theory was perhaps the chief scientific problem of the late 16th and early 17th centuries. To prove this theory was the great passion of Galileo's scientific life. To do so Galileo tried to ignore the naïve inductions from common-sense experience, which were the basis of Aristotle's physics, and to look at things in a new way.

Galileo's new way of looking at the facts of experience represented a change of emphasis which was all important, though each of its two main characteristics had antecedents in an earlier tradition; the proof of it was that it bore fruit in the rapid solution of many different scientific problems. First, he put aside all discussion of the 'essential natures' that had been the subject-matter of Aristotelian physics and concentrated on describing what he observed, that is, on the phenomena. This is seen in his *Dialogue Concerning the Two Principal Systems of the World* (1632) when, during the Second Day, Salviati, representing Galileo himself, replies as follows to the assertion made by Simplicio, the Aristotelian, that everyone knows that what causes bodies to fall downwards is gravity:

You are wrong, Simplicio; you should say that everyone knows that it is called gravity. But I am not asking you for the name, but the essence of the thing. Of this you know not a bit more than you know the essence of the mover of the stars in gyration. I except the name that has been attached to the former and made familiar and domestic by the many experiences we have of it a thousand times a day. We don't really understand what principle or what power it is that moves a stone downwards, any more than we understand what moves it upwards after it has left the projector, or what moves the moon round. We have merely, as I said, assigned to the first the more specific and definite name *gravity*, whereas to the second we assign the more general term impressed power (*virtù impressa*), and the last we call an *intelligence*, either *assisting* or *informing*; and as the cause of infinite other motions we give *nature*.

This attitude to such so-called causes Galileo learnt from the nominalism which had penetrated the Averroïst schools of northern Italy during the 15th century. Such words as 'gravity', he held, were simply names for certain observed regularities, and the first business of science was not to seek unfindable 'essences' but to establish these regularities, to discover proximate causes, that is, those antecedent events which, when other conditions were the same, always and alone produced the given effect. 'Consider what there is that is new in the steelyard,' declared Salviati in the Second Day of the *Two Principal Systems*, 'and therein lies necessarily the cause of the new effect.' He continued, in the Fourth Day, enunciating what J. S. Mill was to call the method of concomitant variations: *

Thus I say that if it is true that one effect can have only one basic cause, and if between the cause and the effect there is a fixed and constant connexion, then whenever a fixed and constant variation is seen in the effect, there must be a fixed and constant variation in the cause. Now since the variations which take place in the tides at different times of the year and of the month have their fixed and constant periods, it must be that regular changes occur simultaneously in the primary cause of the tides. Next, the alterations in the tides at the said times consist of nothing more than changes in their sizes; that is, in the rising and lowering of the water a greater or less amount, and in its running with greater or less impetus. Hence it is necessary that, whatever the primary cause of the tides is, it should increase or decrease its force at the specific times mentioned. ... If then we wish to preserve the identity of the cause, we must find the changes in these

* Francis Bacon called this the method of 'Degrees or Comparison'; cf. below, p. 295.

additions and subtractions that make them more or less potent at producing those effects that depend upon them.

As this passage indicates, Galileo's whole method presupposed measurement. He gave another, more qualitative, illustration of it in his witty reply in *Il Saggiatore*, question 45:

If Sarsi wishes me to believe, on the word of Suidas, that the Babylonians cooked eggs by whirling them swiftly in a sling, I will believe it; but I shall say that the cause of such an effect is very remote from that to which they attribute it, and to discover the true cause I shall argue as follows: If an effect, which has succeeded with others at another time, does not take place with us, it necessarily follows that in our experiment there is something lacking which was the cause of the success of the former attempt; and, if we lack but one thing, that one thing is alone the true cause; now, we have no lack of eggs, nor of slings, nor of stout fellows to whirl them, and yet they will not cook, and indeed, if they be hot they will cool the more quickly; and, since nothing is wanting to us save to be Babylonians, it follows that the fact of being Babylonians and not the attrition of the air is the cause of the eggs becoming hard-boiled, which is what I wish to prove.

In its business of discovering proximate causes, Galileo held, science began with observations and observations had the last word. In accordance with the logic of science of the later Middle Ages, the method of 'resolution and composition', he showed how to arrive at general theories by analysis from experience, to vary the conditions and isolate causes (as in the previous quotation), and to verify or falsify theories by experiment. Distinguishing the method used by Aristotle for investigation from that used in presenting his conclusions, Galileo said in the First Day of the *Two Principal Systems*:

I think it certain that he first obtained by the senses, by experiments and observations, such assurance as was possible of the conclusions, and that afterwards he looked for means to demonstrate them. For this is the normal course in the demonstrative sciences; and it is followed because, when the conclusion is true, by making use of the resolutive method one may hit upon some proposition already demonstrated or arrive at some principle known *per se*; but if the conclusion is false, one could go on for ever without ever finding any known truth – if indeed one does not encounter some impossiblity or manifest absurdity. And you need have no doubt that Pythagoras, long before he had found the proof for which he offered the hecatomb, was sure that the square on the side opposite the right angle in a right-angled triangle was equal to the squares on the other two sides. The certainty of the conclusion assists not a little to the discovery of the proof, meaning always in the demonstrative sciences. But whatever was Aristotle's method of procedure, whether the reasoning *a priori* came before the sense perception *a posteriori* or the other way round, it is enough that Aristotle, as he said many times, preferred the experience of the senses to any argument.

He went on, in the Second Day: 'I know very well that one single experiment or conclusive proof to the contrary would be enough to batter to the ground ... a great many probable arguments.' *

Clearly in this presentation of the role of experiment Galileo's scientific method resembled that of the scholastic philosophers of Oxford and Padua who had interpreted Aristotle in terms of Plato's dialectic and had applied the *reductio ad absurdum* to empirical situations (see above pp. 24, 33 *et seq.*). In his use of 'thought experiments' – but not of impossible imaginary experiments – Galileo also carried on established practices. But he made one advance of the greatest importance. He insisted, at least in principle, on making systematic, accurate measurements, so that the regularities in phenomena could be discovered quantitatively and expressed in mathematics.

The significance of this advance is made very plain in his own comments on William Gilbert's work on magnetism (cf. below, p. 196 *et seq.*) in the Third Day of the *Two Principal Systems*. 'I am going to explain, with a certain likeness to my own,' he said, 'his method of procedure in philosophizing, in order that I may stimulate you to read it. I know that you understand quite well how much a knowledge of events contributes to an investigation of the substance and essence of things; therefore I wish

* Galileo seems to have thought that science advanced through a series of alternatives each decided by a crucial experiment.

you to take care to inform yourself thoroughly about many events and properties that are found uniquely in the lodestone, and not in other stones or other bodies.' He continued :

I have the highest praise, admiration, and envy for this author, who framed such a stupendous concept concerning an object which in-numerable men of splendid intellect had handled without paying any attention to it ... But what I might have wished for in Gilbert would be a little more of the mathematician, and especially a thorough grounding in geometry, a discipline which would have rendered him less rash about accepting as rigorous proofs those reasons which he put forward as *veræ causæ* for the correct conclusions he himself had observed. His reasons, candidly speaking, are not rigorous, and lack that force which must unquestionably be present in those adduced as necessary and eternal scientific conclusions.

It was by his insistence on measurement and mathematics that Galileo combined his strictly experimental method with the second main characteristic of his new approach to science. This was to try to express the observed regularities in terms of a mathematical abstraction, of concepts of which no exemplaries need actually be observed but from which the observations could be deduced. From its consequences the hypothetical abstraction could then be tested quantitatively. Galileo's method of abstrac-tion was explicitly an adaptation of the postulational method of Archimedes and Euclid. It was of revolutionary importance both for his own work and consequently for the whole history of science. Under the influence of the same Greek tradition, such abstractions had certainly been used in some medieval scientific investigations, for example the 'ideal balance' with weightless arms, the mathematical expressions postulated in dealing with problems of motion, and the geometrical devices postulated to 'save the appearances' in astronomy. Following the precedent of Democritus and Plato, the mathematicization of 'form' and 'sub-stance' found especially in 13th-century optics is another aspect of the postulational method of abstraction that Galileo was to exploit. But because of the strength of Aristotelian influence, most pre-Galilean science was in practice constricted by the dominance of naïve and direct generalizations from common-sense experience. Galileo's use of the method of mathematical abstraction enabled him firmly to establish the technique of investigating a phenomenon by specially arranged experiments, in which irrelevant conditions were excluded so that the phenomenon could be studied in its simplest quantitative relations with other phenomena. Only after these relations had been established and expressed in a mathematical formula did he re-introduce the excluded factors, or carry his theory into regions not readily amenable to experimentation.

In Galileo's eyes, one of the principal assets of the Copernican system was that Copernicus had escaped from the naïve empiricism of Aristotle and Ptolemy and taken a more sophisticated attitude to theories used to 'save the appearances'. 'Nor can I sufficiently admire the eminence of those men's intelligence,' says Salviati, during the Third Day of the *Two Principal Systems*,

who have received and held it [the Copernican system] to be true, and with the sprightliness of their judgements have done such violence to their own senses, that they have been able to prefer that which their reason dictated to them to that which sensible experiences represented most manifestly to the contrary ... I cannot find any bounds for my admiration how reason was able, in *Aristarchus* and *Copernicus*, to commit such a rape upon their senses as in spite of them, to make herself mistress of their belief.

Galileo believed the mathematical theories from which he deduced the observations to represent the enduring reality, the substance, underlying phenomena. Nature was mathematical. This view he owed partly to the Platonism which had been popular in Italy, particularly in Florence, since the 15th century. One essential element of this Pythagorean Platonism, which had been made increasingly plausible by the success of the mathematical method in 16th-century physics, was the idea that the behaviour of things was entirely the product of their geometrical structure. During the Second Day of the *Two Principal Systems*, Salviati replies to Simplicio's assertion that he agreed with Aristotle's judgement that Plato had doted too much upon geometry. 'After all,' says Simplicio, 'these mathematical subtleties do very well in the abstract, but they do not work out when applied to sensible and physical matter.' Salviati points out that the conclusions of mathematics are exactly the same in the abstract and in the concrete.

It would indeed be novel if the computations and ratios made in abstract numbers did not afterwards correspond to the gold and silver coins and the merchandises in concrete. Do you know what does happen, Simplicio? Just as the computer who wants his calculations to deal with sugar, silk, and wool must discount the boxes, bales, and other packings, so the mathematical scientist (*filosofo geometra*), when he wants to recognize in the concrete the effects which he has proved in the abstract, must deduct the material hindrances, and if he is able to do so, I assure you that things are in no less agreement than the arithmetical computations.

The faith that inspired nearly all science until the end of the 17th century was that it discovered a real intelligible structure in objective nature, an *ens reale* and not merely an *ens rationis*. Kepler believed himself to be discovering a mathematical order which provided the intelligible structure of the real world; Galileo said, during the First Day of the *Two Principal Systems*, that of mathematical propositions human understanding was 'as absolutely certain . . . as Nature herself'. In fact, though Galileo rejected the kind of 'essential natures' the Aristotelians had been seeking, he simply brought in another kind by the back door. He asserted that since mathematical physics could not deal with the non-mathematical, what was not mathematical was subjective (see above, p. 99 *et seq.*; cf. below, p. 303 *et seq.*). As he affirmed in *Il Saggiatore*, question 6:

Philosophy is written in that vast book which stands forever open before our eyes, I mean the universe; but it cannot be read until we have learnt the language and become familiar with the characters in which it is written. It is written in mathematical language, and the letters are triangles, circles and other geometrical figures, without which means it is humanly impossible to comprehend a single word.

It was precisely in his attitude to these mathematical 'primary qualities' that Galileo the Platonist differed from Plato himself.

Plato had held that the physical world was a copy or likeness of a transcendent ideal world of mathematical forms; it was an inexact copy and so physics was not absolute truth but, as he put it in the *Timæus*, 'a likely story'. Galileo by contrast asserted that the real physical world actually consisted of the mathematical entities and their laws, and that these laws were discoverable in detail with absolute certainty. In the transitional state of contemporary scientific thought his analysis of scientific method had two main purposes. On the one hand he wanted to show that the Aristotelian explanations were not explanations at all, were in fact answers to the wrong questions and totally inadequate to the problems being considered. By eliminating Aristotle's particular conception of the real essences of the physical world, with their various irreducible natural qualities, natural positions in the universe and natural motions, he would eliminate the whole Aristotelian opposition to the new mathematical physics and dynamics and to Copernicus. On the other hand, he wanted to show how to find the true solutions, the true explanations revealing the actual essence and structure of the physical world, and to show how to give reasons for asserting that such explanations were certainly true. Both aims were necessary for his programme of reframing the questions to be asked in order to construct a true and universal mathematical science of motion.

Galileo's Platonism was thus of the same kind as that which had led to Archimedes being known in the 16th century as the 'Platonic philosopher', and with Galileo mathematical abstractions got their validity as statements about nature by being solutions of particular physical problems. By using this method of abstracting from immediate and direct experience, and by correlating observed events by means of mathematical relations which could not themselves be observed, he was led to experiments of which he could not have thought in terms of the old common-sense empiricism.

His approach to the search for the mathematical laws of phenomena, for example of the acceleration of heavy bodies, the swinging of a pendulum, the trajectory of a cannon ball, or the motions of the planets, was in the traditional 'Euclidean' manner of searching for premises from which to deduce the data of the phenomena. Setting up his theories on the Euclidean pattern, his whole procedure was what he called an *'argomento ex suppositione'*. Galileo was a scientist who was most conscious of problems of method and philosophy. There are many references to the subject in both his main works, the *Two Principal Systems* and the *Mathematical Discourses and Demonstrations concerning Two New Sciences* (1638). He described his method fully also in a letter to Pierre Carcavy in 1637. Since it was impossible to deal at once with all the observed properties of a phenomenon, he first reduced it intuitively to its essentials. After this 'resolution' of the essential mathematical relations involved in a given effect, he set up a 'hypothetical assumption' from which he deduced the consequences that must follow. This second stage he called 'composition'. Finally came an experimental analysis, which he also called 'resolution', of examples of the effect in order to test the hypothesis by comparing its deduced consequences with observation. Abstraction was essential to the whole procedure. Thus, for example, in order to deal with a moving body dynamically it became a quantity of matter concentrated at its centre of gravity traversing a given space in a given time. It was strictly the 'physical object' so abstracted and defined that entered into the dynamical theorems. All questions relating to the 'nature' of the object in the Aristotelian sense were to be ignored. Thus Galileo was able to give precise formulation to a conception of motion first hinted at by Ockham and Buridan; and the methodological significance of his distinction between primary and secondary qualities becomes apparent in his treatment of motion kinematically in terms of velocity.

A good example of Galileo's method is his work on the pendulum. By abstracting from the inessentials of the situation, 'the opposition of the air, and line, or other accidents', he was able to demonstrate the law of the pendulum, that the period of oscillation is independent of the arc of swing and simply proportional to the square-root of the length. This having been proved, he could then reintroduce the previously excluded factors. He argued, for instance, that the reason why a real pendulum, of which the thread was not weightless, came to rest, was not simply because of air resistance, but because each particle of the thread acted as a small pendulum. Since they were at different distances from the point of suspension, they had different frequencies and therefore inhibited each other.

Another good example of his method is his study of freely-falling bodies, one of the foundations of 17th-century mechanics. Disregarding Aristotle's conception of motion as a process requiring a continuous cause, and the Aristotelian categories of movement based on purely 'physical' principles still accepted by such writers as Cardano or Kepler, he looked for a definition that would enable him to measure motion. He said, during the First Day of the *Two Principal Systems*:

Let us call velocities equal, when the spaces passed have the same proportion as the times in which they are passed.

In this he followed such 14th-century physicists as Heytesbury and Richard Swineshead, whose works had been printed at the end of the 15th century and taught to Galileo during his youth at Pisa. He tried to arrange things so that he could study the problem under simple and controlled experimental conditions, for example in balls rolling down an inclined plane. He made a few preliminary observations, and analysed the mathematical relations obtaining between two factors only, space and time, excluding all the others. Then he tried to invent what he called a 'hypothetical assumption', which was a mathematical hypothesis from which he could deduce consequences that could be tested experimentally; and since, as Salviati said during the Second Day of the *Two Principal Systems*, 'Nature ... does not do that by many things, which may be done by few,' he adopted the simplest possible hypothesis. During the Third Day of the *Two New Sciences*, 'On Local Motion', he gave the definition of uniformly accelerated motion as a motion which, 'when starting from rest, acquires during equal time-intervals equal increments of velocity.' This, he said, he adopted for one reason, because Nature employs 'only those means which are most common, simple and easy'. His experimental verification consisted of a series of measurements showing the concomitant variations in space travelled and time passed. If the consequences of his hypothesis were verified, he regarded this hypothesis as a true account of the natural order. If they were not, he tried again, until he reached a hypothesis which was verified; and then the particular instance, for example the observed facts about falling bodies, was explained by being shown to be the consequences of a general law. The object of science for Galileo was to explain the particular facts of observation by showing them to be consequences of such general laws, and to build up a whole system of such laws in which the more particular were consequences of the more general. In all this the role of intuition, even of an Aristotelian type although turned to a different purpose, was paramount. Intellectual intuition, abstraction, and mathematical analysis discovered the hypothetical possibilities; experiment became necessary to eliminate false hypotheses among these and to identify and verify the true one. An hypothesis so verified was a true intuitive insight into the details of the real structure of the physical world.

Galileo's approach to physical problems is clearly seen in the *Two New Sciences*, in his deduction of the kinematical laws of freely falling bodies, when Salviati turns away from the suggestion that certain physical causes might account for the facts and concentrates on the kinematical aspect of the problem.

The present does not seem to be the proper time to investigate the cause of the acceleration of natural motion concerning which various opinions have been expressed by various philosophers, some explaining it by attraction to the centre, others by repulsion between the very small parts of the body, while still others attribute it to a certain stress in the surrounding medium which closes in behind the falling body and drives it from one of its positions to another. Now, all these fantasies, and others too, ought to be examined; but it is not really worth while. At present it is the purpose of our Author merely to investigate and to demonstrate some of the properties of accelerated motion (whatever the cause of this acceleration may be) – meaning thereby a motion, such that the momentum of its velocity goes on increasing after departure from rest in simple proportionality to time, which is the same as saying that in equal time-intervals the body receives equal increments of velocity; and if we find that the properties [of accelerated motion] which will be demonstrated later are realized in freely falling and accelerated bodies, we may conclude that the assumed definition includes such a motion of heavy bodies and that their speed goes on increasing as the time and the duration of the motion.

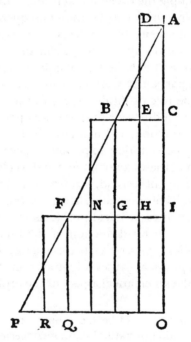

Fig. 4 Diagram used in Galileo's proof that with a body falling with uniform acceleration, in successive equal intervals of time AC, CI, IO, the distances traversed (measured by the areas ABC, CBFI, IFPO) increase as 1, 3, 5, and so on. In modern terminology, assuming v=at, Galileo proved that s=½at². From *Discorsi e dimostrazione matematiche intorno à due nuove scienze*, Bologna, 1655 (1st ed. Leiden, 1638), Third Day.

This passage, indicating a classical turning-point in the history of science, was written in 1638, but Galileo had not always seen so clearly that the acceleration of free fall must be defined and the definition verified as a fact, before there could be any attempt at a dynamical explanation. Galileo's clarification of this distinction measures the progress he made between his early treatment of motion as a young professor at Pisa and his maturer understanding at Padua, to which he went in 1592. It opened the way to his attack on the dynamics itself and to his formulation, incomplete but definite, of the concept of inertial motion. This was the achievement of his period at Florence, to which he returned in 1610 under the special patronage of the Grand Duke of Tuscany.

Earlier discussions of free fall had never separated the kinematic from the dynamical aspects of the problem. The former were always presented as deductions from the latter and thus shared in their inadequacies, a characteristic found even in Soto's correct statement of the kinematic law (see above, p. 108 *et seq.*). No one had thought of simply ascertaining the correct kinematic law independently of any dynamics. In his first original scientific essays, the treatise and dialogue both entitled *De Motu* written at Pisa about 1590, Galileo followed this traditional procedure. The primary aim of these early essays was to refute the dynamical theory and law of motion on which Aristotle had based his arguments against the possibility of motion in a void, the basic assumption being that local motion was the resultant of a proportion between power and resistance to which both were necessary (see above, p. 61 *et seq.*). Galileo made criticisms of Aristotle's dynamics, and especially of his explanations of projectile motion and free fall, similar to those made by Buridan and Albert of Saxony and their followers, but the explanations he offered in their place suggest an attachment rather to the dynamics of Avempace than of Buridan, and to the Pythagorean or Platonic conception of relative gravity. He asserted that a constant power would produce a finite uniform velocity through extended space even without any resistance, as for example in a void; if there were a resistant medium it would simply reduce this finite velocity by a definite amount. Projectile motion would thus be possible in a void; he explained it by the theory of *virtus impressa*. As to free fall, he said that every species of body had a finite natural velocity of fall determined by its intrinsic 'nature'

157

or specific gravity, a velocity that would be realized in a void, where there was no resistance. In a resistant medium this natural velocity would be reduced by a definite amount determined by the relative specific gravities of the body and of the medium; indeed if the latter were the greater the body would rise. This left the problem of why heavy bodies accelerated when they fell from rest. To explain this Galileo supposed that in the case both of a body thrown upwards, and of one at rest above its natural place, a lingering upwards-directed *virtus* was acquired by the displacement from the centre; as the body fell, this *virtus* was gradually reduced, so that the body accelerated downwards until the opposing *virtus* had disappeared entirely, after which the body continued to fall with a constant velocity proper to its gravity. Thus Galileo did not at that time agree with his predecessors like Oresme who held that the acceleration of free fall would continue indefinitely, but rather had independently hit upon a theory proposed in antiquity by Hipparchus.

The physical-causal treatment of motion in these Pisan essays shows Galileo still very far from the kinematical approach for lack of the necessary concept of inertia. While criticizing Aristotle, along somewhat traditional lines, he fully accepted the basic assumptions that a constant velocity required a constant motive power, and that an accelerated velocity required a corresponding increase in effective power. Another example of the same characteristics can be seen in his account in the essays of experiments of dropping different weights from 'a high tower'. These were later associated by Galileo's disciple and biographer Vincenzo Viviani with the Leaning Tower of Pisa, but there is no positive evidence that he actually made any experiments from the Leaning Tower, and his manner of introducing them suggests rather that they were 'thought experiments'. Thus in attacking Aristotle's assumption that speed of fall is proportional to weight, he speaks not only of flinging two stones, one twice as big as the other, from a high tower, but also of dropping two lead spheres, one a hundred times as big as the other, from the moon. He ridicules the notion that one stone will fall twice as fast as the other, and one sphere of lead 100 times as fast as the other. In fact Galileo's basic argument to demonstrate that bodies of the same material but different size would fall with the same speed was exactly that used by Benedetti : the whole cannot fall faster than the part (see above, p. 142). But this did not apply to bodies, such as a piece of lead and a piece of wood, of different material. These fell with velocities proper to their 'natures' and, he wrote in the treatise *De Motu*, 'if they are let go from a high tower, the lead precedes the wood by a long space; and I have often made test of this.... Oh how readily are true demonstrations drawn from true principles!' he exclaimed.

Two other Italian scientists, Giorgio Coresio in 1612 and Vincenzio Renieri in 1641, did actually make such experiments from the Leaning Tower, and they found that even with bodies of the same material the heavier weight reached the ground first, if they were dropped from a sufficient height. Coresio even asserted that the velocity was proportional to the weight, thus confirming Aristotle's 'law'; but Renieri, giving actual figures, showed otherwise. In fact he submitted his results to Galileo, who referred him to his *Dialogue*. Discussing the subject more fully in his *Two New Sciences*, Galileo had pointed out that the actual difference in the velocity in such experiments was widely different from that to be expected from the Aristotelian 'law'. He was also aware that the results disagreed with the expectations of his new dynamics : by that time, having given up the conception of 'natures' as causes of motion, he had come to assume that all bodies of any material would fall with the same velocity. Unimpressed by the disagreement of experiment with theory, Galileo made an abstraction from empirical actuality and said that the theory applied to free fall in a vacuum. In a resistant medium such as air, he said that a lighter body would be retarded more than a heavier one. Same results, different explanations! It has long ago ceased to be possible to regard the Leaning Tower experiment, even supposing Galileo made it, as in any sense crucial, or even new.

The first evidence that Galileo had successfully turned to a kinematical attack on the problem of free fall comes in his famous letter to Paolo Sarpi in 1604, in which he said that he had proved that the spaces passed over by a falling body were to each other as the squares of the times. By this time he must have assumed that the acceleration continued indefinitely, or would do so if it were not for the resistance of the air, which, as he explained in the *Two New Sciences*, tended to limit the velocity of a falling body to a maximum value. He claimed to have deduced his theorem, familiar now as $s = \frac{1}{2}at^2$, from the axiom that the instantaneous velocity was proportional to the *distance* fallen. He used in his demonstration the medieval geometrical method of dealing with varying qualities, taking the integral, Oresme's 'quantity of velocity' (the area A B C in Fig. 2), to represent the distance fallen (Fig. 4). But in fact, as Duhem showed, the axiom, or definition of uniform acceleration, that Galileo, by a curious error, actually assumed in his reasoning was not this impossible one already rejected by Soto, but that instantaneous velocity was proportional to *time*. Indeed the distinction between the two was not one that either the incompletely clear kinematics or mathematics of the period made easy. Exactly the same mistake was made by Isaac Beeckman and Descartes.

It seems likely that Galileo had discovered his mistake and correctly formulated the law of acceleration and the space theorem by 1609, although he published them only in the *Two Principal Systems* in 1632. It is possible that he had already carried out his experiment to test the law with a bronze ball rolling down an inclined plane as early as 1604. This experiment is described in the *Two New Sciences* (1638), where the mathematical demonstration is again set out. In the absence of an accurate clock, he defined equal intervals of time as those during which equal weights of water issued from a small hole in a bucket; he used a very large amount of water relative to the amount issuing through the hole, so that the decrease in head was unimportant. His experiment confirmed his definition and law of free fall, and from it he deduced further theorems.

It was this famous experiment that, on the empirical side, distinguished Galileo's account from all preceding attempts to deal with the problem of free fall, although it is an indication of the contemporary lack of system in presenting scientific results that

Galileo recorded no actual individual measurements, and gave only the conclusions he had drawn from them. In fact Mersenne failed to get the same results when he repeated Galileo's experiment some years later – an indication perhaps of Galileo's confidence in the mathematical and conceptual intuition to which he owed his scientific success quite as much as to his experiments. And it was just because he came to see the law of acceleration and the space theorem within the theoretical structure generated by the new concept of inertial motion that they became the foundations of classical dynamics, and may be considered, as Galileo himself believed them to be, his greatest achievement.

Although the conception of motion developed in his treatise *De Motu* was fundamentally anti-inertial, there are to be found in it applications of the 'Platonic' technique of abstraction in which the conception of inertia is already foreshadowed. For example, in his discussions of a sphere rolling on an infinite horizontal plane, a motion that is neither natural nor violent and therefore can be produced by an infinitely small force, or of the constant finite velocity of a body falling in a void – both cases being abstractions from empirical reality – he abolished by implication the need for a continuing motive power to maintain constant velocity. Later at Padua, just as had happened in the 14th century, he was to give up the theory of *virtus impressa* as an explanation of projectile motion and natural acceleration, in favour of a new theory of *impeto* or *momento*. But Galileo's *impeto* belongs to another conceptual world from Buridan's *impetus*. In Galileo's new dynamics *impetus* as a motive power became redundant: the imprecise idea of the conservation of motion that it contained became analysed into recognizable statements of the laws of inertia (still incompletely generalized by Galileo) and of the conservation of momentum.

In the Second Day of the *Two Principal Systems*, Galileo makes Salviati ask:

whether there is not in the movable, besides the natural inclination towards the opposite direction, another intrinsic and natural quality (*qualità*) which makes it resistant to motion. So tell me once more: Do you not believe that the tendency of heavy bodies to move downwards, for example, is equal to their resistance to being driven upwards? [To which Sagredo replies:] I believe it to be exactly so, and it is for this reason that two equal weights in a balance are seen to remain steady and in equilibrium, the heaviness of one weight resisting being raised by the heaviness with which the other, pressing down, seeks to raise it.

This passage contains in unanalysed form the distinction to be made by Isaac Newton (1642–1727) between weight, the force moving a falling body, and mass, the intrinsic resistance to motion.* It was in fact implied by Galileo's supposition that in a vacuum all bodies would fall with the same acceleration, differences in weight exactly counterbalanced by equal differences in mass (cf. Vol. I, p. 126, note). It was impossible for Galileo to make this distinction clearly, because for him weight was still an intrinsic tendency downwards, not something depending on an *external* relationship with another attracting body, such as had been suggested by Gilbert and Kepler on the analogy of magnetism (see below, p. 197 *et seq.*) and was to be generalized by Newton as the theory of universal gravitation. Nevertheless the conception that there was an intrinsic resistance (*resistanza interna*) to motion, equal to the weight or quantity of matter of a body, gave Galileo his definition and measure of *momento*, and enabled him to take up the problem of the persistence of motion in a manner that made the concept of inertia inevitable.

From the observation that on a balance a large weight placed a short distance from the fulcrum would oscillate in equilibrium

---

* Arising out of the problem of condensation and rarefaction as discussed by Aristotle, the principle was established in the 14th century that the *quantitas materiæ* of a body remained constant in all changes. The term *quantitas materiæ* was used by Giles of Rome. Following the work of Roger Swineshead (who also called it *massa elementaris*), Heytesbury and Dumbleton, Richard Swineshead developed a clear concept of the mathematical measurability of *quantitas materiæ* by the ratio of density and volume. With Buridan it became a dynamical concept (see above, p. 81, note). But weight (*pondus*) remained for the scholastics a property only of 'heavy' bodies, and so they were never able to conceive weight as proportional to mass as Newton did. I am again indebted to Dr Weisheipl for some of this information: cf. above, pp. 101–2, note.

with a small weight placed a proportionately longer distance from the fulcrum, he derived the idea that what persists in motion is the product of weight multiplied by velocity. This product he called *impeto* or *momento*, not a cause of motion like Buridan's *impetus*, but an effect and measure of it. The problem of the persistence of motion was thus the problem of the persistence of *momento*, or momentum. He assumed, in the *Two New Sciences*, Third Day, that the momentum of a given body falling down a frictionless inclined plane was proportional only to the vertical distance, and independent of the inclination; from this he concluded that a body falling down one plane would acquire

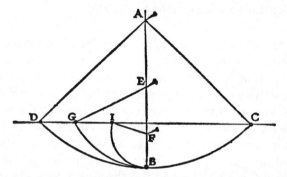

Fig. 5 Galileo's demonstration of inertia with the pendulum. From *Discorsi e dimostrazione matematiche intorno à due nuove scienze*, Bologna, *1655* (*1st ed. Leiden, 1638*), Third day.

momentum that would carry it up another to the same height. The swinging bob of a pendulum was equivalent to such a body, and he showed that if released at C (Fig. 5) it would ascend to the same horizontal line D C, whether it went by the arc B D or, when the string was caught by the nails E or F, by the steeper arcs BG or BI. This result he developed as follows:

Furthermore we may remark that any velocity once imparted to a moving body will be rigidly maintained as long as the external causes of acceleration or retardation are removed, a condition which is found only on horizontal planes; for in the case of planes which slope downwards there is already present a cause of acceleration, while on planes sloping upwards there is retardation; from this it follows that motion along a horizontal plane is perpetual; for, if the velocity be uniform, it cannot be diminished or slackened, much less destroyed. Further, although any velocity which a body may have acquired through natural fall is permanently maintained so far as its own nature is concerned, yet it must be remembered that if, after descent along a plane inclined downwards, the body is deflected to a plane inclined upward, there is already existing in this latter plane a cause of retardation; for in any plane this same body is subject to a natural acceleration downwards. Accordingly we have here the superposition of two different states, namely, the velocity acquired during the preceding fall which, if acting alone, would carry the body at a uniform rate to infinity, and the velocity which results from a natural acceleration downwards common to all bodies.

As he had already argued in the *Two Principal Systems*, perpetual motion was the limiting case, reached in an ideal world without friction, as the acceleration and retardation given respectively by downwards and upwards sloping planes each gradually tended to zero with the approach of the planes to the horizontal. The *impeto*, or momentum, impressed on a body by its movement then persisted indefinitely. Thus motion was no longer conceived of as a process requiring a cause commensurate with the effect but, as Ockham had foreshadowed, was simply a state of the moving body persisting unchanged unless acted on by a force. Force could therefore be defined as that which produced, not velocity, but a *change* of velocity from a state either of rest or of uniform motion. Further, when a body was acted on by two forces, each was independent of the other. Galileo assumed for practical purposes that the uniform motion preserved in the absence of an external force would be rectilinear, and this enabled him to calculate theoretically the trajectory of a projectile. In the *Two New Sciences*, Fourth Day, he showed that the path of a projectile, which moved with a constant horizontal velocity received from the gun and a constant acceleration vertically downwards, was a parabola, and that the range on a horizontal plane was greatest when the angle of elevation was 45 degrees. There could be no better proof than this theorem of the superiority of the theoretician able to foresee yet unobserved results, over the pure empiricist who could see only the facts already observed. As he said:

The knowledge of a single fact acquired through the discovery of its causes prepares the mind to ascertain and understand other facts without need of recourse to experiments, precisely as in the present case, where by argument alone the Author proves with certainty that the maximum range occurs when the elevation is 45°. He thus demonstrates what perhaps has never been observed in experience, namely, that of other shots those which exceed or fall short of 45° by equal amounts have equal ranges.

Even more emphatic was Salviati's assertion in the Second Day of the *Two Principal Systems*: 'I am certain, without observation, that the effect will happen as I tell you, because it must so happen.'

Galileo certainly arrived by implication at the conception of inertial motion, which was the illumination of mind that made it possible for Newton to complete the terrestrial and celestial mechanics of the 17th century; but Galileo himself did not state the law of inertia completely. He was investigating the geometrical properties of bodies in the real world, and in the real world it was an empirical observation that bodies fell downwards towards the centre of the earth. Thus, adapting the Pythagorean theory, he regarded gravity as the natural tendency of bodies to proceed to the centre of the collection of matter in which they found themselves, and weight as an innate physical property possessed by bodies; this was the source of movement or *impeto*. Galileo remained faithful all his life to the basic assumption, already found in the dialogue *De Motu*, that gravity was the essential and universal physical property of all material bodies. Confining his physical investigations to terrestrial bodies, he could take the centre of the earth to determine favoured directions in space, even though space itself was empty, homogeneous extension. The only 'natural' properties he left to bodies were their weight and their equivalent inertial 'internal resistance' to change in a motion. 'Natural gravity' was the only force he considered. It was thus in a form taking account of these assumptions that he expressed his version of the law of inertia. As he wrote in the Third Day of the *Two New Sciences*:

Just as a heavy body or system of bodies cannot move itself upwards, or recede from the common centre towards which all heavy things tend, so it is impossible for any heavy body of its own accord to assume any motion other than one which carried it nearer to the aforesaid common centre. Hence, along the horizontal by which we understand a surface, every point of which is equidistant from this same common centre, the body will have no momentum (*impeto*) whatever.

In the real world, therefore, the 'plane' along which movement would continue indefinitely was a spherical surface with its centre of the earth. As he said in the *Two Principal Systems*, Second Day:

A surface which is neither declining nor ascending ought in all its parts to be equally distant from the centre. ... Then a ship moving over a calm sea is one of those movables which run along a surface that is neither declining nor ascending, and, if all external and accidental obstacles were removed, it would thus be disposed to move incessantly and uniformly from an impulse once received? I conclude, he said in the First Day, that only circular motion can naturally suit bodies which are integral parts of the universe as constituted in the best order, and that the most that can be said for rectilinear motion is that it is assigned by nature to the bodies and their parts only where these are disposed outside their natural places, in a bad order, and therefore in need of being restored to their natural state by the shortest path. From which it seems to me that it may reasonably be concluded that for the maintenance of perfect order among the parts of the universe, it is necessary to say that movable bodies are movable only circularly; and if there are any that do not move circularly, these are necessarily immovable, there being nothing but rest and circular motion apt to the conservation of order.

This conception of motion enabled Galileo to say that the circular motion of the heavenly bodies, once acquired, would be retained. Moreover, he said that it was impossible to prove whether the space of the real universe was finite or infinite. His universe thus contained bodies with independent physical properties, which affected their movements in real space. The same line of thought can be seen in the remark in the *Two Prin-cipal Systems* that a cannon ball without weight would continue horizontally in a straight line, but that in the real world, where bodies had weight, the movement which bodies conserved was in a circle. For practical purposes of calculation he assumed, as in his work on the trajectory of a projectile, that it was rectilinear motion that was conserved. But his conception of motion en-abled him to say that in the heavenly bodies circular motion would be conserved. He did not have to explain their movements by gravitational attraction.

The intellectual revolution which had cost the 'Tuscan artist' such an anguish of effort, and yet left him still just short of reducing physics completely to mathematics, made it possible for his followers to take the geometrization of the real world as evident. Cavalieri got rid of gravity as an innate physical property, and said that like any other force it was due to external action. Evangelista Torricelli (1608–47) regarded gravity as a dimension of bodies similar to their geometrical dimensions. Giordano Bruno (1548–1600), continuing the scholastic discus-sion of plural worlds and the infinity of space, had realized that Copernicus, in making it plausible to take any point as the centre of the universe, had abolished absolute directions (see below, p. 176 *et seq.*). He had popularized the idea that space was actually infinite and therefore without favoured natural directions. The French philosopher and mathematician, Pierre Gassendi (1592–1655), whose predecessors in the 16th century had, unlike the Italians, sometimes tended to identify the con-tinuous quantity of geometry with physical extension, identified the space of the real world with the abstract, homogeneous, infinite space of Euclidean geometry. He had learnt from Demo-critus and Epicurus to conceive of space as a void, and from Kepler to regard gravity as an external force (see below, p. 197 *et seq.*). He therefore concluded, in his *De Motu Impresso a Motore Translato* published in 1642, that since a body moving by itself in a void would be unaffected by gravity, and since such a space was indifferent to the bodies in it, as Aristotle's space and its remnants in Galileo were not, the body would continue in a straight line forever. Gassendi thus first published the explicit statement that the movement which a body tended to conserve indefinitely was rectilinear, and that a change in either velocity or direction required the operation of an external force. He also first consciously eliminated the notion of *impetus* as the cause of motion. With the complete geometrization of physics, the principle of inertial motion thus became self-evident.

Gassendi had been anticipated in the expression of this principle, though not in its publication, by René Descartes (1596–1650) in his book *Le Monde*, begun some time before 1633. But if Descartes can thus be claimed as the first to have given expression to the complete principle of inertia, one fundamental and in the end fatal distinction between his and Galileo's methods of procedure must be emphasized. Whereas Galileo reached his incomplete inertial principle as a deduction from a principle of conservation of momentum supported by physical reasoning, Descartes based his complete principle on an entirely metaphysical assumption of God's power to conserve movement. Descartes had intended *Le Monde* to be a system of celestial mechanics based on the Copernican theory, but, discouraged by the condemnation of Galileo in 1633 for the similar excursion made in the *Two Principal Systems* (see below, p. 206 *et seq.*), he dropped the project, and the incompleted work was not published until 1664, when its author was already dead. The mechanical ideas contained in *Le Monde* he again resumed in the *Principia Philosophiæ* (1644). Carrying to an extreme, which Galileo had been unable to realize, the notion that the mathematical was the only objective aspect of nature, he said that matter must be understood simply as extension (see below, pp. 308–9). In creating the universe of infinite extension God also gave it motion. All sciences were thus reduced to measurement and mathematics,* and all change to local motion. Motion, being something real, could neither increase nor decrease in total

---

* 'In order to be able to prove by demonstration everything that I will deduce, I do not accept any principles in physics that are not also accepted in mathematics; these principles are sufficient because all the phenomena of nature can be explained by means of them,' *Principia Philosophiæ*, II,

amount, but could only be transferred from one body to another. The universe therefore continued to run as a machine, and each body persisted in a state of motion in a straight line, the geometrically simplest form in which God set it going, unless acted on by an external force. Only a void was indifferent to the bodies in it and, since Descartes accepted the Aristotelian principle that extension, like other attributes, could exist only by inhering in some substance, he held that space could not be a void, which was nothing, but must be a *plenum*. Only a *tendency* to continuous velocity in a straight line would therefore be possible in the real world. For Descartes the real world was simply geometry realized; movement he conceived of simply as a geometrical translation, with time as a geometrical dimension like space. The great mistake that resulted from this treatment was that Descartes completely failed to understand how to measure quantity of motion and thus failed to grasp the essential concept of the conservation of momentum. The movement that was always in a straight line was motion at an instant, conceived purely kinematically without any non-geometrical properties of inertia.

This theory left Descartes with the problem of the curvilinear motion of the planets. Having rejected action at a distance and all causes of deflection from inertial motion except mechanical contact, he could not accept a theory of gravitational attraction. He therefore tried to explain the facts by vortices in the *plenum*. He considered the original extension to have consisted of blocks of matter, each of which revolved rapidly about its centre. The consequent attrition then produced three kinds of secondary matter, characterized by luminosity (sun and stars), transparency (inter-planetary space, i.e., ether), and opacity (earth). The particles of these matters were not atomic but divisible to infinity, and their geometrical shapes accounted for their various properties. They were all in contact, so that motion

---

64. When mathematics was used to explain physical events the necessary requirement was that 'all the things which are deduced should agree completely with experience', *Princ. Philos.*, III, 46. Descartes' position in the Augustine-Platonist tradition was thus similar to that of Grosseteste or Roger Bacon.

could occur only by each successively replacing the next and thus producing a vortex, in which motion was transmitted by mechanical pressure (Plate 5). Such vortices carried the heavenly bodies round. Mechanical pressure was also the means of propagation of such influences as light and magnetism. The *plenum*, or ēther, which owed some of its characteristics to Gilbert and Kepler, was thus loaded with the physical properties, among them what was later called 'mass', that could not be reduced to geometry.

The vortex theory shows Descartes empirically at his weakest, and Newton was to demonstrate in the *Principia Mathematica* (1687) that it would not in fact yield Kepler's laws of planetary motion and so was falsified by observation (cf. below, pp. 204–5).

In spite of his great contributions to mathematics and to the mathematical techniques of physics, Descartes developed his cosmology to a considerable extent on entirely non-mathematical lines, and certainly it makes a striking contrast with Galileo's approach to physical problems. Starting from a background of scholastic physics, Galileo achieved his successes by eliminating the physical-causal elements from the problem of motion; his approach to dynamics was through kinematics, and although his passionate concern with the new astronomy gave him a general cosmological objective, his method was to try to solve each individual problem separately, to discover empirically what laws were in fact exhibited by the natural world, before facing the task of reassembling them into a whole. While appreciating Galileo's individual kinematic descriptions, Descartes found his work lacking in a total view of physics and his method of abstraction defective exactly at the point where Galileo had made it so effective: its turning away from the problem of physical causes. Commenting in 1638 on Galileo's recently published *Discourses concerning Two New Sciences*, Descartes characterized his own position by contrast, writing to Mersenne:

I will begin this letter with my observations on Galileo's book. I find that in general he philosophizes much better than the average, in that he abandons as completely as he can the errors of the schools, and attempts to examine physical matters by the methods of mathematics. In this I am in entire argreement with him, and I believe that there is absolutely no other way of discovering the truth. But it seems to me that he suffers greatly from continued disgressions, and that he does not stop to explain all that is relevant to each point; which shows that he has not examined them in order, and that, without having considered the first causes of nature, he has merely sought reasons for certain particular effects; and thus he has built without a foundation. A month later he wrote again : As to what Galileo has written about the balance and the lever, he explains very well what happens (*quod ita fit*), but not why it happens (*cur ita fit*), as I have done in my *Principles*.

Descartes was not alone in not accepting Galileo's methods as covering the complete range of physical problems; many physicists especially in France, for example Fermat, Mersenne and Roberval, shared his hesitations. It was just because Descartes himself took the opposite course of inquiring beyond the mathematical descriptions into physical causes and the nature of things, and of boldly constructing an entire system of science ranging from psychology and physiology through chemistry to physics and astronomy, writing a new *Timæus*, that his ideas became in many ways by far the greatest single influence on the history of science in the 17th century. They established the general line of thought even of those who, like Newton, were most critical of the Cartesian system in detail. Descartes approached physics as a philosopher. It must not be supposed that for that reason he did not appreciate the function of experiments or make them himself; certainly he did (cf. below, pp. 244 *et seq.*, 258 *et seq.*). But it was through his philosophical method and the universality claimed for its most fundamental results that he came to dominate the scientific thought of the period and to provide in one bold sweep at least something comprehensive and consistent to disagree with. Descartes saw the object of his philosophical method as the search by rational analysis for the simplest elements making up the world, 'simple natures' that could not be reduced to anything simpler and so had no logical definitions (see below, p. 308 *et seq.*). So far as the physical world was concerned he found these in extension and motion. 'If I am not deceived,' he wrote in *Le Monde*, 'not only these four qualities [heat, cold, wetness, dryness], but also all the others, and even all the forms of inanimate bodies, can be explained, without having to suppose anything else in their matter but motion, size, shape, and the arrangement of their parts.' From these 'simple natures', and from purely metaphysical principles, partly relating to the perfection and goodness of God, he then proceeded to deduce the laws that the actual world must follow. He admitted that his conclusions might be wrong in detail, and he gave up the attempt to reduce the complicated observed world, with its many unknown variables, to mathematical laws; hence the largely qualitative character of *Le Monde* and the *Principia Philosophiæ*. But of the correctness of his general aims and general conclusions he never had any doubts.

It was the most fundamental general conclusion of Descartes' mechanistic philosophy that all natural phenomena could eventually, when sufficiently analyzed, be reduced to a single kind of change, local motion; and that conclusion became the most influential belief of 17th-century science. This, and the consequent doctrines of universal corpuscularity and the universality of action by physical contact, provided the 17th century with a new conception of nature in place of Aristotle's qualitative 'forms' or 'natures'; they provided scientists with a 'regulative belief' determining the form given to physical and physiological theories. The Cartesian philosophy of nature was the immediate subject of most of the controversies in which Newton and Newtonianism became involved; the *Principia Mathematica* (1687) itself, while pursuing the same very general aims as the *Principia Philosophiæ*, was written partly as a polemic against the details of the Cartesian system and the methods of arriving at them. Moreover it was not only in the philosophy of science that Descartes' influence was felt. Christian Huygens (1629–95) owed his scientific awakening to Descartes and never entirely deserted his point of view; and in the conception of kinetic energy found obscurely in Leibniz's conception of *vis viva* and fully developed in the 19th century, Descartes could be claimed to have originated a substantial contribution to dynamics.

The history of Cartesianism begins only in the middle of the 17th century and belongs to this volume only in reminding us that the direction of thought culminating in Galileo's method of abstraction and descriptive analysis of motion was balanced by another less willing to see physics alienated, even temporarily, from the search for the nature and causes of things. So far as the inertial principle was concerned, it was not Descartes but Galileo who provided the conception of motion on which Huygens, Newton and others were to build the classical mechanics of the 17th century. The dynamical inquiries of these mathematicians, though leading to the enunciation of a number of separate principles whose connexion with each other was not at the time always clearly understood, such as the law of falling bodies, the concepts of inertia, force and mass, the parallelogram of forces and the equivalence of work and energy, really involved only one fundamental discovery. This was the principle, established experimentally, that the behaviour of bodies towards one another was one in which accelerations were determined, the ratio of the opposite accelerations they produced being constant and depending only on a characteristic of the bodies themselves, which was called mass. It was a fact which could be known only by observation that two geometrically equivalent bodies would move differently when placed in identical relations with the same other bodies. Where Galileo had halted before the real world and Descartes, geometrizing from abstract principles, hid this physical property in vortices, Newton made an exact mathematical reduction of mass from the facts of experience. The relative masses of two such bodies were measured by the ratio of their opposite accelerations. Force might then be defined as that which disturbed a body from a state of rest or uniform rectilinear motion, and the force between two bodies, for example that of gravitation, was the product of either mass multiplied by its own acceleration. Inertial motion was an ideal limit, the state of motion of a body acted on by no other. The problem that had been so puzzling to those who first questioned the Aristotelian law of motion, why, excluding the resistance of the medium, bodies of different masses fell to the earth with the same acceleration, then found its solution in the distinction between mass, a property of the body providing intrinsic resistance, and weight, caused by the external force of gravitation acting on the body. Differences in weight might be considered as exactly counter-balanced by proportional differences in mass. And the same mass had a different weight according to its distance from the centre of the earth. When these conceptions were generalized by Newton, the old problems of the acceleration of freely falling bodies and of the continued motion of projectiles were finally solved; and when the same principles were carried once more into the sky in the theory of universal gravitation, Buridan's aspiration was realized, and the movements of the heavens, which Kepler had correctly described, were united with these homely phenomena in one mechanical system. This not only brought about the final destruction of the hierarchically-ordered finite world of irreducibly-different 'natures', which had formed the Aristotelian cosmos; it was a vast illumination of mind. The principles, first effectively established by Galileo, on which the new mechanics were constructed then seemed finally justified by their success.

2. ASTRONOMY AND THE NEW MECHANICS

Though, after its arrival in Western Christendom in the 13th century, the Ptolemaic system had been commonly regarded as simply a geometrical calculating device, the need was felt for an astronomical system which would both 'save' the phenomena and also describe the 'actual' paths of the heavenly bodies through space. Since the 13th century, observation and the revision of tables had gone on in connexion with the chronic desire to reform the calendar and with the practical demands of astrology and navigation. Regiomontanus had been summoned to Rome for consultation on the claendar in 1475, the year before he died, and his work was used by the Portuguese and Spanish ocean navigators. Some medieval writers, for instance Oresme and Nicholas of Cusa, had suggested alternatives to the geo-static system as a description of physical 'fact' and, in the early years of the 16th century, the Italian Celio Calcagnini (1479–1541) put forward in a vague form a theory based on the earth's rotation. His countryman, Girolamo Fracastoro (1483–1553), attempted to revive the system of concentric spheres without epicycles. It was left for Nicholas Copernicus (1473–1543) to elaborate a system which could replace Ptolemy's as a calculating device and yet represent physical 'fact', and also 'save' additional phenomena, such as the diameter of the moon, which according to Ptolemy's system should have undergone monthly variations of nearly a hundred per cent.

Copernicus was educated first at the University of Cracow and then at Bologna, where he studied law but also worked with the professor of astronomy, Domenico Maria Novara (1454–1504). Later he proceeded to Rome, to Padua where he studied medicine, and to Ferrara where he completed his law. The remainder of his life was spent at Frauenberg, a cathedral town in East Prussia, where he performed the functions of a cleric, doctor and diplomat and produced a scheme which was the basis of a reform of the currency. In the midst of this busy life he proceeded to reform astronomy. Here, though he made a few observations, his work was that of a mathematician. He is a supreme example of a man who revolutionized science by looking at the old facts in a new way. He took his data mainly from the *Epitome in Almagestum* (printed 1496) of Peurbach and Regiomontanus and from Gerard of Cremona's Latin translation of the *Almagest*, which was printed at Venice 1515. Novara, a leading Platonist, had taught him the desire to conceive of the constitution of the universe in terms of simple mathematical relationships. Inspired by this, he set about producing his new system.

Martianus Capella had preserved for the succeeding centuries Heraclides' theory that Mercury and Venus, whose orbits are peculiar in their restricted angular ranges from the sun (the other planets may be seen at any angular distance, or 'elongation', from the sun), actually revolved round the sun, while the sun with the remaining heavenly bodies revolved round the earth. Heraclides was also reported to have let the earth revolve daily on its axis. Copernicus not only gave the earth a daily rotation but made the whole planetary system, including the earth, revolve round a static sun in its centre. His reluctance to publish this theory, of which the manuscript was complete by 1532, seems to have depended largely on the fear that it would be considered absurd. He had been satirized on the stage near Frauenberg in 1531, and his anxiety would certainly have been confirmed had he lived to hear the comments of such diverse personalities as the Italian mathematician Francesco Maurolico and the German revolutionary Martin Luther (1483–1546). 'The fool,' said Luther, 'would overturn the whole science of astronomy.' Copernicus eventually drew up a short summary (*Commentariolus*), which seems to have become known to the Pope, and in 1536 he was asked by Cardinal Nicolaus von Schönberg to make his theory known to the learned world. Georg Joachim (Rheticus), a professor at Wittenberg (who is notable for having introduced the improvement of making trigonometrical functions depend directly on the angle instead of on the arc), had journeyed to Frauenberg in 1539 to study Copernicus' manuscript, and in 1540 Rheticus published his *Narratio Prima de Libris Revolutionum* concerning it. Copernicus' work was thus well advertised when, having been seen through the press by Rheticus, it appeared at Nuremberg in 1543, dedicated to Pope Paul III under the title *De Revolutionibus Orbium Cœlestium*. Its practical value was demonstrated when Erasmus Reinhold used it to calculate the *Prussian Tables* (1551), though these suffered from the inaccuracy of Copernicus' data, and when the figure for the length of the year given in *De Revolutionibus* was proposed, though not used, as the basis of the reform of the calendar instituted by Pope Gregory XIII in 1582. In spite of the cautious preface by Andreas Osiander, stating the contrary, Copernicus certainly considered the revolution of the earth as a physical fact and not a mere mathematical convenience. *De Revolutionibus* thus posed the problems that occupied the greater part of physics down to Newton.

The Copernican revolution was no more than to assign the daily motion of the heavenly bodies to the rotation of the earth on its axis and their annual motion to the earth's revolution about the sun, and to work out, by the old devices of eccentrics and epicycles, the astronomical consequences of these postulates (Fig. 6, Plate 6).

It was in postulating the annual motion of the earth that Copernicus made his great strategic advance in theory over the medieval discussions of a reformed astronomy, and opened the way for the full mathematical development of a new system. For example, though Oresme had made the earth spin on its axis, his system remained geocentric. There were certain peculiarities in the mathematics of the geocentric system that Copernicus may have noticed: the constants of epicycle and deferent were reversed between the lower planets (Mercury and Venus) and the upper ones; and the sun's period of revolution appeared in the calculations for each of the five planets (see Fig. 6). Of the steps by which he arrived at the conception of a heliocentric system Copernicus has left no detailed account. He described simply in the preface to *De Revolutionibus* how he was urged to think out a new way of calculating the motions of the spheres because he found that the mathematicians disagreed among themselves, and used different devices: concentric spheres, eccentric spheres, epicycles. He concluded that there must be some basic mistake.

Then when I pondered over this uncertainty of traditional mathematics in the ordering of the motions of the spheres of the orb, I was disappointed to find that no more reliable explanations of the mechanism of the universe, founded on our account by the best and most regular Artificer of all, was established by the philosophers who have so exquisitely investigated other details concerning the orb. For this reason I took up the task of re-reading the books of all the philosophers which I could procure, exploring whether any one had supposed the motion of the spheres of the world to be different from those adopted by the academic mathematicians.

In this way he came across Greek theories of the double motion of the earth, on its axis and round the sun, and these he developed, following the example of his predecessors who had not scrupled to imagine whatever circles they required to 'save the appearances'.

177

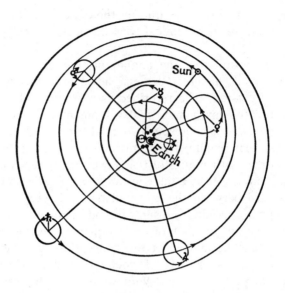

Fig. 6A, B Comparison of the Ptolemaic (A) and Copernican (B) systems (cf. Vol. I, Figs 2 and 3). Although his system was essentially a collection of independent devices for each heavenly body, the relative periods of revolution had established a traditional order of the orbits which Ptolemy adopted. By reversing the positions of the earth and the sun, Copernicus was able to use these periods to assess the relative mean distances of the planets from the sun and to rationalize the relationships between the epicycles and deferents of the lower (Mercury and Venus) and upper planets (see table below). The earth's motion round its orbit in the Copernican system is in fact reproduced in the Ptolemaic system not only by the sun's orbit but also by the deferent of each lower planet (the planet's orbit being reproduced by Ptolemy's epicycle) and by the epicycle of each upper planet (the planet's orbit here being reproduced by Ptolemy's deferent). It is not possible to show these points clearly in the diagram by drawing to scale. The positions of the centres of the planetary orbits relative to that of the sun's orbit in the Ptolemaic system, and to the sun itself in the Copernican system, are shown by the dots at the inner ends of the radii of the deferents; i.e., the large circles. Copernicus regarded his greatest technical achievement as the elimination of the objectionable Ptolemaic equant (cf. Vol. I, p. 97), which he achieved by referring the planetary motions not to the central sun but to the centre (D) of the earth's orbit, which itself revolved round the sun on two further circles. This device introduced inaccuracies in the planetary latitudes, especially of Mars, and

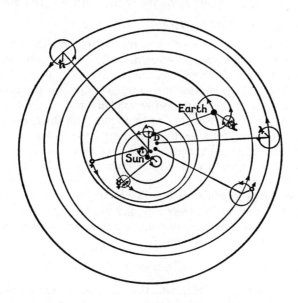

it was Kepler who in fact made the sun the point of reference for the planetary orbits (see Plate VII). Mercury received special treatment from Ptolemy, who made the centre of its deferent rotate slowly round another circle. Copernicus retained this device and in addition introduced the unique treatment of making the planet oscillate, or 'librate,' on the diameter of its epicycle instead of travelling round it. By a simple geometrical construction (not given here) it can be shown that any complexity introduced into one system in order to 'save the appearances' can be matched in the other, so that the two systems can be made equivalent in representing the angle at which a planet appears when seen from the earth. But the two systems differ in their ranges of theoretical possibilities for the lower planets (Mercury and Venus), and these differences can provide an empirical test for choosing between them. According to the Copernican system, but not to the Ptolemaic, the lower planets can appear on the side of the sun remote from the earth (they cannot do so in the Ptolemaic system since they are inside the sun's orbit); their greatest angular distances from the sun are reached when earth-planet-sun form a right angle; and only they should show complete phases like the moon. Galileo confirmed these Copernican conclusions with his telescope (see pp. 191, 210 *et seq.*). The Ptolemaic system can, however, be made to yield the same conclusions by making the epicycles of Mercury and Venus rotate round the sun, a suggestion made by Heraclides of Pontus (see Vol. I, p. 101) and adopted by Tycho Brahe for the whole planetary system (see p. 187). (Diagrams re-

179

'Occasioned by this,' he wrote,

I also began to think of a motion of the earth, and although the idea seemed absurd, still, as others before me had been permitted to assume certain circles in order to explain the motions of the stars, I believed it would readily be permitted me to try whether, on the assumption of some motion of the earth, better explanations of the revolutions of the heavenly spheres might not be found. And thus, assuming the motions which in the following work I attribute to the earth, I have finally found, after long and careful investigation, that when the motions of the other planets are referred to the circulation of the earth and are computed for the revolution of each star, not only do the phenomena necessarily follow therefrom, but the order and magnitude of the stars and all their orbs and the heaven itself are so connected that in no part can anything be transposed without confusion to the rest and to the whole universe.

'Therefore,' he continued in book 1, chapter 10,

we are not ashamed to maintain that all that is beneath the moon, with the centre of the earth, describe among the other planets a great orbit round the sun which is the centre of the world; and that what appears to be a motion of the sun is in truth a motion of the earth; but that the size of the world is so great, that the distance of the earth from the sun, though appreciable in comparison to the orbits of the other planets, is as nothing when compared to the sphere of the fixed stars. And I hold it to be easier to concede this than to let the mind be distracted by an almost endless multitude of circles, which those are obliged to do who detain the earth in the centre of the world. The wisdom of nature is such that it produces nothing superfluous or useless but often produces many effects from one cause. If all this is difficult and almost incomprehensible or against the opinion of many people, we shall, please God, make it clearer than the sun, at least to those who know something of mathematics. The first principle therefore remains undisputed, that the size of the orbits is measured by the period of revolution, and the order of the spheres is then as follows, commencing with the uppermost. The first and highest sphere is that of the fixed stars, containing itself and everything and

---

drawn after William D. Stahlman's diagrams in Galileo Galilei, *Dialogue on the Great Systems of the World*, revised translation by Giorgio de Santillana, Chicago, 1953, pp. xvi–xvii.)

(a) Ptolemaic System

	Ratio of radii (corresponding to mean distance from sun in Copernican system)	Angular velocity (degrees per day)	Modern value of sidereal mean motion (degrees per day)
	Epicycle/Deferent	*Epicycle*	
Earth ⊕			
Moon ☽			
Mercury ☿	0·3708	4·09233	4·09234
Venus ♀	0·7194	1·60214	1·60213
Sun ☉		0·98563 (sun's orbit)	0·98561 (earth's orbit)
	Deferent/Epicycle	*Deferent*	
Mars ♂	1·5206	0·52406	0·52403
Jupiter ♃	5·2167	0·08312	0·08309
Saturn ♄	9·2336	0·03349	0·03346

(b) Copernican System

	Mean distance from sun expressed as ratio of distance of earth	Modern value	Period of revolution round sun (days)
Sun ☉		0·00257 (from earth)	27·33 (round earth)
Mercury ☿	0·3763	0·3871	88
Venus ♀	0·7193	0·7233	225
Earth ⊕	1·0000	1·0000	365·25
Mars ♂	1·5198	1·5237	687
Jupiter ♃	5·2192	5·2028	4,332
Saturn ♄	9·1743	9·5389	10,760

therefore immovable, being the place of the universe to which the motion and places of all other stars are referred. For while some think that it also changes somewhat [this refers to precession], we shall, when deducing the motion of the earth, assign another cause for this phenomenon. Next follows the first planet Saturn, which completes its circuit in thirty years, then Jupiter with a twelve years' period, then Mars, which moves round in two years. The fourth place in the order is that of the annual revolution, in which we have said that the earth is contained with the lunar orbit as an epicycle. In the fifth place Venus goes round in nine months, in the sixth Mercury with a period of eighty days. But in the midst of all stands the sun. For who could in this most beautiful temple place this lamp in another or better place than that from which it can at the same time illuminate the whole? Which some not unsuitably call the light of the world, others the soul or the ruler. Trismegistus calls it the visible God, the Electra of Sophocles the all-seeing. So indeed the sun, sitting on the royal throne, steers the revolving family of stars.

The consequences of Copernicus' postulates were of two kinds, physical and geometrical. The daily rotation of the earth encountered the Aristotelian and Ptolemaic physical objections, based on the theory of natural motions, concerning 'detached bodies', an arrow or a stone sent into the air, and the strong east wind (see above, p. 90 *et seq.*). To these Copernicus replied in the same way as Oresme, making circular movement natural and saying that the air shared that of the earth because of their common nature and also perhaps because of friction. He held that falling and rising bodies had a double motion, a circular motion when in their natural place, and a rectilinear motion of displacement from, or return to, that place. The objection to this argument was that if bodies had a natural circular movement in one direction they should have a resistance, analogous to weight, to motion in the other. The answer to this, like that to the argument that the earth would be disrupted by what is now sometimes called 'centrifugal' force, which Copernicus merely said would be worse for the enormous celestial sphere if it rotated, had to await the mechanics of Galileo.

To the annual motion of the earth in an eccentric circle round the sun Copernicus' critics objected on three scientific grounds

First, it conflicted with the Aristotelian theory of natural movements, which depended on the centre of the earth being at the centre of the universe. To this Copernicus replied, with Oresme and Nicholas of Cusa, though abandoning Cusa's theory of balancing heavy and light elements, that gravity was a local phenomenon representing the tendency of the matter of any astronomical body to form spherical masses. The second objection arose from the absence of observable annual stellar parallaxes, or differences in position of the stars. Copernicus attributed this to the enormous distance of the stellar sphere from the earth compared with the dimensions of the earth's orbit. The third objection continued to be a stumbling-block till Galileo changed the whole conception of motion, when it ceased to be relevant. The Aristotelians maintained that each elementary body had a single natural movement, but Copernicus gave the earth three motions: the two mentioned above which accounted, respectively, for the rising and setting of the heavenly bodies and for the passage of the sun along the ecliptic and the retrogradations and stations of the planets, and a third which was intended to account for the fact that the axis of the earth, notwithstanding the annual motion, always pointed to the same spot on the celestial sphere. This third motion was also made to account for the precession of the equinoxes and their illusory 'trepidations'.

With the sun and the celestial sphere, the boundary of the finite universe, at rest, Copernicus proceeded to provide the usual eccentrics, deferents and epicycles to account for the observed movements of the moon, sun and planets by means of perfect uniform circular motion. On the mathematical aspects of the result, Neugebauer in his *Exact Sciences in Antiquity* (1957, p. 204) comments as follows: 'The popular belief that Copernicus' heliocentric system constitutes a significant simplification of the Ptolemaic system is obviously wrong. The choice of the reference system has no effect whatever on the structure of the model, and the Copernican models themselves require about twice as many circles as the Ptolemaic models and are far less elegant and adaptable.' Copernicus' main mathematical contributions, according to Neugebauer, were three in number. He clarified the steps from observations to parameters, thus making a methodological improvement. He introduced with his system a criterion for assigning relative distances to the planets. And he suggested the proper solution of the problem of latitudes. But his belief in the imaginary trepidations of the equinoxes led to unnecessary complications and, by taking the centre of the earth's orbit as the centre of all the planets' motions, his treatment of Mars had considerable errors. Further, he relied on ancient and inaccurate data. This last defect was remedied by Tycho Brahe (1546–1601), who showed that the trepidations were due solely to errors in observation; and Johann Kepler (1571–1630), while considering Tycho's results, was to build his system from the orbit of Mars.

Copernicus had produced a mathematical system at least as accurate as Ptolemy's, with both mathematical advantages and disadvantages. Theoretically and qualitatively it was certainly simpler, in that he could give a unified explanation of a number of different features of planetary motion which in Ptolemy's system were arbitrary and disconnected. He could account for the retrogradations and stations of the planets as mere appearances due to a single movement of the earth, and could give a simple explanation of various motions peculiar to individual planets. In the 16th century it was also counted in his favour that he had reduced the number of circles required; he used 34. Copernicus had also argued that the postulated movements of the earth did not conflict with physics, that is, with Aristotle's physics. These arguments in favour of the heliostatic system were negative, and moreover in order to effect the reconciliation he had to interpret Aristotle's physics, just as Oresme had done, in a sense different from that accepted by most of his contemporaries. It is not surprising that many of them remained unconvinced. How then did Copernicus justify his innovation, both to himself and publicly, and why did it make so strong and so emotional an appeal later to Kepler and Galileo? A large part of the answer certainly lies in the Neoplatonism they all shared. In the passage already quoted from *De Revolutionibus*, book 1, chapter 10, Copernicus justifies the new system he sets out by an appeal to its simplicity (qualitative, not quantitative) and to the special position it gives to the sun. The intellectual biographies of Kepler and Galileo, and the manner in which they used these and similar arguments, show that they too had committed themselves to the heliocentric system because of their metaphysical beliefs, before they had found arguments to justify it physically.

The Copernican system appealed first to three types of interest. The *Alfonsine Tables* had caused dissatisfaction both because they were out of date and no longer corresponded to the observed positions of stars and planets, and because they differed from Ptolemy on the precession of the equinoxes and added other spheres beyond his 9th, deviations offensive to humanists who believed that the perfection of knowledge was to be found in the classical writings. All practical astronomers, whatever their views on the hypothesis of the earth's rotation, thus turned to the 16th-century *Prussian Tables* calculated on Copernicus' system though, in fact, these were scarcely more accurate. Some humanists regarded Copernicus as the restorer of the classical purity of Ptolemy. Another group of writers, such as the phycisist Benedetti, Bruno, and Pierre de la Ramée, or, as he was called, Petrus Ramus (1515–72), saw in the Copernican system a stick with which to beat Aristotle. Finally, scientists like Tycho Brahe, William Gilbert (1540–1603), Kepler and Galileo, came to face the full meaning of *De Revolutionibus* and attempted to unify observations, geometrical descriptions and physical theory. It was because of the absence of such a unity that until the end of the 16th century, while everyone used the *Prussian Tables*, no one advanced astronomical theory. Tycho Brahe's contribution was to realize that such an advance demanded careful observation, and to make that observation.

Tycho's main work was done at Uraniborg, the observatory built for him in Denmark by the king. His first task was to improve the instruments then in use. He greatly increased their size, constructing a quadrant with a 19-foot radius and a celestial globe 5 feet in diameter, and he improved methods of sighting and graduation. He also determined the errors in his instruments, gave the limits of accuracy of his observations, and took account of the effect of atmospheric refraction on the apparent positions of heavenly bodies. It had been customary before Tycho to make observations in a somewhat haphazard manner, so that there had been no radical reform of the ancient data. Tycho made regular and systematic observations of known error, which revealed problems hitherto hidden in the previous inaccuracies.

Tycho's first problem arose when a new star appeared in the constellation Cassiopeia on 11 November 1572, and remained until early in 1574. Scientific opinion received a marked shock from this object. Tycho attempted to determine its parallax and showed that this was so small that the star must be beyond the planets and adjacent to the Milky Way. Although he himself never fully accepted it, the mutability of celestial substance had thus been definitely demonstrated. Also, though comets had been regularly observed since the days of Regiomontanus, Tycho was able to show, with his superior instruments, that the comet of 1577 was beyond the sun and that its orbit must have passed through the solid celestial spheres, if these existed. He also departed from the Platonic ideal and suggested that the orbits of comets were not circular but oval. Further, Aristotelian theory held that comets were manifestations in the air. It is significant that, although it would have been possible with instruments available from antiquity to show that comets penetrated the unchanging world beyond the *moon*, such observations were not in fact made until the 16th century. In 1557 Jean Pena, royal mathematician at Paris, had maintained on optical reasoning that some comets were beyond the moon and hence had rejected the spheres of fire and of the planets. He held that air extended to the fixed stars. Tycho went further and abandoned both the Aristotelian theory of comets and the solid spheres. At the same time, the discovery of land scattered all over the globe led other natural philosophers, such as Cardano, to abandon the theory of concentric spheres of earth and water based on the Aristotelian doctrine of natural place and motion. Land and sea they held to form one single sphere.

While Tycho provided the observations on which to base an accurate geometrical description of heavenly motions, he was led by physical as well as by Biblical difficulties to reject the rotation of the earth. He did not consider that Copernicus had answered the Aristotelian physical objections. Further, before the invention of the telescope had revealed the fact that the fixed stars, unlike the planets, appear as mere luminous points and not as discs, it was usually held that they shone by reflected light, and their brightness was taken as a measure of their magnitude. Tycho therefore deduced, from the absence of observable annual stellar parallax, that the Copernican system would involve the conclusion that the stars had diameters of incredible dimensions. He produced a system of his own (1588), in which the moon, sun and fixed stars revolved round a stationary earth while all the five planets revolved round the sun. This was geometrically equivalent to the Copernican system, but escaped what he considered to be the latter's physical defects and included the benefits of his own observations. It remained an alternative to Copernicus (or Ptolemy) during the first half of the 17th century, and when Tycho bequeathed his observations to Kepler, who had come to work with him, he asked him to use it in the interpretation of his data.

Kepler did more than this. Michael Mästin (1550–1631), under whom he had first studied, had, like Tycho, also calculated the orbit of the comet of 1577, and he declared the Copernican system alone capable of accounting for it. Kepler persisted in this opinion. He was also strongly influenced by Pythagoreanism. The vision of abstract harmony, according to which he believed the world to be constructed, sustained him through the drudgery of arithmetical computation to which he was consigned both by his astronomical researches and by his work as a professional astrologer. Throughout his life he was inspired by the search for a simple mathematical law which would bind together the spatial distribution of the orbits and the motions of the members of the solar system. After numerous trials he arrived at the idea published in his *Mysterium Cosmographicum* (1596), that the spaces between the planetary orbits each corresponded, from Saturn to Mercury, to one of the five regular solids or 'Platonic bodies': cube, tetrahedron, dodecahedron, icosahedron and octahedron. His object was to show the necessity of there being six and only six planets and of their orbits being of the relative sizes they are, as calculated from their periods round the sun. He tried to show that the five regular solids could be fitted to the six orbits so that each orbit was inscribed in the same solid about which the next outer orbit was circumscribed. He then went to Tycho Brahe, who had moved to Prague, from whom alone he could get the correct values of the mean distances and eccentricities that would confirm this theory. He was forced instead to set it aside, but his mathematical vision came to perceive in Tycho's data the foundations of celestial harmony. Having worked out the orbit of Mars on each of the three current theories, the Ptolemaic, the Copernican and the Tychonic, he saw that Copernicus had unnecessarily complicated matters by not allowing the planes of all the planetary orbits to pass through the sun. Even when this assumption was made there was still an error of 8 or 9 minutes in the arc of the orbit of Mars; and this could not be attributed to inaccuracy in the data. This forced him to abandon the assumptions that planetary orbits were circular and the movements of the planets uniform, and led him to formulate his first two laws: (1) Planets move in ellipses with the sun in one focus; (2) each planet moves, not uniformly, but so that a line joining its centre to that of the sun sweeps out equal areas in equal times (*Astronomia Nova aitiologetos, seu Physica Cœlestis tradita commentariis de motibus stellæ Martis ex observationibus G. V. Tychonis Brahe*, 1609– Plate 7).

It was actually the second of these laws that Kepler discovered first. One of the difficulties encountered was the considerable variation in the velocity of Mars round its orbit, so that it was faster nearer the sun than remote from it. He first tried to render this variation mathematically by reintroducing the equant, which Copernicus had rejected. But he found that there was no equant that permitted the accurate calculation of all the observations. His proof that the same variations occurred in the earth's orbit demonstrated mathematically the similarity of its motion to those of the planets. He saw the problem then as that of finding a theorem relating the velocity of a planet's rotation at any point to its distance from the sun in an eccentric orbit. This he solved by a method of integration by which he showed that the duration of the planet on a very small arc of its trajectory was proportional to its distance from the sun. Guided in his approach to this problem by his physical conception of a power or *virtus* extending from the sun and moving the planets, it followed that this motive power was inversely proportional to the distance from the sun. Thus the motive power was inversely proportional to the duration of the planet on an arc of its orbit – a conclusion entirely in keeping with the Aristotelian dynamical assumption that velocity requires a motive power.

It was in the course of these calculations and the checking of the predicted positions against Tycho Brahe's data that Kepler began to have his revolutionary doubts whether the planetary orbits were really circular. He had decided to give up the circular movements in 1604. As he wrote in his *Astronomia Nova*, part 3, chapter 40:

My first error was to take the planet's path as a perfect circle, and this mistake robbed me of the more time, as it was taught on the authority of all philosophers, and is consistent in itself with metaphysics.

The fact that Kepler came to break what Koyré has called the 'spell of circularity', while Galileo did not, makes an interesting contrast in the character of their Platonism. Galileo denied the Platonic ontological distinction between geometrical figures and material bodies; so far as he could he saw the physical world as geometry realized; and this made it difficult for him to deny the privileged status of circularity in physics and astronomy while accepting it in mathematics and, as has been shown recently, in aesthetics (cf. above, pp. 150–51, 166). Kepler, on the other hand, by retaining the ontological distinction between ideal form and material realization, was able without violence to his Platonic metaphysics to accept a deviation from circularity forced on him by the empirical data. He argued that celestial bodies, *qua* bodies, were bound to deviate from the perfectly circular course because their motions were not the work of mind but of nature, of the 'natural and animal faculties' of the planets, which followed their own inclinations, as he said in his *Epitome Astronomiæ Copernicanæ*, book 4, part 3, chapter 1 (1620).

Again guided by his conception of the physical causes of planetary motion, Kepler at first supposed that the non-circular orbit was an ovoid resulting from two independent motions, one caused by the sun's *virtus* and the other by a uniform rotation of the planet on an imaginary epicycle produced by a *virtus* of its own. Kepler found himself unable to deal mathematically with the various ovoid curves he tried, so he decided to use as an approximation the ellipse, of which the geometry had been fully worked out by Apollonius. He discovered that the ellipse fitted his law of areas perfectly, an empirical conclusion for which he later tried to give a physical explanation by means of an oscillating motion or 'libration' of the planet on the diameter of its epicycle (cf. Fig. 6, Mercury).

After ten years' further labour he arrived at his third law, published in 1619 in the *Harmonice Mundi*: (3) the squares of the periods of revolution (p_1, p_2) of any two planets are proportional to the cubes of their mean distances (d_1, d_2) from the sun (C), that is, $\dfrac{p_1{}^2}{p_2{}^2} = \dfrac{d_1{}^3}{d_2{}^3}$. This was a law for which Kepler had searched from the beginning of his career, but he made his discovery in the end almost accidentally. Following the method of trial and error, he made a series of comparisons of the instantaneous velocities and the periods and the distances of the different planets, but reached no significant formula. Eventually he tried comparisons of powers of these numbers, and found that those of his 'third law' gave an exact empirical fit.

These laws could scarcely have been formulated without the work of Greek geometers, especially Apollonius, on conic sections. This subject had been developed by Maurolico and, in a commentary on Witelo (1604), by Kepler himself. In deducing his second law, Kepler made a contribution to mathematics, introducing the innovation, from which considerations of strict logic had restrained the Greeks, of considering an area as made up of an infinite number of lines generated by revolving a given curve about an axis (cf. above, p. 139). For the integration required for his second law he used a method similar to that by which Archimedes had determined the value of π. The work of the practical astronomer was also greatly assisted by improvements in methods of computation, first by the systematic use of decimal fractions introduced by Stevin, but above all by the publication in 1614 of the discovery of logarithms by John Napier (1550–1617). Following this, other mathematicians calculated tables for trigonometrical functions and accommodated logarithms to the natural base e. The slide rule was invented by William Oughtred in 1622. Kepler made use of some of these innovations in reducing to order the practical results of his own and Tycho's work for the *Rudolphine Tables*, published in 1627.

Kepler's three laws provided a final solution of the ancient problem of discovering an astronomical system which would both 'save' the phenomena and describe the 'actual' paths of the bodies through space. Copernicus' 'third motion' of the earth was abandoned since, there being no celestial spheres, the phenomena it was supposed to explain were attributed simply to the fact that the earth's axis remained parallel to itself in all positions. The independent invention of the telescope (with magnifications of up to about thirty) by Galileo added confirmation for the 'Copernican' theory. One of Tycho's objections to this theory was removed when Galileo was able to show, by finding the distance at which a stretched cord of known thickness would just eclipse them, that the fixed stars were not of the incredibly enormous dimensions Tycho had supposed they would have to be, on the assumption that brightness was proportional to magnitude, in order to be as bright as they are at a distance sufficient for them to show no parallax. Galileo also resolved parts of the Milky Way into individual stars, and he confirmed Copernicus' deduction that Venus, because of the position he held it to have inside the earth's orbit, would have complete phases like the moon. The other lower planet, Mercury, also has complete phases, whereas Mars has only partial phases (cf. Fig. 6). In 1631 Pierre Gassendi observed the transit, which Kepler had predicted, of Mercury across the sun's disc, and established that it described an orbit between the sun and the earth. The transit of Venus was observed in 1639 by the English astronomer Jeremiah Horrocks (1619–41), Galileo, in his *Sidereus Nuncius* (1610), described the mountains on the moon and the four satellites of Jupiter, which he took as a model of Copernicus' solar system (cf. Plate 8). Later he observed Saturn as misshapen (his telescope would not resolve the rings), and he was able to show that the variations in the apparent sizes of Mars and Venus corresponded with the distances of these bodies from the earth according to the Copernican hypothesis. His observation of spots on the sun, by which he claimed to estimate its rate of rotation, also added evidence against the Aristotelian theory of immutability. Sun-spots were also described by Johann Faber and by the Jesuit Father Christopher Scheiner (1611), who soon afterwards constructed a telescope embodying improvements suggested by Kepler.

The astronomical theory of the early years of the 17th century was thus the achievement of the practical alternation of hypothesis and observation which had proceeded since Copernicus. Kepler gave an account of his conception of the philosophy and methods of astronomy in the first book of his text-book, *Epitome Astronomiæ Copernicanæ* (1618). He conceived astronomy to begin with observations, which were translated, by means of measuring instruments, into lengths and numbers for treatment by geometry, algebra and arithmetic. Next, hypotheses were formed which brought the observed relations together in geometrical systems which 'saved the appearances'. Finally, physics studied the causes of the phenomena related by an hypothesis, which must also be consistent with metaphysical principles. The whole inquiry sought to discover the true planetary motions and their causes, at present hidden in 'God's pandects' but to be revealed by science.

Kepler's achievement was in fact much more than simply to discover the true descriptive laws of planetary motion; he also made the first suggestions towards a new physical cosmology into which they would fit. That he did not succeed in this attempt is in part a measure of the extreme difficulty of the problem, which was solved only when Newton united Kepler's planetary laws with the completion of Galileo's terrestrial dynamics

1. Nicole Oresme with an armillary sphere. From *Le Livre du Ciel et du Monde*, *Bibliothèque Nationale*, Paris, MS français 565 (14th century)

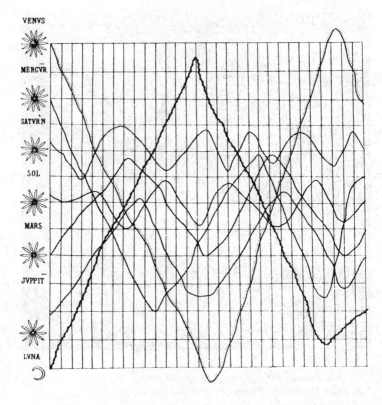

2. The earliest-known graph; showing the changes in latitude (vertical divisions) of the planets relative to longitude (horizontal divisions). From MS Munich 14436 (11th century)

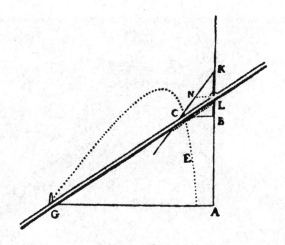

Aprés cela prenant vn point a difcretion dans la courbe,
comme C, fur lequel ie fuppofe que l'inftrument qui fert
a la defcrire eft appliqué, ie tire de ce point C·la ligne
C B parallele a G A, & pourceque C B & B A font deux
quantités indeterminées & inconnuës , ie les nomme
l'vne y & l'autre x. mais affin de trouuer le rapport de
l'vne à l'autre; ie confidere auffy les quantités connuës
qui determinent la defcription de cete ligne courbe,
comme G A que ie nomme a, K L que ie nomme b, &
N L parallele a G A que ie nomme c. puis ie dis, comme
N L eft à L K, ou c à b, ainfi C B, ou y, eft à B K, qui eft
par confequent $\frac{b}{c} y$: & B L eft $\frac{b}{c} y - b$, & A L eft x +
$\frac{b}{c} y - b$. de plus comme C B eft à L B, ou y à $\frac{b}{c} y - b$, ainfi
a, ou G A, eft à L A, ou x + $\frac{b}{c} y - b$. de façon que mul-

Sſ   tipliant

3. A page from Descartes, *La Géométrie* (1637), in which he
discusses the algebraic equation of a parabola

4. The mathematical disciplines and philosophy. The student is
met by Euclid at the outer gate. Inside he finds Tartaglia
surrounded by the mathematical disciplines: Arithmetic,
Geometry, Astronomy, Astrology, etc. A cannon is firing,
showing the trajectory of the projectile. At the far gate stand
Aristotle and Plato, to welcome the student into the presence of
Philosophy. Plato holds a scroll with the inscription 'Let no one
untrained in geometry enter here' (cf Vol. I, p. 25). From N.
Tartaglia, *Nova Scientia*, Venice, 1537

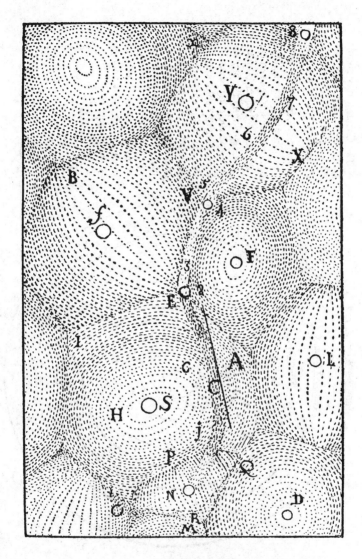

5. Diagram of vortices, from Descartes, *Principia Philosophiæ*, Amsterdam, 1644. Planets are carried in the whirlpool of subtle matter round the sun S. A comet, escaped from a vortex, is seen descending by an irregular path from the top right. Descartes thought that it would be impossible to reduce the motion of comets to a law

6. The Copernican system. From Copernicus, *De Revolutionibus Orbium Coelestium*, Nuremberg. 1543

7. *(below)* Kepler's demonstration of the elliptical orbit of Mars. If the sun is at one focus (*n*) of the ellipse (the curve shown by the broken line) and the planet at *m*, then according to Kepler's second law the radius *nm* sweeps out equal areas in equal times. The small diagram on the right is part of Kepler's proof of the equivalence of motion on an ellipse to that on a deferent and epicycle. From *Astronomia Nova*, Prague, 1609

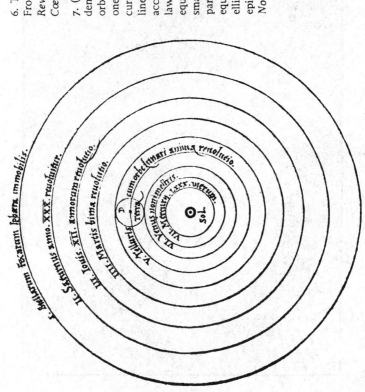

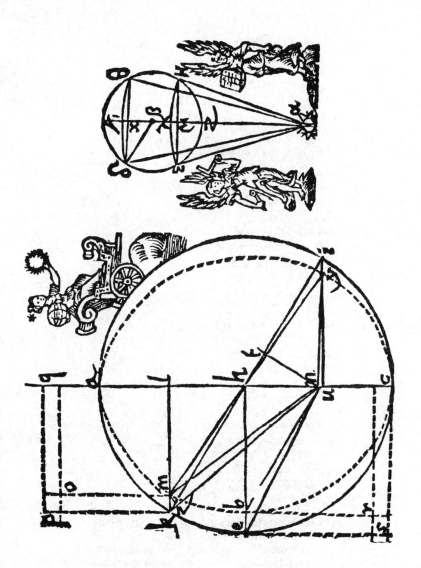

8. Page from Thomas Harriot's papers at Petworth House, describing his observations on Jupiter's satellites made at Syon House, on the Thames near Isleworth, and from the roof of a house in London. Harriot knew of Galileo's discovery of the satellites on 7 January 1610, but as early as July 1609 he had himself been observing the moon with a telescope. The upper part of the page is a rough entry of his first observations, and the lower part is the beginning of a fair copy he afterwards wrote out. See note on p. 337

9. Telescope and other instruments in use, and an apparatus for showing sun spots by projection on to a screen. From C. Scheiner, *Rosa Ursina*, Bracciani, 1630

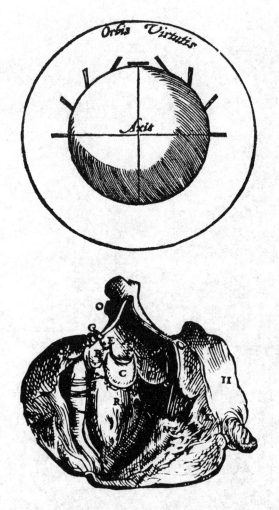

10. (*above*) The earth as a magnet, and magnetic dip. From Gilbert, *De Magnete*, London, 1600

11. (*below*) The heart and its valves. From Vesalius, *De Humani Corporis Fabrica*, Basel, 1543

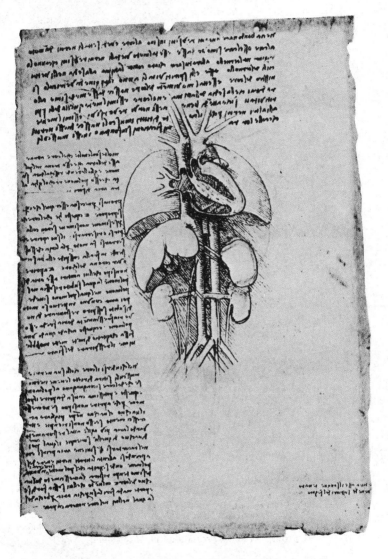

12. Leonardo's drawing of the heart and associated blood vessels. From *Quaderni d'Anatomia 4*, Royal Library, Windsor, MS; by Gracious permission of H.M. the Queen. See note on p. 337

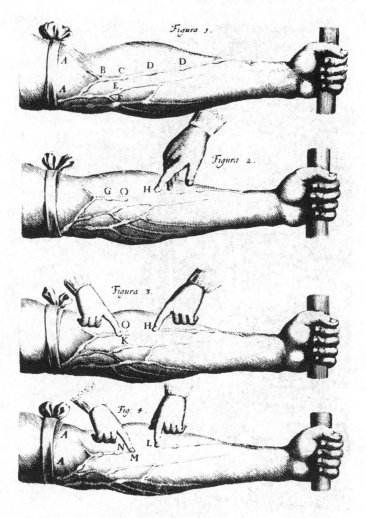

13. Harvey's experiments showing the swelling of nodes in veins at the valves. From *De Motu Cordis*, London, 1639 (1st ed. 1628)

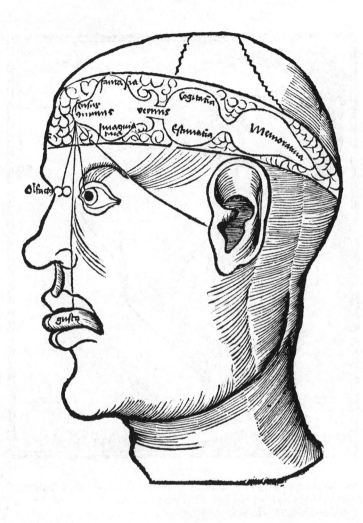

14. The *sensus communis* and the localized functions of the brain. From G. Reisch, *Margarita Philosophica*, Heidelberg, 1504

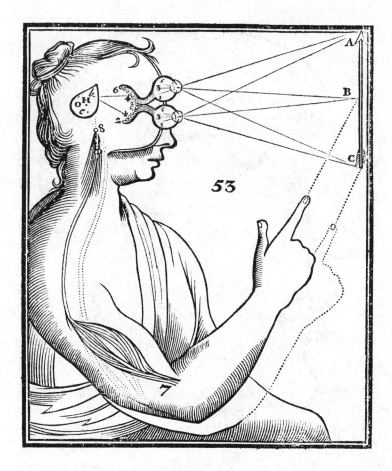

15. Descartes' theory of perception, showing the transmission of the nervous impulse from the eye to the pineal gland and thence to the muscles. From *De Homine*, Amsterdam, 1677 (1st ed. Leiden, 1662)

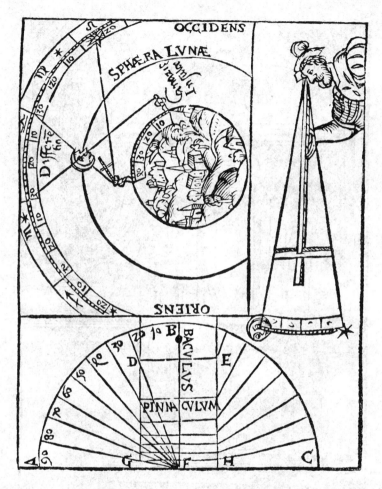

16. A cross-staff in use for surveying. From Petrus Apianus, *Cosmographia*, Antwerp, 1539

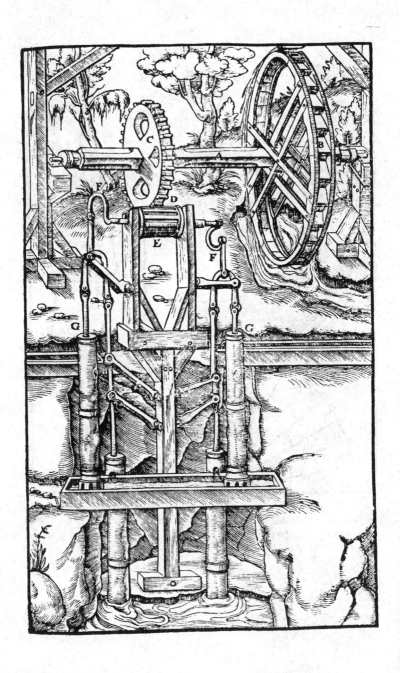

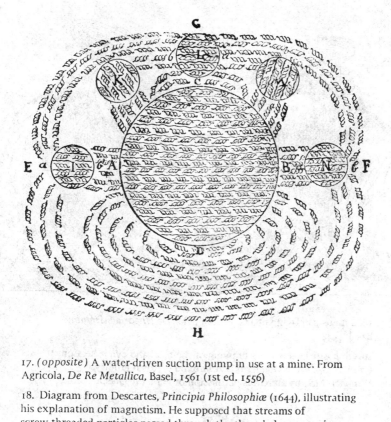

17. (*opposite*) A water-driven suction pump in use at a mine. From
Agricola, *De Re Metallica*, Basel, 1561 (1st ed. 1556)

18. Diagram from Descartes, *Principia Philosophiæ* (1644), illustrating
his explanation of magnetism. He supposed that streams of
screw-threaded particles passed through the threaded passages in
the earth and in iron, thus causing the alignment seen in the effect
of a magnet on a piece of iron or of the earth on a compass needle

19. Botanists drawing plants. From Fuchs, *De Historia Stirpium*, Basel, 1542

20. *(opposite)* Leonardo's drawing of the head and eye in section (cf Plate 15). From *Quaderni D'Anatomia* 5. Royal Library, Windsor, MS; by Gracious permission of H.M. the Queen

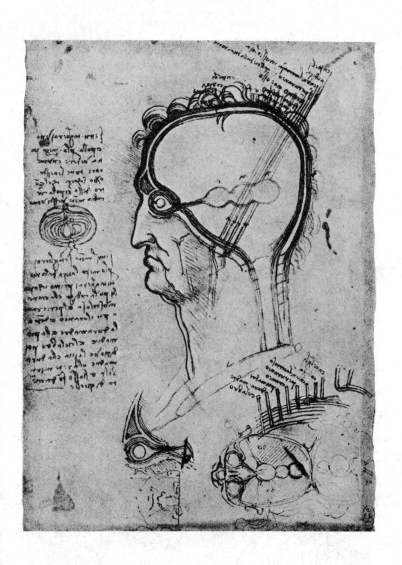

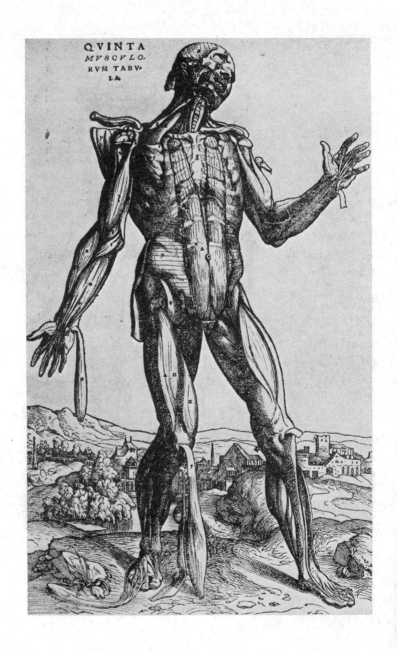

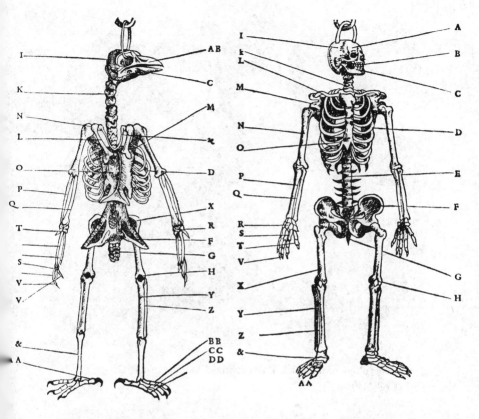

21. (opposite) A dissection of the muscles. From Vesalius, *De Humani Corporis Fabrica*, 1543

22. Diagrams illustrating the comparison between the skeletons of a man and a bird, from Belon, *Histoire de la nature des oyseaux*, Paris, 1555

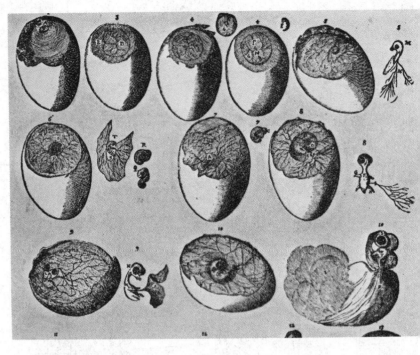

23A. Embryology of the chick. From Fabrizio, *De Formatione Ovi et Pulli*, Padua, 1621

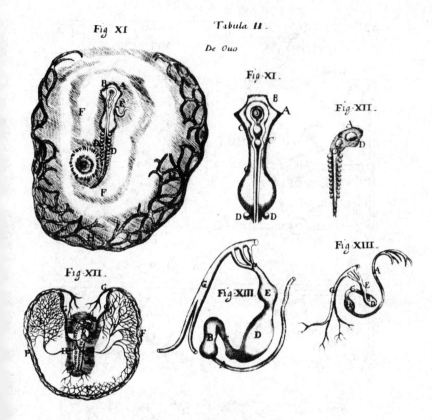

23B. Embryology of the chick showing the use of the microscope
(cf Plate 23A). From Malpighi, *De Formatione Pulli in Ovo* (first
published 1673), in *Opera Omnia*, London, 1686

24: The comparative anatomy of the ear ossicles from Casserio,
De Vocis Auditisque Organis, Ferrara, 1601

by means of the bridging law of universal gravitation. Towards that bridging law Kepler supplied both a positive contribution and a direction of inquiry. Following the preface to *De Revolutionibus*, the view had become common that, as Francis Bacon expressed it in his criticism of Copernicus in his *Novum Organum* (book 2, aphorism 36), the heliostatic system had been 'invented and assumed in order to abbreviate and ease the calculations', but was not literally and physically true. 'There is no need for these hypotheses to be true, or even to be at all like the truth,' Osiander had written in this preface; 'rather, one thing is sufficient for them: that they should yield a calculus which agrees with the observations.' It was Kepler who first detected that Copernicus had not written these words. He disagreed with them strongly. The goal of the inquiry, he insisted, was to discover how the planets actually moved, and not only how but why they moved as they did and not otherwise: 'so that I might ascribe the motion of the Sun to the earth itself by physical, or rather metaphysical reasoning, as Copernicus did by mathematical,' as he said in the preface to the *Cosmographic Mystery*.

In fact Kepler made his discoveries of the three laws of planetary motion in a search for something much more, in the course of a metaphysical inquiry behind the visible appearances into the underlying harmonies expressed in purely numerical relations which he held to constitute the nature of things: the *harmonice mundi* which became manifest in the planetary motions and in music – an actual 'music of the spheres'. A reader unprepared for the individualities of Kepler's processes of thought may find the bulk of his difficult writings, concerned as much with questions like the nature of the Trinity, of celestial harmony, and of the relation of divine to human knowledge as with astronomy, an almost unintelligible matrix into which gems of science somehow got embedded. This would be to misunderstand the organization of his thought entirely; and it would be to miss an obvious clue to perhaps the most important element in any original scientific thinking: the bridge of intuition and imagination by which he crossed the logical gap from the immediate results of observation to the theory by means of which he explained these results. All the evidence points to the bridge in Kepler's mind being made by the preconceptions of the metaphysical inquiries of which his science formed a part. Developed first in analogy to the relations between the persons of the Trinity, his conception of the structure of the universe became part of a theological creed. But it was also part of Kepler's presuppositions – a point that came out vividly in a controversy on the subject with the English Rosicrucian Robert Fludd (cf. below, p. 253) – that the true structure and harmonies of the universe were those verified in observation. After his first visit to Tycho Brahe in 1600 he wrote in a letter to his friend Herwart von Hohenburg:

I would have concluded my research on the harmonies of the world, if Tycho's astronomy had not fascinated me so much that I almost went out of my mind; still I wonder what could be done further in this direction. One of the most important reasons for my visit to Tycho was the desire, as you know, to learn from him more correct figures for the eccentricities in order to examine my *Mysterium* and the just mentioned *Harmonice* for comparison. For these speculations *a priori* must not conflict with experimental evidence; moreover they must be in accordance with it.

In developing this criterion of empirical confirmation he took account of the range of confirmation, asserting for example that the Copernican hypothesis was 'truer' than the Ptolemaic because, of the two, it alone could put the planets in an order round the sun according to their periods. Kepler's laws of planetary motion and his attempts to explain them were thus, so to speak, carved out of his preconceived Neoplatonic metaphysics by as strict an application as possible of quantitative methods and of the principle of the empirical test. It is this that makes him so interesting an example of scientific thinking, so different from that to be expected from a too literal application of the austerities of a positivist or 'operationalist' interpretation or of the canons of J. S. Mill.

Kepler's central metaphysical conception was of the existence from eternity in the mind of God of archetypal ideas, which were reproduced on the one hand in the visible universe and on the other in the human mind. Of these geometry was the archetype of the physical creation and was innate in the human mind. As he wrote in 1599 to Herwart von Hohenburg:

To God there are, in the whole material world, material laws, figures and relations of special excellency and of the most appropriate order. ... Let us therefore not try to discover more of the heavenly and immaterial world than God has revealed to us. Those laws are within the grasp of the human mind; God wanted us to recognize them by creating us after his own image so that we could share in his own thoughts. For what is there in the human mind besides figures and magnitudes? It is only these which we can apprehend in the right way, and if piety allows us to say so, our understanding is in this respect of the same kind as the divine, at least as far as we are able to grasp something of it in our mortal life. Only fools fear that we make man godlike in doing so; for God's counsels are impenetrable, but not his material creation.

To this conception he joined the ancient doctrine of the *signatura rerum*, of the signs of things, according to which the external form of a thing was held to point to properties and a level of reality that were not directly visible. In the *Cosmographic Mystery* he described at length the visible universe as a sign or image of the Trinity, having the most perfect form of the sphere: the Father was represented by the centre, the Son by the outer surface, and the Holy Ghost by the radius having an equality of relationship between centre and surface.* In creating the visible universe in accordance with this geometrical symbolism, God placed at the centre a body to represent the Father by its radiation of power and light: this was the sun. Following the precedent of earlier Neoplatonic cosmologies, for example that of Grosseteste (see Vol. I, pp. 88–9), Kepler conceived all natural powers as flowing out from bodies to assume a spherical form; and so by analogy with the power emanating from the Father, the sun became the instrument giving visible shape and life to the cosmos and everything in it, a universe in which everything was animate. It was the *anima motrix* or 'motive soul' of the sun that hurried the planets round in their circular orbits, and also the comets,

* An analogous symbolism, differently arranged, is found in the medieval Celtic cross.

with a velocity depending on its power after it had reached their respective distances. It has been suggested that it was because Kepler approached the problem of the planetary motions with this archetypal image in mind that he became a convinced Copernican.* Certainly he never abandoned the *animæ motrices* as the 'physical' motive power, even after he had been forced by the observational data he obtained from Tycho Brahe to give up the circular orbits. He was encouraged in his continued use of these causal conceptions as a guide to his mathematical inquiries by the explanations William Gilbert had given of his recent dis-coveries in magnetism.

Gilbert was court physician to Queen Elizabeth, who gave him a pension to pursue his research. He took a considerable interest in astronomy, but his main achievement was to work systematic-ally through an entire field of scientific inquiry, the field of mag-netism and electricity as then capable of study. Gilbert's *De Magnete* (1600), though containing some measurements, was en-tirely non-mathematical in treatment, and is the most striking illustration of the independence of the experimental and mathe-matical traditions in the 16th century (cf. above, p. 148). He derived his methods largely from Petrus Peregrinus, whose work had been printed in 1558, and from practical compass-makers such as Robert Norman, a retired mariner whose book, *The Newe Attractive* (1581), contains the independent discovery of the mag-netic dip. This had been observed first by Georg Hartmann in 1544. Gilbert extended Peregrinus' work to show that the strength and range of a uniform lodestone was proportional to size. He also showed that the angle of dip of a freely suspended needle varied with latitude. Peregrinus had likened the needle-lines traced on a spherical magnet to meridia, and called the points where they met poles. Gilbert inferred from the orienta-tions in which magnets set with respect to the earth that the latter was itself a huge magnet with its poles at the geographical poles. This he confirmed by showing that iron ore was magnet-ized in the direction in which it lay in the earth. The properties

* See C. G. Jung and W. Pauli, *The Interpretation of Nature and the Psyche*, London, 1955.

of lodestones and of the compass were thus included in a general principle (Plate 10).

Gilbert also made a study of electrified bodies, which he called *electrica*. He showed that not only amber, but also other substances such as glass, sulphur and some precious stones, attracted small things when rubbed; he identified a body as being 'electric' by using a light metallic needle balanced on a point. He pointed out that while the lodestone attracted only magnetizable substances, which it arranged in definite orientations, and was unaffected by immersion in water or screens of paper or linen, electrified bodies attracted everything and heaped them into shapeless masses and were affected by screens and immersion. Niccolo Cabeo (1585–1650) later observed that bodies flew off again after being attracted; Sir Thomas Browne said they were *repelled*.

Gilbert's empiricism extended only as far as the facts he had established. He used a balance to disprove the old story, accepted by Cardano, that the magnet fed on iron, but his explanations of magnetism and electricity, though not inconsistent with the facts, did not arise out of them. His explanation was really an adaptation of Averroës' theory of 'magnetic species' in a setting of Neoplatonic animism. Beginning with the principle that a body could not act where it was not, so that all action by means of matter must be by contact, he asserted that if there appeared to be action at a distance there must be a material 'effluvium' responsible for it. Such an effluvium, he held, was released from electrified bodies by the warmth of friction. He excluded magnetic action from this explanation, because, since it could pass through matter, it could not be due to a material effluvium; the motion of iron towards a magnet was more like that of a self-moving soul. But he extended the theory of effluvia to explain the earth's attraction for falling bodies, the effluvium here being the atmosphere. Without going into details, he attributed the diurnal rotation of the earth, which he accepted, to magnetic energy, and the orderly movements of the sun and planets to the interaction of their effluvia.

Kepler was himself interested in magnetism, and Gilbert's work stimulated him to use this phenomenon to explain the physics of the universe. Here he accepted the current Aristotelian conception of motion as a process requiring the continuous operation of a motive power. As a young man reading Scaliger he had adopted the Averroïst doctrine of Intelligences moving the heavenly bodies, but he had afterwards abandoned it because he wanted to consider only mechanical causes. He explained the continued daily rotation of the earth on its axis by the *impetus* which God impressed on it at the creation. But, like Nicholas of Cusa, he identified this *impetus* with the earth's soul (*anima*), thus reinstating the equivalent of an Intelligence. This *impetus*, he held, did not corrupt, for, with the Pythagorean theory of gravity which he accepted, circular motion could, without contradiction, be considered the natural motion of earth. To answer the traditional objections to the earth's daily rotation he developed Gilbert's suggestions. He considered lines, or elastic chains of force, which he held to be magnetic, to emanate radially from the earth's *anima motrix* and carry round the moon, clouds and all bodies thrown above the surface of the earth. Similar lines from the *animæ motrices* of Jupiter and Saturn carried round their satellites, and lines from the sun carried round the whole planetary system as the sun turned on its axis. It was this theory of a magnetic force, which diminished as distance increased so that the velocity of a planet in its orbit varied inversely as the distance from the sun, that led him to his second law. The rotation of the sun swinging its magnetic lines in a vortex would move the planets in a circle; the deviation from this to produce an elliptical orbit he tried to explain by the oscillations caused by the attraction and repulsion of their poles. Further, just as the motive force of the sun was magnetic, so there was an analogy between magnetism and gravitation. Gravitation was the tendency of cognate bodies to unite and, if it were not for the motive power carrying the moon and earth round their orbits, they would rush together, meeting at an *intermediate* point. This last was an entirely new idea.

It was Kepler's idea that a satellite was kept on its orbit by *two* forces, one the mutual radial attraction with the central body and the other the motive power of the *anima motrix* impelling it laterally, that made his physical system the opening into the unification of terrestrial and celestial dynamics by Newton. The beginning of Kepler's achievement in this direction was his development of the Pythagorean conception of gravity. Oresme, Copernicus, Gilbert and Galileo had all rejected Aristotle's conception of gravity as a tendency to move towards a special place, the centre of the universe, and substituted for it gravity as the tendency of cognate bodies to unite; and the analogy with magnetism had been made by more than one medieval writer before being exploited again by Gilbert. Kepler considered this tendency to be caused by a real attraction (*virtus tractoria*) exerted externally by one body on another. His innovation was to make the attraction (in gravitation as in magnetism) *mutual*, and then to express it in a dynamical form. He wrote in the introduction to his *Astronomia Nova*:

> If two stones were placed close together in any place in the universe outside the sphere of the power (*virtus*) of a third cognate body, they would, like two magnetic bodies, come together at an intermediate point, each moving such a distance towards the other, as the mass (*moles*) of the other is in proportion to its own.

Postulating that the earth and the moon were cognate bodies, like two stones, he continued:

> If the moon and the earth were not retained, each in its orbit, by their animal and other equivalent forces, the earth would ascend towards the moon one fifty-fourth part of the distance between them, and the moon descend towards the earth about fifty-three parts; and they would join together; assuming, however, that the substance of each is of one and the same density.

That the attractive force of the moon did actually extend to the earth he concluded from the ebb and flow of the tides, which he supposed to be caused by the moon pulling the water of the seas towards itself: a theory which Grosseteste had foreshadowed, reminding us once more of the persistence of the complex ideas that went with Neoplatonism (cf. Vol. I, p. 137). Kepler supposed it likely that a much stronger force extended from the earth to the moon, and beyond.

Kepler developed this conception of gravitation only in application to the earth and the moon; he did not suppose, for example, that the sun and the planets were cognate bodies attracting each other. Similarly he failed to grasp the cosmological significance of the inverse-square law which he formulated as a photometric law relating the intensity of light to the distance from its source, for example the sun. Again displaying both his consistently 'realist' philosophy of science, and the complex of Neoplatonic associations that clung to all developments of the 'cosmology of light' (cf. Vol. I, pp. 88–9, 112–13, 136), he described, in the introduction to his *Astronomia Nova*, the course of his inquiries into the motive forces swinging the planets round.

I have begun by saying that in this work I shall treat astronomy not on the basis of fictional hypotheses (*hypotheses fictitiæ*), but on the basis of physical causes, and that for this purpose I have found it necessary to proceed by stages. The first stage was the demonstration that the eccentrics of the planets meet in the body of the sun. Next, by deductive reasoning, I proved that since, as Tycho Brahe showed, the solid orbs do not exist, it follows that the body of the sun is the source and the seat of the force which makes all the planets revolve round the sun. I showed likewise that the sun performs this in the following manner : while remaining in the same place, the sun nevertheless rotates as if on a tower and in fact emits through the breadth of the world an immaterial species (*species*) from its body, analogous to the immaterial species of its light.

This species, because of the rotation of the solar body, revolves in the form of a very fast vortex, which extends through the whole immensity of the universe and carries the planets with it, drawing them round in a circle with a vehemence (*raptus*) which is intenser or weaker according to whether the density of this *species*, in accordance with the law of its flow (*effluxus*), is greater or less.

The interaction of the individual motors of the planets with this common motor then produced the deviation from the circle. So far, so good. Kepler had raised for the first time the question of what moved the planets, since the spheres did not exist.

In his *Ad Vitellionem Paralipomena* (1604) Kepler had shown that if, as he held, light and other powers (*virtus*, *species*) expanded in a sphere from their source, then their strength would decrease as the area of the surface of the sphere, that is in proportion to the square of the radius. But in his *Epitome Astronomiæ Copernicanæ* (book 4, part 2, chapter 3; 1620) he specifically denied that this photometric law applied to the motive force of the sun, which he said decreased in *simple* proportion to distance. He tried to argue that the inverse-square law applied only to the sun's light. His argument was that whereas the sun's light expanded in a sphere, so that its intensity decreased according to the areal increase in the surface of the sphere, the sun's motive force extended only in the *plane* of each planetary orbit and decreased with the linear increase in the circumference. Certainly he was very far from applying it to *attraction* between the sun and the planets.

Kepler in fact resembles Galileo in supplying elements towards a unifying principle of cosmology whose need he saw clearly but of whose realization he stopped short. Their omissions are curiously complementary and have a curious symmetry in preparation for the Newtonian synthesis. Neither Galileo nor Kepler had really grasped the dynamical problem presented by the planets. Galileo believed like Copernicus that the planetary revolutions were a 'natural' motion; that is, they required no external mover and could be accepted on grounds of order alone. Galileo was able to hold this because he ignored Kepler's demonstration of the elliptical orbits, which he certainly knew. Whether he did so for metaphysical or aesthetic reasons, or simply as he said in 1614 because Kepler's writing was 'so obscure that apparently the author did not know what he was talking about', the result was that he continued to regard the planets as revolving in circles (cf. above, p. 166). In any case he did not admit that the planets required any forces, either lateral or centripetal, to keep them in their orbits. Thus by ignoring Kepler's descriptive laws Galileo failed to see that the actual geometry of the heavens vitiated any spherical model, and so he missed the problem of how the planets were retained in their elliptical orbits.

Kepler's attempt to solve this problem, on the other hand,

was vitiated by his failure to grasp the full meaning of the inertial principle which had been clearly though incompletely stated by Galileo in his second *Letter on the Sunspots* in 1612.* Continuing to suppose that continuous uniform velocity required a continuous motive power, Kepler saw this supplied by the *species motrix* or *virtus motoria* supposed to emanate from the sun; and since these swung the planets round laterally, he did not suppose that a centripetal force was required to keep them in their orbits instead of flying off tangentially. He failed to see the universal significance of the model he had himself supplied of the earth and the moon.

The uncertainty which Kepler himself seems to have felt in his inquiries into the vast problems which he had undertaken are shown by the changes he made, after each failure, in his approach to scientific explanation.† After discovering that the planetary theory proposed in the *Mysterium Cosmographicum* would not fit the facts, he turned from a conception of satisfactory explanation as one in which mathematical harmonies are discovered in the chaos of observations, to a mechanical conception of the universe as a regulative and heuristic guide to the investigations published in *Astronomia Nova*. The title of this work is itself revealing: *The New Astronomy or Physics of the Heavens explored on the Basis of the Law of Causality and developed in Analyses of the Movements of Mars based on Observations by Tycho Brahe.* While preparing this work he wrote to Herwart von Hohenburg in 1605:

I am much occupied with the investigation of the physical causes. My aim in this is to show that the celestial machine is to be likened not to a divine organism but rather to a clockwork ..., insofar as nearly all the manifold movements are carried out by means of a single, quite simple magnetic force, as in the case of a clockwork all

* Galileo's *Letter* was written in 1612 and published in 1613. It was Kepler who introduced the word *inertia* into physics but he used it to mean an intrinsic resistance to motion and inclination to rest if in motion.

† Cf. Gerald Holton, 'Johannes Kepler's universe: its physics and mathematics,' *American Journal of Physics*, xxiv (1956); A. Koyré, 'L'œuvre astronomique de Kepler,' *XVIIe Siècle*, 1956, No. 30.

motions [are caused] by a simple weight. Moreover I show how this physical conception is to be presented through calculation and geometry.

In the end the physical theory of the *species motrix* emanating from the sun, put forward in *Astronomia Nova*, also proved an empirical failure, for it was observed that the apparent speed of the sun's rotation, then believed to be measured by that of the sun-spots, did not agree with that of the planets. For his next work Kepler allowed his conception of mathematical harmony to satisfy him as a criterion of satisfactory explanation, and in the *Harmonice Mundi* he announced his Third Law without any attempt to deduce it from mechanical principles. Two quite distinct meanings were involved in this conception of 'harmony'. According to the first, the Second Law, for example, was harmonious because it showed the areal velocity as constant; and it is worth noting that just as Ptolemy's constant angular velocity was more abstract and removed from immediate observation than Aristotle's directly observable constant linear velocity, so Keplers' areal velocity was a discovery of constancy or uniformity at a further stage of abstraction. The second meaning of Kepler's harmony applied to the 'fitness' or 'rightness' of the structure of the universe, for example the sun's 'rightful place' in the centre. The two meanings seem to have no logical connexion but both performed heuristic and regulative functions in all Kepler's work.

Because they could see only sections of the total picture that was to emerge later, the attempts that both Kepler and Galileo made not only to answer the traditional objections to the earth's motion but also to produce conclusive arguments in favour of it failed to convince most of their contemporaries. For example, the magnetic chains adopted by Kepler to explain the motion of the moon would have made all movement of projectiles impossible. Galileo was better placed for the negative operation of meeting objections to the earth's motion. For example, with his conceptions of *impeto* and of the composition of *impeti* he was able to show that the argument from 'detached bodies' lost its premisses. In his *Dialogue Concerning the Two Principal Systems of the*

World, the Ptolemaic and the Copernican (a revealing title indicating his indifference to Tycho Brahe and to Kepler) he pointed out that such bodies would retain the velocity received from the rotating earth unless forced to do otherwise. The remaining mechanical objection to the 'Copernican' theory was from 'centrifugal force'. Galileo argued that this depended, not on the linear velocity of a point on the earth's surface, but on the angular velocity of rotation, and therefore that it was no greater on the earth's surface than on a smaller body rotating once in the 24 hours. This would be negligible compared with gravity. Actually centrifugal force depends on both the linear and the angular velocity, as Huygens was the first to show. Although the proof of the earth's motion remained one of the principal goals of Galileo's dynamical work, in the end he was unable, in spite of all his determined efforts, to do more than show that this was at least as plausible as the supposition that it was at rest.

It was by an explicitly unrestricted and universal comparison of bodies on the earth with those in the heavens that Newton, with the indispensable aid of some intermediate writers, finally produced the synthesis of his *Principia Mathematica* (1687). Newton united Galileo's kinematic laws of falling bodies and projectiles and the completion of his principle of inertia, with Kepler's descriptive laws of planetary motion and the completion of his conception of gravitation (cf. above, pp. 162–74). Comparing a planet with a projectile, he was then able to attribute the forward motion of each to inertia, and the deflection from a rectilinear trajectory to gravitation. A planet was thus a projectile whose velocity prevented it from falling on to the earth, so that its orbit formed an ellipse instead of a parabola.* Newton showed that the acceleration of fall of the moon in its elliptical orbit round the earth was equal to that required by Galileo's law of acceleration of free fall; the same applied to the planets' orbits round the sun. He deduced Kepler's Third Law from his inverse-square law of universal gravitation.† He showed that it was

* This orbit was achieved experimentally with the launching of Sputnik on 4 Oct. 1957.

† And *vice versa*; see below, p. 215, note.

dynamically impossible for the enormous sun to turn round the diminutive earth, but that a central body and its satellite must revolve round their common centre of gravity, which in the solar system was inside the surface of the sun. Thus he was able to succeed where Galileo and Kepler had failed, not merely in refuting the arguments against the earth's movement, but in showing that the arguments in favour of it were compelling. They were compelling within a universal system of dynamics confirmed in all other tested fields of observation. For the first time since, in Hellenistic times, the observations had forced astronomers to abandon Aristotle's concentric spheres in favour of the physically inexplicable mathematical devices of epicycles and eccentrics and had produced the dichotomy between the physical explanation of the celestial motions and the mathematical means of predicting them – the dichotomy between Aristotelian physical cosmology and Ptolemaic mathematical astronomy that had existed throughout the Middle Ages – a conclusive criterion became available for choosing one calculating system in preference to another that was equally accurate in making predictions within the limits of astronomy. The choice operated by showing that only one of the alternative systems was compatible with a wider field of observations. It was Newton's achievement to put the dynamical criterion, foreseen and prepared by Galileo and Kepler, into operation, and to unite for the first time the explanation with the means of prediction. Beginning with the same fundamental physical axioms of the laws of motion and of gravitation, the steps followed in setting out the *explanation* of the motions of bodies became exactly those made in *predicting* their motions. So cosmology as a science of 'natures' independent of calculation and the means of prediction disappeared within a genuine synthesis of mathematical-physics (cf. Vol. I, pp. 83 *et seq.*, 99 *et seq.*; also below, p. 303 *et seq.*).

Another difficulty remaining for the heliocentric system from the side of observation, and one which Galileo had been unable to solve, was the absence of stellar parallax. This was observed for the first time in 1838 by F. W. Bessel in the star 61 Cygni, though James Bradley, when looking for parallax, had observed, in 1725, that the fixed stars described small ellipses within exactly the duration of the terrestrial year, and that stars from the poles of the ecliptic to the ecliptic described figures which were increasingly less circular and more approaching straight lines. This was convincing evidence for the movement of the earth in an ellipse round the sun, but Bradley recognized that what he had observed were not parallactic ellipses, but aberrational ellipses due to the earth's approach, on one side of its orbit, to the light coming from the stars, and recession from it on the other.

It was because of his vision of a unified mathematical-physical cosmology that Galileo came into philosophical conflict with certain contemporary theologians; the other aspects of his troubles with the Roman Inquisition and the course of his trial belong rather to the history of Roman ecclesiastical policy and judicial procedure – in this case obscure enough – than to the history of science. Yet it is significant that it was a theological problem, the relationship between astronomical theory and Scripture, between the cosmology discovered by scientific reasoning and that presented as revealed by God, that should have made the truth of the 'realist' view of science shared by Galileo and Kepler the great question of the day in the philosophy of science. Oresme had already considered and then withdrawn before the same passages of Scripture, which must be *literally* false if the new cosmology were literally and physically true (see above p. 94 *et seq.*). For example Joshua's command on the evening of the battle of Gibeon : 'Sun, stand thou still upon Gibeon; and thou, Moon, in the valley of Ajalon; and the Sun stood still, and the moon stayed . . .' (Joshua, x, 12, 13), implied that the sun was normally in motion. Other passages contradicted the other essential Copernican postulate, that the earth moved : for example, from Psalm 93, 'the world also is stablished, that it cannot be moved'. Granting the various mathematical and practical advantages of the new astronomy, as everyone was prepared to, there were two ways of avoiding this conflict. One was to abandon the literal interpretation of Scripture, a course that had been followed, albeit with proper caution, by the Fathers themselves when the occasion had called for it. The other way was to weaken the truth of natural science, to treat astronomical theory not as a discovery of the real physical world, a world of abstract laws perhaps but knowable as true, but as a convenient fiction for making the calculations, 'merely a poetical conceit, a dream', 'a chimæra', as Galileo wrote ironically in a letter to Leopold of Austria in 1618.

After some preliminaries, Galileo finally stated his position publicly in 1615 in his open *Letter to Madame Cristina of Lorraine, Grand Duchess of Tuscany*, written on the advice of some clerical friends, partly to clear himself of a malicious rumour that he was an unbeliever, and also to try, unsuccessfully, to prevent the ecclesiastical authorities from making the fatal mistake of condemning the Copernican system on theological grounds. Citing the authority of St Augustine, Galileo argued that God was the author not only of one great book but of two, of Nature as well as the Scriptures. Truth was to be studied in both, but with different results. The book of nature was to be read in the language of mathematical science and the results expressed in physical theory; the Scriptures, on the other hand, contained no physical theory, but revealed to us our moral destiny. When they referred to natural phenomena they used the language of ordinary understanding, conforming to popular ideas, without implying that their literal meaning was to be taken as referring to physical fact. Indeed he pointed out that the Scriptures had always been understood to use figurative language at many points, as when they mentioned the eye or hand or anger of God, where a literal interpretation would be directly heretical. It was against both reason and tradition to use a literal interpretation of Scriptures to throw doubt upon the truth of statements expressing either the direct evidence of the senses or necessary conclusions from that evidence.

'It seems to me,' Galileo wrote in his *Letter to the Grand Duchess:*

that in discussing natural problems we ought not to start from the authority of the texts of the Scriptures, but from the experience of the senses and from necessary demonstrations (*dalle sensate experienze e dalle dimostrazioni necessarie*). For, Holy Scripture and Nature alike proceeding from the Divine Word, the former as the dictation of the Holy Ghost, the latter as the most observant executrix of God's commands; and moreover, it being convenient in the Scriptures (by way of condescension to the understanding of all men) to say many things different, in appearance and so far as concerns the naked signification of the words, from the absolute truth; but nature, on the other hand, being inexorable and immutable and never passing beyond the bounds of the laws assigned to her, as if she did not care whether her abstruse reasons and modes of operation were or were not within the capacity of men to understand; it is clear that those things concerning natural effects which either the experience of the senses sets before our eyes or necessary demonstrations prove to us, ought not to be called in question on any account, much less condemned on the basis of texts of Scripture which may, in the words used, seem to mean something different. For every expression of Scripture is not tied to strict conditions like every effect of nature; nor does God reveal himself less admirably in the effects of nature than in the sacred words of Scripture.

Clearly, he concluded, it was not the intention of the Holy Ghost to teach us physics or astronomy, or to show us whether the earth moved or was at rest. These questions were theologically neutral, although certainly we should respect the sacred text, and where appropriate use the conclusions of science to help to discover its meaning. The purpose of the Holy Ghost in the Scriptures, as he expressed it wittily in a remark which he attributed to Cardinal Baronio, was to teach us 'how to go to heaven, not how the heavens go'.

'This granted,' he continued:

and it being true, as has been said, that two truths cannot be contrary to each other, it is the office of a judicious interpreter to try to penetrate to the true senses of sacred texts, which undoubtedly will agree with those natural conclusions which manifest sense and necessary demonstrations have first made sure and certain. Indeed, it being the case, as has been said, that the Scriptures, for the reasons stated, admit in many places of interpretations far from the sense of the words; and, moreover, we not being able to affirm that all interpreters speak by divine inspiration (for, if it were so, then there would be no diversity between them concerning the meanings of the same texts); I should think that it would be an act of great prudence to forbid anyone to usurp the texts of Scripture and as it were to force them to maintain this or that natural conclusion for truth, of which the senses and demonstrative and necessary reasons may one time or another assure us the contrary. For who will prescribe bounds to man's intelligence and invention? (*E chi vuol por termine alli umani ingegni?*) Who will assert that all that is sensible and knowable in the world is already discovered and known? Perhaps those who on other occasions avow (and with great truth) that *ea quæ scimus sunt minima pars earum quæ ignoramus* [those things that we know are very few in comparison with those we do not know]. Indeed, if we have it from the mouth of the Holy Ghost himself that *Deus tradidit mundum disputationi eorum, ut non inveniat homo opus quod operatus est Deus ab initio ad finem* [God offers the world for their disputation, so that no man may find out the work that he makes from the beginning to the end – *Ecclesiastes*, iii, 11], we ought not, as I conceive, contradicting such a sentence, stop the way to free philosophizing about the things of the world and of nature, as if they were already certainly found and all clearly known.

A man of the world as well as a convinced Catholic and a dedicated natural philosopher, a guest cherished at aristocratic tables for his genial intelligence and witty conversation, Galileo knew well the weight that political decisions, both ecclesiastical and secular, attached, of their nature, to convenience and administrative peace. With a prophetic foresight into his own future troubles he pointedly emphasized the distinction between the conditions for a change of legal or commercial and of scientific opinion.*

* Cf. Francis Bacon, *Advancement of Learning* (1605): 'Yet to those that seek truth and not magistrality, it cannot but seem a matter of great profit, to see before them the several opinions touching the foundations of nature: not for any exact truth that can be expected from those theories; for as the same phenomena in astronomy are satisfied by the received astronomy of the diurnal motion, and the proper motions of the planets, with their eccentrics and epicycles, and likewise by the theory of Copernicus, who supposed the earth to move (and the calculations agree indifferently to both), so the ordinary face and view of experience is many

I would entreat those wise and prudent Fathers that they should with all diligence consider the difference that exists between demonstrative knowledge and knowledge where opinion is possible : to the end, that weighing well in their minds, with what force necessary conclusions compel acceptance, they may the better ascertain for themselves that it is not in the power of those who profess the demonstrative sciences to change their opinions at pleasure and to apply themselves now on one side and now on the other; that there is a great difference between commanding a mathematician or a philosopher and disposing of a merchant or a lawyer; and that the demonstrated conclusions touching the things of nature and of the heavens cannot be changed with the same facility as opinions about what is legal or not in a contract, rent, or bill of exchange.

On the basis of the observations and of the new dynamics Galileo believed that it would be possible to demonstrate that the heliocentric system was a necessary conclusion from the data. He had seen with his telescope a model of the solar system in Jupiter and his satellites, and he had measured the great annual variation in the apparent diameters of Venus and Mars. His observations of the phases of Venus had confirmed, so far as he had gone, the prediction from the Copernican system that the inner planets, and they alone, would show complete phases like the moon when observed from the earth (cf. Fig. 6). There were, he said, 'many other sensible observations which can never by any means be reconciled with the Ptolemaic system, but are most weighty arguments for the Copernican'. There were some natural propositions of which human science and discourse could furnish us only rather with 'some probable opinion and plausible conjecture, than with any certain and demonstrated knowledge'. But 'there are others, of which either we have or we may confidently believe that it is possible to have, by experiments, prolonged

---

times satisfied by theories and philosophies; whereas to find the real truth requireth another manner of severity and attention.' He added : 'So many say that the opinion of Copernicus touching the rotation of the earth, which astronomy itself cannot correct, because it is not repugnant to any of the phenomena, yet natural philosophy may correct.'

observations and necessary demonstrations, an indubitable certainty; as for instance, whether the earth or the sun move or not, and whether the earth is spherical or otherwise'.

If the Copernican theory, or the particular opinion of the earth's mobility, were prohibited and declared contrary to the Catholic faith without prohibiting astronomy as a whole, Galileo continued his earnest advocacy, it could only cause great scandal. It could only be to the detriment of souls 'to give them occasion to see a proposition proved which it would afterwards become a sin to believe. And what other would the prohibiting of the whole science be than an open contempt of a hundred texts of the Holy Scriptures, in which we are taught that the glory and the greatness of Almighty God are admirably discerned in all his works, and divinely read in the open book of heaven?' It would be to contradict all the evidence of God's intention in endowing man with his admirable intelligence and inquiring reason. Galileo warned theologians against putting the faithful in the embarrassing position of having to believe as true what their senses, and scientific demonstrations, might show them to be false, or of committing a sin when they believed what their reason convinced them to be true. Moreover he pointed out that even the geostatic system disagreed with the literal words of Scripture. For example, if Joshua's command to the sun was meant to have been taken literally, according to this system he should have addressed it to the Prime Mover, for by stopping the sun and moon alone he would have deranged the whole celestial system, yet there is no evidence that he did so. The association of Aristotelian cosmology and Ptolemaic astronomy with the language of theology was not only purely accidental but far from complete.

Galileo wrote in the language of uncompromising scientific realism. He believed in an objective world of unchanging law existing independently of the inventions of men, a true world which it is the business of science to discover, to be sure by subtle theoretical reasoning, but nevertheless with certainty. 'Nothing ever changes in nature to accommodate itself to the comprehensions or motions of men,' he wrote to his friend Elia Diodati in

1633. While dedicated to a mathematical approach to the natural world, he agreed with the medieval 'physicists' rather than the 'mathematicians' in astronomy, and was not content to stop simply at 'saving the phenomena'. Like Aquinas he presupposed a true physical theory, a real physical substance causing the phenomena (cf. Vol. I, pp. 96–7). But if the real physical world was an abstract structure of the real mathematical 'primary qualities' and their laws, qualities determining the nature of physical substance, then the system of theories stating these laws must necessarily be formulated consistently throughout the entire range of physical phenomena according to uniform mathematical principles. It was precisely the discontinuities in the existing science of motion, for example between Ptolemaic astronomy and Aristotelian cosmology and between the qualitatively different kinds of motion within the latter, that Galileo found so unsatisfactory. It was quite true, as Salviati said in the Third Day of the *Two Principal Systems*, that 'the principal aim of pure astronomers is to give reasons only for the appearances in celestial bodies, and to fit to these and to the motion of the stars such structures and compositions of circles that the motions following from these calculations correspond with those same appearances, having few scruples about admitting anomalies which might in fact prove troublesome in other respects'. But a criticism he made of the Ptolemaic system was just that 'although it satisfied an astronomer merely arithmetical (*puro calcolatore*), yet it did not afford satisfaction or content to the astronomer philosophical', that is, who was also a natural scientist. But, he added, Copernicus 'had very well understood that if one might save the celestial appearances with assumptions false in nature, it might much more easily be done with true suppositions'.

The characteristic of Galileo's philosophy of science that came to dominate his side of the controversy over the Copernican theory was the particular form of his conviction that his new mathematical science was a method of reading the real book of nature. It was his belief that 'natural propositions' could be 'demonstrated necessarily', that the experimental verification of a theory could establish it with 'indubitable certainty'. De-

scribing the opening of an investigation by means of an 'hypothetical assumption', he said in *Two New Sciences* that this could
be accepted conditionally 'as a postulate, the absolute truth of
which will be established when we find that the inferences from
it correspond to and agree with experiment'. He used such language not only when establishing the kinematic law of free fall
as a fact, but also in speaking of the Copernican theory. So, when
he repeated the argument that this was more economical than
the Ptolemaic theory, he was not using it in any conventionalist
sense. It was Nature herself that 'does not do that by many
things, which may be done by few', as he said in the Second Day
of his *Two Principal Systems*.

Apparently not clearly distinguished in his own mind from this
conviction that irrefutable verification was possible in science,
Galileo's fundamental contribution to the cosmological debate
was to see that a new and precise physical criterion, such as had
long been accepted as appropriate to decide between rival mathematical theories in astronomy (see Vol. I, p. 99 *et seq.*), was at
hand in the new inertial dynamics. Treating all motion, celestial
and terrestrial alike, as explicable by a single system of dynamics, he wanted to unite, in this system, the explanation with
the means of prediction of the various motions. In the law of
inertia he saw the possibility of a higher theory with which the
geocentric theory was incompatible and only the heliocentric
theory compatible. He failed in his own attempt to use this
dynamical criterion because he failed both to generalize the law
of inertia completely and to appreciate the true geometry of the
heliocentric system as set out by Kepler, but it was by means of
this criterion that the decision was eventually made.

But in 1615 Galileo had not yet begun to stress the dynamical
argument for the Copernican theory, and it was rather with the
difficulty of establishing necessary truths about the things of
experience in any particular case that the principal actor on the
ecclesiastical side of the debate made his riposte. This was Cardinal Robert Bellarmine (1542–1621). A student of astronomy in
his youth, it had been Bellarmine's unhappy task to frame the
decision that led Giordano Bruno to his death at the stake in

1600.* Undoubtedly his policy over Galileo was based on a determination never to let that episode be repeated. Over seventy years old, he aimed at administrative peace, and his method of achieving it was to take the alternative way to Galileo's in order to escape the conflict between astronomy and Scripture. His policy was to weaken the conclusions of natural science and to accept the new astronomy as in no sense established with 'indubitable certainty' but only as 'probable opinion and plausible conjecture', to accept it only in a form that would leave undisturbed the literal interpretation of Scripture and the Aristotelian cosmology which historical accident had married with it. He shut his eyes to the respects in which the union was less like marriage than living in sin. Yet although primarily administrative in their aim and limited in their application, it cannot be denied that Bellarmine's arguments succeeded in making a philosophical point against Galileo. Their two philosophies represent a classical polarization of opposites, an antithesis in the conception of the discoveries and inventions of theoretical science that is at once ancient, persistent, and easily misunderstood.

The principle had been well known to scholastic logicians, that the phenomena cannot uniquely determine the hypotheses that must 'save', or explain them, when the same conclusions can be deduced from different premises; to assert that agreement with observation proved an hypothesis true was to commit the fallacy of 'affirming the consequent'. This principle, developed at Oxford in the 13th and 14th centuries, had been a commonplace of the logical school of Padua in the early 16th century (cf. above, p. 40 *et seq.*). A typical statement is that by Agostino Nifo. In his commentary on Aristotle's *Physics*, Nifo had distinguished between the logical processes of discovery and of demonstration,

---

* It seems that Bruno was not charged with his advocacy of the Copernican system. According to Lynn Thorndike, *History of Magic and Experimental Science*, vol. 6, p. 427: 'Except that on March 24, 1597, he was admonished to give up such idle notions of his as that of a plurality and infinity of worlds, what counted most against him was his apostasy from his Order, his long association with heretics, and his questionable attitude as to the Incarnation and Trinity.'

and had contrasted the certainty of mathematics, where premisses and conclusion are reciprocating, with the conjectural character of our knowledge of causes in natural science.* Going on to consider astronomical hypotheses, Nifo wrote in his *De Cælo et Mundo Commentaria*, published at Venice in 1553, book 2:

> In a good demonstration, the effect necessarily follows from the assumed cause, and this must necessarily be assumed in view of the observed effect. Now the eccentrics and epicycles being admitted, it is true that the appearances are saved. But the converse of this is not necessarily true, namely that given the appearances, the eccentrics and epicycles must exist. This is true only provisionally until a better explanation is discovered which both necessitates the phenomena and is necessitated by them. Accordingly those men are in error who, taking a natural phenomenon, the occurrence of which might flow from many causes, conclude in favour of one cause.

The occasion that led Bellarmine to use this logical doctrine to draw the teeth of Galileo's arguments for the new astronomy was a letter written by a countryman of Galileo's, the Carmelite friar Paolo Antonio Foscarini, who had followed Galileo in suggesting that the Copernican system should be considered as a physical truth, not as a mere calculating device, and had shown how the relevant passages of Scripture could be reconciled with it. Bellarmine's reply, also written in 1615, rejected Foscarini's proposal. 'It seems to me,' he wrote:

> that your Reverence and Signor Galileo act prudently when you content yourselves with speaking hypothetically (*ex suppositione*) and not absolutely, as I have always understood that Copernicus spoke. To say that on the supposition of the Earth's movement and the Sun's quiescence all the celestial appearances are explained better than by the theory of eccentrics and epicycles [!], is to speak with excellent good sense and to run no risk whatever. Such a manner of speaking is enough for a mathematician. But to want to affirm that the Sun, in very truth, is at the centre of the universe and only rotates on its axis without going from east to west, and that the earth is in the third

---

* Many scientists, including Descartes and Newton, have shared the ideal of trying to make natural science as nearly as possible like mathematics in this respect; cf. below, pp. 308, 326.

heaven [i.e., sphere – see Plate 6] and revolves with the greatest speed round the sun, is a very dangerous attitude and one calculated not only to arouse all scholastic philosophers and theologians but also to injure our holy faith by contradicting the Scriptures. Your Reverence has clearly shown that there are several ways of interpreting the Word of God, but you have not applied these methods to any particular passage; and, had you wished to expound by the method of your choice all the texts which you have cited, I feel certain that you would have met with the very gravest difficulties.

As you are aware, the Council of Trent forbids the interpretation of the Scriptures in a way contrary to the common opinion of the holy Fathers.... It will not do to say that this is not a matter of faith, because though it may not be a matter of faith *ex parte objecti* or as regards the subject treated, yet it is a matter of faith *ex parte dicentis,* or as regards him who announces it....

If there were a real proof that the Sun is in the centre of the universe, that the Earth is in the third heaven, and that the Sun does not go round the Earth but the Earth round the Sun, then we should have to proceed with great circumspection in explaining passages of Scripture which appear to teach the contrary, and rather admit that we did not understand them, than declare an opinion to be false which is proved to be true. But, as for myself, I shall not believe that there are such proofs until they are shown to me. Nor is it a proof that, if the Sun be supposed at the centre of the universe and the Earth in the third heaven, the celestial appearances are thereby saved, equivalent to a proof that the sun actually is in the centre and the Earth in the third heaven. The first kind of proof might, I believe, be found, but as for the second kind, I have the gravest doubts, and in the case of doubt we ought not to abandon the interpretation of the sacred text as given by the holy Fathers.

Evidently Bellarmine had not mastered the technical details of *De Revolutionibus,* but he had read Osiander's cautious preface. The Copernican system was to be treated simply as a mathematical hypothesis for making calculations; it had been used as such for the Gregorian calendar in 1582. Galileo's views on the interpretation of Scripture, explicitly an exposition of the doctrines of St Augustine and the Fathers, were in themselves well received in Rome. The only query was the prudence of a layman setting out to teach the theologians their business. But it was Bellarmine's philosophical policy, the policy of Osiander, a Lutheran pastor, that prevailed in the deliberations of the Congregation of the Holy Office, before which the Copernican affair had come. No doubt the Roman authorities were partly concerned to preserve the text of Holy Scripture against private interpretations on the Protestant model, to which there seemed no limit. In any case they played for safety. Galileo's personal intervention in Rome failed to convince anyone that the Copernican theory was physically true, though it was useful in clearing him personally of a quite unfounded and maliciously inspired suspicion of heresy and blasphemy. On 24 February 1616 the theological experts, or Qualifiers, of the Holy Office delivered their famous report. They reported that the proposition that 'the sun is the centre of the world and altogether devoid of local motion' was 'foolish and absurd philosophically, and formally heretical, in as much as it expressly contradicts the doctrines of Holy Scripture in many places, both according to their literal meaning, and according to the common exposition and meaning of the holy Fathers and Doctors'; and that the proposition that 'the earth is not the centre of the world nor immovable, but moves as a whole, and also with a daily motion' was worthy 'to receive the same censure in philosophy and, as regards theological truth, to be at least erroneous in faith'.

On 5 March the Congregation of the Index issued its decree prohibiting Copernicus' *De Revolutionibus* until it had been corrected. Partly because of the intervention of Cardinal Maffeo Barberini, the future Pope Urban VIII, the Congregation made a distinction between scientific hypothesis and theological interpretation, and refused to prohibit *De Revolutionibus* altogether. The 'corrections' amounted to very minor changes, but made it appear that it was presenting only an hypothesis. In 1620 the book was once more permitted to be read. Moreover, the prohibition was not issued in such a way that the Copernican theory ever became formally heretical, although many contemporaries, unaware of the niceties of the distinction, understandably thought it was. The book by Foscarini on the interpretation of Scripture was at the same time prohibited absolutely. Galileo was not mentioned by name, although he was really the central character in the drama and the principal victim. Nothing if not forthright, he had been unsparing in his advocacy of the new astronomy during the whole of that Roman winter. 'We have here Sig. Galileo, who often, in gatherings of men of curious mind, bemuses many concerning the opinion of Copernicus which he holds for true', wrote a certain urbane Monsignor Querengo (in a letter included in the National Edition of Galileo's Works, published in Florence).

He discourses often amid fifteen or twenty guests who make hot assaults upon him, now in one house, now in another. But he is so well buttressed that he laughs them off; and although the novelty of his opinion leaves people unpersuaded, yet he convicts of vanity the greater part of the arguments with which his opponents try to overthrow him. On Monday in particular, in the house of Federico Ghisilieri, he achieved wonderful feats; and what I liked most was that, before answering the opposing reasons, he amplified them and fortified them himself with new grounds which appeared invincible, so that in demolishing them subsequently he made his opponents look all the more ridiculous.

Certainly the simple facts of personalities were an important influence in this drama on which so much philosophical ink has been spilt. After the decree Querengo wrote again expressing the view of the uncommitted man of the world.

The disputes of Signor Galileo have dissolved into alchemical smoke, since the Holy Office has declared that to maintain this opinion is to dissent manifestly from the infallible dogmas of the Church. So here we are at last, safely back on a solid Earth, and we do not have to fly with it as so many ants crawling around a balloon.

There are two documents alleging to describe what was said to Galileo after the Congregation of the Holy Office had reached its decision. According to a certificate given to him by Bellarmine, he was simply notified of the decree declaring that the Copernican theses were contrary to Scripture 'and consequently must not be either held or defended'. But according to a minute, possibly false, inserted into the Inquisition record of the proceedings, Galileo was warned by Bellarmine 'of the error of the aforesaid opinion and admonished to abandon it; and immediately thereafter' was ordered by the Commissary General of the Holy Office, in the presence of Bellarmine and other witnesses, 'in the name of His Holiness the Pope and the whole Congregation of the Holy Office, to relinquish altogether the said opinion that the Sun is the centre of the world and immovable and that the Earth moves; nor further to hold, teach or defend it in any way whatsoever, verbally or in writing; otherwise proceedings would be taken against him by the Holy Office; which injunction the said Galileo acquiesced in and promised to obey'. The difference between these two versions was to become material in Galileo's trial in 1633.

Galileo waited for an opportunity to prove an opinion he had good, though not conclusive, reasons for holding to be true. This came with the election in 1623 of Maffeo Barberini as Pope Urban VIII, a Florentine, a friend of the arts, and with Galileo a member of the Accademia dei Lincei. Galileo had disposed of all the arguments put forward *against* the earth's motion. Moreover he came to the conclusion that only by assuming the double motion of the earth on its axis and round the sun was it possible to explain the ebb and flow of the tides. Disbelieving in attraction at a distance, he did not accept Kepler's gravitational theory. He proposed instead an explanation based on the conservation of the sea's momentum. His object was to show that the movements of the tides could be demonstrated from the assumption of the earth's daily and annual revolutions, and that the existence of those revolutions was demonstrated by the existence of the tides. It was this capital dynamical proof that eventually formed the culmination of the *Dialogue concerning the Two Principal Systems of the World* (1632), in the Fourth Day, to which all the preceding dynamical discussion led up. It did not carry much conviction with Galileo's contemporaries, but it was only with the later work of Huygens and Newton that it became possible to get to the bottom of the matter and to see the fallacy in Galileo's ingenious argument.

Galileo's hopes for a genuine reopening of the Copernican question were not fulfilled. Urban agreed to his publishing a further discussion of the subject only on condition that it should remain hypothetical. The ecclesiastical point of view had not in fact changed since the time of Bellarmine. Its difference from Galileo's can be measured by the speech at the end of the *Dialogue* in which Galileo put into the mouth of Simplicio the opinions with which the Pope had instructed him to conclude. In discussing the assertion that it was possible to prove the motion of the earth conclusively, Simplicio asked whether God by his infinite power and wisdom could not have caused the tides by some other means than that envisaged by Galileo. 'Keeping always before my mind's eye a most solid doctrine that I once heard from a most eminent and learned person, and before which one must fall silent ...,' he declared, 'I know that you would reply that He could have, and that He would have known how to do this in many ways, which are beyond the reach of our minds. From which I forthwith conclude that, this being so, it would be an extravagant boldness for anyone to limit and confine the divine power and wisdom to some particular fancy (*fantasia particolare*) of his own.' Salviati responds: 'An admirable and truly angelic doctrine, and well in accord with another one, also divine, which, while it grants to us the right to argue about the constitution of the universe (perhaps in order that the working of the human mind shall not be curtailed or made lazy) adds that we cannot discover the work of His hands.'

The argument from God's omnipotence that had been used to free natural science from the restrictions of Aristotelianism in the 13th century had proved a boomerang.* Galileo's point of view was that while this argument was undoubtedly true, he

---

* Cf. Leibniz's letter to the Abbé Conti, Nov. or Dec. 1715, referring to Newton's natural theology, on which he was engaged in a controversy with Samuel Clarke: 'And because we do not yet know perfectly and in detail how gravity is produced or elastic force or magnetic force, this does not give us any right to make of them scholastic occult qualities or miracles; but it gives us still less right to put bounds to the wisdom and power of God and to attribute to him a *sensorium* and such things.' (*Recueil de diverses Pièces sur la Philosophie, la Religion Naturelle, L'Histoire, les*

was interested in discovering the way God had *actually* acted in making the world.

Thus if he were to publish a proof of the Copernican theory at all without going directly against ecclesiastical authority, it was impossible for Galileo to avoid some prevarication. The general order contained in the decree of 1616 still stood. It was his miscalculation of the risk that led to his disaster, although it may be justly argued that this in no way justified the action that was taken against him. He took every precaution, aided by his friends, the Master of the Holy Palace, the chief official responsible for licensing, and the Pope's own secretary, to make sure that the *Dialogue* would appear with every proper official sanction. It received the *imprimatur* of the Archbishop of Florence, although there seems to have been some genuine confusion between the different authorities, all of them well disposed. Following the Pope's instructions Galileo had added a preface and a conclusion disclaiming that his arguments were anything more than probable and hypothetical. But since the whole burden of the discussion in the pages between preface and conclusion had the very opposite intention, the casuistry of this disclaimer was the more obvious. Urban, with some justice, accused Galileo of breaking a personal promise made to himself. The Roman Inquisition then charged him with disobeying the injunction recorded in the minute of 1616 and of only pretending to present the condemned opinion 'as an hypothesis' (*hypothetice*). Galileo strongly denied any knowledge of the injunction. After proceedings that were anything but straightforward, he was found guilty, three out of the ten Cardinal judges withholding their signatures, and on 22 June 1633, in the Dominican Convent of Santa Maria Sopra Minerva, he was obliged to abjure his belief in the condemned Copernican theses. The *Dialogue* was prohibited. The earliest appearance of the famous phrase *Eppur si muove* seems to have been in the in-

---

Mathématiques, etc. par Mrs Leibnitz, Clarke, Newton, & autres Auteurs célèbres, ed. Des Maiseaux, Amsterdam, 1720, ii, 9.) 'Credulity is hurtful, so is incredulity: the business therefore of a wise man is to try all things, hold fast to what is approv'd, never to limit the power of God, nor assign bounds to nature.' (Boerhaave, *A New Method of Chemistry*, London, 1741.)

scription on a portrait of Galileo painted in the year of his death. It is unlikely, after a submission so humiliatingly complete, that in fact he murmured these words on rising from his knees. As to his physical treatment during the trial, all the evidence shows that the worst he suffered was confinement in comfortable quarters. It was a more serious inconvenience to be confined for the rest of his life to his farm at Arcetri, in the hills to the south overlooking Florence. His real suffering was of a different kind. Experience had taught Galileo to distinguish between truth and the behaviour of those claiming to act in its interest. But it was almost past bearing to suffer humiliation at the hands of the authorities of the Church in whose doctrines he believed and whom it was his desire to serve. The triumph of 'ignorance, impiety, fraud and deceipt', as he described the trial in later life, was as unnecessary as it was unwelcome a conclusion to the intelligent inquiries of Christian philosophers of science.

The decree against the Copernican theses and Galileo's condemnation placed Catholics in a false position for more than a century, without preventing excellent work being done in practical astronomy in Italy and other Catholic countries and the uninhibited development of other sciences there. Galileo himself, though an old man, went on with his work on mechanics and completed what was really his most important contribution to the subject, his *Discourses Concerning Two New Sciences*. But he had it published in Holland, in 1638. Even in theoretical astronomy excellent work went on behind the façade of ingenious quibbles. For example Alfonso Borelli, in 1660, observed the letter of the decree by limiting to Jupiter and his satellites the suggestive theory of celestial mechanics which he obviously intended to apply to the earth and the moon. Another curious result of the decree was the edition of Newton's *Principia* published in 1739–42, with a commentary by the Minim Fathers Le Seur and Jacquier presenting the Newtonian system of the world 'hypothetically'; the *Principia* had been announced originally in the *Philosophical Transactions of the Royal Society* as a mathematical demonstration of the Copernican system. Certainly the atmosphere was embarrassing for the 'free philosophizing about the things of the world and nature' which Galileo had pleaded so earnestly to keep open. Richelieu instigated an attempt to have the Copernican theses condemned at the Sorbonne, but without success : it was decided that the question was one for philosophy, not for authority. It was on hearing of Galileo's condemnation that Descartes, already a nervous philosopher and living in Holland, explicitly adopted his policy of dissembling in philosophy and became, in Maxime Leroy's phrase, the *'philosophe au masque'*. In November 1633 he wrote in alarm to Mersenne, who was arranging the publication of *Le Monde*, asking for news of the affair of the Copernican theory : 'and I confess that if it is false, then so are the whole foundations of my philosophy, because it is demonstrated from them beyond doubt'. When he discovered what had happened he sent Mersenne further letters withdrawing *Le Monde*, writing in 1634 :

No doubt you know that Galileo was arrested a short time ago by the Inquisitors of the Faith, and that his opinion concerning the movement of the Earth has been condemned as heretical. Now I would like to point out to you that all the things that I explained in my Treatise, among which was this opinion about the movement of the Earth, depend so much upon the other that it is enough to know that one of them is false, to know that all the reasons which I used are invalid; and although I thought that they were based on very certain and evident demonstrations, I would not wish for anything in the world to maintain them against the authority of the Church. I know well that it could be said that everything that the Inquisitors of Rome have decided does not thereby become an incontinent article of faith, and that it would first be necessary for it to be accepted by the Council. But I am not so much in love with my thoughts as to want to make use of such qualifications in order to maintain them; and I want to be able to live in peace and to continue the life I have begun in taking as my motto : *bene vixit, bene qui latuit* [he lives well who keeps well out of sight], accepting the fact that I am happier to be delivered from the fear that I would make more acquaintances than I want through my book than sorry to have wasted the time and trouble I have spent in writing it.... I have seen a notice about the condemnation of Galileo, printed at Liège on the 20th of September 1633, in which these words occur : *quamvis hypothetice a se illam proponi simularet* [although he pretended that it was being proposed by him hypothetically], so that they even seem to forbid the use of this hypo-thesis in astronomy; ... having seen nowhere that this censure has been authorized by the Pope or by the Council, but only by a particular Congregation of Cardinal Inquisitors, I do not lose all hope that the same thing will happen with it as with the Antipodes, which were more or less condemned at one time, and so that my *Monde* will be able to see the light of day in the course of time; towards which circumstance I will have to use my wits.

When Descartes eventually published his cosmology in the *Principia Philosophiæ* in 1644, it was under the cover of his presentation of physical theories as fictions (cf. below, p. 322). 'I want what I have written to be taken simply as an hypothesis,' he wrote, 'which is perhaps far removed from the truth.' With the definition that he had developed of motion as *simply* trans-lation from proximity to one set of bodies to proximity to an-other set, he was able to suppose that all motion was completely relative, any set of bodies being able to be chosen arbitrarily as the point of reference at rest. This enabled him to declare form-ally that the earth could be considered at rest. The conventiona-lism and fictionalism forced on physicists by the anti-Copernican decree had entered deeply into Descartes' soul, and earned him the polemics of Newton. The decree and the theological ambience in which it was issued had a greater responsibility for the more 'positivist' aspects of 17th-century thought than may sometimes be realized (cf. below, p. 316 *et seq.*).

Descartes had seen the important point that without papal ratification the Copernican theses had not been formally declared contrary to faith and heretical. The same point was made by Gassendi. The Commissioner General himself, Vincenzo Macu-lano da Firenzuola, who had conducted the proceedings against Galileo, admitted to Galileo's Benedictine pupil and friend Bene-detto Castelli that astronomical questions could not be decided by Scripture, which was concerned only with matters relating to salvation. In the decades that followed a number of Jesuit writers made the same point as Descartes and Gassendi. For example the French Jesuit Honoré Fabri, writing in 1661 in defence of the geocentric passages of Scripture, added that if conclusive reasons were ever found he did not doubt that the Church would say that they should be understood 'figuratively'. It was not until 1757 that Pope Benedict XIV annulled the anti-Copernican decree. At length in 1893 Pope Leo XIII made the *amende honorable* to Galileo's memory by basing his encyclical *Providentissimus Deus* on the principles of exegesis that Galileo had expounded, and rejected the fundamentalism of Bellarmine and of the Qualifiers of the Holy Office.

Not to be outdone, Pierre Duhem in 1908 made his famous declaration, in his *Essai sur la notion de théorie physique de Platon à Galilée* (*Annales de philosophie chrétienne*, 1908, vol. 6, pp. 584–85, 588), that more recent developments in physics had shown that 'logic was on the side of Osiander, Bellarmine and Urban VIII and not on that of Kepler and Galileo; that the former had grasped the exact significance of the experimental method, while the latter had been mistaken'. ... 'Suppose the hypotheses of Copernicus were able to explain all known appearances, what can be concluded is that they may be true, not that they are necessarily true, for in order to legitimate this last conclusion, it would have to be proved that no other system of hypotheses could possibly be imagined which could explain the appearances just as well; and this last proof has never been given.'

Duhem was making the valid point, fully developed in his *La Théorie Physique: son objet, sa structure* (1914), that experiment can never *establish* a theory irrefutably. But by introducing the dynamical criterion for choosing between two theories equally accurate in 'saving the appearances' of the heavens Galileo was in effect introducing a test of a theory by its range of applicability, as indeed Duhem realized. By this test it may be claimed that Galileo and Kepler showed how to go about *refuting* an astronomical theory and that in fact Newton did refute the geocentric hypothesis.* In this way the experimental falsification of the consequent could make it necessary to *deny* the antecedent, even though its verification would not enable the antecedent to be affirmed. Leaving aside Duhem's misapprehension of Bellar-

* Cf. Karl R. Popper, 'Three views of human knowledge' in *Contemporary British Philosophy: Personal Statements* (Third Series), London, 1956.

mine's very restricted application only to astronomical theories of the 'positivist' interpretation of science advocated by Duhem himself, the view that the two rival theories were merely alternative calculating devices certainly does not survive Galileo's test.

Discussing this controversy in 1844, J. H. Newman, the future cardinal, wrote in his *Sermons chiefly on the Theory of Religious Belief*: 'If our sense of motion be but an accidental result of our present senses, neither proposition is true and both are true, neither true philosophically, both true for certain practical purposes in the system in which they are respectively found.' Newman was not of course trying to make a revolution in logic but to deal with a difficulty in a theological controversy, but a similar point is sometimes made by those who say that Einstein's principle of general relativity has made Galileo's problem meaningless because motion and rest can be defined only by reference to a conventional standard, so that it is equally legitimate to take a stationary earth or a stationary sun for the frame of reference. But for general relativity it makes just as much sense to say that the earth rotates, as it did for Galileo and for Newton. To take a medieval example, it can be said to rotate just as a millstone can be said to rotate: it rotates with reference to all local inertial systems. It was in this sense that the earth's motion was at issue. A sophisticated logical interpretation of science is unavoidably faced with the fact that theoretical scientific analysis can make genuine physical discoveries, even though Galileo's assertion that a theory empirically verified according to his principles is a 'necessary' truth must be regarded as evidence that he himself was still the prisoner of a too simple Euclidean model for physics.

3. PHYSIOLOGY AND THE METHOD OF EXPERIMENT AND MEASUREMENT

Experimental physiology was another branch of science in which the quantitative approach, which Galileo had used with such success in mechanics and which was to achieve such astonishing triumphs in astronomy, was used with great effect in the 17th century.

Galileo himself had shown, while studying the strength and cohesion of materials, that whereas weight increased as the cube, the area of cross-section, on which strength depended, increased only as the square of the linear dimensions. There was thus a definite limit to the size of a land animal which its limbs could bear and its muscles move, but animals living in water, which supported the weight, might reach enormous dimensions.

One of the first to apply Galileo's methods to physiological problems was his colleague, the professor of medicine at Padua, Santorio Santorii (1561–1636). He described a number of instruments such as a pulsilogium, or small pendulum for measuring pulse-rate, and a clinical thermometer. He used the latter to estimate the heat of a patient's heart by measuring the heat of the expired air, which was supposed to come from the heart. He also made instruments to measure temperature in the mouth and others to be held in the hand. His method of measurement was to observe the distance which the liquid in the thermometer fell during 10 beats of a pulsilogium. As this depends not only on the patient's temperature but on the speed of his peripheral circulation, which increases with fever, Santorio's measurement of the rate of rise of temperature was probably an excellent indication of fever. In another work, *De Medicina Statica* (1614), he described an experiment which laid the foundation of the modern study of metabolism. He spent days on an enormous balance, weighing food and excrement, and estimated that the body lost weight through 'invisible perspiration'.

It was to William Harvey (1578–1657) that the revolution in physiology was chiefly due. After graduating at Cambridge, Harvey spent five years at Padua under Hieronymo Fabrizio of Aquapendente (c. 1533–1619), who was Galileo's colleague and personal physician. There Harvey learnt from his revered teacher to value the comparative method (see below, pp. 284–5). Most of his own researches into comparative anatomy were lost during the English Civil War, but in the two books which contain his extant contribution to science he emphasized the importance of comparative anatomy, both for its own sake and for elucidating the structure and physiology of man. He examined the hearts of a large number of vertebrates, including lizards, frogs and fish, and of invertebrates such as snails, a small transparent shrimp and insects. In insects he observed the pulsating dorsal vessel with a magnifying glass. Although his period at Padua coincided with Galileo's professorship there, there is no evidence that they ever met, nor did Harvey ever mention Galileo in his works. Nevertheless Harvey's method of restricting his research into biological processes to problems which could be solved by experiment and measurement might well have been learnt from the great mechanist. At any rate he breathed the same atmosphere and, although his references to logic were almost entirely to Aristotle, he also resembles Galileo in that his most important work was a perfect practical exhibition of the methods of 'resolution' and 'composition'.

Harvey's first statement of his theory of the general circulation of the blood occurs in his notes for lectures given at the Royal College of Physicians in London from 1616 to 1618 (published in 1886 as the *Prelectiones Anatomiæ Universalis*) but this seems to have been a later addition. Several of the constituents of his theory had already been discovered by his predecessors, but no one before him had seen that the difficulties raised by Galen's account of the motion of the blood were such as to require a revision of the whole theory. In fact Harvey's originality, no less than Galileo's, sprang from his ability to see familiar facts from an entirely novel point of view. The essential anatomy of the vascular system had been known since Galen's time and was as familiar to Harvey's immediate predecessors as to himself. It was not on purely anatomical grounds that he was able to reject Galen's complete separation between the venous and arterial systems (cf. Vol. I, p. 172 *et seq.*). He made his reinterpretation on the basis of a total shift in physiological theory; once this was accepted, the anatomical structures all fell into place in the new scheme.

The chief points in Galen's theory that became problems for Harvey were his assertions (i) that venous blood was produced continuously in the liver from the food, (ii) that it flowed in the veins out from the liver to all parts of the body, (iii) that only a small fraction of it entered the heart itself, and found its way from the right to the left ventricle to be converted into the arterial blood (this raised the questions of the existence of the pores in the interventricular septum and of the pulmonary circulation), (iv) his assertion that the arterial blood was drawn out of the heart in diastole and his account of the arterial pulse, and (v) his account of the two-way motion of air and waste in the venous artery. The first point raised the question of the amount and speed of the blood travelling in the vessels, and the others those of the direction of flow and of the action of the heart. None of these was considered except in isolation by any of Harvey's predecessors.

Leonardo da Vinci had maintained that the heart was a muscle, and made admirable drawings of it which included the discovery of the moderator band in the sheep. He had also followed the movements of the heart in the pig by means of needles thrust through the chest wall into its substance, and constructed models to illustrate the action of the valves. His views on the movements of the blood were, however, almost entirely Galenic and, moreover, it is not known whether his anatomical manuscripts had an influence similar to those on mechanics (Plate 12). The French physician, Jean Fernel, seems to have been the first to have observed, in 1542, that, in contradiction to current teaching, when the ventricles contracted (systole) the arteries *increased* in size, and to have stated that this was because of the blood (and compressed spirits) entering them. But in general Fernel expressed the ideas accepted before Harvey, relating the motion of the heart primarily to the supposed cooling function of respiration, and arguing in favour of Galen against Aristotle on the cause of its action and of the pulse. In 1543, Vesalius published his observations showing that he had been able to discover no pores through the interventricular septum; he had probed the pits in the septum, and found that 'none of these pits penetrate (at least according to sense) from the right ventricle to the left' (cf. Plate 11). In the second edition (1555) of his *De Fabrica* (see below, pp. 278–9) he was even more definite about the absence of pits, remarking: 'I am not a little hesitant concerning the heart's function in this respect.'*

A similar doubt, together with the view that the heart was a muscle and had two and not three ventricles, had already been asserted by the 13th-century Egyptian (or Syrian) physician, Ibn al-Nafis al-Qurashi. Ibn al-Nafis had maintained, as against Avicenna and Galen, that, since there was no passage through the septum, venous blood must pass from right to left ventricle via the arterial vein (pulmonary artery), through the lungs, where it spread through their substance and mixed with the air in them, and then back to left side of the heart in the venous artery (pulmonary vein). This work seems to have been unknown in the West†; the first Western writer to publish the theory of the pulmonary circulation (1553) was the Catalan scholar Miguel Serveto (1511–53), who mentioned, in the course of a theological discussion, that some blood passed from right to left ventricle via the lungs, where it changed colour. He supposed some also to pass through the interventricular septum. Serveto's interests were primarily theological and it is probable that he derived these ideas from some other source, although he had in fact studied anatomy, having been a pupil of Johannes Günther of Andernach

* It was the accepted interpretation in the 16th century that Galen had held that the blood passed from the right to the left side of the heart through these pores. This was also the view of Avicenna (cf. *Canon medicinæ*, III. xi. i. 1, Venice, 1608, i. 669–70), although Galen's own writings seem to leave open the possibility that some blood passed through the lungs (see Vol. I, pp. 173–4). Harvey himself may have interpreted Galen in the latter sense, although his remarks are equivocal: 'From Galen, that great Prince of Physicians, it seems clear that the blood passes through the lungs from the arterial vein [pulmonary artery] into the minute branches of the venous artery [pulmonary vein], urged to this both by the beating of the heart and by the movements of the lungs and thorax.' (*De Motu Cordis*, Ch. 7.) At least Harvey did pay Galen the tribute of having provided clear evidence for the pulmonary circulation by his description of the cardiac valves and of the anastomosis of arteries and veins in the lungs; but he ridiculed the view that a current of 'sooty wastes' could flow back through the mitral valves from the left ventricle to the lungs.

† A Latin translation by Andrea Alpago of Ibn al-Nafis' great commentary on Avicenna's *Canon* was published in Venice in 1547, but, curiously enough, this omitted the section on the pulmonary circulation.

in Paris at the same time as Vesalius himself. There is at present no evidence that either he or the Paduan anatomist Realdo Colombo (c. 1516–59) knew of Ibn al-Nafis, and some scholars have suggested that it was Serveto who inspired Colombo in his views on the lesser circulation. In view of the curious context in which Serveto announced the discovery, others have suggested that the borrowing was more probably the other way round; it is even possible that Colombo derived the idea of the lesser circulation from Vesalius himself, whose pupil he had been at Padua. Colombo himself, in his *De Re Anatomica* (1559), not only put forward the idea of the pulmonary circulation but also supported it with experiments. He noted, as Fernel had done, that cardiac systole (contraction) coincided with arterial *expansion* and cardiac diastole (expansion) with arterial contraction; and he showed, further, that the complete closure of the mitral valve prevented pulsation in the pulmonary vein. When he opened this vein he found not fumes, as the Galenists would have expected, but blood, and he concluded that blood passed from the lung (where a change in colour was observed) through the pulmonary vein back to the left side of the heart. Like Serveto, he believed that some blood passed also through the interventricular septum. Both writers held also to the Galenic view that blood was made in the liver. Thus neither had any idea of the true nature of blood, and although Colombo had observed that the pulsation of the brain was synchronous with that of the arteries, he did not arrive at the idea of the general or systemic circulation.

The same may be said for Colombo's Catalan pupil, Juan Valverde, who gave an account of the lesser circulation in 1554. Valverde seems to have claimed no originality for himself, and some scholars have argued, on the grounds that he stated, like Serveto, that the pulmonary vein contains both blood and air, that it was Serveto who influenced him. Against this others have argued that it was from Colombo's teaching that Valverde learnt of the idea of the lesser circulation; Colombo's treatise, published posthumously in 1559, may well have been written before Valverde's. Certainly Colombo himself claimed the new idea as his own, and hitherto unknown.

The Dutch anatomist Volcher Coiter (1534–*c.* 1576) also made some experiments on the heart. He made a comparative study of the living hearts of kittens, chickens, vipers, lizards, frogs and eels, and observed that in the excised organ the auricles contracted before the ventricles and that the heart was lengthened in systole and shortened in diastole. He also showed that a small detached piece of heart muscle would continue to beat.

Some observations on the movements of the blood were made also by the Italian physiologist and botanist, Andrea Cesalpino (1519–1603). He said, in his *Quæstionum Peripateticarum* (1571), that when the heart contracted it forced blood into the aorta and when it expanded it received blood from the *vena cava*. In his *Quæstionum Medicarum* (1593), book 2, question 17, he said:

The passages of the heart are so arranged by nature that from the *vena cava* a flow takes place into the right ventricle, whence the way is open into the lung. From the lung, moreover, there is another entrance into the left ventricle of the heart, from which there is a way open into the aorta artery, certain membranes being so placed at the mouths of the vessels that they prevent return. Thus there is a sort of perpetual movement from the *vena cava* through the heart and lungs into the aorta artery, as I have explained in my *Peripatetic Questions*. If we take into account that in the waking state there is a movement of natural heat towards the exterior, that is to say, towards the organs of sense, while in the sleeping state there is, on the contrary, a movement towards the interior, that is, towards the heart, we must judge that in the waking state much of the spirit and blood become engaged in the arteries, since it is by them that access is had to the nerves, while, on the other hand, in sleep the animal heat comes back through the veins to the heart, but not by the arteries, since the access provided by nature to the heart is through the vena cava and not through the aorta ... For in sleep the native heat passes from the arteries into the veins through the process of communication called anastomosis, and thence to the heart.

He used this to explain the observation that when a vein was ligatured it swelled up on the side away from the heart. But his views on the subject lacked clarity and decision, and in his last work in 1602–3 he formally stated that blood went *forth* from the heart through the veins as well as the arteries. Though he used the word *circulatio*, he understood this to mean a to-and-fro movement as in the rising and falling of fluid, evaporation followed by condensation, in chemical distillation. Thus he did not understand the general circulation any more than Colombo, Serveto or Valverde, or Carlo Ruini who, in 1598, also published a description of the pulmonary or lesser circulation in his treatise on the anatomy of the horse, or Fabrizio who, in 1603, gave the first clear and adequate figures of the valves in the veins but believed that their function was to counteract the effect of gravity and prevent the blood accumulating in the hands and feet. (These valves had been described by Charles Estienne in 1545 – see below, p. 276 – and after that they were studied by several anatomists, none of whom understood their action.)

The theory of the general circulation of the blood was, in fact, first advanced by William Harvey and published in 1628 in his *Exercitatio Anatomica de Motu Cordis et Sanguinis in Animalibus*. The earliest knowledge of his great theory comes, however, from the *Prelectiones*, and these provide valuable evidence of how he reached it.

There is a conversation recorded by Robert Boyle in 1688 but relating to over thirty years earlier, although still nearly twenty years after the publication of *De Motu Cordis*, in which Harvey himself seems to connect his theory with the results of the great Italian tradition of anatomical studies. 'I remember,' Boyle wrote,

that when I asked our famous Harvey, in the only discourse I had with him (which was but a while before he died), what were the things that induced him to think of the circulation of the blood, he answered me that when he took notice that the valves in the veins of so many parts of the body were so placed, that they gave free passage to the blood towards the heart, but opposed the passage of the venal blood the contrary way, he was invited to imagine that so provident a cause as nature had not so placed so many valves without design; and no design seemed more probable than that since the blood could not well, because of the interposing valves, be sent by the veins to the limbs, it should be sent through the arteries, and return through the veins, whose valves did not oppose its course that way. (Boyle, *Works*, 1772, vol. 5, p. 427)

More recently it had been suggested that Harvey's theory of the general circulation was a natural development of the work of his predecessors on the pulmonary circulation. Neither of these suggestions receives much support from his own writings, but at another level, that of method, the Italian tradition is clear indeed. Harvey himself shows us that his great illumination came from his use of the comparative method; his ability to follow out its consequences came from his clear grasp of the use of experiment and of measurement. All this was the teaching of Padua, but it was the use to which he put those methods that raised him to an altogether higher level of originality.

This can be seen in the contrast between him and the anatomists who had discussed the pulmonary circulation. These had never questioned the basic Galenic assumption that the veins and the right side of the heart formed a system, centred on the liver, which was quite distinct in function and structure from the system formed by the arteries and the left side of the heart. Between the two lay the lungs, receiving nutriment from the venous blood sent from the right ventricle, providing from the air the principle of its conversion into arterial blood in the left ventricle, and cooling and cleansing the heart itself. They had looked for the solution of one particular problem: how the blood passes from the right to the left side of the heart in man, a problem that arose and was solved within Galen's system itself. Looking beyond man to a whole range of red-blooded animals and even to animals like shrimps, insects and snails, Harvey saw that this was only part of the more general problem of the movement of the blood in the body as a whole. In fishes, which have no lungs, in frogs, toads, snakes and lizards which resemble fish in having only a single ventricle, and also in the embryos of lunged animals, the first problem did not in fact arise at all. 'The common practice of anatomists,' he wrote in chapter 6 of *De Motu Cordis*, 'in dogmatizing on the general make-up of the animal body, from the dissection of dead human subjects alone, is objectionable. It is like devising a general system of politics, from the study of a single state, or pretending to know all agriculture from an examination of a single field. It is fallacious to attempt to draw general conclusions from one particular proposition.' 'Had anatomists only been as conversant with the dissection of the lower animals as they are with that of the human body, the matters that have hitherto kept them in a perplexity of doubt would, in my opinion, have met them freed from every kind of difficulty.'

Far from being simply a continuation of the work of his predecessors, the main object of Harvey's whole argument was to put forward, and to demonstrate by experiment and accessory evidence, a conclusion diametrically opposed to their basic Galenic assumptions about the course of the blood and the action of the heart. The question of the pulmonary circulation plays a very secondary role in his whole argument; indeed he discussed it fully only in a letter written in 1651, to Paul Marquard Slegel of Hamburg. Harvey's originality was altogether greater than the sum of the contributions of his predecessors. What he did was to be the first, since Galen, to attempt 'a general system of politics' in questions of anatomy and physiology. He was the first to produce a theory which, as he insisted both in *De Motu Cordis* and in the controversies to which it gave rise, comprehended all the particular circulatory systems of different animals and embryos in a general scheme. By demonstrating an alternative to the central doctrine of Galen's system, he raised an entirely new range of questions about physiology in general.

Harvey's discussion, both in the *Prelectiones* and in *De Motu Cordis*, indicates that it was Galen's assertion that the blood left the heart in diastole and his account of the arterial pulse that led to his first doubts. The argument in the *Prelectiones* closely follows that in the first eight chapters of *De Motu Cordis*. Both begin with a 'resolution' of the problem into its parts, so that the cause might be discovered through its effects. After analysing the difficulties for Galen's theory, citing many observations made by others, he concentrated on demonstrating the action of the heart in systole, the nature of the pulse, and the consequent continuous flow of the blood through the heart, in various animals and in the foetus, as a result of its continuous beat. The *Prelectiones* conclude with a statement of the hypothesis of the general circulation similar to that in chapter 8 of *De Motu*

Cordis. Probably the discussion in his lectures stopped there, for he was demonstrating the anatomy of the thorax as a whole, and this had to be completed in one day because there were no preservatives. The remaining chapters of *De Motu Cordis* clearly form a second section corresponding to the 'compositive' part of the argument. He described the testing of his hypothesis by three consequences that follow from it; stated it definitely in chapter 14; and added further accessory evidence.

He began his demonstration by showing that the contraction of the heart was a muscular contraction beginning with the auricles and passing to the ventricles, whose contraction then caused the expansion of the arteries. In contradiction to the conceptions of its action held by Aristotle and by Galen, he was to conclude that the heart was a force pump. This suggested that there was a flow of blood from the veins through the heart into the arteries, and the arrangement of the cardiac valves would prevent its return. He then showed that if either the pulmonary artery or the aorta alone were punctured, the contraction of the right ventricle was followed by a jet of blood from the former, and the contraction of the left ventricle by a jet of blood from the latter; the two ventricles contracted and dilated in unison. In the fœtus he pointed out that the structure of the heart and vessels was designed to bypass the lungs, which were not yet functioning. He said that the blood from the *vena cava* passed through an opening, the *foramen ovale*, into the pulmonary vein and so via the left ventricle into the aorta. (Actually the *foramen ovale* opens directly into the left auricle.) The blood entering the pulmonary artery was carried into the aorta by the fœtal *ductus arteriosus*. The two ventricles thus operated as one, and the condition in the embryo of animals with lungs corresponded to that in the adults of animals such as fish which had no lungs. In the adults of the animals with lungs the blood could not pass through the two fœtal passages, which were closed, but had to go from the right to the left side of the heart via the tissue of the lungs themselves.

From the structure and continuous beat of the heart Harvey concluded that the flow of the blood through it was not only in one direction but also continuous. It would follow from this that unless there were some passage from the arteries back to the veins in the body at large, as well as in the lungs, the veins would soon be drained and the arteries would be ruptured from the quantity of blood flowing into them. There was, therefore, no escape from the hypothesis which he enunciated in chapter 8 of *De Motu Cordis*:

I began to think whether there might not be *a motion as it were, in a circle*. Now this I afterwards found to be true; and I finally saw that the blood was forced out of the heart and driven by the beating of the left ventricle through the arteries into the body at large and into its several parts, in the same way as it is sent by the beating of the right ventricle through the arterial vein [pulmonary artery] into the lungs, and that it returns through the veins into the *vena cava* and so to the right ventricle, in the same way as it returns from the lungs through the venous artery [pulmonary vein] to the left ventricle.

Proceeding to the testing of this hypothesis, Harvey next made a number of deductions which, if experimentally verified, would both confirm it and finally eliminate the rival hypothesis of Galen that the blood was continuously produced in the liver from the ingested food. First, he demonstrated that, with the blood flowing continuously in one direction through the heart, it could be calculated from the heart's capacity and rate of beat that it pumped through itself in an hour, from the veins to the arteries, more than the whole weight of the body. That the blood did flow continuously through the heart only in the direction from veins to arteries, he confirmed by further experiments. In a serpent, whose vessels were conveniently arranged for experimental investigation, when the *vena cava* was pinched with forceps the heart drained and became pale, whereas when the aorta was similarly closed the heart became distended and purple. This was in keeping with the arrangement of the valves. Secondly, he showed, by experiments with ligatures, that this same large amount of blood that passed through the heart was forced through the arteries to the peripheræ of the body, and that there the blood flowed in the same continuous stream in one direction only, but, in those regions, from arteries to veins. In the limbs the arteries are deeply placed, while the veins are near the surface. A moderately tight ligature round the arm would constrict the latter but not the former, and he found that this produced a distension of the hand with blood. A very tight ligature stopped the pulse and flow of blood into the hand altogether and no distension was observed. Finally, he showed that the blood returned to the heart in the veins. Anatomical investigations showed that the valves were arranged in the veins so that the blood could flow only towards the heart, a fact which Fabrizio had not realized. Harvey showed that when the arm was ligatured moderately tightly so that the veins swelled up, 'nodes' were formed at the position of the valves (Plate 13). If the blood were pushed out of the vein below the valve by running the finger along it in the peripheral direction, the emptied section remained flat, and he concluded that this was because the valve prevented the blood from running back into it. This explanation he confirmed by further experiments of the same kind. He therefore arrived at the definitive conclusion, in *De Motu Cordis*, chapter 14:

Since all things, both argument and ocular demonstration, show that the blood passes through the lungs and heart by the action of the ventricles, and is sent for distribution to all parts of the body, where it makes its way through the pores of the flesh into the veins, and then flows by the veins from the circumference on every side to the centre, from the lesser to the greater veins, and is by them finally discharged into the *vena cava* and right auricle of the heart, and this in such a quantity, with such an outflow through the arteries, and such a reflux through the veins, as cannot possibly be supplied by the ingesta, and is much greater than can be required for mere purposes of nutrition; it is therefore necessary to conclude that the blood in the animal body is impelled in a circle, and is in a state of ceaseless motion; that this is the act or function which the heart performs by means of its pulse; and that it is the sole and only end of the motion and contraction of the heart.

Published at Frankfurt, the scene of an annual book fair, Harvey's treatise became widely distributed. In spite of criticism by some established professors such as Jean Riolan of Paris, his theory was fairly soon adopted, especially by younger anatomists, an example of the fact that often only a new generation can appreciate a fundamental revolution, partly because to it the new doctrine has ceased to be revolutionary. John Aubrey wrote in his vignette of Harvey :

I have heard him say, that after his booke of the Circulation of the Blood came out, that he fell mightily in his practize, and that 'twas be-leeved by the vulgar that he was crack-brained; and all the physitians were against his opinion, and envyed him; many wrote against him, as Dr Primige Paracisanus, etc. With much adoe at last, in about 20 or 30 years time, it was received in all the Universities in the world; and, as Mr Hobbes sayes in his book 'De Corpore,' *he is the only man, perhaps, that ever lived to see his owne doctrine established in his life time.*

Harvey's theory was an immense illumination to physiology, to which it directed the interest of all biologists. His treatise provided a model of method. After him, abstract discussion of such questions as the nature of life or of 'innate heat' gradually gave way to the empirical investigation of how the body worked. He himself had left somewhat vague the passage of the blood from the arteries to veins, and the demonstration of his theory was finally completed when, in 1661, Malpighi observed, under the microscope, the flow of blood through the capillaries in the frog's lung. About the same time Jean Pecquet and Thomas Bartholin worked out the lymphatic system, beginning with Pecquet's observations, at the end of Harvey's life, on the lacteals, the vessels that carry the chyle (emulsified fat) from the small intestines to the veins by way of the thoracic duct – an important complement to Harvey's theory which the aged physiologist himself rejected on the very grounds of comparative anatomy that had guided his own work. He could find no trace of lacteals in birds and fish. 'Nor,' he wrote to Dr R. Morrison, 'do I see any reason why the route by which the chyle is carried in one animal should not be that by which it is carried in all animals whatsoever; nor indeed, if a circulation of the blood be necessary in this matter, as it really is, that there is any need for inventing another way.' Those very aptitudes for theoretical generalization to which he owed his greatest discoveries were to blind him to the apparent inconsequentiality of fact.

The study of the blood, the bearer of food and oxygen, was, in fact, well placed to lay the foundations of physiology, and Harvey's elucidation of its mechanics was followed later in the 17th century by the researches especially of Boyle, Hooke, Lower and Mayow into the chemical problem of respiration, which they related for the first time to the general problem of combustion.

But Harvey himself never understood the function of respiration, and when we come to consider his views on the purpose of the circulation in general we must put ourselves into the framework of a philosophy of nature very different from that of modern physiology, a framework of questions extending beyond the range of those to which Harvey's elucidation of the mechanics of the circulation was the positive answer incorporated into modern science.

The philosophy of nature of a period different from our own, the whole complex of presuppositions and conceptions that a particular explanation eminently satisfies at a given time, is often more clearly indicated by secondary writers than by the great innovators whose originality inevitably transforms the background of ideas into which they were born. One of the first of contemporaries to accept Harvey's theory was the London physician, alchemist and Rosicrucian Robert Fludd, many of whose own writings had been issued by the same Frankfurt publisher. But Fludd saw in the great discovery of 'his friend, colleague and compatriot, well versed not only in anatomy but also in the deepest mysteries of philosophy', as he called Harvey in his *Integrum Morborum Mysterium* in 1631, not the beginning of a new physiology, but a demonstration of something quite different: of the correspondence between the microcosm of the body and the macrocosm of the celestial spheres; a demonstration that the spirit of life retained an impression of the planetary system and of the zodiac, an impression of the cicular motion of the heavenly bodies that ruled the world below.

Cool, clear and rational as he was, an empirical scientist to the depths of his mind, it is clear that Harvey himself would not have been unwilling to accept this compliment. At the end of the passage already quoted from chapter 8 of *De Motu Cordis*, in which he described how the idea of circulation came into his mind, Harvey related the motion of the blood to a general view of the world. A true pupil of Padua, his view is basically Aristotelian : 'The authority of Aristotle has always such weight with me that I never think of differing from him inconsiderately,' as he said later in *De Generatione Animalium* (exercitatio 11). It was basic to Aristotle's philosophy of nature that circular motion is the noblest form of motion and that the circular motion of the heavenly bodies forms the pattern to which the motions of sublunary bodies, and especially of the microcosm of living organisms, aspire. Aristotle had made the heart the principle organ of the body and the origin of the blood and vessels. After his account of the mechanical pumping of the blood round the body by the heart, Harvey likens its circular motion to the cycle of water evaporating under the sun's heat from the moist earth and returning again as rain, thus producing the generations of living things, and to the annual cycle of the weather with the approach and retreat of the sun; both 'as Aristotle says ... emulate the circular motion of the superior bodies.'

And so in all likelihood does it come to pass in the body, through the motion of the blood. All the parts may be fed, warmed and quickened by the warmer and more perfect vaporous, spirituous and, as I may say, nutritive blood; and this, on the contrary, may become, in contact with the parts, cooled, thickened, and so to speak effete, so that it returns to its origin, the heart, as to its source, the inmost temple of the body, to recover its perfection and virtue. Here it is again liquified by natural heat – potent, burning, a kind of treasury of life, and it is impregnated with spirits and as it might be said with balsam; and thence again it is dispersed; and all this depends on the motion and beating of the heart. Consequently the heart is the beginning of life, the sun of the microcosm, just as the sun in his turn might well be called the heart of the world; for it is by the potency (*virtus*) and beating of the heart that the blood is moved, perfected and quickened to life (*vegetatur*) ... for the heart indeed is the perfection of life, the source of all action.

This view of the cosmological pattern in which the circulation of the blood took its place, Harvey shared with another Aristotelian,

Cesalpino. Like Harvey, Cesalpino had regarded the renewed 'perfection' of the blood as the immediate purpose of its passage through the heart; and like Harvey he described a cyclical process of heating and evaporation in the heart, followed by cooling and condensation in the parts of the body, corresponding to the alchemical cycle of *distillatio*. These notions, the analogy of the microcosm and the macrocosm, the prevalence of cycles in nature, the excellence of the circle, were in fact commonplaces and occur in various forms in all the Aristotelian, alchemical, Paracelsist, and Neoplatonic writings of the period. They appear, for example, in the symbolic embryology of Peter Severinus (1571) and of Johann Marcus Marci of Kronland (1635). Harvey himself returned to them in his *De Generatione Animalium* (1651) as the analogy of the coming and going of new generations, especially in the cycle of change, described in his theory of 'epigenesis', from the undifferentiated seed to the first differentiated matter, the blood, thence to the fully differentiated adult, and back to the seed forming the new generation.

It is this philosophical conception of cycles that united the two great fields of Harvey's work (see below, pp. 286-7), and this is a good illustration of the fact that if we are to understand the appearance of a discovery or a new explanation, and the particular form it takes, we must look beyond the purely empirical grounds on which it rests. The latter indeed are never alone in determining the scientist's expectations and the direction of his attention and his vision; these are inevitably to some extent the products of a theory, and certainly in Harvey's case of unverified ontological assumptions about the world which constituted his philosophy of nature. But the difference between a scientist like Harvey and the mere speculators like Fludd, with whom he may have shared many such assumptions, was that he put his theories to effective empirical tests. In this he stood in the same relation to Fludd as did Kepler. To the end of his life Harvey denied that the blood underwent any essential change in the lungs; he held that the blood was cooled in the body generally and he thought that the traditional view might be correct, that breathing cooled it especially. But he distinguished this problem from the *fact* of the circulation : 'I own I am of opinion,' he wrote in the *Second Dis-quisition to Jean Riolan* (1649), 'that our first duty is to inquire whether the thing be or not, before asking wherefore it is.' Harvey's great strength as a master of the experimental method, and his superiority over all other contemporary biologists, was that he had both the gifts of imagination that made him a great discoverer and a great theoretician, and the gifts of reason that showed him how to test his theories by precise and quantitative experiments.

It was the theoretical gifts that were uppermost in the mind of the co-founder, with Harvey, of modern physiology, Descartes. In his *Discours de la Méthode* (1637) Descartes had expressed the hope of arriving at rules which would reform medicine in the same way as he had attempted to reform the other sciences. He was one of the first to accept Harvey's discovery of the circulation of the blood, though he did not understand the pumping action of the heart, which he still regarded as producing its action through its vital heat. Although he gave the credit for the discovery of the circulation to '*un médécin d'Angleterre*' (*Discours*, part 5) Descartes claimed for himself the elucidation of the mechanism of the heart. He thought that it was the vital heat of the heart that caused it to expand by vapourizing the blood drawn into it on contraction, and that it was this expansion in diastole that sent the blood along the arteries and into the body and the lungs, where it cooled and liquified and returned to the heart, where the cycle began again. Descartes was in fact reviving Aristotle's explanation, in opposition to both Galen and Harvey (cf. Vol. I, p. 175, below, p. 311 *et seq.*). It is indeed curious that a man who claimed to have divested himself of all former prejudices should have repeated the old error, already detected a century before, that the blood left the heart in diastole, and that his physiological system as a whole should so much have resembled those of Galen and Aristotle. But it is not by such details that Descartes' achievement should be judged; indeed had they caused him any hesitation perhaps he would never have made it. His contribution was to grasp and assert one big theoretical idea : that the body is a machine, and that all its operations are to be explained by the same physical principles and law as apply in the inanimate world. Though he still used terms like 'spirits', these were simply material, and they obeyed general mechanical laws; the special spirits and principles charged in the old physiology with each particular function had been eliminated. Whereas the philosophy of nature, the system of analogies with cycles of nature and with the sun, within which Harvey worked out how the heart and blood moved was of little use in suggesting further inquiries, Descartes' mechanism was immediately fruitful. In spite of his misunderstanding, he had made a point against Harvey in pressing the question of the *cause* of the heart's beating. He wanted to show that this would follow from known mechanical laws, and so appear as a phenomenon expected within the general system of mechanics.

'But lest those who are ignorant of the force of mathematical demonstrations,' he wrote in part 5 of the *Discours*, 'and who are not accustomed to distinguish true reasons from mere verisimilitudes, should venture, without examination, to deny what has been said, I wish it to be considered that the motion which I have now explained follows as necessarily from the very arrangements of the parts, which may be observed in the heart by the eye alone, and from the heat which may be felt with the fingers, and from the nature of the blood as learned from experience, as does the motion of a clock from the power, the position and the shape of its counterweights and wheels.'

In the presentation of his mechanistic theory Descartes made explicit an even greater contribution to physiology, for he did so in terms of one of the most fruitful *methods* known to science: the method of the theoretical model. A theoretician of scientific method as well as of physics and physiology, Descartes was perfectly conscious of what he was doing; it was he who made the method of the physical and chemical model the powerful tool of analysis it has ever since been in physiological research. His *'homme-machine'* was a theoretical body, which he tried to construct from the known principles of physics in such a manner that he could deduce from it the physiological phenomena observed in actual living bodies. In his *Primæ Cogitationes circa*

Generationem Animalium he even faced the basic question of machines begetting machines. His physiology was Galenic and Aristotelian, but it was Galen and Aristotle *more geometrico demonstrata.*

Moreover Descartes was not ignorant of the subject at first-hand; he had spent several years studying anatomy, and in *La Dioptrique*, published together with the *Discours* as part of his exemplification of method, himself made a fundamental contribution to the physiology of vision.

'I am resolved to leave all the people here to their disputes,' he said in part 5 of the *Discours*,

and to speak only of what would happen in a new world, if God were now to create somewhere in the imaginary spaces matter sufficient to compose one, and were to agitate variously and confusedly the different parts of this matter, so that there resulted a chaos as disordered as the poets ever feigned, and after that did nothing more than lend his ordinary concurrence to nature, and allow her to act in accordance with the laws which he had established.

Of the mechanistic theory of the living body, which he claimed to be able to derive from these laws, he said :

Nor will this appear at all strange to those who are acquainted with the variety of movements performed by the different automata, or moving machines, fabricated by human industry, and with the help of but few pieces compared with the great multitude of bones, muscles, nerves, arteries, veins, and other parts that are found in the body of each animal. Such persons will look upon this body as a machine made by the hands of God, which is incomparably better arranged, and adequate to movements more admirable than in any machine of human invention.

He gave a detailed account of this theoretical body in his treatise *L'Homme*, which formed part of *Le Monde ou Traité de la Lumière* (completed in 1633 but published posthumously in 1664).

'I assume that the body is nothing more than a statue or machine of clay,' he wrote;

we see clocks, artificial fountains, mills, and other similar machines which, although made by man, yet have the power of moving themselves in several different ways; and it seems to me that I could not imagine as many kinds of movement in it as I suppose to have been made by the hands of God, nor attribute to it so much artifice that you could not think it could have more ... I want you to consider next that all the functions which I have attributed to this machine, such as the digestion of food, the beating of the heart and arteries, the nourishment and growth of the members, respiration, waking, and sleeping; the impressions of light, sounds, odours, tastes, heat and other such qualities on the organs of the external senses; the impression of their ideas on the common sense and the imagination; the retention of imprinting of these ideas upon the memory; the interior motions of the appetites and passions; and, finally, the external movements of all members, which follow so suitably as well the actions of objects which present themselves to sense, as the passions and impressions which are formed in the memory, that they imitate in the most perfect manner possible those of a real man; I desire, I say, that you consider that all these functions follow naturally in this machine simply from the arrangement of its parts, no more and no less than do the movements of a clock, or other *automata*, from that of its weights and its wheels; so that it is not at all necessary for their explanation to conceive in it of any other soul, vegetative or sensitive, nor of any other principle of motion and life, than its blood and its spirits, set in a motion by the heat of the fire which burns continually in its heart, and which is of a nature no different from all fires in inanimate bodies.

In Descartes' theory the body of a human being was occupied by a rational soul. Since the mind was an unextended thinking substance while the body was an unthinking extended substance, some of his critics and followers, such as Gassendi and Malebranche, held that these two substances could have no point of contact. But Descartes held that they interacted through one organ and one only, the pineal gland in the brain (Plate 15; cf. Plate 14; Vol. I, p. 171; below, p. 315 *et seq*.). One reason for his choice of the pineal gland was that it was the only organ in the brain that was single and not divided into right and left sides. Thus it was adapted to interact with all parts of the body. He held that the cerebral cavity, in which the pineal gland was suspended, contained animal spirits distilled in the heart from the blood, and that through pores in the internal surface of this cavity animal spirits entered the nerves, which he thought were fine hollow tubes. Inside each nerve he held that there were numerous very fine threads, one end of each being attached to a part of the sense organ to which the nerve ran, and the other to a small door at the pore where the nerve reached the internal surface of the brain. The whole nervous function in this machine depended only on the control of the flow of the purely material animal spirits in the brain and nerves, just, he said, as organ music depended only on the control of the air in the pipes.

For example, when light coming from an external object was focussed on to the retina, it pushed a corresponding set of threads in the optic nerve. These in turn opened the corresponding pores on the internal surface of the brain, acting like the wires of a bell-pull. The image formed on the retina was thus reproduced in the pattern of pores opened, and so was traced in the spirits on the surface of the pineal gland. There it was immediately apprehended by the rational soul, which thus received a sensation of the external object. The mind was thus presented with a token of the external world, not the thing in itself.

When, on the other hand, the soul willed a certain action, it acted on the body by moving the pineal gland so that it deflected the animal spirits into the pores opening into the nerves leading to the muscles concerned. The animal spirits acted on the muscle at the end of a nerve by flowing into it and making it swell up, thus causing it to move the limb or part of the body to which it was attached.

By means of this hypothetical model Descartes was able to offer mechanical explanations of many common neurological and psychological phenomena, for example of the coordinated control of an action such as walking where many different muscles are involved, of emotions, of images formed without external objects, of falling to sleep and waking up, of dreams, and of memories, which he held to be the physical traces of the paths of the animal spirits. His explanation of vision and the eye is especially remarkable for its close control by observation and experiment, combined with mathematical analysis of the optical phenomena concerned.

In contrast with man the brutes were simply *automata* and nothing more. Though animals were considerably more compli- cated, there was no difference in principle between them and the *automata* constructed by human ingenuity. 'There is,' wrote Des- cartes in a letter to the Marquis of Newcastle, on 23 November, 1646, 'no one of our external actions which can assure those who examine them that our body is anything more than a machine which moves itself, but which also has in it a mind which thinks – excepting words, or other signs made in regard to whatever subjects present themselves, without reference to any passion.' He had said the same thing in the *Discours*. The noises made by animals indicated no such controlling mind and we should not be deceived by their apparently purposive behaviour.

I know, indeed, that brutes do many things better than we do, but I am not surprised at it; for that, also, goes to prove that they act by force of nature and by springs, like a clock, which tells better what the hour is than our judgement can inform us. And, doubtless, when swallows come in the spring, they act in that like clocks. All that honey-bees do is of the same nature.

The mechanical principles that Harvey had adopted as a method were thus converted by Descartes into a complete philo- sophy of nature, and just as he had ignored the empiricism of Galileo so he did that of the English physiologist. All three men, however, inspired their successors to bring about the mechaniza- tion of biology. The iatro-mechanical school adopted the prin- ciple that biological phenomena were to be investigated entirely by 'mathematical principles'. The stomach was a retort, the veins and arteries hydraulic tubes, the heart a spring, the viscera sieves and filters, the lung a bellows and the muscles and bones a system of cords, struts and pulleys. The adoption of such conceptions certainly exposed many problems for investigation by the now established mathematical and experimental methods, a particu- larly successful application being the study of the mechanics of the skeleton and muscular system by Giovanni Alfonso Borelli in his book, *On the Motion of Animals* (1680). But they were soon carried to naïve extremes which oversimplified the complexity and variety of physiological processes, especially of biochemical processes. Moreover the exhaustiveness of Cartesian mechanism entirely obliterated biological phenomena that could not be immediately reduced to them, especially the apparent purposiveness of animal behaviour (for example, in the nest-building of birds) and the whole question of the adaptation of the parts and functions of the body to each other and of the whole to the environment. These problems continued to interest naturalists like John Ray (1627–1704) and they became an important element in natural theology, proving, not only for Ray but also for physical scientists like Boyle and Newton, as the title of Ray's book expressed it, *The Wisdom of God manifested in the Works of the Creation* (1693). In physiology they prompted a return to more vitalistic explanations, but it is a tribute to the power of Descartes' theoretical genius that the question of vitalism and mechanism continued until the 20th century to be argued (sometimes unconsciously) in the philosophical terms established by him and his 17th-century critics.

4. THE EXTENSION OF MATHEMATICAL METHODS TO INSTRUMENTS AND MACHINES

As the 17th century progressed, experiment and the use of mathematics became so intimately linked that such a case as that of William Gilbert, who had carried on his experimental studies of magnetism almost without mathematics, would by the end of the century have been almost inconceivable. If causal relations such as those discovered by Gilbert remained incapable of expression in mathematical terms even by Galileo himself, it was generally believed that it was only a matter of time before the problem would be overcome and that this would depend largely on the development of more accurate instruments for measuring.

One of the instruments which Galileo did much to perfect was the clock. At the end of the 15th century the first clocks driven by a spring instead of by weights had been introduced in Nuremberg and this made possible the invention of the portable watch, as, for instance, the 'Nuremberg eggs'. The use of a spring introduced a new problem, for the force it exerted decreased as it became unwound. Various devices were designed to overcome this difficulty, the most successful being the so-called 'fusee' introduced in the middle of the 16th century by the Swiss Jacob Zech. The main principle of this device was to make the driving barrel taper gradually, so that, as the spring became unwound, the loss of force was compensated by an increase in leverage provided by making the spring act on successively wider sections of the barrel. It was still not possible, however, to get a clock that would keep accurate time over a long period. This was becoming a necessity for several purposes, but particularly for the ocean-going navigation that had been expanding since the end of the 15th century. The only practical method of determining longitude depended on the accurate comparison of the time (by the sun) on the ship with that at some fixed point on the earth's surface, for instance Greenwich. Such a clock became possible when a pendulum was introduced as a regulating mechanism. In watches a balance-spring served the same purpose. In 1582 Galileo had discovered that a pendulum swung isochronously, and later saw that this fact might be used in designing a clock. The first accurate clock was invented in 1657 by Huygens, quite independently of Galileo's suggestion. But it was not until the 18th century that the navigational problem was finally overcome, when devices were introduced to compensate for the irregular motion of a ship and for changes in temperature.

Another form of measurement in which the demands of navigation and travel led to great improvements in the 16th and 17th centuries was the method of making maps. The sensational voyages of Bartholomew Diaz round the Cape of Good Hope in 1486, of Christopher Columbus who reached America in 1492, of Vasco da Gama who reached India in 1497, and of many other sailors who searched for the North-West or the North-East passage, not only added a new world to European consciousness but also made accurate maps and methods of fixing position a fundamental necessity. The essential requirement for mapping the terrestrial globe was a linear measure of the arc of the meridian, for there were few astronomical estimations of latitude, and practically none for longitude, until the 18th century. Various improvements on medieval estimations of the degree were made during 16th and early 17th centuries, but the first accurate figure was given by the French mathematician, Jean Picard, in the second half of the 17th century. In spite of inaccurate figures for the degree, cartography improved greatly from the end of the 15th century. This was in the first place due to a renewed interest in the maps of Ptolemy's *Geography* (see Vol. I, pp. 215–16). Ptolemy had emphasized the need for the accurate fixing of position, and his maps were drawn on a complete network of parallels and meridians. In the 16th century charts were produced showing much more restricted areas than the medieval charts, and on these rhumb-lines were shown in a simplified form. The compass was used to establish the meridian line, the fact of magnetic variation with longitude being known and taken into consideration. Petrus Apianus, or Bienewitz, whose map, published in 1520, was one of the earliest to show America, in 1524 wrote a treatise on cartographical methods and in another work, *Cosmographicus Liber*, gave a list of latitudes and longitudes of many places in the known world, illustrated with maps (Plate 16). Another 16th-century cartographer, Gerard de Cremer or, as he was called, Mercator, of Louvain, in 1569 produced the well-known projection that is still in use showing the spherical earth on a two-dimensional paper. He also experimented with other kinds of projection and he took care to base his maps either on personal surveys, as in his map of Flanders, or on a critical comparison of the information collected by explorers. The same care was shown by other 16th-century cartographers such as Ortelius, who was geographer to the King of Spain, and Philip Cluvier, who published works on the historical geography of Germany and Italy.

It was in these questions that the governments and administrators of the period showed the greatest interest in science and that the most contact occurred between scientists and mathematicians from the universities, on the one hand, and practical craftsmen – instrument-makers and navigators – on the other.

Undoubtedly the most advanced institution concerned with these problems was the long-established *Casa de Contratación*, the great school of navigation at Seville which so much impressed one of the ship's masters of the English explorer Richard Chancellor. But even in a country like England, where in the mid-16th century instrument-makers and pilots were being brought over from the Continent to make up for native backwardness, private enterprise helped to achieve what lack of government patronage left undone. From the second half of the century mathematicians like Robert Recorde, John Dee, Leonard and Thomas Digges, Thomas Hood (employed by Queen Elizabeth's government), Henry Briggs (at Gresham College in London), and Thomas Harriot made efforts to improve mathematical education, especially that of the master-craftsmen, and even gave practical instruction in the new methods of navigation. John Dee, for example, was commissioned to instruct Martin Frobisher's sailing-master before he set out on his first voyage in 1576; Thomas Digges spent several months at sea demonstrating the new methods; and Thomas Harriot accompanied Sir Walter Raleigh's colonists to Virginia in 1585 as 'mathematical practitioner' and adviser.

Essential to accurate cartography on land were accurate surveying methods, and these were improved in the 16th and 17th centuries. The use of the astrolabe, quadrant and cross-staff to measure height and distance was known in the Middle Ages, and in the 16th century Tartaglia and others showed how to fix position and survey land by compass-bearing and distance. In the late 15th and early 16th centuries very accurate maps were made of Alsace, Lorraine and the Rhine Valley, notably by Waldseemüller of Strassburg (1511), in which roads were marked off in miles and a compass rose was shown. It is thought that these maps were made with a primitive theodolite known as the polimetrum. The method of triangulation, by which a whole country could be surveyed from an accurately measured base line but otherwise without direct measurement, was first published in print by the Flemish cartographer, Gemma Frisius, in 1533. In England, the first accurate maps were made by Saxton, at the end of the 16th century, and Norden, early in the 17th. An outstanding question which was not settled for some years was the adoption of a common prime meridian. English cartographers adopted Greenwich in the 17th century, but it was not generally accepted until 1925.

The first instrument for measuring temperature seems to have been invented by Galileo some time between 1592 and 1603, but three other investigators seem independently to have designed a thermometer, thermoscope, calendar-glass or weather-glass, as it was variously called, at about the same time. Galen had represented heat and cold by a numerical scale and, by the 16th century, though the senses were the only means of estimating temperature, the idea of degrees of these qualities had become a commonplace in medical and natural-philosophical literature (see above, p. 110). The scales of degrees there described, such as that of eight degrees of each quality, were among those used for the earliest thermometers. These instruments were themselves adaptations of ancient Greek inventions. Philo of Byzantium and Hero of Alexandria had both described experiments based on the expansion of air by heat (see above, p. 51, note), and Latin versions of their works existed. That of Hero's *Pneumatica* was printed twice in the 16th century. The first thermometers, which were adaptations of some of their apparatus, consisted of a glass bulb with a stem dipping into water in a vessel. Air was driven out of the bulb by heat and, on cooling, water was drawn back into the stem. The stem was marked in degrees and, as the air in the bulb contracted and expanded, the movement of the water up and down it was held to measure temperature, although, as we now understand, the water would move also with changes in atmospheric pressure.

The attribution of the first invention of this instrument to Galileo rests solely on the testimony of his contemporaries, for it is described in none of his extant works. The first published account of it was given in 1612 in a commentary on Avicenna by the physiologist Santorio Santorio, who used it for clinical purposes. A similar instrument, which seems to have been a modification of Philo's apparatus, was used a few years later by Robert Fludd to demonstrate, according to him, the cosmic effects of light and darkness and heat and cold, to indicate or predict weather conditions, and to measure temperature changes. Another type of thermometer, consisting of a tube with a sealed bulb at each end, seems to have been invented by another contemporary, the Dutchman Cornelius Drebbell (1572–1634). This instrument depended for its operation on the difference in temperature between the air in each bulb, which moved coloured water up or down the stem.

These air thermometers were used for various purposes in the 17th century, though mostly for medical purposes. J. B. van Helmont (1577–1644), for example, used a modification of the open type to take body temperature. They were very inaccurate and the open type was particularly sensitive to changes in atmospheric pressure. The French chemist Jean Rey adapted it in 1632 to form a water thermometer which measured temperature by the expansion and contraction of water instead of air; but technical difficulties prevented the construction of an accurate thermometer until the 18th century.

The desire to measure prompted the invention of an instrument which would give some idea of the weight of the atmosphere, again an instrument for which Galileo was also initially responsible. Such observations as that water would not run out of a water clock while the hole at the top was closed were usually explained, after the 13th century, either by Roger Bacon's 'continuity of universal nature' or in terms of the void (see above, pp. 54–5). Galileo did not, like the Aristotelians, regard a void as an impossibility. He produced the earliest recorded artificial vacuum by drawing a piston from the bottom of an air-tight cylinder and, like Giles of Rome, he attributed the resistance encountered to the 'force of the vacuum'. When he learnt that a pump would not lift water above 32 feet, he assumed that this was the limit of the force. He did not connect these phenomena with atmospheric weight. In 1643 it was shown, at Torricelli's suggestion, that when a long tube with one end sealed was filled with mercury and inverted with its open end under mercury in a vessel, the length of the column of mercury standing in the tube was less than that of the water raised by a pump in proportion to the greater density of mercury. The empty space above the mercury became known as the 'Torricellian vacuum', and Torricelli attributed the effect to the weight of the atmosphere. Torricelli's apparatus was adapted to form the familiar J-tube barometer. His conclusions were confirmed when, under Pascal's direction, a barometer was carried to the top of the Puy de Dôme and it was found that the height of the mercury decreased with altitude, that is with the height of atmosphere above it.

The possibility of creating a vacuum led a number of scientists during the 16th and 17th centuries to try to devise a practical steam engine. The earliest of these were driven, in fact, not by the force of expansion of steam but by atmospheric pressure operating after steam in the cylinder had been condensed, though some writers, for instance de Caus in 1615 and Branca in 1629, suggested using the turbine device described by Hero of Alexandria, a jet of steam directed on to a wheel with blades. The most important practical problem for which steam engines were suggested was the pumping of water. The problem of keeping the ever-deepening mines free from water became increasingly serious throughout the 16th and 17th centuries. Agricola in his *De Re Metallica* described several types of device used for this purpose in the early 16th century: a chain of dippers worked by a crank turned by hand; a suction pump worked by a water wheel, with a cam to work the piston and with pipes made of hollow tree trunks clamped with iron bands (Plate 17); a force pump worked by a crank; and a rag-and-chain device in which the buckets were replaced by balls of horsehair and the motive power was provided by men walking a treadmill or a horse driving a whim. Pumps were needed also to provide water for fountains, and for town supplies. Augsburg was supplied with water by a series of Archimedean screws turned by a driving shaft which raised the water to the tops of towers, from which it was distributed in pipes; London was supplied after 1582 by a force pump driven by a tide-wheel set up near London Bridge by the German engineer, Peter Morice, and later by other horse-driven pumps; and pumps were used to supply Paris and other towns, and to work the fountains at Versailles and Toledo. As early as the mid-16th century Cardano had discussed methods of producing a vacuum by condensing steam, and in 1560 G. B. della Porta (1536–1605) suggested using a device based on this principle for raising water. This suggestion was put forward again in 1663 by the Marquis of Worcester. The earliest steam engine operating with a cylinder and piston was designed by the French engineer Denis Papin, who had worked with Boyle and invented the condensing pump and also the pressure cooker, or 'steam digester', as he called it, with a safety-valve. He also designed a steam-driven carriage. A practical steam engine based on the condensation of steam was patented in 1698 by Thomas Savery; it was used in at least one mine and to supply water to several country houses. Hearing of this, Papin in 1707 designed a high-pressure boiler with an enclosed fire-box, and a steam-boat propelled by paddle-wheels. It was his design that Thomas Newcomen a little later successfully adapted for his engine worked by atmospheric pressure; even James Watt's engines were still primarily atmospheric. Towards the end of the 18th century engines were invented which were driven by the expansive force of steam at high pressure.

The Torricellian vacuum was taken as a final refutation of Aristotle's arguments against the existence of void which, according to some of his followers, 'nature abhorred'. The arguments against the void, drawn from the Aristotelian law of motion, had already been disposed of by Galileo. But Aristotle himself had sometimes confused arguments against the existence of void, in the sense of 'non-being', with physical arguments against, for instance, the absence of a resisting medium. Many of his 17th-century critics did the same. The Torricellian vacuum was not an ontological void such as Descartes, among others, could not accept. It was a space which, at least theoretically, contained no air or similar matter. Indeed, although later physicists were not so sensitive to metaphysical niceties as Descartes, they found it necessary to postulate a *plenum* of some sort, and this continued to play a variety of physical roles down to the 20th century. Torricelli showed that light was transmitted through a vacuum and, beginning with Gilbert's effluvia, 17th-century physicists filled up the void with a medium, the ether, capable of propagating all the known influences such as gravity, magnetism and light. Descartes himself attempted to explain magnetism by vortices which, like Averroës' *species magnetica*, entered by one pole of the magnet and left by the other. He held that these acted on iron because the resistance of its particles to the flow drew it to the magnet (Plate 18). Non-magnetizable substances did not offer such resistance.

Instruments designed for closer observation as well as for more accurate measurement were also constructed during the 17th century, the most important being the telescope and the compound microscope. The propagation of light was still explained by most 16th-century opticians in terms of the 'species' theory, which they related to their knowledge of geometrical optics. Notable attempts were made by Leonardo da Vinci, Maurolyco and Porta to give an account of the operation of the eye by means of an improved knowledge of lenses and the comparison of the eye with a *camera obscura*. But all three still believed that the eye's lens was the sensitive organ and that the image had to be erect and correctly orientated. The retina was first recognized as the sensitive organ by the anatomist Felix Plater (1536–1614). Realdo Colombo and Hieronymo Fabrizio drew the lens in the front part of the eye and not, as had been done previously, in the middle. Kepler in his commentary on Witelo (1604) first demonstrated that the rays focused by the cornea and lens formed a real inverted image on the retina.

A convenient method of isolating stars by observing them through a tube had already been introduced by the Arabs and, with the spread of spectacles, the lens-grinding industry had developed in a number of centres. Pioneer work on combinations of mirrors and perhaps lenses was carried out, apparently under the inspiration of Roger Bacon, by the English mathematicians Leonard Digges (d. *c.* 1571) and his son Thomas, but they set up their apparatus on frames, without tubes. It seems that some sort of telescope with lenses in a tube was constructed in Italy about 1590. In any case it is recorded that a Dutch spectacle-maker named Janssen in 1604 copied an Italian model marked with that date, and the record gives point to Porta's obscure account in

1589 of a combination of convex and concave lenses. For some reason Galileo only heard of the Dutch instruments, and he then constructed his telescope and compound microscope from his scientific knowledge of refraction.* He did not fully understand this phenomenon and Kepler, in his *Dioptrica* (1611), gave a more intelligible theory. Galileo's combination of concave and convex lenses was replaced by combinations of convex lenses, and in the course of time rules were worked out for determining focal lengths and apertures. The true law of refraction, that the ratio of the sines of the angles of incidence and refraction is a constant depending on the media concerned, was discovered in 1601 by Harriot and rediscovered a few years before 1626 by Willibrord Snell (1591–1626). The law was also formulated, probably in the first instance independently, by Descartes, who gave it its first publication in his *Dioptrique* in 1637.

Descartes attempted to conceive of the physical nature of light in a more strictly mathematical form than his predecessors. In accordance with his own mechanical principles he held that light consisted of particles of the *plenum* and that it was transmitted instantaneously by mechanical pressure from one particle to the next. Colour he held to depend on the different rotary velocities of the particles. When giving 'Snell's law' he presented it as a deduction from this conception of the mechanical nature of light, and in his *Météores* (1637) he tried to use this law to explain the two phenomena exhibited by the rainbow, the bright circular bow and the colours. Theodoric of Freiberg's diagrams of the formation of the primary and secondary bows, showing the essential fact of the internal reflection of the sunlight in the raindrops, had been published in Erfurt in 1514, and Antonio de

---

* When the Frenchman, Jean Tarde, called on Galileo in 1614, he said, 'Galileo told me that the tube of a telescope for observing the stars is no more than 2 feet in length; but to see objects well, which are very near, and which on account of their small size are hardly visible to the naked eye, the tube must be two or three times longer. He tells me that with this long tube he has seen flies which look as big as a lamb, are covered all over with hair, and have very pointed nails, by means of which they keep themselves up and walk on glass, although hanging feet upwards.' Galileo, *Opere*, Ed. Naz., Vol. 19, p. 589.

Dominis had given a somewhat inaccurate report of a similar explanation in 1611 (see Vol. I, pp. 122–4). This was almost certainly known to Descartes, if he did not know Theodoric's own diagrams. But Descartes' knowledge of the law of refraction and of optics in general made his treatment of the subject altogether superior to that of his predecessors. He not only gave a complete account of the refraction and reflection of the rays in the water drops causing the rainbow, but also showed that those coming to the eye at an angle of about 41 degrees from their original direction from the sun were much more dense than those coming from other directions and so produced the primary bow. He definitely associated the colours with differential refrangibility, which he explained by his theory of rotating particles. Some time later Johann Marcus Marci of Kronland (1595–1667) showed that rays of a given colour were dispersed no further by a second prism. Neither Descartes nor Marcus was able to produce an adequate theory of colour, which had to wait until their experiments with prisms had been repeated and extended by Newton, with again an altogether superior theoretical understanding of the question. This 17th-century work of Descartes, Newton, Hooke, Huygens and others on light made it possible for serviceable microscopes and telescopes to be constructed, but the usefulness of both these instruments was somewhat reduced by the failure to overcome the chromatic aberration, which became serious with powerful lenses. With telescopes the problem of getting a large magnification was overcome by using concave mirrors instead of lenses, but a really powerful microscope became possible only in the 19th century.

5. CHEMISTRY

In chemistry, such progress as was achieved by the middle of the 17th century was the result rather of experiment and observation alone than of interpretation of facts in terms of mathematical generalizations. The expansion of alchemy and the pursuit of more strictly practical ends, such as painting and mining, had led, during the 14th and 15th centuries, to a fairly wide familiarity with ordinary chemical apparatus. Although this had included the balance, this instrument had not, as Cusa had suggested, been combined with *inventio*, or discovery, and the 'art of latitudes' for the development of a quantitative chemical theory. Mineral drugs had begun to come into pharmaceutical and medical practice, and through an extended study of them chemistry was given a marked impetus during the early decades of the 16th century by the bizarre Philippus Aureolus Theophrastus Bombastus von Hohenheim, or Paracelsus (1493–1541). Paracelsus was an accomplished experimenter and added a few facts to chemical knowledge, for instance the observation that while the vitriols were derived from a metal the alums were derived from an 'earth' (metallic oxide). He also contributed the *tria prima*, sulphur, mercury and salt, to chemical theory. The Arabs had held that sulphur and mercury were the chief constituents of metals, but Paracelsus made sulphur (fire, the inflammable principle), mercury (air, the fusible and volatile principle) and salt (earth, the incombustible and non-volatile principle) the immediate constituents of all material substances. The ultimate constituents of matter, of which these *tria prima* were themselves composed, were the four Aristotelian elements. He illustrated his theory by burning wood, which gave off flames and fumes and left ash.

The chief influence that Paracelsus had on chemistry was through his assertion that its main business was not with the transmutation of metals, though he held this to be possible, but with the preparation and purification of chemical substances for use as drugs. After him, chemistry became an essential part of medical training, and for nearly a century doctors were divided into paracelsists (or 'spagyrists') and herbalists, who kept to the old herbal drugs. The former were often very incautious in their remedies but, however disastrous for the patient, the contribution of iatro-chemistry (medical chemistry) to chemistry itself is well illustrated by the clear and systematic account of techniques and substances given in the *Alchymia* (1597) of Andreas Libavius (1540–1661). Like the practical manuals of Vanoccio Biringuccio (1480–1539), Agricola and Bernard Palissy (1510–c.

1590) in other aspects of the subject, Libavius' book shows the progress of the 16th century in the collection of fact.

The first serious improvements in method, aimed at the chemical analysis of the nature of matter, were made by Johann Baptista van Helmont. After graduating in medicine at Louvain, van Helmont made a wealthy marriage and settled down to the charitable practice of his profession and research in his laboratory. His writings, which he left unpublished, were collected after his death and published by his son under the title *Ortus Medicinæ*. An English translation, *Oriatrike or Physick Refined*, appeared in 1662. Van Helmont's empiricism showed the influence both of the practical chemists who had preceded him and, in spite of his attacks on the schools, of nominalism and Augustinian-Platonism. He held that the sources of human knowledge were both Divine illumination and sensory experience. 'The meanes of obtaining Sciences, are onely to pray, seek and knock,' he said in the tract *Logica Inutilis* which forms chapter 6 of the *Oriatrike*. In the study of nature there was no true *inventio*, or discovery, but 'by bare observation' of concrete and measurable objects.

For when anyone sheweth me *lapis Calaminaris*, the preparing of *Cadmia* or *Brasse Oare*, the content of, or what is contained in Copper, the mixture and uses of *Aurichalcum*, or *Copper* and *Gold*, which things I knew not before, he teacheth, demonstrateth, and gives the knowledge of that, which before there was ignorance of.

But the logic of the school philosophers could not lead to such discoveries. By itself 'Logical invention is a meer re-taking of that which was known before.' After observation had been made, the investigator was led by *ratio*, that is formal logic and mathematics, to a knowledge of the active principles, which in effect were analogous to the Aristotelian substantial form, and were the source of the observed behaviour. But, van Helmont said, unless such reasoning was accompanied by intuition or illumination its conclusions were always uncertain.

Van Helmont made this theory of knowledge the basis of a suggested reform of education. 'Certainly I could wish,' he said in the *Oriatrike,* chapter 7, referring to the schools' teaching of Aristotle and Galen,

that in so short a space of life, the Spring of young men, might not be hereafter seasoned with such trifles, and no longer with lying Sophistry. Indeed they should learn in that unprofitable three years space, and in the whole seven years, Arithmetick, the Science Mathematical, the Elements of Euclide, and then Geographie, with the circumstances of Seas, Rivers, Springs, Mountains, Provinces, and Minerals. And likewise, the properties, and Customs of Nations, Waters, Plants, living Creatures, Minerals, and places. Moreover, the use of the Ring, and of the Astrolabe. And then, let them come to the Study of Nature, let them learn to know and separate the first Beginnings of Bodies ... And all those things, not indeed by a naked description of discourse, but by handicraft demonstration of the fire. For truly, nature measureth her works by distilling, moystening, drying, calcining, resolving, plainly by the same meanes, whereby glasses do accomplish those same operations. And so the Artificer by changing the operations of nature, obtains the properties and knowledge of the same.

Van Helmont held that there were two 'first beginnings' of bodies. He had performed Cusa's experiment with the willow (see above, p. 111), and this convinced him that the ultimate inert constituent of material substances was water. The active principle which disposed the water and constructed the specific concrete thing was a 'ferment or seminal beginning', which was generated in matter by the Divine light (or celestial influence). This last brought the 'archeus', the efficient cause enabling the ferment to construct the 'seed' which developed into a stone, metal, plant or animal. 'For,' as he said in the *Oriatrike,* chapter 4,

the seminal efficient cause containeth the Types or Patterns or things to be done by itself, the figure, motions, houre, respects, inclinations, fitnesses, equalizings, proportions, alienation, defect, and whatsoever falls in under the succession of dayes, as well in the business of generation, as of government.

Such bodies were constructed in accordance with the 'idea' of the archeus. In the generation of animals the *archeus faber* of the male seed epigenetically constructed the embryo out of the materials provided by the female. Seeds of organic origin were not, however, indispensable for generation, and perfect animals might be produced when the archeus acted on a suitable ferment. Indeed, van Helmont held that the parent was only equivocally the efficient cause of the offspring. It was the 'natural occasion' on which the seed was produced, but the effective efficient cause was God. This theory was similar to that of the 'occasionalists' (see below, pp. 315–16). He held that there were only two causes operating in natural events, the material and the efficient.

Van Helmont held that there were specific ferments and archei in the stomach, liver and other parts of the body; these controlled their functions, on which his views were in general Galenic. He held also that a disease was an alien entity imposing its way of life, or archeus, on that of the patient; and in developing this idea he became a pioneer in ætiology and morbid anatomy. By putting into practice the doctrine that knowledge of the ferments was to be derived from observation of their material effects, he was able also to assign specific functions to many of the Galenic and other principles. He demonstrated the acid digestion, or 'fermentation', in the stomach, and its neutralization by the bile. These, he said, were the first two fermentations of the food passing through the body. The third took place in the mesentry; the fourth was in the heart, where the red blood became more yellow by the addition of vital spirits; the fifth was the conversion of arterial blood into vital spirit, mainly in the brain; the sixth was the elaboration of the nutritive principle in each part of the body from the blood. Van Helmont also anticipated something like the principle of the specific energy of nerves when he said that vital spirit conveyed to the tongue accounted for the perception of taste, but would not cause taste in the finger.

In pure chemistry, van Helmont made systematic use of the balance and demonstrated the conservation of matter, which, he held, secondary causes could not destroy. He showed that if a certain weight of silica were converted into waterglass and the latter were then treated with acid, the precipitated silicic acid would on ignition yield the same weight of silica as that originally taken. He showed also that metals dissolved in the three main mineral acids could be recovered again, and realized that when one metal precipitated another from a solution of a salt this did not, as Paracelsus had thought, imply transmutation. Perhaps his most important work was on gases. He himself coined the name 'gas' from the Greek *chaos*. Several medieval and later writers had recognized the existence of aqueous and earthy 'exhalations' as well as air, but van Helmont was the first to make a scientific study of different kinds of gases. Here his research was made much more difficult by the lack of a convenient apparatus for collecting gases. The different kinds of gas he mentioned included a *gas carbonum* given off by burning charcoal (usually carbon dioxide but also carbon monoxide); a *gas sylvester* given off by fermenting wine, by spa water, and by treating a carbonate with acetic acid, and also found in certain caves, which put out a flame (carbon dioxide); a red poisonous gas, which he also called *gas sylvester*, given off when aqua fortis acted on such metals as silver (nitric oxide); and an inflammable *gas pingue* formed by dry distillation of organic matter (a mixture of hydrogen, methane and carbon monoxide). Van Helmont also took an interest in respiration, of which he maintained the purpose to be not, as Galen had said, to cool but to maintain animal heat; this it did by a ferment in the left ventricle which changed the arterial blood into vital spirit.

Several other chemists made experiments with gases during the early decades of the 17th century in connexion with the phenomena of combustion. According to the accepted theory, combustion involved the decomposition of compound substances with the loss of the inflammable 'oily' principle present in the 'sulphur'. Burning would thus result in a decrease in weight. Several observations were made, however, which led to the development of new ideas on this subject. The experiment of 'enclosed combustion', in which a candle was lighted in a glass upturned in a basin of water, had been desbribed by Philo (see above, p. 51, note), and Francis Bacon referred to it as a common experiment. It was repeated by Robert Fludd (1617), and when the water rose as the air was consumed he described the latter as 'nourishing' the flame. It had also been known by both Arab and 16th-century chemists that during calcination metals increased in weight. In 1630 Jean Rey gave reasons for believing that the definite and limited 'augmentation' in weight, which he observed in the calx of lead and tin, could have come only from the air, which he said mixed with the calx and became attached to its most minute particles. He maintained, further, that all elements, including fire, had weight and that this weight was conserved throughout chemical changes. These facts and ideas were clearly incompatible with the theory of the 'oily' principle, and when this principle was developed as 'phlogiston' it had to be considered as having negative weight. But it was not until towards the end of the 18th century that combustion was firmly associated with oxidation, when it became the central question of the Chemical Revolution initiated by Lavoisier and his contemporaries.

The universal mechanism which accompanied the successes of mathematical physics entered chemistry through the development of the atomic theory. Such natural philosophers as Bruno, who had argued for the actual existence of natural or physical *minima*, continued the scholastic discussions of this problem; and it was given prominence by Francis Bacon who, though he changed his mind later, began with a favourable opinion of atoms and also said that heat was a condition produced by the vibration of corpuscles. Galileo said that change of substance 'may happen by a simple transposition of parts'. The first application of the atomic theory to chemistry was made by the Dutchman Daniel Sennert (1572–1637). Sennert maintained that substances subject to generation and corruption must be composed of simple bodies, from which they arose and into which they were resolved. These simple bodies were physical and not merely mathematical *minima*, and were in fact atoms. He postulated four different kinds of atoms, corresponding to each of the four Aristotelian elements, and elements of the second order (*prima mixta*) to which the Aristotelian elements gave rise when combined. He held that atoms, for example of gold in solution in acid and of mercury in sublimation, retained their individuality in combination, so that the original substances could be regained from compounds. Similar ideas were expressed by Joachim Jung (1587–1657), through whom they later became known to Robert Boyle (1627–91).

Contributions to the atomic theory were made also by Descartes, for although he did not believe in indivisible physical *minima*, he tried to extend his mechanistic principles to chemistry by attributing the properties of various substances to the geometrical shapes of their constituent earthy particles. For instance, he supposed the particles of corrosive substances such as acids to be like sharp-pointed blades, while those of oils were branching and flexible. These ideas were used later by John Mayow (1643–79), and they became familiar to chemists through the *Cours de Chymie* (1675) by Nicholas Lémery (1645–1715). Another geometer, Gassendi, popularized the atoms of Epicurus (1649), maintaining, however, that they had not been in existence since eternity but had been created with their characteristic powers by God. He based his belief in the existence of void on Torricelli's experiments and, like Descartes, connected chemical properties with the shapes of the atoms. He also attributed combination into *moleculæ* or *corpuscula* to mechanisms such as hooks and eyes. Gassendi's system was the subject of an English work by Walter Charleton (1654), physician to Charles II and an early fellow of the Royal Society. The microscope had lent an interest to discovering the side of atoms and Charleton argued, from such phenomena as volatilization and solution, that the smallest discernible microscopic particle contained ten hundred thousand million invisible particles. Through Charleton the atomic theory became well known in mid-17th-century England. When it was adopted by Boyle and Newton, the empirical conceptions of van Helmont and the earlier practical chemists were transformed in accordance with mechanical principles, and chemistry, like physics, finally set out on its course of being reduced to a mathematical science. After the discovery of 'combination weights' and Dalton's generalization of the results in his atomic theory early in the 19th century, the fulfilment of that process became inevitable.

6. BOTANY

Botanical studies up to the middle of the 17th century were confined principally to the business of collecting and classifying facts, and were left almost untouched by the mathematical revolution in scientific thought. In fact, even in the 20th century, botany, like many other branches of biology, remains singularly intractable to mathematical treatment. The theory in which the animate world eventually found a universal explanation, the theory of organic evolution, was based on logical rather than mathematical abstractions.

The dual interest of medical men in descriptive botany and anatomy, which continued into the 16th century, brought it about that these were the first aspects of biology to develop and that this was almost entirely the work of medical men. It was customary in some places, such as Montpellier, to take up botany in summer and anatomy in winter. The first books on scientific botany to be printed were nearly all herbals. The best of these, such as the *Latin Herbarius* (1484), which had probably already existed in manuscript, and the *German Herbarius* (1485), besides being compilations from classical, Arabic and medieval Latin authors, also included descriptions and illustrations of local, for instance German, plants. Rufinus, the best of the known medieval Latin herbalists, seems, however, to have been forgotten.

Besides the medical interest in identifying plants for use as drugs, 16th-century doctors shared with lexicographers the humanist interest in identifying the plants mentioned in the recently printed Latin editions of Pliny (1469), Aristotle (1476), Dioscorides (1478), and Theophrastus (1483). More than one humanist naturalist, of whom the Swiss Conrad Gesner (1516–65) is a typical example, began by trying to find and identify in his own country, for purposes of textural criticism, the plants and animals mentioned by classical authors; and out of this developed an interest in local fauna and flora for their own sake. The extraordinary interest which animals, plants and rocks were arousing among such people by the middle of the 16th century is shown by the enormous correspondence on the subject, with

descriptions of local expeditions and the transmission of speci-
mens, drawings and descriptions, carried on by Gesner and other
naturalists. It was soon realized, as indeed Albertus Magnus and
Rufinus had known well, that there were other creatures in existence besides those known to the ancients. The classical limitations were finally destroyed by the new fauna, flora, foods and
drugs coming to Europe from the New World and the East.
Plants and animals were then described and drawn for their own
sakes and called by their common vernacular names to a large
extent without reference to the classics.

The first result of this 16th-century botanical activity, which
was greatest in Germany, the Netherlands, southern France and
Italy, was to increase the number of individual plants known.
Lists of local flora and fauna were drawn up for various districts.
Botanical gardens, which had long been kept by monasteries and,
from the 14th century, had been planted by some medical
schools, were established in various further university towns
such as Padua (1545), Bologna (1567) and Leyden (1577). The last
two were presided over respectively by Aldrovandi and Cesalpino, and by de l'Ecluse. Others were established later at Oxford
(1622), Paris (1636), and other places. The practice of preserving
dried plants, 'dry gardens', which began in Italy, also allowed
botany to go on in the winter months. At the same time the
Portuguese herbalist Garcia da Orta published a book on Indian
plants at Goa (1563), and the Spaniard Nicolas Monardes published the first descriptions of 'el tabaco' and other American
plants (1569–71).

In the northern school, whose interest was purely floristic, a
continuous development of botanical ideas may be traced from
the four 'fathers' of German botany to Gaspard Bauhin. For all
the members of this school the primary intention was simply to
make it possible to identify individual wild and cultivated
plants and distinguish them from those resembling them. This led
to concentration on accurate illustrations and descriptions. The
illustrations, which in the herbal of Otto Brunfels (1530), the first
of the German fathers, were made by Hans Weiditz, an artist of
the school of Albrecht Dürer (1471–1528), were at first greatly

superior to the pedantic traditional descriptions. With Jerome Bock (1539) and Valerius Cordus (1561) the latter began gradually to improve. The object of both illustrations and descriptions was simply to depict the most easily recognizable aspects of external appearance, such as the form and disposition of roots and branches, the shape of the leaves, and the colour and shape of the flowers (Plate 19). There was no interest in the comparative morphology of the parts. For instance, the glossary of terms given by the third German father, Leonard Fuchs (1542), referred almost entirely to such general characters; and the earlier attempts at classification, for instance those made by Bock and the Netherlander Rembert Dodœns (1552), were based for the most part on artificial characteristics such as edibility, odour or medicinal properties.

Since the task of describing individual forms necessarily involved distinguishing them from near relations, some appreciation of 'natural' affinity was inevitable. Gesner, whose botanical work unfortunately was not published until long after his death and thus apparently had little or no influence on his contemporaries, distinguished different species of a given genus, for example Gentian, and also seems to have been the first to draw attention to the flower and fruit as diagnostic characters. Other writers, such as Dodœns and Charles de l'Ecluse (1576), though primarily interested simply in giving order to their work, placed together within each artificial division plants belonging to what are now recognized as natural groups. This practice had been carried even further by Mathias de Lobel (1571), like de l'Ecluse a graduate of Montpellier, who had based his classification mainly on leaf structure. It reached its final stage in Gaspard Bauhin (1560–1624), professor of anatomy at Basel. Bauhin's descriptions are precise and diagnostic, as may be seen from that of the beet, which he called *Beta Cretica semine aculeato*, given in his *Pro-domus Theatri Botanici* (1620).

From a short tapering root, by no means fibrous, spring several stalks about 18 inches long: they straggle over the ground, and are cylindrical in shape and furrowed, becoming gradually white near the root with a slight coating of down, and spreading out into little sprays. The plant has but few leaves, similar to those of *Beta nigra*, except that they are smaller, and supplied with long petioles. The flowers are small, and of a greenish yellow. The fruits one can see growing in large numbers close by the root, and from that point they spread along the stalk, at almost every leaf. They are rough and tubercled and separate into three reflexed points. In their cavity, one grain of the shape of an *Adonis* seed is contained; it is slightly rounded and ends in a point, and is covered with a double layer of reddish membrane, the inner one enclosing a white, farinaceous core.

The number of plants described by Bauhin had increased to about 6,000, as compared with the 500 or so given by Fuchs. He systematically used a binomial nomenclature, though he did not invent this system for it had occurred in a 15th-century manuscript of the *Circa Instans*. In his *Pinax Theatri Botanici* (1623), he gave an exhaustive account of the synonyms used by earlier botanists. In enumerating the plants described, he proceeded, as de Lobel had done, from supposedly less perfect forms, such as grasses and most of the Liliaceæ, through dicotyledonous herbs to shrubs and trees. Both he and de Lobel thus made a practical distinction between monocotyledons and dicotyledons and, as some of their predecessors had done in varying degrees, put together plants belonging to such families as the Cruciferæ, Umbelliferæ, Papilionaceæ, Labiateæ, Compositæ, etc. Such grouping was, however, based entirely on an instinctive appreciation of likeness in form and habit. There was no conscious recognition of comparative morphology, and no system was set out based on the understanding and analysis of morphological features. The main effort of the northern school was in fact towards the accumulation of more and more empirical descriptions, until by the end of the 17th century John Ray (1682) was able to cite 18,000 species.

The man who made it possible to reduce this mass of information to some sort of rational order was the Italian Andrea Cesalpino, professor of medicine first in Pisa and then in Rome, where he was also physician to Pope Clement VIII. Cesalpino brought to the study of botany not only the floristic knowledge of herbalists, but also an interest in the detailed morphology of the separate parts of the plant and an Aristotelian mind capable of forming generalizations. He based his attempt, set out in the *De Plantis* (1583), to explain the 'real' or 'substantial' affinities between plants on the Aristotelian principle that the final cause of vegetative activity was nutrition, of which the reproduction of the species was simply an extension. In his day the role played by the leaves in nutrition was still unknown, and the nutritive materials were supposed to be absorbed by the roots from the soil and carried by the veins up the stem to produce the fruit. The centre of vital heat, corresponding to the heart in animals, was the pith, and Cesalpino held that it was also from the pith that the seeds were produced. The co-operation of the male and female parts of flowers in reproduction had not yet been discovered, and he supposed that the flower was simply a system of protecting envelopes round the seed, comparable to the fœtal membranes of animals. On these principles he divided plants first, according to the nature of the stem conducting the nutritive materials, into woody and herbaceous plants and again, within these groups, according to the organs of fructification. Here he began with plants such as fungi, which he held had no seed but were spontaneously generated from decaying substances, and passed through others such as ferns, which propagated by a kind of 'wool', to plants with true seeds. He then classified these last according to the number, position and shape of the parts of the fruit, with sub-divisions based on root, stem and leaf. Characteristics such as colour, odour, taste or medicinal properties he considered to be mere accidents.

Cesalpino's attempt to deduce a 'natural' classification from the principles he had assumed was in result deplorable. The distinction between monocotyledons and dicotyledons was less clear than with the herbalists and, out of the 15 classes he made, only one, the Umbelliferæ, corresponds to what would now be recognized as a natural group. Nevertheless, his system was based on considerable knowledge and clear principles which, however wrong, were the first to be introduced by botanists of the time into the study of plants. His followers had something to work on. The first to criticize and develop Cesalpino's ideas was Joachim Jung (1587–1657), a German professor of medicine who probably came across his ideas while studying at Padua. Jung accepted the idea that nutrition was the fundamental vegetative function and, like Cesalpino, based his idea of species on reproduction. He made what was then a great advance by discussing morphology as far as possible in independence of physiological questions.

Theophrastus, whose *Historia Plantarum* had been translated into Latin by Theodore of Gaza (1483), had given morphological descriptions of the external parts of plants from roots to fruit. He had also set forth the 'homology' of the perianth members of flowers, watched the development of seeds, and to some extent distinguished between monocotyledons and dicotyledons. His interests had been by no means confined to morphology. He had made an attempt to understand the relation between structure and function, habits and geographical distribution, had described the fertilization of the date palm, and had tried to understand the caprification of the fig, though the flowers were distinguished only by Valerius Cordus. Theophrastus had also established the first rudiments of plant nomenclature, and there was practically no further development in the subject until similar morphological descriptions and distinctions were made by Jung.

Jung's precise definitions of the parts of plants, for which he made use of the logical refinements developed by the later scholastics and of his own mathematical gifts, were the foundation of subsequent comparative morphology. For instance, he defined the stem as that upper part of the plant above the root which stretched upwards in such a way that back, front and sides could not be distinguished, while in a leaf the bounding surfaces of the third dimension (apart from the length and breadth) in which it was extended from its point of origin were different from one another. The outer and inner surfaces of a leaf were thus differently organized and this, as well as the fact that they fell off in autumn, enabled compound leaves to be distinguished from branches. Botanists were not yet ready to follow this lead, and neither Jung nor Cesalpino had much effect on their contemporaries, who continued to devote their energies to empirical descriptions. It was only at the end of the 17th century that botanists once more recognized the need for a 'natural' system of classification and attempted to base it on comparative morphology. The culmination of their effort was the system of Linnæus (1707–78), who acknowledged his debt to both Cesalpino and Jung. When the 'natural' classification came itself to call for an explanation this was supplied by the theory of organic evolution.

7. ANATOMY AND COMPARATIVE ANIMAL MORPHOLOGY AND EMBRYOLOGY

The great advances made in anatomy and zoology during the 16th and early 17th centuries were, like those in botany, due simply to a new precision of observation and remained largely untouched by mathematics. Just as 16th-century botany began with the object of identifying medicinally useful plants, so anatomy began with such aspects as would facilitate the work of surgeons and artists. What the practical needs of the surgeon chiefly required were good topographical descriptions; comparative morphology had little interest for him. The painters and sculptors of whom several, such as Andrea Verrocchio (1435–88), Andrea Mantegna (d. 1516), Leonardo da Vinci, Dürer, Michelangelo (1475–1564) and Raphæl (1483–1520), are known to have used the scalpel, required little more than surface anatomy and a knowledge of bones and muscles. As the century went on, however, a greater practical interest was taken both in functional questions and in the structure and habits of animals. In both developments by no means the least important factor was the brilliant revolution brought about by the artists themselves in anatomical illustration.

The artist who has left most evidence of his anatomical exercises is Leonardo da Vinci and, as in mechanics, his researches went far beyond the practical needs of his craft. He even planned a text-book of anatomy in collaboration with the Pavian professor Marcantonio della Torre (c. 1483–1512), who died before the book was written. Leonardo was guided by earlier text-books and repeated some of the old mistakes, such as drawing the lens in the centre of the eye (cf. Plate 20). His claim always to have followed experience may be accepted in the same spirit as the same claim made by many of his predecessors. He made several original observations on both human and comparative anatomy, and carried out physiological experiments which were often fruitful and always ingenious. He was one of the first to make use of serial sections. The animals which he mentioned as the subjects of his researches include *Gordius*, moths, flies, fish, frog, crocodile, birds, horse, ox, sheep, bear, lion, dog, cat, bat, and monkey. His best figures were of bones and muscles, those of the hand and shoulder being clear and substantially accurate. Others exhibited the action of muscles. He made models with bones and copper wire, and pointed out that the power of the biceps brachii depended on the position of its insertion with respect to the hand. He compared the limbs of man and horse, showing that the latter moved on the tips of its phalanges. He studied the wing and foot of the bird, the mechanics of flight, and the operation of the diaphragm in breathing and defæcation. He studied the heart and blood vessels. He also made good drawings of the placenta of the cow, but was uncertain whether the maternal and fœtal blood streams were connected or not. One of his most ingenious feats was to make wax casts of the ventricles of the brain. He also carried out experiments on the spinal cord of the frog, and concluded that this organ was the 'centre of life'.

Leonardo made a further contribution to biology, as well as to geology, when he used inland shells to support Albert of Saxony's theory of the formation of mountains (Vol. I, pp. 138–9). 'Why,' he asked, 'do we find the bones of great fishes and oysters and corals and various other shells and sea-snails on the high summits of mountains by the sea, just as we find them in low seas?' *

There had been a continuous interest in local geology in Italy since the 13th century, and in his speculations on geology Leonardo made use of his own observations on the sea coast, the

* J. P. Richter, *The Literary Works of Leonardo da Vinci*, 2nd ed., Oxford, 1939, vol. 2, p. 175.

Alps and its streams, and Tuscan rivers such as the Arno. He rejected the theories that fossils were not the remains of living things but were accidents or 'sports' of nature or had been spontaneously produced by astral influence, and that they were organic remains which had been transported from elsewhere by the Flood. He accepted instead Avicenna's theory of fossil formation which he had learnt from Albertus Magnus. He then maintained that the arrangement of shells in strata, with gregarious forms such as oysters and mussels in groups and solitary forms apart just as they were found living on the seashore, and with crabs' claws, shells with those of other species fastened to them, and bones and teeth of fish all mixed up together, suggested that fossils were the remains of animals which had formerly lived in the same place just as contemporary marine animals did. The mountains on which the shells were found had formerly formed the sea floor, which had been, and was still being, gradually raised by the deposit of river mud.

The shells, oysters and other similar animals which originate in sea-mud, bear witness to the changes of the earth round the centre of our elements. This is proved thus: Great rivers always run turbid, owing to the earth, which is stirred by the friction of their waters at the bottom and on their shores; and this wearing disturbs the face of the strata made by the layers of shells, which lie on the surface of the marine mud, and which were produced there when the salt waters covered them; and these strata were covered again from time to time with mud of various thicknesses, or carried down to the sea by the rivers and floods of more or less extent; and thus these shells remained walled in and dead underneath these layers of mud raised to such a height that they came up from the bottom to the air. At the present time these bottoms are so high that they form hills or high mountains, and the rivers, which wear away the sides of these mountains, uncover the strata of these shells, and thus the softened side of the earth continually rises and the antipodes sink closer to the centre of the earth, and the ancient bottoms of the sea have become mountain ridges.*

The surgical developments of the 15th century, which received fresh impetus from the printing of *De Medicina* of Celsus in 1478,

* Richter, vol. 2, pp. 146–47.

first issued an anatomical discovery with Alexander Achillini's (1463–1512) description, in his commentary on Mondino, of 'Wharton's duct', of the entry of the bile duct into the duodenum, and of the hammer and anvil bones of the middle ear. The clear influence of naturalistic art on anatomical illustration is first seen in the Italian work, *Fasciculo di Medicina* (1493), while Berengario da Carpi (d. 1550), professor of surgery in Bologna, was the first to print figures illustrating his text. In his commentary on Mondino (1521), Berengario also described a number of original observations. He demonstrated experimentally that the kidney is not a sieve, for when injected with hot water from a syringe it merely swelled up and no water passed through. He showed in a similar way that the bladder of a nine months' unborn child had no opening other than the urinary pores. He also denied the existence of the *rete mirabile* in man, gave the first clear accounts of the vermiform appendix, the thymus gland and other structures, had some idea of the action of the cardiac valves, and coined the term *vas deferens*. Another surgeon of the same period who had a good practical knowledge of anatomy was Nicholas Massa, who published a work on the subject in 1536. The first to publish illustrations showing whole venous, arterial, nervous or other systems (1545) was Charles Estienne (1503–64), of the well-known family of French humanist printers. He also traced the blood vessels into the substance of the bone, noted the valves in the veins, and studied the vascular system by injecting the vessels with air. Another work which illustrates the advances in anatomy made during the early decades of the 16th century is the tract published by Giambattista Canano (1515–79) in 1541, in which he showed each muscle separately in its relations with the bones.

Besides these improvements in knowledge of anatomy, a number of purely empirical advances were made in practical surgery in the 16th century. One of the greatest problems for an army surgeon was how to treat gunshot wounds. At first these were believed to be poisonous and were treated by scalding with oil of elders with terrible results. One of the first doctors to abandon this practice was Ambroise Paré (1510–90), who described in his fascinating *Voyages en Divers Lieux* how he had had so many men to treat after the attack on Turin, in 1537, when he was in the service of King Francis I of France, that he ran out of oil. Next morning he was amazed to find that the men who had been left untreated were much better than those whose wounds had been scalded with oil, and thereafter he gave up this practice. Paré also gave a good account of the treatment of fractures and disclocations, and of herniotomy and other operations. Surgery in northern Europe was still largely in the hands of comparatively uneducated barbers and cutters, though some of these showed considerable skill. The itinerant lithotomist Pierre Franco, for example, was the first to perform suprapubic lithotomy for removing stone in the bladder. In Italy, surgery was in the hands of anatomists with a university training, like Vesalius and Hieronymo Fabrizio, and so it could benefit from the improvement of academic knowledge. The work in plastic surgery which had begun in the 15th century was carried on in the 16th century by the Bolognese Gaspere Tagliacozzi, who restored a lost nose by transplanting a flap of skin from the arm, leaving one end still attached to the arm until the graft on the nose had established itself.

While these anatomists and surgeons were extending the practical achievements of their predecessors, medical men of another group were endeavouring, as in other sciences, to return to antiquity. The first humanist doctors, such as Thomas Linacre (*c.* 1460–1524), physician to Henry VIII, tutor to Princess Mary and founder and first president of the College of Physicians, or Johannes Günther (1487–1574), who at Paris numbered Vesalius, Serveto and Rondelet among his pupils, were literary men rather than anatomists. They encouraged, and co-operated in making the new Latin translations of Galen and Hippocrates which, along with the old, were printed in numerous editions from the end of the 15th century. They devoted their energy to establishing the text of these authors rather than to observation, and Mondino was objectionable to them not so much because he disagreed with nature as because he disagreed with Galen. They also began a violent attack on the old Latinized Arabic terminology of

Mondino, which they 'purified' by substituting classical Latin or Greek for Arabic words and transformed into the anatomical terminology still in use.

It was in this atmosphere of both practical observation and humanist prejudice and literary research that the so-called father of modern anatomy, the Netherlander André Vésale of Brussels, or Andreas Vesalius (1514–64), began his work. In it he exhibits both features. The *De Humani Corporis Fabrica* (1543) may be regarded as the outcome of an attempt to restore both the letter and the standards of Galen. In it Vesalius followed Galen, as well as other authors to whom he did not acknowledge his debt, in many of their mistakes as well as in their true observations. He placed the lens in the middle of the eye, repeated Mondino's misunderstanding of the generative organs, represented the kidney as a sieve, and adduced some conclusions about human anatomy from the study of animals, a practice for which he criticized Galen. Further, he differed in no important respect from Galen in physiology. He shared his Greek master's eye for the exhibition of living function in anatomical structure. The function of an organ, according to Galen, was the final cause of its structure and mechanical action and thus the explanation of its presence. The inspiration of the anatomical research which he stimulated was strongly teleological, and Vesalius himself regarded the human body as the product of Divine craftsmanship. This must be accounted an important factor in the passion with which he pursued his dissections. But it was the illustrations that were the really revolutionary feature of *De Fabrica* (Plate 21). No anatomical drawings can compare with them except the unpublished ones by Leonardo himself; together they make the most brilliant demonstration of how close the relations were between descriptive biology and naturalistic art. But the illustrations of *De Fabrica* achieve more than mere naturalism; the astonishing series representing the dissection of the muscles are at once an exhibition in detail of the relations between the structure and function of the muscles, tendons, bones and joints, and a dance of death, a drama played out by a corpse suspended from a gibbet against the background of a continuous landscape in the

Euganean hills. Whose work the illustrations of *De Fabrica* and its companion volume the *Epitome* (published with it at Basel in 1543) were has not been finally determined, but it is practically certain that they emanated from the atelier of Titian, and that among the artists who worked on them under the supervision of the master was Vesalius himself.

The work of Vesalius contained by far the most detailed and extensive descriptions and illustrations yet published of all the systems and organs of the body. Though his account of the other organs usually does not compare with that of the bones and muscles, whose relations he illustrated very well, he nevertheless made a large number of new observations on veins, arteries and nerves, greatly extended the study of the brain though without entirely rejecting the *rete mirabile*, and showed that bristles could not be pushed through the supposed pores in the inter-ventricular septum of the heart. He also repeated several of Galen's experiments on living animals and showed, for instance, that cutting the recurrent laryngeal nerve caused loss of voice. He showed that a nerve was not a hollow tube, though physiologists continued to believe the contrary until the 18th century. He showed also that an animal whose thoracic wall had been pierced could be kept alive by inflating the lungs with bellows.

A contemporary of Vesalius who, had his anatomical illustrations been published when they were completed in 1552, instead of in 1714, might have ranked with him as one of the founders of modern anatomy, was the Roman, Bartolomeo Eustachio (1520–74). He introduced the study of anatomical variations, particularly in the kidney, and gave excellent figures of the ear ossicles, the relations of the bronchi and blood vessels in the lung, the sympathetic nervous system, the larynx, and the thoracic duct.

As events turned out, Vesalius, and not Eustachio, set his mark on anatomy. He made the centre of the subject Padua, where he was professor from 1537 until he became physician to the Emperor Charles V in 1544, and a large part of the subsequent history of anatomy down to Harvey is the story of Vesalius' pupils and successors. The first of these was his assistant

Realdo Colombo (*c.* 1516–59), who experimentally demonstrated the pulmonary circulation of the blood (see above, p. 231). He was followed by Gabriel Fallopio (1523–62), who described the ovaries and the tubes called after him, the semi-circular canals of the ear, and several other structures. Fallopio's own pupils extended the Vesalian tradition at Padua into the study of comparative anatomy, but in the meantime similar interests had begun to develop elsewhere.

Many of those who were attracted by the printed editions of Pliny or of the Latin translations of Aristotle's zoological works developed from being humanist lexicographers into naturalists. A good example of this is William Turner (*c.* 1508–68) whose book on birds (1544), while being largely a compilation and accepting some legends such as that of the barnacle goose, also contained some fresh observations. Sixteenth-century zoology thus began as a gloss on the classics written increasingly from nature. The system of classification recognized by Albertus Magnus in Aristotle's writings, which the Oxford scholar and doctor, Edward Wotton, attempted to restore (1552), was the framework of the subject.

Besides birds, the first animals to attract attention were fish. Accounts of several local fish fauna, those of the sea at Rome and Marseilles and of the river Moselle, were written during the first half of the 16th century, but the scientific study of marine animals really began with the *De Aquatilibus* (1553) of the French Naturalist, Pierre Belon (1517–64). Belon had already become well known for his account of a voyage to the eastern Mediterranean, during which he made some interesting biological observations (1533). He took an ecological view of his group; his 'aquatiles' were the fish of 'cooks and lexicographers' and included cephalopods and cetacea as well as *pisces*. He made the first modern contributions to comparative anatomy. He dissected and compared three cetacean types, realized that they breathed air with lungs, and compared the heart and skeleton to those of man. He depicted the porpoise attached by the umbilical cord to the placenta, and the dolphin with its new-born young still surrounded by fœtal membranes. He also made a comparative study of fish anatomy, and in another small book, *Histoire Naturelle des Oiseaux* (1555), in which he intuitively recognized certain natural groups of birds, he depicted the skeletons of a bird and a man side by side to show the morphological correspondences between them (Plate 22). Another Frenchman, Guillaume Rondelet (1507–66), who became professor of anatomy at Montpellier and may have been 'the Physitian our honest Master Rondibilis' of Rabelais (who had also studied medicine there), included a similar heterogeneous collection of aquatic animals in his *Histoire Naturelle des Poissons* (1554–55). This was also a valuable work. In it he pointed out the anatomical differences between the respiratory, alimentary, vascular and genital systems of gill- and lung-breathing aquatic vertebrates, and depicted the viviparous dolphin and the ovoviviparous shark. He endeavoured to discover the morphological correspondence between the parts of the mammalian and piscine hearts. He discussed the comparative anatomy of gills, which he considered to be cooling organs, but he also showed that fish kept in a vessel without access to air would suffocate. He considered the teleostean swim-bladder, which he discovered, to be a kind of lung. Another heterogeneous work on aquatic animals published about the same time (1554), which is of interest in showing the influence of contemporary art in its excellent zoological illustrations, is that of I. Salviani (1514–72).

Another contemporary of these writers was the polyhistor and naturalist, Conrad Gesner. He attempted to draw up, on the lines of Albertus Magnus or Vincent of Beauvais, whom he quoted, an encyclopædia containing the observations of all his predecessors from Aristotle to Belon and Rondelet. In the course of this he also made observations of his own and, through his vast correspondence, was a stimulus to others. In the zoological part of this work, the *Historia Animalium* (1551–58), he seems to have been so uncertain about classification that he arranged the animals in alphabetical order. In other works, containing extracts from the *Historia*, he set them out according to the Aristotelian system, omitting only the insects. The material for the insects, which had been compiled by Gesner, Wotton and

Thomas Penny (*c.* 1530–88), was eventually published as Thomas Mouffet's *Theatrum Insectorum* (1634). Mouffet's 'insects' were those of Aristotle, and included myriapods, arachnids and various sorts of worms as well as the modern group of insects. His book contained a number of fresh observations, most of them the work of Penny. Gesner's work as an encyclopædist and zoologist was contained by Ulysses Aldrovandi (1522–1605), professor of natural history at Bologna, who among other things wrote the first book on fishes which did not include other aquatic forms.

Both Gesner and Aldrovandi included in their encyclopædic labours catalogues of fossils, or 'figured stones', of which several collections had been made in the 16th century, including one by Pope Sixtus V at the Vatican. The fossils included in these collections were mainly echinoderms, mollusc shells and fish skeletons, and considerable interest attached to their origin. On this matter opinion in fact remained divided until the 18th century, and it was not easy to recognize the organic origin of some fossils. Those who held that fossils were not of organic origin explained them by such theories as astral influence or generation by subterranean vapours. Even among those who held that fossils were organic remains some believed them to have been transported to the mountains by the Flood. The theory that organisms had been fossilized where they had once lived and were found had persisted in the writings of Albertus Magnus. Girolamo Fracastoro (1483–1553) accepted this view and so did Agricola, who held that the process of mineral-formation and fossilization was due to a *succus lapidescens*, which may have meant precipitation from solution. Another writer, the French potter Bernard Palissy, who had learnt through Cardano of Leonardo's ideas on these questions, went further and arrived at some understanding of the significance of fossil forms for comparative morphology. He regretted that Belon and Rondelet had not described and drawn fossil fish as well as living forms; they would then have shown what kind of fish had lived in those regions at the time when the stones in which they were found had congealed. He himself made a collection of fossils, recognized the identity of a number of forms, such as sea-urchins and oysters, with their living relatives, and even distinguished marine, lake and river varieties. In contrast with these bold ideas, Gesner admitted some fossils as petrified animals but regarded others as products *sui generis* of the earth itself. He made an attempt to classify them, taking their shape, the things they resembled and so on, as his criteria. Aldrovandi regarded fossils not as the remains of normal living forms but as incomplete animals in which spontaneous generation had failed full accomplishment.

Another aspect of biology which received fresh attention during the 16th century was embryology, the study of which was revived by Aldrovandi, who was inspired by Aristotle and Albertus Magnus to follow the development of the chick by opening eggs at regular intervals. Into this he initiated his Dutch pupil, Volcher Coiter, who, before finally settling at Nuremberg, studied also under Fallopio, Eustachio and Rondelet. He was thus an intellectual descendant of Vesalius, and the first of them to adopt the comparative method. In the chick, on which his observations were on Aristotelian lines, he discovered the blastoderm, but he left it for Aldrovandi to explain how the eggs passed from the ovary into the oviduct, and failed to recognize that the avian ovary was homologous with the mammalian 'female testis'. He made a systematic study of the growth of the human fœtal skeleton and pointed out that bones were preceded by cartilages. He also made a systematic study of the comparative anatomy of all vertebrate types except the fishes. His emphasis on points of difference, rather than homology, shows that he did not fully grasp the significance of the comparative method, but his comparisons, beautifully illustrated by himself, greatly extended the range of the subject. He was most successful in his treatment of skeletons, of which he compared those of many different types, from frog to man. He also made a comparative study of living hearts. He tried to interpret the structure of the mammalian lung in terms of the simpler organs of frogs and lizards, and understood the difference in their respiratory mechanisms. He made a number of particular anatomical discoveries, of which that of the dorsal and ventral nerve roots was perhaps the most important, and he tried to classify mammals on an anatomical basis.

The comparative method was systematically extended to embryology by Fallopio's successor at Padua, Hieronymo Fabrizio, who was professor there at the same time as Galileo. Fabrizio made a number of contributions to anatomy. His embryological theory like that of his pupil Harvey, was in principle entirely Aristotelian. But he held that the majority of animals were generated not spontaneously but from 'eggs', gave good figures of the later stages of development of the chick (Plate 23A; cf. Plate 23B) and made a careful study of the embryology of a large number of vertebrates. In the last he paid particular attention to the fœtal membranes and confirmed the assertion of Julius Cæsar Arantius (1564), that although the fœtal and maternal vascular systems were brought into close contact with the placenta there was no free passage between them. He gave a clear account of other already known structures associated with the fœtal blood system, such as the *ductus arteriosus* and the *foramen ovale* (discovered by Botallus, 1564). The valves in the veins had been observed by a number of anatomists, but Fabrizio published the first clear and adequate pictures of them (1603), which Harvey afterwards used to illustrate his book. In his comparative studies Fabrizio attempted to assess the points common to the various vertebrates and those defining specific differences. He held that each sense organ had its own special function and could perform no other, but although he drew the lens in its correct position in the eye he still believed that it was the seat of vision. He attempted to analyse the mechanics of locomotion, and compared the actions of the internal skeleton of the vertebrate and the external skeleton of the arthropod. He observed that the worm moved by the alternate contraction of its longitudinal and circular muscles, and examined the relation of the centre of gravity to posture in the bird. It was not, however, until Borelli (1680) was able to make use of Galileo's mechanics that these problems received an adequate solution.

Fabrizio's comparative method was carried still further by his former servant and pupil, Giulio Casserio (1561–1616), who succeeded him at Padua. Casserio has been described as a great craftsman, who endeavoured to explain the fabric of man by reference to that of the lower animals. He divided his investigation, as Galen had done, into structure, action and uses (function). His method was first to describe the human condition in fœtus and adult and then to follow it through a long series of other animals. This is well illustrated in his study of the organs of the voice and hearing, during which he described the sound-producing organs of the cicada and the auditory ossicles of a large number of land vertebrates, and discovered the internal ear of the pike (Plate 24).

Casserio's successor, Adriaan van der Spieghel (1578–1625), whose chief work was to improve anatomical terminology, was the last of the great Paduan line, and after his time animal biology itself developed in a different direction. His contemporary at Pavia, Gasparo Aselli (1581–1626), discovered the lacteal vessels while dissecting a dog which had just had a meal containing fat. These are lymphatic vessels which conduct into the blood stream, at the jugular vein, fatty substances absorbed in the intestine, but Aselli thought they led from the intestine to the liver. Another contemporary, Marc Aurelio Severino (1580–1656), a pupil at Naples of the anti-Aristotelian philosopher Campanella, compiled on comparative anatomy a treatise entitled *Zootomia Democritœa* (1645) out of respect for his master's views. In this he recognized the unity of the vertebrates, including man, but he regarded man as the basic 'archetype', determined by Divine design, and divergences from this as due to differences in function. He discovered the heart of the higher crustacea, dissected but misunderstood that of cephalopods, recognized the respiratory function of fish gills, invented the method of studying blood vessels by injection with a solidifying medium, and recommended the use of the microscope. Though he wrote after Harvey, he suffered from the same defects as his predecessors.

The effort of the 16th-century anatomists had been to explore, describe and compare the structure of the human and the animal body, to make some attempt to relate the results by a zoological classification and to understand the variety of animal forms. They laid the foundation of work which was to lead to the theory of organic evolution, but, not only were their conceptions of physiology vague, inaccurate and uncoordinated, but also their inferences did not arise out of a critical and comprehensive consideration of the facts. Their conceptions of biological function were, in fact, largely inherited from the past, and as yet remained unrelated to their discoveries of structure. These matters were being brought into relation by another son of Padua, William Harvey (see above, p. 227 *et seq.*).

In embryology, Harvey made a number of advances. Although he has been criticized for his work on this subject, in fact he carried into this difficult field the same principles as he had used with such success in analysing the simpler problem of the movement of the blood. Among his positive contributions to comparative embryology were a number of particular observations on the placenta and other structures, the final recognition of the cicatricula on the yolk membrane as the point of origin of the chick embryo, and a clear discussion of growth and differentiation. Another contribution was implied by his remark in his *Exercitationes de Generatione Animalium* (1651), exercitatio 62 : 'The egg is the common beginning for all animals.' Albertus Magnus, who had made a similar remark (see Vol. I, p. 163), certainly also accepted the spontaneous generation of the eggs or *ova* themselves; and since Harvey was not unequivocal on the point, especially in *De Motu Cordis*, opinions differ about whether he did too. Some passages do definitely suggest that he held all plants and animals to originate from 'seeds' arising from parents of the same species, though these 'seeds' might sometimes be too small to be seen. As he declared in *De Generatione Animalium* : 'many animals, especially insects, arise and are propagated from elements and seeds so small as to be invisible (like atoms flying in the air), scattered and dispersed here and there by the winds; and yet these animals are supposed to have arisen spontaneously, or from decomposition, because their ova are nowhere to be seen.' Francesco Redi, who first experimentally disproved spontaneous generation in insects (1668), read

Harvey's views in this sense. Thus, although Harvey did not understand the nature of the *ovum*, which he still identified in insects with the larva or pupa, and in mammals, with small embryos surrounded by their membrane or chorion, his ideas, which crystallized into the *omne vivum ex ovo* that appeared on the frontispiece of his book, stimulated research into the subject by his followers.

Harvey's own observations led him to reject both the Aristotelian and Galenic theories of fertilization. According to Aristotle, the uterus of a fertilized female should have contained semen and blood, according to Galen a mixture of male and female semen. In the king's deer, which Harvey dissected at Hampton Court, he could find no such visible proof of conception for some months after mating. He was unfortunate, because in this respect deer are peculiar; but he could also see nothing for several days in more normal animals such as dogs and rabbits. He therefore concluded that the male contributed an immaterial influence like that of the stars or of a magnet, which set the female egg developing. Although the production of eggs in ovarian follicles was not discovered until after Harvey, he may thus be considered the originator of the 17th-century 'ovist' theory according to which the female contributed the whole of the embryo. After Leeuwenhoek, with his microscope, had discovered the spermatozoon (1677), the opposite school of 'animalculists' made the same claim for the male, and the resulting controversy continued throughout most of the 18th century.

The other great embryological controversy over which Harvey's followers spent their energies was that of epigenesis and preformation. Harvey himself had clearly reaffirmed Aristotle's preference for the former, at least in sanguineous animals; he held development to be the production of structures *de novo* as the embryo approached that final adult form. Again the later ovists and animalculists alike held that the adult was formed by the 'evolution', or unfolding, of parts already completely present in the germ. This was more in keeping with the mechanism of the age, and the year after Harvey's death Gassendi published a theory of panspermatic preformationism based on his atomic theory. Some time earlier, an even completer mechanistic theory of biology had been worked out by Descartes (see above, p. 243 *et seq.*).

This work on reproduction was to lead to the formulation of the germ theory of disease, though that was not fully understood until the time of Pasteur in the 19th century. In the early 16th century a theory that diseases were caused by the transference of *seminaria*, or seeds, was put forward by Fracastoro. He is famous for introducing the name syphilis and for describing that disease, which had first appeared in a virulent form in 1495 in Naples, then occupied by Spanish troops, during the siege by the soldiers of Charles VIII of France. He set forth his theory of disease in his *De Contagione*, published in 1546, in which he reiterated the already known facts that disease could be transmitted by direct contact, by clothing and utensils, and by infection at a distance as with smallpox or plague (see Vol. I, pp. 234–6). To explain such action at a distance, he made use of a modification of the old theory of the 'multiplication of species'; he said that during the putrefaction associated with disease minute particles of contagion were given off by exhalation and evaporation, and that these 'propagated their like' through the air or water or other media. When they entered another body, they spread through it and caused the putrefaction of that one of the four humours to which they had the closest analogy. To such *seminaria* Fracastoro attributed the spread of contagious phthisis, rabies and syphilis.

Fracastro seems also to have been the first to recognize typhus, and the habit of carefully recording case-histories, which had been seen in the *consilia* and plague tracts made since the 13th century, produced a number of good accounts of diseases in the 16th century, for example the clear description of the sweating sickness published by John Caius in 1552. This practice increased in the 17th century and produced such excellent clinical records as Francis Glisson's account of infantile rickets in 1650, Sir Theodore Turquet of Mayerne's medical history of King James I, and the careful descriptions of measles, gout, malaria, syphilis, hysteria, and other diseases made by Thomas Sydenham (1624–

89). This insistence on observation, and suspicion of the all-too-facile theories which had prevented new approaches to the facts, led to a great increase in empirical knowledge and in empirical methods of treatment; indeed even now, in the 20th century, medicine is still largely an empirical art. As early as the early 16th century, if not still earlier, mercury was used for syphilis, and from the early 17th century cinchona bark, the source of quinine, was used for malaria. This had been introduced into Europe from Peru by Jesuit missionaries after whom it became known as 'Jesuits' bark.' A clear understanding of infectious diseases, as indeed the understanding of the causes of the functional and organic disorders of the body, had to await the gradual acquisition of the fundamental knowledge of biology and physiology during the 18th and 19th centuries.

8. PHILOSOPHY OF SCIENCE AND CONCEPT OF NATURE IN THE SCIENTIFIC REVOLUTION

By the middle of the 17th century European science had gone a long way since Adelard of Bath had demanded explanations in terms of natural causation, and since the experimental and mathematical methods had begun to develop within the predominantly Aristotelian system of scientific thought of the 13th and 14th centuries. Certainly in experimental and mathematical technique revolutionary progress had been made by the 17th century, and this was to go on with breathtaking speed throughout that century. To take only one science as an example, astronomy in 1600 was Copernican, and not even completely so; in 1700 it was Newtonian, and was supported by the impressive structure of Newtonian mechanics. Yet the statements on aims and methods expressed by the spokesmen of the new 17th-century science were remarkably similar to those expressed by their predecessors in the 13th and 14th centuries, who were, in fact, also spokesmen of modern science at an earlier stage in its history. They were remarkably similar – with a difference.

The utilitarian ideal, for example, was given expression by Francis Bacon in words very like to those of his 13th-century namesake, even down to the particular value he placed on the inductive method. 'I am labouring to lay the foundation,' said Bacon in the preface to his *Great Instauration*, 'not of any sect or doctrine, but of human utility and power.' The purpose of science was to gain power over nature. The object of the Great Instauration, or new method, was to show how to win back that dominion which had been lost at the Fall. In the past, science had been static, while the mechanical arts had progressed, because in science observation had been neglected. It was only through observation that knowledge of nature could be gained; it was only knowledge that led to power; and the knowledge that the natural scientist was to look for was knowledge of the 'form', or causal essence, whose activity produced the effects observed. Knowledge of the form gave mastery over it and its properties, and so the positive task of Bacon's new method was to show how to obtain knowledge of the form. As he declared in the *Novum Organum* (1620), book 1, aphorism 3: 'Human knowledge and human power are one; for where the cause is not known the effect cannot be produced. Nature to be commanded must be obeyed; and that which in contemplation is as the cause is in operation as the rule.' What he meant by the 'form' of a body or a phenomenon he explained further in book 2, aphorism 2: 'For though in nature nothing really exists besides individual bodies, performing pure individual acts according to a fixed law, yet in philosophy this very law, and the investigation, discovery, and explanation of it, is the foundation of knowledge as well as of operation. And it is this law, with its clauses, that I mean when I speak of Forms; a name which I rather adopt because it has grown into use and became familiar.'

The parenthetical conclusion to this quotation is a warning that Bacon may be concealing in his deceptively scholastic language concepts far removed from the 'substantial form' and real qualities in the sense of the scholastic 'natures'. It also serves as a reminder that the historian of scientific method must necessarily include in his field of consideration not only the logical procedures described and used by a natural philosopher, but also – and without these he will understand nothing – the actual problems to which the procedures were applied and the assumptions made concerning the kind of explanation they should yield. For example it is impossible to see the point of Grosseteste's or Ockham's discussions of scientific method without the context of the philosophy of nature to which they applied. Galileo and Kepler aimed their analyses of scientific method at the particular kinematic and dynamical problems they were trying to solve; their point can be seen only in relation to these, and to the kinds of laws they expected to discover.

The procedures of science are methods of answering questions about phenomena; the questions give definition to the phenomena and constitute them into problems. Much of what is asked about such data will be determined simply by the technical procedures, mathematical and experimental, in current use or being developed. But the form the questions take, the direction and extent to which they are pressed in the search for an explanation, will inevitably be strongly influenced by the investigator's philosophy or conception of nature, his metaphysical presuppositions or 'regulative beliefs', for it is these that will determine his conception of the real subject of his inquiry, of the direction in which the truth hidden in the appearances will be found. It is these that will often determine what a scientist regards as significant in a problem; they may inspire his scientific imagination, as they did with Kepler and Galileo; and they may set limits to what he regards as admissible in an explanation, as the objection to action at a distance did for the critics of Newton's theory of gravitation. These philosophical assumptions may of course themselves be profoundly modified in the course of a scientific investigation. They may be falsified by observation, as Newton's falsified the assumption of the circularity of all celestial motion. Or they may be in themselves not empirically falsifiable, like the scholastic conception of 'natures' or the belief that all phenomena can be reduced to matter and motion. Such conceptions are abandoned or modified only by re-thinking. But there has never been natural science with no preconception at all of theoretical objectives of a philosophical kind.

In the actual history of science many of the most fruitful

theories have been developed from preconceived ideas of the kinds of laws or theoretical entities that will be discovered to explain the phenomena. The history of the inquiry has to a large extent consisted of using the sharp tools of mathematics and experiment to carve out of these preconceptions a theory exactly fitting the data. A good example of this is the atomic theory, first seen as scientific material of this kind in the 17th century and eventually reduced to exact empirical form by John Dalton in 1808. So far as scientific method is concerned, the whole period from the 13th to the 17th century can be seen as one in which the functions both of the experimental principles of verification and falsification and correlation, and of mathematical techniques, were understood and applied with increasing effect to reduce philosophies of nature to exact science (cf. above, p. 26 *et seq*.). For example the Neoplatonic philosophy of nature, with its geometrical conception of the ultimate 'form' of things, first became scientifically significant with Grosseteste's philosophy of light. But in spite of his analysis of the logic of experimental science Grosseteste himself was capable of leaving the explanations he derived from his Neoplatonism not only very loosely connected with the data but sometimes actually contradicting them. It was the more technical and less philosophical mathematical and experimental investigators of the period, inspired by Euclid and Archimedes more than Plato and Aristotle, who were more empirically accurate in practice; and it was only when the technical procedures were fully exploited by Galileo and Kepler that Neoplatonism yielded exact science.

It was precisely in such a critical role that Francis Bacon conceived his inductive method for 'the discovery of forms'. By 'form' Bacon meant something quite specific: geometrical structure and motion. The common idea of him as a pure empiricist, starting with no preconceived ideas or hypotheses, is by no means borne out by his principal work on scientific method, the *Novum Organum*, although it is nearer the mark in the endless tables of instances forming the 'Natural and Experimental Histories' of the *Sylva Sylvarum*. Bacon's achievements are those of a philosopher with a clear grasp of the function of the empirical principle but almost none at all of the technical procedures necessary, not only to solve problems, but even to formulate them in a scientifically significant manner.

In his *Novum Organum* Bacon of course explicitly set out to replace the *Organum* of Aristotle, but when compared with the various conceptions of scientific method held in classical and early modern times it is clear that Bacon's method has far more in common with Aristotle's than do, for example, the postulational methods of Archimedes and Galileo. He based his method on the analysis of matter rather than the idealizations of mechanics; it was aimed at discovering the composition of bodies, and it is significant that a large number of his examples were taken from chemistry. But if one is looking for the ancestry of his method, it is easy to see it in the postulational method of Democritus and in Plato's dialectic (cf. above, pp. 24, 148–9).

The current view against which Bacon and other contemporary advocates of the 'new philosophy' were writing was that the explanation of phenomena could be given in terms of the qualitative substantial forms and real qualities forming the 'natures' of the scholastics. Finding these unhelpful, the natural philosophers of the period assimilated their philosophy of nature to the new science by developing a more mathematical conception of the 'form' based on the atomism of Democritus and Epicurus and of Hero of Alexandria (see Vol. I, p. 46, note; above, p. 51, note), while Galileo and Kepler came to distinguish between the primary, real, geometrical qualities actually belonging to bodies and the secondary, subjective qualities produced by the action of these on the organs of sense (see below, p. 305). Bacon was one of the earliest modern writers to propose the complete reduction of all events to matter and motion. In his *Cogitationes de Natura Rerum* he had written: 'The doctrine of Democritus concerning atoms is either true, or useful for demonstration.' His proposal for 'the discovery of forms' in the *Advancement of Learning* (1605) was an inquiry into the explanation of the properties of bodies, but he asserted that this had got too far from experiment. His object was to base the inquiry not on the atoms of the philosophers but on induction. Then, as he said in the *Novum Organum*, book 2, aphorism 8, 'We shall be led only to real particles, such as really exist.' These constituted the 'latent configuration' of the form, hidden from sight but discoverable by inductive reasoning. Their movement constituted the 'latent process', variation in motion producing different manifest effects in the 'nature', by which he meant any type of observable occurrence, such as heat, light, magnetism, planetary motion, fermentation. Thus his preconception of the kind of entities his inductive analysis would yield was just as definite as that of the scholastic writers on scientific method who discussed the 'resolution' of bodies into the four Aristotelian elements and causes or of a disease into one of a set of preconceived species of a genus (cf. above, pp. 29, 40–43). And Bacon described the form, as he conceived it, in language similar to that used by the scholastics of the four Aristotelian causes, the conditions necessary and sufficient to produce the observed effect. 'For,' he said in book 2, aphorism 4, 'the Form of a nature is such, that given the Form the nature infallibly follows.' This led him to base the inquiry for the form on the methods of agreement or presence, difference or absence, and concomitant variation (cf. above, p. 146).

Bacon's method followed the pattern of the inductive and deductive processes already seen in his medieval predecessors. His chief contribution to the theory of induction was to set out very clearly and in great detail both the method of reaching the definition of a 'common nature', or form, by collecting and comparing instances of its supposed effects, and the method of eliminating false forms (or what would now be called hypotheses) by what he called 'exclusion'. This was analogous to Grosseteste's method of 'falsification' (*falsificatio*). Bacon said in the *Novum Organum*, book 1, aphorism 95:

Those who have handled sciences have been either men of experiment or men of dogmas. The men of experiment are like the ant; they only collect and use: the reasoners resemble spiders, who make cobwebs out of their own substance. But the bee takes a middle course, it gathers its material from the flowers of the garden and of the field, but transforms and digests it by a power of its own. Not unlike this is the true business of philosophy; for it neither relies solely or chiefly on the powers of the mind, nor does it take the matter which it gathers from natural history and mechanical experiments and lay it up in the memory whole, as it finds it; but lays it up in the understanding altered and digested. Therefore from a closer and purer league between these two faculties, the experimental and the rational (such as has never yet been made) much may be hoped.... Now [he went on in book 2, aphorism 10] my directions for the interpretation of nature embrace two generic divisions; the one how to educe and form axioms from experience; the other how to deduce and derive new experiments from axioms.

The first step towards the discovery of a form was to make a purely empirical collection of instances of the phenomenon or 'nature' being investigated. As an illustration of both his method and the kinds of things that should be investigated, he gave his well-known example of the 'form of heat'. As he said in the *Novarum Organum*, book 2, aphorism 10: 'We must prepare a *Natural and Experimental History*.' The next step was made by what he claimed to be a new kind of induction, hitherto used in part only by Plato. The current kind of induction 'by simple enumeration' was, he said, in book 1, aphorism 105, generally based on too few facts and 'exposed to peril from a contradictory instance. ... But the induction which is to be available for the discovery and demonstration of sciences and arts, must analyse nature by proper rejections and exclusions; and then, after a sufficient number of negatives, come to a conclusion on the affirmative instances.' To make this 'true and legitimate' induction the observations must be classified into three 'Tables and Arrangements of Instances'. The first was a table of 'Essence and Presence' or agreement, which included all events where the form sought (e.g., heat) was present; the second was a table of 'Deviation or of Absence in Proximity' which included all events where the effects of the form sought were not observed; the third was a table of 'Degrees or Comparison' which included instances of variations in the observed effects of the form sought either in the same or in different subjects. Induction then consisted simply of the inspection of these tables. 'The problem is,' said Bacon in the *Novum Organum*, book 2, aphorisms 15 and 16, upon a review of the instances, all and each, to find such a nature as is always present or absent with the given nature, and always increases and decreases with it ... The first work therefore of true induction (as far as regards the discovery of Forms) is the rejection or exclusion of the several natures which are not found in some instance where the given nature is present, or are found in some instance where the given nature is absent, or are found to increase in some instance when the given nature decreases, or to decrease when the given nature increases. Then indeed after the rejection and exclusion had been duly made, there will remain at the bottom, all light opinions vanishing into smoke, a Form affirmative, solid and true and well defined.

On the basis of this unelimited residue the investigator then embarked upon what he called in aphorism 20 'an essay of the Interpretation of Nature in the affirmative way'. The first stage in this process led only to the 'First Vintage' or a working hypothesis. So, he concluded: 'From a survey of the instances, all and each, the nature of which Heat is a particular case appears to be Motion ... Heat itself, its essence and quiddity, is Motion and nothing else.' From this hypothesis new consequences were deduced and tested by further observations and experiments until eventually, by repeated and varied observation followed by elimination, the 'true definition' of the form was discovered, and this gave certain knowledge of the reality behind the observed effects, knowledge of the true law in all its clauses. 'The Form of a thing,' he said in *Novum Organum*, book 2, aphorism 13, 'is the very thing itself, and the thing differs from the form no otherwise than as the apparent differs from the real, or the external from the internal, or the thing in reference to man from the thing in reference to the universe.'

The form for Bacon was always some mechanical disposition; induction eliminated the qualitative and the sensible leaving geometrical fine structure and motion. The form of heat was thus motion of particles; the form of colours a geometrical disposition of lines. In fact, by Bacon's time the word 'nature' itself had come to mean mechanical properties, the *natura naturata* of the Renaissance. The spontaneous animating principle, *natura naturans*, of such writers as Leonardo da Vinci or Bernadino Telesio (1508–88) had practically disappeared.

The discovery of the form was the end of the 'experiments of Light' which occupied the essential first stage in science but, as Bacon put it in the *Great Instauration* :

those twin objects, human Knowledge and human Power, do really meet in one; and it is from ignorance of causes that operation fails.

The final purpose of science was power over nature. Moreover, he said in the *Novum Organum*, book 1, aphorisms 73 and 124 :

fruits and works are as it were sponsors and sureties for the truth of philosophies ... Truth therefore and utility are here the very same things : and works themselves are of greater value as pledges of truth than as contributing to the comforts of life.

Thus, when Bacon excluded final causes from science it was not because he did not believe in them, but because he could not imagine an applied teleology as there was an applied physics. By following his 'experimental philosophy' he held that future humanity would achieve an enormous increase in power and material progress. As he expressed it in the *Novum Organum*, book 1, aphorism 109 :

There is therefore much ground for hoping that there are still laid up in the womb of nature many secrets of excellent use, having no affinity or parallelism with anything that is now known, but lying entirely out of the beat of the imagination, which have not yet been found out.

And he believed that the final achievement of the branch of science which he described in the *Advancement of Learning* as 'Natural Magic' would be the transmutation of the elements.

It was through his utilitarianism and his empiricism rather than the actual canons of his inductive method that Bacon chiefly influenced his followers, although his ideas on method certainly had some effect in England. Even Harvey declared in his *De Generatione*, exercitatio 25: 'in the words of the learned Lord Verulam to "enter upon our second vintage" ...' His most important influence was in the Royal Society. Bacon's description of the research institute, Solomon's House, in his *New Atlantis*, published posthumously in 1627, was the real inspiration of the various schemes for scientific institutions or colleges that were finally realized in the foundation of the Royal Society. Under Bacon's influence the Fellows dedicated themselves from the beginning to experimental inquiries, and they aimed at promoting not only 'Natural Knowledge' but knowledge that would be useful in trades and industries. In the *Advancement of Learning* Bacon declared the true end of scientific activity to be the 'glory of the Creator and the relief of man's estate'. Echoing this, the second charter of the Royal Society, which received the Great Seal on 22 April 1663 and by which the Society is still governed, laid down that the investigations of its Fellows 'are to be applied to further promoting by the authority of experiments the sciences of natural things and of useful arts, to the Glory of God the Creator, and the advantage of the human race'. The Fellows were asked by the English government to investigate such problems as the practices used in navigation and in mining, and they themselves saw in technology a means of improving the empirical basis of science (cf. above, p. 133). It was this emphasis on the usefulness of science as well as his empiricism, that made Bacon the hero of d'Alembert and the French encyclopædists of the 18th century.

Thomas Sprat in his *History of the Royal Society* (1667) expressed a typical opinion of Bacon in describing his writings as the best 'defence of Experimental Philosophy, and the best Directions, that are needful to promote it,' and in saying at the same time that Bacon's Natural Histories were not only sometimes inaccurate but also that he seemed 'rather to take all that comes, than to choose, and to heap, rather than to register'. A typical example is the inquiry into the form of heat, where the instances ranged from warm feathers to the sun's rays, and from 'hot' pepper to the 'burning' of the hands by snow. Bacon's influence certainly sometimes led to a blind empiricism, but more typical was that on a man like Robert Hooke, who was one of those who actually made use of Bacon's methods, expounding them in his *General Scheme* published in the *Posthumous Works*

(1705), but he was too good an experimentalist, mathematician, and deviser of hypotheses to be in any way restricted by what Bacon had laid down.

The only scientist of the period who saw himself as a complete Baconian was Boyle: 'designed by Nature to succeed' to the fame of the great Verulam as the *Spectator* described him in 1712. 'By innumerable experiments He, in great Measure, filled up those Plans and Out-Lines of Science, which his Predecessor had sketched out.' Boyle was extremely influential in handing on Bacon's empiricism, his distaste for systems, his insistence on the primacy of experiments over theory, to Newton himself and to the 18th century. For example the significant *Prœmial Essay* in his *Physiological Essays* (1661) was aimed at reinforcing Baconian empiricism as against Cartesian rationalism and the speculative development of systems far beyond the experimental evidence. As he wrote: 'It has long seemed to me none of the least impediments of the real advancement of true natural philosophy, that men have been so forward to write systems of it, and have thought themselves obliged either to be altogether silent, or not to write less than an entire body of physiology.' But Boyle's work and contemporary reputation are revealing just because they show the influence of the side of Bacon that has so often been neglected: his philosophy of nature. No more than Bacon was Boyle a completely anti-theoretical experimentalist; he is more truly seen, as his 18th-century editor Peter Shaw described him, as the 'restorer of the mechanical philosophy' in England.* As he himself wrote in the *Producibleness of Chymical Principles* (1679) appended to the second edition of the *Sceptical Chymist*: 'For though sometimes I have had occasion to discourse like a Sceptick, yet I am far from being one of that sect; which I take to have been little less prejudicial to natural philosophy, than to divinity itself.'

In fact, far from being a sceptical empiricist, Boyle was very ready to make use of hypotheses as aids to research. Arguing in favour of the 'Corpuscularian doctrine' in the preface of his

* Cf. M. Boas, 'The establishment of the mechanical philosophy,' *Osiris*, 1952, vol. 10.

Mechanical Origin ... *of* ... *Qualities* (1675), he wrote: 'For, the use of an hypothesis being to render an intelligible account of the causes of the effects, or phænomena proposed, without crossing the laws of nature, or other phænomena; the more numerous and the more various the particles are, whereof some are explicable by the assigned hypothesis, and some are agreeable to it, or, at least, are not dissonant from it, the more valuable is the hypothesis, and the more likely to be true. For it is much more difficult to find an hypothesis that is not true, which will suit with many phænomena, especially if they be of various kinds, than but with a few.' But he concluded: 'I intend not therefore by proposing the theories and conjectures ventured at in the following papers, to debar myself of the liberty either of altering them, or of substituting others in their places, in case a further progress in the history of qualities shall suggest better hypotheses or explications.' In an unfinished and unpublished tract entitled *Requisites of a Good Hypothesis*, he made a further distinction between a 'good hypothesis', which explained the largest number of facts without contradiction, and an 'excellent hypothesis', which was the unique explanation, or, at least, was uniquely good. Such an hypothesis must not only yield predictions, but such predictions as will enable it to be put to experimental test. The fragment is worth quoting as a whole:

The Requisites of a good Hypothesis are:
  That it be Intelligible.
  That it neither Assume nor Suppose anything Impossible, unintelligible, or demonstrably False.
  That it be consistent with itself.
  That it be fit and sufficient to Explicate the *Phænomena*, especially the chief.
  That it be, at least consistent, with the rest of the *Phænomena* it particularly relates to, and do not contradict any other known *Phænomena* of nature, or manifest Physical Truth.
The Qualities and Conditions of an *Excellent Hypothesis* are:
  That it be not *Precarious*, but have sufficient Grounds in the nature of the Thing itself or at least be well recommended by some Auxiliary Proofs.

That it be the *Simplest* of all the good ones we are able to frame, at least containing nothing that is superfluous or Impertinent.

That it be the *only* Hypothesis that can Explicate the Phænomena; or at least, that do[e]s Explicate them so well.

That it enable a skilful Naturalist to foretell future Phænomena by their Congruity or Incongruity to it; and especially the events of such Experim'ts as are aptly devis'd to examine it, as Things that ought, or ought not, to be consequent to it.*

Boyle's problem was the same as Bacon's and of other contemporaries faced with the scientific uselessness of the Aristotelian doctrines of 'natures'. As he wrote in the preface of the *Mechanical Origin ... of ... Qualities*: 'if, by a bare mechanical change of the internal disposition and structure of a body, a permanent quality, confessed to flow from its substantial form, or inward principle, be abolished, and, perhaps, also immediately succeeded by a new quality mechanically producible; if, I say, this come to pass in a body inanimate, especially, if it be also, as to sense similar, such a phænomenon will not a little favour that hypothesis, which teaches, that these qualities depend upon certain contextures, and other mechanical affections of the small parts of the bodies, that are endowed with them, and consequently may be abolished when that necessary modification is destroyed.' The diverse and prolix collection of essays that form the product of his forty years' devotion to natural philosophy had a single aim : to discover through experiment an explanation of the properties of bodies, to develop a universal theory of matter on the same intelligible principles as the new science of mechanics. By his analysis of 'the origin of forms and qualities' Boyle meant just what Bacon meant by 'the discovery of forms'. The object of his 'corpuscular philosophy', neither atomist nor Cartesian but developed along the lines suggested by Bacon, was to explain all the manifest properties of bodies by the two prin-

* *Boyle Papers*, vol. 37, Miscellaneous, in the Library of the Royal Society of London. There are several versions, with minor variations; see M. Boas, 'La méthode scientifique de Robert Boyle', *Revue d'histoire des sciences*, 1956, vol. 9; R. S. Westfall, 'Unpublished Boyle papers relating to scientific method', *Annals of Science*, 1956, vol. 12.

ciples of matter and motion, by the size, shape and motion of particles as indicated by extensive experiments. This form of the mechanical philosophy was reinforced by Boyle's experimental production of a vacuum and his experiments on the air. The strongly empirical aspect of his thinking is shown for example by his unwillingness to commit himself as to the cause of the air's elasticity, of which he stated the quantitative characteristics in 'Boyle's Law'. There is a parallel to this in the attitude taken by Edme Mariotte, who also formulated this law, and by Pascal. Boyle was always only too careful to test and illustrate by experiment the many particular hypotheses he formed in the course of his researches. But the form of these particular hypotheses and the kind of theoretical entities they contained was determined by a philosophy of nature that was not submitted to falsification but was a 'regulative belief' *assumed* in all his scientific thinking. This was the belief in universal mechanism which was held by Bacon no less than by Descartes and was soon to become predictively fruitful in the world-machine of Newton. As Boyle wrote in his *Excellency and Grounds of the Mechanical Hypothesis* (1674): 'By this very thing that the mechanical principles are so universal, and therefore applicable to so many things, they are rather fitted to include, than necessitated to exclude, any other hypothesis, that is founded in nature, as far as it is so'.

The desire for certain knowledge of nature, which inspired Francis Bacon's work on method, and which in fact since St Augustine or indeed since Plato had inspired the whole rationalist tradition of European thought, with its belief that what is certain is true of reality, was the principal motive behind all 17th-century science; it is what made the 17th century so conscious of method. Until the end of the 17th century, when this Aristotelian form of predication of attributes as inhering in real persisting substances began to be criticized in the new empiricism of John Locke (1632–1704), all scientists were inspired by the faith that they were discovering through and behind the particular observed phenomena the intelligible structure of the real world. And so it was supremely important to have a method that would facilitate this discovery of real nature behind the appearances and that would guarantee the certainty of the result. The same emphasis on method is seen in all science, whether in the numerous 'methods' put forward by botanists in search of a 'natural' as opposed to a merely artificial system of classification, or in the experimental method and the mathematical method of chemists and physicists.

Except for some biologists to whom organisms still presented a problem, by the middle of the 17th century the assumption made by nearly every natural philosopher who set out to discover this real physical world was that what they would discover would be something mathematical in form. It was Galileo who laid down the methodological desiderata for this mechanical philosophy by his explicitly kinematic treament of motion and firm rejection of any consideration of Aristotelian 'natures' and causes, for example in the *Two New Sciences* (see above, pp. 155–7; cf. p. 98 *et seq.*). He described the concept of nature which his methods had in view very clearly in 1623 in *Il Saggiatore*, both in question 6 (see above, p. 151) and in his famous distinction between primary and secondary qualities in question 48. Discussing Aristotle's remark in *De Cælo* (book 2, chapter 7) that 'motion is the cause of heat', he wrote:

But first I want to propose some examination of that which we call heat, whose generally accepted notion comes very far from the truth if my serious doubts be correct, in as much as it is supposed to be a true accident, affection, and quality really residing in the thing which we perceive to be heated. No sooner do I form a conception of a piece of matter or a corporeal substance, than I feel the need of conceiving that it has boundaries which give it this or that shape; that relative to others it is large or small; that it is in this or that place, in this or that time; that it is moving or still; that it touches or does not touch another body; that it is single, few, or many; nor can I, by any effort of imagination, dissociate it from those qualities (*condizioni*). But I feel no need to apprehend it as necessarily accompanied by such conditions as to be white or red, bitter or sweet, sounding or silent, pleasant or evil smelling. On the contrary, if the senses had not perceived these qualities, perhaps the reason and imagination alone would never have arrived at them. Therefore I hold that these tastes, odours, colours, etc. on the part of the object in which they seem to reside, are nothing more than pure names, and exist only in the sensitive body, so that if the animate being (*animale*) were removed, these qualities would themselves vanish. But yet, having given them special names different from those of the other primary and real qualities (*accidenti*), we would persuade ourselves that they also exist just as truly and really as the latter. I can explain my conception more clearly with an example. I pass a hand, first over a marble statue, then over a living man. As to the hand's own action, this is the same with respect to both bodies – that is, the primary qualities, motion and touch, for we call them by no other names. But the animate body which suffers such operations feels different sensations (*affezioni*) according to the different parts touched. For example, when touched under the soles of the feet, on the kneecaps, or under the armpits, it feels, besides the common feeling of being touched, another to which we have given a particular name, calling it tickling. This feeling is all ours, and does not belong to the hand at all; and it seems to me that it would be a grave mistake to say that, besides motion and touch, the hand has in itself another faculty, different from these, namely the tickling faculty, so that tickling would be a quality residing in the hand. A small piece of paper, or a feather, lightly drawn over any part of our body you wish performs, in itself, the same action everywhere, that is it moves and touches; but in us, touching between the eyes, on the nose, or under the nostrils, it excites an almost unbearable tickling, though in other parts we can hardly feel it at all. Now this tickling is all in us, and not in the feather, and if the animate and sensitive body were removed, it would be no more than a mere name (*un puro nome*). I believe that many qualities (*qualità*) which are attributed to natural bodies, such as tastes, odours, colours, and others, have a similar but no greater existence.

He went on to relate each of four senses to the four traditional elements, in a corpuscular theory of matter. Touch corresponded to earth, taste to water, smell to fire, hearing to air. The fifth sense, vision, corresponded to light, ether. Thus earth was continually being resolved into 'minimal particles' (*particelle minime*) of different kinds. Some of these, having been 'lodged on the upper surface of the tongue, and penetrating its tissue after being dissolved in its moisture, produce tastes that are pleasant or unpleasant according to the diversity of contact provided by the different shapes of these particles, and according to whether they are few or many and more or less rapidly in motion.' Similarly for smell and hearing. 'But,' he concluded, 'I hold that there exists nothing in external bodies for exciting in us tastes, odours and sounds other than sizes, shapes, numbers, and slow or swift motions; and I conclude that if the ears, tongue and nose were removed, shape, number and motion would remain but there would be no odours, tastes or sounds, which apart from living beings I believe to be nothing but names, exactly as tickling is nothing but a name if the armpit and the skin inside the nose be removed.' As to the relation of vision to light, he concluded: 'Of this sensation and the things connected with it I do not pretend to understand more than very little, and since I have not much time to explain, or rather to sketch that little, I shall remain silent.'

In this famous passage Galileo outlined a true mechanical philosophy of nature. Combining Democritus' distinction between the perceptual world of sensory appearance (which Aristotle took to be real) and the conceptual real world of the primary qualities, with a corpuscular conception of matter derived from Hero of Alexandria (see Vol. I, p. 46, note; above, p. 51, note), he offered an explanation of the manifest physical properties of bodies in terms of the characteristics of their component particles. These moreover he conceived dynamically, taking into account the variation of their motion, and seeming to envisage the extension to the particles of mathematical laws such as had proved so successful in dealing with the motions of macroscopic bodies.

Galileo's ultimate scientific aim of discovering the real structure of the physical world, of reading the real book of nature in mathematical language, is clearly shown not only in his controversies over the Copernican theory but in everything he wrote about the philosophy of science (see above, pp. 144 *et seq.*, 207 *et seq.*). Certainly this envisaged the establishing of a quantitative and empirically-verified connexion between the real but unobservable entities defined by the primary qualities and the observed properties of which these entities were the cause. Galileo himself also provided, in his 'resolutive-compositive' method, the effective means of exploring and establishing such a connexion. But the tactics exemplified in his kinematic approach to motion, his method of breaking up a problem into separate questions and proceeding step by step, meant that Galileo himself never in fact developed his mechanical philosophy into a scientific explanation, a theory deductively related to the prediction of the data. In fact with the current state of scientific knowledge it would have been rash speculation to attempt such a development systematically. Galileo preferred to keep it as the ultimate goal of his empirical progress.

It was Descartes who first not only claimed that the mechanical philosophy was the universal explanation of all physical phenomena, but also attempted to carry out the explanations in detail. Lacking Galileo's scientific finesse and sense of empirical fact, Descartes criticized Galileo's treatment of motion for providing mathematical descriptions without philosophical basis and therefore without explanation (see above, p. 170). Descartes' confident philosophical rationalism, his clear conception of a universal philosophy of nature as the goal of science, swept him into regions of speculation before which much better scientists hesitated. But just this speculative rashness was the source of his uniquely important contribution to the scientific movement. His bold unifying conception of the universe as an integrated whole explicable by universal mechanical principles applicable equally to organisms and to dead matter, to the microscopic particles and to the heavenly bodies, provided the succeeding generations of natural philosophers – astronomers, physicists, chemists, physiologists – with a programme. He gave them an hypothesis, a model whose properties they could exploit. Becoming the prevailing philosophy of nature by the mid-17th century, Cartesianism also brought out into the open philosophical problems inherent in the mechanical philosophy regarded as the whole truth and nothing but the truth. Even when Descartes' epistemology and metaphysics were rejected, his physics had a dominant influence, in the Royal Society as much as in the Académie des Sciences. Any new system had to make its way against it, and even the most celebrated alternative, the Newtonian system, to which Cartesian resistance in France was overcome only by Maupertuis (1698–1759) and Voltaire (1694–1778), was based on the same general programme of discovering the unifying laws of cosmology. It succeeded by establishing this Cartesian objective with greatly superior empirical precision. Even when proved wrong in detail, the general programme of Cartesian mechanism remained a guide to inquiry, and its general concepts also showed themselves admirably and fruitfully adaptable to the requirements of experimental results, as for example in physiology, in the theories of light of Hooke and Huygens, and in the later history of Descartes' *matière subtile* or ether filling space (cf. above, p. 169).

The basis of Descartes' philosophy of nature was his division of created reality (i.e. as distinct from God) into two mutually exclusive and collectively exhaustive essences or 'simple natures', extension and thought, and his conception of the method which was designed to give him certain knowledge of this reality. It is significant that Descartes should have resembled a medieval natural philosopher like Grosseteste or Roger Bacon in presenting his first published scientific results as examples of the application of a conception of scientific method. The epoch-making volume of treatises published in 1637 had the full title: *Discours de la Méthode pour bien conduire sa raison, et chercher la vérité dans les sciences. Plus la Dioptrique, les Météores et la Géometrie, Qui sont des essais de cette Méthode.* The fact that two of these treatises should have dealt with optics and that his earliest cosmological essay should have had the sub-title *Traité de la Lumière* is also an indication of at least part of Descartes' intellectual ancestry. But before any of these works he had already, between 1619 and 1628, written his fullest treatise on method, his *Regulæ ad Directionem Ingenii*, published posthumously in 1701. Such an order of composition could scarcely show more strongly his confidently rationalist approach to science.

'By method,' Descartes wrote in Rule iv of the *Regulæ*, 'I mean a set of certain and easy rules such that anyone who obeys them exactly will first never take anything false for true and secondly, will advance by an orderly effort, step by step, without waste of mental effort, until he has achieved the knowledge of everything that does not surpass his capacity of understanding.' He went on in Rule v: 'The whole of method consists in the order and disposition of the objects to which the mind's attention must be turned, that we may discover some truth. And we will exactly observe this method, if we reduce involved and obscure propositions step by step to simpler ones, and then, from an intuition of the simplest ones of all, try to ascend through the same steps to the knowledge of all others.'

A distinction must be made between Descartes' method as applied to philosophy and as applied to science. So far as philosophy is concerned, the rules he gave for analysing the data of experience were to prepare the mind for an intuitive act, similar to that described by Aristotle at the end of the *Posterior Analytics*, by which the 'simple natures' were grasped. These were, for example, thought, extension, number, motion, existence, duration – self-evident 'clear and simple ideas' which could not be reduced to anything simpler and so had no logical definitions. The purpose of the rules was to choose and arrange the data for this act of intuition, and they included a form of induction involving the principle of elimination. Descartes' philosophical aim was to reduce the 'involved and obscure propositions', with which we began from experience, to propositions that were either self-evident (simple natures) or had been already shown to follow from self-evident propositions. Having done this, he would then be able to explain the whole of the data of experience by showing that they could be deduced from the discovered 'simple natures'. In his search for the 'simple natures' constituting the created world he held that he had been successful. The ultimate substance of everything was either *res extensa* or *res cogitans*. As he wrote in the *Principia Philosophiæ*, part I, principle 53:

Although any one attribute is sufficient to give us knowledge of substance, there is always one principal property of substance which constitutes its nature and essence, and on which all the others depend. Thus extension in length, breadth and depth constitutes the nature of

bodily substance; and thought constitutes the nature of thinking substance. For all else that may be attributed to body presupposes
extension, and is merely a mode of this extended thing; and in the
same way everything that we find in mind is merely so many diverse
forms of thinking. Thus, for example, we cannot conceive of shape
except in an extended thing, nor of movement except in an extended
space; and similarly imagination, feeling and will exist only in a thinking thing, and we cannot conceive of them without it. But we can, on
the contrary, conceive of extension without shape and movement and
of thinking thing without imagination or feeling, and similarly for the
other attributes.

In part 2, section 4 he asserted the identity of matter and
extension even more emphatically, writing : 'The nature of matter, or of body in general, does not consist in its being a thing
which is hard or heavy or coloured or which affects our senses
in some other way, but only in its being a substance which is
extended in length, breadth and depth. . . . Its nature consists
simply in this, that it is a substance with extension.' Thus the
secondary qualities were subjective; only extension and motion
had any objective existence; and all the properties that we observed in matter were due to the diversification of the original
matter, under the influence of motion, into particles of different
sizes and shapes and motions and their subsequent aggregation
into bodies of various kinds. So anxious was Descartes to banish
the substantial forms and all innate real qualities – 'occult
properties' that he excluded even the idea that bodies were naturally endowed with weight. It was for assuming gravity to be an
innate quality, and for not attempting to explain it, that Descartes criticized Galileo to Mersenne (cf. above, p. 170). His own
attempt to explain gravity was by the *matière subtile* or ether
acting mechanically in the *plenum* of matter identified with
extension. In this *plenum* all action was by contact; it excluded
the possibility of a vacuum and was the basis of his theory of
vortices; and it enabled him to exclude the 'occult force' of
attraction at a distance.

When Descartes first discussed the application of his method
to natural science he was as confident of success as he was in

philosophy. The 'Universal Mathematics' adumbrated in the *Regulæ* was to repeat the structure of his philosophical system depending on the 'simple natures'. It was to embrace the whole physical world and to subordinate to itself all the particular sciences, and within this scheme science would discover the invariable cause, the invariable connexion between the *datum* of experience and the *quæsitum* of theory. Here indeed would be a complete union of prediction and explanation, if only it could be proved.

Descartes' account of scientific method in the *Regulæ* was a variant on the familiar double procedure of analysis and synthesis or resolution and composition. The object of scientific inquiry was to reduce the complex problems, as presented by experience, which he described in somewhat Aristotelian language as 'composite *a parte rei*', to specific constituent problems for quantitative solution, so that the complex situation could then be reconstituted theoretically and explained by deduction from the discovered elements and laws that produced it. The first stage of the analysis led to a classification of the data, and on the basis of these the investigator then set up hypothetical 'conjectures' of the cause. These were required because the complexity of nature necessitated an indirect route to the truth, and the next stage was to deduce the empirical consequences that followed from them, and to eliminate false conjectures by applying the Baconian method of the *experimentum* or *instantia crucis*, using the methods of agrement, difference and concomitant variation. The 'composite' of theory showed the true cause when it corresponded perfectly with the 'composite' of things. So the theory explained the facts, and the facts proved the theory (cf. above, pp. 41, 214, below, p. 327). Descartes described this reciprocal movement as a 'demonstration', writing in the *Discours*, part 6 :

If some of the matters of which I have spoken in the beginning of the *Dioptrics* and *Meteors* should offend at first sight, because I call them hypotheses and seem indifferent about giving proof of them, I request a patient and attentive reading of the whole, from which I hope those hesitating will derive satisfaction; for it appears to me that the reasonings are so mutually connected in those treatises, that, as the last are demonstrated by the first which are their causes, the first are in their turn demonstrated by the last which are their effects. Nor must it be imagined that I here commit the fallacy which logicians call a circle; for since experience renders the majority of the effects most certain, the causes from which I deduce them do not serve so much to establish their existence as to explain them; but on the contrary, the existence of the causes is established by the effects.

An 'Augustinian-Platonist' in the same way as Grosseteste and Roger Bacon, just as they found certainty only in Divine illumination, so Descartes found it only in the belief that the most perfect of all Beings would not deceive him. Backed by that guarantee, he asserted, in a letter to Mersenne written on 27 May 1638, 'There are only two ways of refuting what I have written: one is to prove by some experiments or reasoning that the things I have assumed are false; and the other, that what I deduce from them cannot be deduced.' Unfortunately, as Newton delighted to show, on all too many occasions Descartes exposed himself to refutation on just these grounds (cf. above, p. 170 *et seq.*).

Descartes' whole process of inquiry by means of conjectures presupposed the mechanical philosophy as the basis of the explanation, as distinct from the mere prediction or summary of the facts. For Descartes such explanations must always be the ultimate goal of scientific inquiry, because it was they that connected the particular phenomena of experience to the 'simple natures' that ultimately constituted the world and so provided the ultimate explanation of all phenomena. So by putting natural science into this philosophical framework Descartes made it necessary in some degree to answer the final question before asking the first one.

The same point of view appeared in his attitude to Harvey. In his description, in the *Traité de l'Homme*, of how the body could act according to purely mechanical laws, Descartes acclaimed Harvey's discovery of the circulation of the blood but refused to accept his account of the systole and diastole of the heart on the grounds that, even if Harvey's facts proved correct, he had not explained the *reason* for the heart's contraction. Descartes' own explanation of the heart-beat in fact rejected those of both Harvey and Galen alike and was a revival of Aristotle's conception of the heart as the centre of vital heat which caused the expulsion of the blood from the heart by making it boil and expand (see above, p. 243 *et seq.*). Later, in his *Description du Corps Humain* (1648; published 1664), Descartes admitted that *'une expérience fort apparente'*, such as one he suggested on the vivisection of a rabbit's heart, might confirm Harvey's account of the heart's motion, but he added: 'Nevertheless that only shows that the observations can often even lead us into being deceived, when we do not sufficiently examine all the causes which they could have.' Harvey's theory might be shown to agree with many of the phenomena, but 'that did not exclude the possibility that all the same effects might follow from another cause, namely from the dilatation of the blood which I have described. But in order to be able to decide which of these two causes is true, we must consider other observations which cannot agree with both of them.' The choice between the rival hypotheses must be made by an *experimentum crucis* which would falsify one of them.

The ultimate objective of Descartes' method, in science as in philosophy, was thus in the final analysis to display the connexion by 'long chains of deduction' between the ultimate ontological reality, as discovered in the 'simple natures', and the many particulars of experience. In this conception of an ultimately ontological goal of scientific discovery Descartes in fact agreed with Platonizing mathematical physicists like Galileo and Kepler, who had introduced such empirical conviction into the identification of the substance of the real world with the mathematical entities contained in the theories used to predict the 'appearances'. It was not in this ultimate ontological goal, but in the smaller degree of empirical caution with which he moved towards it, that Descartes differed from these more empirical contemporaries.

It was in the extreme and systematic form given to it by Descartes, offering a comprehensive metaphysical and cosmological alternative to the Aristotelian philosophy, that the mechanical philosophy raised the philosophical problems that came to shape the character not only of the epistemology and metaphysics of the period but also of the philosophy of science. For example, the doctrine of the subjectivity of the 'secondary' qualities was taken up by Locke and incorporated into his new theory of knowledge, according to which the proper objects of our knowledge are not things in an external world but the data of experience received through the sense organs, and organized by the mind. This is not the place to discuss Locke's epistemology, but it is interesting that it should have been the 'restorer' of the mechanical philosophy himself, Robert Boyle, who pointed out that the primary qualities or geometrical concepts in terms of which mathematical physics organized and interpreted experience were no less mental than the secondary qualities, and that if either group had any claim to reality then both had equal claims. George Berkeley (1685–1753) was to make a similar criticism.

A whole range of problems was raised by Descartes' absolute identification of matter with extension, aimed at the uncompromising exclusion from bodies of any innate properties whatsoever. In physics the difficulties this made in accounting for gravitation and in determining what was conserved in the conservation of motion became the main subjects of controversies between Huygens, Leibniz and the Newtonians. These are a good illustration of the metaphysical origin of many scientific concepts which were only later tailored to the requirements of quantitative precision (cf. above, p. 172). The total exclusion of the active principles in things corresponding to the scholastic 'natures' created a general difficulty for the whole doctrine of causation. Strictly speaking all 'secondary' causation (that is, causation apart from God's direct intervention) became impossible, as some of Descartes' followers pointed out. Some writers, for example Gassendi and Sir Kenelm Digby (1603–65), tried to deal with this general problem by returning to a form of atomism and, with some confusion, attributed efficient causality to the atoms themselves. A somewhat different solution to the whole problem of interaction was proposed by Leibniz with his theory of monads. These solutions came to have a considerable influence in biology, where the Cartesian doctrine of matter had caused great embarrassment by altogether excluding organisms. For example, when Maupertuis and Buffon (1707–88) tried to explain on mechanical principles such phenomena as the adaptation of the functions of the parts of living things to the needs of the whole and the teleological appearances of embryological development and of animal behaviour, they turned these particles in which causality was lodged into the '*molécules organisées*'. Maupertuis pointed out very clearly that mechanical concepts formulated to explain only a restricted range of inorganic phenomena must be expected to prove inadequate, when applied to other phenomena for which they were not designed. Since biological phenomena seemed to demand both active principles and teleology, his solution was to offer an explanation of them in terms of the antecedent movement of particles, whose behaviour anticipated the ends towards which they moved and the functions to be served by the organs they formed. In developing this form of explanation Maupertuis came to put forward the first systematic theory of organic evolution, and to discuss for the first time in this context the production of order out of disorder by the operation of chance.

It was in the question of interaction between body and mind, between the absolutely distinct extended substance and thinking substance, that the Cartesian system brought out into the open the most intractable problem for the mechanical philosophy, and one that has profoundly affected the whole philosophy of nature that has been developed by scientists, especially physiologists, since the 17th century. For Aristotelian philosophy there was strictly speaking no mind-body problem, since the soul, the *animus* of the scholastics, which included the mind (cf. Vol. I, p. 171, note), was the 'form' of the human being, and determined the nature of the psycho-physical unity just as the form of an inanimate body determined its nature. The problem arose with the mechanistic conception of the body. Joseph Glanvill wrote rhetorically in *The Vanity of Dogmatizing* (1661): 'How the purer spirit is united to this Clod, is a knot too hard for fallen Humanity to unty.'

Descartes discussed the question principally in his *Traité de l'Homme, Les Passions de l'Âme,* and the *Principia Philosophiæ.* His procedure in formulating it was clear and intelligent. Accepting the distinction between mind (sensation, feeling, thinking) and matter (as conceived mechanically), he decided on philosophical grounds that there was interaction between them in the human body. The main philosophical grounds for this conclusion were that we could not deny the reality, for example, of the apparent power of the body to generate in us sensations and feelings, without regarding God as a deceiver, which would be incompatible with his perfection. Moreover there was no good reason to deny it. Consequently he looked for a connexion between mind and body in an appropriate physiological mechanism, which he located in the pineal gland (cf. above, p. 245 *et seq.*).

Beginning with Gassendi, the critics of Descartes' theory of interaction pointed out that any point of contact between the mutually exclusive extended unthinking substance and unextended thinking substance was ruled out by definition. This led to a re-examination of the terms of Descartes' formulation of the theory of interaction and the development of three other solutions, parallelism, materialism, and phenomenalism. Between these four possibilities the problem has oscillated ever since.

Historically the first alternative to Cartesian interactionism was the form of parallelism known as 'occasionalism'. Developed principally by Geulincx (1625–69) and Nicolas Malebranche (1638–1715), this doctrine attributed all causal action immediately to God. When an event A seemed to produce another event B, they held that what really happened was that A furnished the occasion for God voluntarily to produce B. Thus although a physical event happening in the body might seem to produce a sensation in the mind, and an act of will might seem to produce a movement of the body, there was in fact no causal link between two such events except in God who produced them both. In his activities God usually followed fixed rules, so it was possible for natural philosophers to formulate general scientific laws. This was a position similar to that of Ockham (see above, p. 46).

The materialist solution of the mind-body problem was an attempt to reach the unity of theory at which science aims by showing that mental phenomena could be exhaustively derived from, or reduced to, the laws governing the behaviour of matter. The first modern author to put forward a materialist theory of this kind was Thomas Hobbes (1588–1679). It is natural that from the beginning materialism should have been associated with the motive of turning one half of the Cartesian duality into a system of anti-theological metaphysics, flying the banner of science. In the hands of the 'physiologists' of the French *Encyclopédie* like La Mettrie, D'Holbach, Condorcet and Cabanis, man became nothing but a machine; consciousness became a secretion of the brain just as bile was a secretion of the liver; and physical and physiological laws as they conceived them were taken as the norm of the laws not only of mind but also of history and the historical progress of society. Directly descended from the Cartesian mechanical philosophy and Newtonian physics, these conceptions developed by the 18th-century French natural philosophers and sociologists became the direct ancestors of the materialist doctrines associated with Charles Darwin's theory of evolution and its sociological extensions in the 19th-century doctrine of progress.

The phenomenalist, or idealist, solution aimed at getting rid of the Cartesian dualism by taking as the primary objects of knowledge not things in an external world known by means of sensation, but the data of sensation themselves. The physical world was then regarded as a mental construction from these data, existing only in a mind, although, as Berkeley argued, the only mind in which it could properly be said to exist was God's mind. It is characteristic of this doctrine that, in opposition to materialism, it was widely associated with the motive of saving theology from the conclusions that were being drawn from science and from the mechanical philosophy by writers motivated in the opposite direction.

Indeed the whole development of philosophy in relation to

science, and of the philosophy of science, since the 17th century is properly intelligible only within the wider context of the beliefs, especially the theological beliefs, of the period. Undoubtedly the dualism of the mechanical philosophy led to a feeling of bleak isolation of the human spirit, knowing beauty, conscience and the simple pleasures of the secondary qualities, in an inhuman infinity of matter-in-motion. 'Thus is Man that great and true Amphibium,' Sir Thomas Browne pointed the contrast in *Religio Medici* (1643) in his vivid baroque, 'whose nature is disposed to live, not only like other creatures in divers elements but in divided and distinguished worlds.' This reflects an effect on the sensibility that certainly forms part of the so-called 'crisis of conscience' to which the Scientific Revolution gave rise. But there were also specific theological doctrines whose practical influence on contemporary philosophy was probably much more important. For example Descartes, acting with unquestionable sincerity, kept a sharp eye on the doctrine of transubstantiation when developing his theory of matter and of material change. When he heard of Galileo's condemnation on the strength of certain Scriptural texts, he was prepared, with perhaps less unquestionable sincerity, to change his whole philosophy (cf. above, p. 222 *et seq.*).

Considerable light is thrown on the position in which Galileo and Descartes found themselves in relation to contemporary theology by recalling the moves that followed the introduction of Aristotelian philosophy into the West in the 13th century (cf. Vol. I, pp. 71–9, above, 48–9). The Aristotelian system came into circulation accompanied by the Averroïstic doctrines that the universe was a necessarily determined emanation from God's reason, instead of a free creation of his will as Christian theology taught; that the ultimate rational causes of things in God's mind could be discovered by the human reason; and that Aristotle had in fact discovered those causes, so that the universe must necessarily be constituted as he had described it, and could not be otherwise. By means of the Christian doctrines of the inscrutability and absolute omnipotence of God, the 13th-century theologians and philosophers liberated rational and empirical inquiry

into the laws that nature in fact exhibits from this absolute subjection to a metaphysical system. The price of this liberation,
however, was a much less exacting subjection to the revealed
Christian doctrines, and especially to that of the truth of the
word (literal or interpreted) of Scripture. Galileo no less than
Oresme was prepared willingly to pay this price, though not in
the currency pressed into his hand. What he rejected was in fact
the currency of Ockham, who, in his anxiety to save the content
of revelation from any possibility of threat from the side of
reason, had made radical further use of the doctrine of God's
absolute omnipotence to destroy the rational content of science
altogether. The observed regularities of the world became mere
regularities of fact, and the laws expressing them became at their
strongest mere possibilities, at their weakest simply conventional devices for correlation and calculation.

The currency that Galileo flung aside when it was offered to
him by Bellarmine and by Pope Urban VIII, Descartes was quick
to make his own. At the outset of his philosophical and scientific
inquiries Descartes had written with the greatest confidence of
being able to discover true and ultimate explanations. But after
1633 he became the *'philosophe au masque'*. He withdrew *Le
Monde,* and in the revised version of his system published in
Principia Philosophiæ in 1644 he made his famous declaration of
scientific theories as mere fictions. 'I want what I have written
to be taken simply as an hypothesis, which is perhaps far removed from the truth; but yet that having been done, I believe
that it will have been well done if everything deduced from it
agrees completely with the observations. For if that happens it
will be no less useful in practice than if it were true, because we
can use it in just the same way to set out the natural causes to
produce the effects we want.' (Part 3, section 44.) He continued
(in section 45): 'I shall assume here some things which I believe
to be false.' For example, he believed that, as the Christian religion required, God had created the world complete at the beginning, and with God's omnipotence this was reasonable. But we
could sometimes understand the general natures of things better
by supposing hypotheses which we did not believe to be literally

true, for example that all organisms came from seeds, 'although we know that they have not been produced in this way, if we are to describe the world only as it is, or rather as we believe that it was created.' He concluded, in section 47: 'Their falsity does not prevent that which may be deduced from them from being true.'

The policy indicated in these passages, the policy of Ockham, of Osiander, of Bellarmine, was aimed primarily not at interpreting the theoretical formulations of science but at tolerance between them and Christian theology. It was aimed at showing not only that the development of an anti-theological metaphysics was not a necessary consequence of the mechanical philosophy of science, but that science was in fact unable to yield any metaphysics at all. Adopted out of prudence, it is oddly placed in Descartes' philosophical outlook as a whole. It provided an escape clause allowing the practice of science to go on even in the face of theological propositions it might seem to contradict.

Many other aspects of 17th-century thought reflect the same tendency to avoid difficulty by separating scientific problems from theological and metaphysical entanglements as completely as possible. An example of this can be seen in occasionalism, for since God's will is inscrutable the occasionalist is left in fact only with observation and correlation as the proper objects of scientific inquiry.

It became a characteristic of many scientists of the period, of Mersenne, Pascal, Roberval, Mariotte, to refuse to discuss 'causes' in their physical inquiries; and likewise the Royal Society, consciously avoiding contentious subjects, became heavily experimental. The same policy of separating natural science from questions of ultimate causes was expressed by Boyle when he wrote in *The Excellency and Grounds of the Mechanical Hypothesis* (*Works*, abridged by Peter Shaw, 1725, vol. i, p. 187): 'The philosophy I plead for, reaches but to things purely corporeal; and distinguishing between the first origin of things, and the subsequent course of nature, teaches that God ... establish'd those rules of motion, and that order amongst things corporeal, which we call the laws of nature. Thus, the universe being once fram'd by God, and the laws of motion settled, and all upheld by his perpetual concourse and general providence ... the phenomena of the world are physically produced by the mechanical properties of the parts of matter.'

As events turned out, none of these moves to avoid theological trouble succeeded in their objectives. The advance of science did in fact give rise to materialist metaphysics, naïve certainly but to become nevertheless influential in the 18th and 19th centuries, and by definition anti-theological. The God of the scientists, of Boyle, the 'intelligent and powerful being' praised by Newton in the *Principia*, when taken over by the 18th-century Deists, no longer gave any primacy or uniqueness to Christianity among the religions. Most corrosive of all, the 'fictionalist' or 'conventionalist' policy adopted by Descartes and pressed forward by Berkely, became in the hands of secular philosophers like David Hume (1711–76), and of Immanuel Kant (1724–1804), the source of a doctrine that was anti-rational and anti-theological alike. Applied universally, as it inevitably was, it ceased to be a defence of theology against science and became a threat to all knowledge, whether rational or revealed. The way was open to the explicitly anti-theological and anti-metaphysical positivism of Auguste Comte (1798–1857) and John Stuart Mill (1806–73), and to the agnosticism of T. H. Huxley, which became so characteristic a part of the philosophical ambience of science in the 19th century. This was a consequence of the influence of their intellectual careers in which neither Galileo nor Descartes would have taken any pleasure, yet in some degree both foresaw it.

It would be misleading to leave the impression that all discussion of the philosophy of science in the 17th and 18th centuries was directed only at taking an attitude to theology. Dropping the crudely theological objective of Bellarmine and Descartes, the problem for philosophers became the relation of scientific knowledge to the possibilities of knowledge in general. From the time of Descartes the justification of the assumptions, procedures and conclusions of the new science became an essential part of the general problem of knowledge, which included the questions both of finding explanations (as distinct from mere predictions) in science and of the possibility of rational theology. All the great philosophers following Descartes, especially Leibniz, Berkeley, Kant, and Mill, contributed profoundly to the philosophy of science and were themselves profoundly influenced by their analyses of scientific thought.

No less important, both for the general philosophical atmosphere generated by science and for the philosophy of science, were the discussions of problems in this field by scientists themselves. Although these can be properly understood only within the wider philosophical context, they had in fact a distinctive objective. Where philosophers were primarily interested in science in relation to the general problem of knowledge, scientists usually became interested in the philosophy of science primarily in relation to specific problems encountered in the course of their scientific work. Many of these were not essential to a purely scientific solution. For example it is not necessary to discuss the mind-body problem in order to investigate the physiology of the brain and sense-organs, or to discuss the admissibility of action at a distance in order to investigate the laws of planetary motion. Nevertheless it was necessary that investigators looking for explanations from science should discuss such problems. No doubt because of their different objectives, the 20th-century dichotomy between the philosophy of science of scientists and that of philosophers can be seen in embryo even in the 17th century. Each tending more and more to ignore the writings of the other, the division solidified in practically all European educational systems in the 19th century, to the increasing disadvantage of both sides.

The discussions of the philosophy of science by scientists that most profoundly influenced the development of scientific thought in the 17th century all concerned the relationship between specific theories formulated for the purpose of predicting particular phenomena, and the mechanical philosophy of nature in terms of which it was assumed that all explanations in physics must be given. In fact the problem was similar to that existing between the predictive theories of the 13th and 14th centuries and the Aristotelian philosophy of nature. By the time the Royal Society had received its first charter in 1662 and the Académie des Sciences had been established in 1666, the attitudes to the problem had tended to polarize around the two dominant philosophies of science of the period, the empiricism and experimentalism inspired by Bacon and Galileo with its inveterate dislike of systems, and the Cartesian rationalism with its unifying conception of universal principles applying to every aspect of the physical world. The former was favoured by the majority of the English and the latter had its strongest supporters in France and Holland, but in fact no natural philosopher of the period escaped the influence of both. It was from the English experimental school, especially from Boyle and Newton, that the philosophy of science of the scientists as distinct from the philosophers received its most characteristic expression. Boyle and Newton were as convinced as Galileo that science discovered in its theories genuine knowledge about a real and objective natural world. But while the discovery of explanations and real causes remained their ultimate goal, they pursued a tough-minded policy of distinguishing sharply between experimentally established laws that gave accurate predictions, and the assumptions of the accepted philosophy of nature. Details of this last, especially those added speculatively by Descartes, they were always prepared to shelve. Thus they objected equally to the idea that scientific theories were mere fictions or calculating devices, and to the new scholasticism into which Descartes' lesser followers had crystallized his mechanical system. Their real contribution to contemporary and, indeed, to all succeeding philosophy of science was their systematic use of the experimental principle of verification and falsification to distinguish clearly between the different kinds of statements involved in a scientific system. The attitude taken up by this experimental school was well characterized by William Wotton in 1694 in his *Reflections upon Ancient and Modern Learning* : 'And therefore,' he wrote in chapter 20, 'that it may not be thought that I mistake every plausible Notion of a Witty Philosopher for a new Discovery of Nature, I must desire that my former Distinction between *Hypo-*

theses and *Theories* may be remembered. I do not here reckon the several *Hypotheses* of *Des Cartes*, *Gassendi*, or *Hobbes*, as Acquisitions to real Knowledge, since they may only be Chimæras, and amusing Notions, fit to entertain working Heads. I only alledge such Doctrines as are raised upon faithful Experiments, and nice Observations; and such Consequences as are the immediate Results of, and manifest Corollaries drawn from these Experiments and Observations: Which is what is commonly meant by *Theories*.'

It was Newton, becoming the acknowledged master of the experimental philosophy, who achieved the clearest appreciation of the relation between the empirical elements in a scientific system and the hypothetical elements derived from a philosophy of nature. Newton wrote no systematic philosophy of science, but like Galileo he was forced into discussions of scientific method by the controversies to which both his theory of colour and his theory of gravitation gave rise. Both were said by Cartesian critics, and especially by Huygens and Leibniz, to be descriptive and predictive but not explanatory. Presented in the context of controversy, and always in relation to specific problems, his statements have led to considerable misunderstanding. But they clearly indicate a consistent policy throughout. Forced into discussion by Huygens' criticism of his 'New Theory about Light and Colours,' published in the *Philosophical Transactions of the Royal Society* in 1671–72, it was in the subsequent controversy that Newton first took up his characteristic position. He pointed out first that his inquiry into the phenomenal laws was independent of any inquiry into the causes or mechanical processes producing them; secondly that it was only after the phenomenal laws had been established experimentally as the data to be explained that the inquiry for the explanation could begin with hope of success; and thirdly that no experimentally established law could be refuted because it was contradicted by an hypothesis about the causes of the phenomena. As he wrote on 2 June 1672 to Henry Oldenburg, the secretary of the Royal Society, in a letter printed in Samuel Horsley's edition of Newton's *Opera* (1782, vol. 4, pp. 314–15):

For the best and safest method of philosophizing seems to be, first diligently to investigate the properties of things and establish them by experiment, and then to seek hypotheses to explain them. For hypotheses ought to be fitted merely to explain the properties of things and not attempt to predetermine them except in so far as they can be an aid to experiments. If any one offers conjectures about the truth of things from the mere possibility of hypotheses, I do not see how any thing can be determined in any science; for it is always possible to contrive hypotheses, one after another, which are found rich in new tribulations. Wherefore I judged that one should abstain from considering hypotheses as from a fallacious argument, and that the force of their opposition must be removed, that one may arrive at a maturer and more general explanation.

He was to make these points again, in defence of his theory of gravitation, in query 31 of the *Opticks* (1706) and in the Rules of Reasoning in Philosophy, especially Rule iv (1726), at the beginning of the third book of the *Principia*.

From this eminently reasonable position Newton brought clarity into the whole subject of scientific method and logic, and established a policy that was both critical and fruitful for dealing with the relation between the data and phenomenal laws on the one hand, and hypotheses about causes on the other. By means of this policy he showed how mechanical hypotheses could be a fruitful guide to research without becoming misleading. Indeed, possibly because he was not deceived about their hypothetical status, where others would propose one explanation and defend it against all objections, his fertile mind would suggest a whole range of hypotheses, for example of the ether as an explanation of the phenomena of light, gravitation, cohesion, electric and magnetic attraction. Far from excluding from the competence of science the discovery of the real processes in nature causing the phenomenal laws, Newton in fact took them so seriously as the ultimate objective of scientific inquiry that he insisted that the investigation of causes must be conducted as rigorously as that of the laws themselves. 'There are therefore Agents in Nature able to make the Particles of Bodies stick together by very strong Attractions,' he exclaimed in query 31 of the *Opticks*, 'And it is the Business of experimental Philosophy to find them out.' The famous aphorism, *hypotheses non fingo*, in the General Scholium at the end of Book 3 in the second edition of the *Principia* (1713), was directed, as Koyré has pointed out, not against hypotheses about real causes, but against Cartesian fictions and fictionalism. Indeed it is likely that he chose the title *Principia Mathematica* in order to give direct point to his polemic against Descartes' *Principia Philosophiæ*. Thus Newton reversed Descartes' rebuke to Galileo for not providing explanations, and did so by Galileo's own methods of science, which he brought to completion.

Newton certainly did not regard scientific laws as mere predicting devices. They were written in the phenomena, though they were not open to direct inspection and had to be discovered or 'inferred' or 'deduced' from the phenomena by appropriate mathematical and experimental analysis. In the sense that he was searching for true explanations, Newton had the same objective as Aristotle and all his intellectual descendants. But the Aristotelian 'natures' offered explanations divorced from predictive laws. It was this divorce that had occasioned the whole discussion between prediction and explanation since the 13th century and had led to the replacement of Aristotelian physics by the mathematical and mechanical philosophy of nature. As Newton wrote of the Aristotelian 'natures' in query 31 of the *Opticks*, echoing Galileo:

Such occult Qualities put a stop to the Improvement of natural Philosophy, and therefore of late Years have been rejected. To tell us that every Species of Things is endow'd with an occult specified Quality by which it acts and produces its manifest Effects, is to tell us nothing: But to derive two or three general Principles of Motion from Phænomena, and afterward to tell us how the Properties and Actions of all coporeal Things follow from those manifest Principles, would be a very great step in Philosophy, though the Causes of those Principles were not yet discover'd: And therefore I scruple not to propose the Principles of Motion above-mention'd, they being of very general Extent, and leave their Causes to be found out.

By applying the same rigorous quantitative methods to hypotheses about causes as to laws, Newton wanted to point the way towards the goal of the whole experimental school of natural philosophy : the union of explanatory theory and predictive laws in a single theoretical system. Thus, having solved, by means of his laws of motion and of gravitation, the problem of the dynamics of macroscopic bodies on earth and in the heavens, he wrote in the preface to the first edition of the *Principia* : 'I wish we could derive the rest of the phenomena of Nature by the same kind of reasoning from mechanical principles, for I am induced by many reasons to suspect that they may all depend upon certain forces by which the particles of bodies, by some causes hitherto unknown, are either mutually impelled towards one another, and cohere in regular figures, or are repelled and recede from one another. These forces being unknown, philosophers have hitherto attempted the search of Nature in vain; but I hope the principles here laid down will afford some light either to this or some truer method of philosophy.'

Two further passages indicate the continuity of the logical structure of his science with the long tradition stretching back through Galileo and the medieval writers on the 'resolutive-compositive' method to the Greek geometers (cf. above, p. 29). In query 31 of the *Opticks* he wrote :

As in Mathematicks, so in Natural Philosophy, the Investigation of difficult Things by the Method of Analysis, ought ever to precede the Method of Composition. This Analysis consists in making Experiments and Observations, and in drawing general Conclusions from them by Induction, and admitting of no Objections against the Conclusions, but such as are taken from Experiments, or other certain Truths. For Hypotheses are not to be regarded in experimental Philosophy.* And although the arguing from Experiments and Observations by Induction be no Demonstration of general Conclusion; yet it is the best way of arguing which the Nature of Things admits of, and may be looked upon as so much the stronger, by how much the Induction is more general. And if no Exception occur from Phænomena, the Conclusion may be pronounced generally. But if at any time afterwards any Exception shall occur from Experiments, it may then begin to be pronounced with such Exceptions as occur. By this way of Analysis we may proceed from Compounds to Ingredients, and from Motions to the Forces producing them; and in general, from Effects to their

* That is, hypothesis in the sense of explicit fictions.

Causes, and from particular Causes to more general ones, till the Argument end in the most general. This is the Method of Analysis: And the Synthesis consists in assuming the Causes discover'd, and establish'd as Principles, and by them explaining the Phænomena proceeding from them, and proving the Explanations.

Replying in 1712 to Roger Cotes, who was seeing the second edition of the *Principia* (1713) through the press, Newton wrote to clarify further his conception of the distinctions to be made between the different propositions of a scientific system. His purpose was to explain the phrase *hypotheses non fingo* in the General Scholium. He wrote:

... as in Geometry the word Hypothesis is not taken in so large a sense as to include the Axioms and Postulates, so in Experimental Philosophy it is not to be taken in so large a sense as to include the first Principles or Axioms which I call the laws of motion. These Principles are deduced from Phænomena and made general by Induction: which is the highest evidence that a Proposition can have in this philosophy. And the word Hypothesis is here used by me to signify only such a Proposition as is not a Phænomenon nor deduced from any Phænomena but assumed or supposed without any experimental proof.

In one case Newton seems to have meant that laws (or 'principles') were 'deduced from phenomena' in the strict and literal sense, for he showed that just as Kepler's planetary laws could be deduced from the laws of motion and the inverse-square law of gravitation, so the last could be deduced from Kepler's Third Law, describing the phenomena. What he had done in fact was to demonstrate a reciprocal implication between a more and a less general law; his other statements show that he recognized clearly that this does not apply to the relationship between a law and the phenomenal data. In the search for certainty in science, the reciprocal relationship represented an ideal derived from mathematics (cf. above, pp. 42, 200, 205). Showing clearly the 'Euclidean' conception of the structure of theoretical science established by the long tradition which he had inherited, the purpose of Newton's distinction was to state explicitly the extent to which the first principles of a science and of an explanation could be said to have been verified. In the controversies on this question into which his explanations of colour and of planetary motion had drawn him, his policy was to reject, on the one hand, hypotheses proposed as explicit fictions and, on the other, the use of hypotheses of any kind as objections to experimentally established laws, against which the only objections could be contrary experimental evidence or proof of logical inconsistency. So he concluded finally in Rule iv in book 3 of the third edition of the *Principia* (1726): 'In experimental philosophy we are to look upon propositions inferred by general induction (*per inductionem collectæ*) as accurately or very nearly true, notwithstanding any contrary hypotheses that may be imagined, till such time as other phenomena occur, by which they may either be made more accurate, or liable to exceptions. This rule we must follow, that the argument of induction may not be evaded by hypotheses.'

A further well-known passage, from the preface to Huygens' *Traité de la Lumière* (1690), shows how far the method of reasoning in the new physics of the 17th century had moved from the Greek conception of geometrical demonstration. Instead of the justification of conclusions by showing them to be the necessary consequences deduced from first principles accepted as axiomatic, attention is now transferred to justification of the theoretical principles themselves by their observable consequences. It is asserted that the test by consequences achieves not certainty but only probability. The probability of a theory being true is said to increase with the number and range of confirmations, especially in predicting new phenomena. And it is claimed that this method enables us to discover the causes of events. 'There is to be found here,' Huygens wrote, 'a kind of demonstration that does not produce so great a certainty as that of geometry, and is indeed very different from that used by geometers, since they prove their propositions by certain and incontestable principles, whereas here principles are tested by the consequences derived from them. The nature of the subject permits no other treatment. Nevertheless it is possible to reach in this way a degree of probability that is often scarcely less than complete certainty. This happens when the consequences of our assumed principles agree perfectly with the observed phenomena, and especially when such confirmations are numerous, but above all when we can imagine and foresee new phenomena which should follow from the hypotheses we employ and then find our expectations fulfilled. If in the following treatise all these evidences of probability are to be found together, as I think they are, this ought to be a very strong confirmation of the success of my inquiry, and it is scarcely possible that things should not be almost exactly as I have represented them. I venture to hope, therefore, that those who enjoy finding out the causes of things and can appreciate the wonders of light will be interested in these various speculations about it.'

For two centuries it was widely held by scientists that Newton had himself provided just such a union between prediction and explanation as all had been searching for, but already among Newton's earliest critics there were philosophers who did not share his optimism that science could discover 'causes' at all. Newton himself had stressed the sharp empirical distinction that *in fact* existed between knowledge of laws and of causes as envisaged by the current philosophy of nature. Reviving the conclusion reached by the scholastic logicians from Grosseteste to Nifo and Zabarella, that the data of observation cannot uniquely determine the theory that explains them, some 18th-century philosophers began to see the results of scientific inquiry less as discoveries about nature than as products of the methods of thought used.

The most acute of the contemporary critics of the Newtonian system was Berkeley, who in his *De Motu* (1721) anticipated much of Mach's famous analysis of Newton's basic assumptions. Developing arguments similar to those used by the medieval logicians, Berkeley came to the conclusion that neither the Newtonian system nor any other scientific theory could give an account of 'the nature of things' or establish the causes of phenomena. Such a physical system was a 'mathematical hypothesis'; it established simply the 'rules' by which phenomena were found to be connected, and by means of which they could be predicted. Berkeley claimed that there was no justification for Newton's conceptions of absolute space and time and that all motion was relative.

Hume, the 18th-century Ockham, went even further than Berkeley in claiming that science was irrational and that explanation was strictly speaking impossible. Since the empirical data did not carry their own explanation or give grounds for belief in causality, and since he could see no other grounds, he concluded that there was nothing objective in causal necessity beyond regular concomitance and sequence. 'In a word, then,' he declared in section 4 of his *Inquiry Concerning Human Understanding*, 'every effect is a distinct event from its cause. It could not, therefore, be discovered in the cause; and the first invention or conception of it, *a priori*, must be entirely arbitrary.'

An analogous 'nominalist' view of biological categories above that of species was developed by Buffon (1707–88) and other biologists in their critique of Linnæus' 'realist' system of classification. Buffon declared that nature contained only individuals; that the species, defined as the succession of individuals capable of interbreeding, was a real category; but that the 'family' and the higher categories were mere names.

Awakened by Hume's critique, yet firmly believing in the truth of the Newtonian system, to the extension of which in fact he made a contribution as a physicist, Kant found himself able to admit Newtonian science as true only at the price of denying that it had discovered a real world of nature behind the world of appearance. Similarly he found himself obliged to deny the possibility of rational knowledge of God, in whom he also firmly believed. Kant could admit Newtonian science as a true science of nature precisely because he came to regard nature itself as the world of phenomena, the world as it appeared to our assimilating minds, and because he came to regard scientific theories as products of the methods of organizing experience, methods dictated by the structure of our minds. Because of that structure Kant believed that the scientist approached nature with certain necessary principles in mind, of which Euclid's propositions were explicit formulations, and that he necessarily presupposed these principles in all his knowledge and in all the theories with which he attempted to organize his experience. It was this view of science, the reflection of a philosophical situation produced by the success of the scientific revolution itself, as seen by a mind acutely aware of the processes of theoretical construction, that Kant described in his brilliant preface to the second edition of the *Critique of Pure Reason* (1787):

When Galileo let balls ot a particular weight, which he had determined himself, roll down an inclined plane, or Torricelli made the air carry a weight, which he had previously determined to be equal to that of a definite volume of water; or when, in later times, Stahl changed metal into calx, and calx again into metal, by withdrawing and restoring something; a new light flashed on all students of nature. They comprehended that reason has insight into that only, which she herself produces on her own plan, and that she must move forward with the principles of her judgements, according to fixed law, and compel nature to answer her questions, but not let herself be led by nature, as it were in leading strings, because otherwise accidental observations, made on no previously fixed plan, will never converge towards a necessary law, which is the only thing that reason seeks and requires. Reason, holding in one hand its principles, according to which concordant phenomena alone can be admitted as laws of nature, and in the other hand the experiment, which it has devised according to those principles, must approach nature, in order to be taught by it: but not in the character of a pupil, who agrees to everything the master likes, but as an appointed judge, who compels the witnesses to answer the questions which he himself proposes. Therefore even the science of physics entirely owes the beneficial revolution in its character to the happy thought, that we ought to seek in nature (and not import into it by means of fiction) whatever reason must learn from nature, and could not know by itself, and that we must do this in accordance with what reason itself has originally placed into nature. Thus only has the study of nature entered on the secure method of a science, after having for many centuries done nothing but grope in the dark.*

* This passage has a suggestive position in the parallel development of conceptions of nature and of thought. At least since Francis Bacon, philosophers and scientists had been reducing nature to matter in motion and, somewhat later, thought to the association of impressions and ideas. The

All the subsequent philosophies of science that have developed in the 19th and 20th centuries have taken their shape in one way or another from the doctrines developed from Francis Bacon, Galileo and Descartes to Kant. It was for example an easy step from Kant's view that theories are read not in but *into* nature, to Auguste Comte's assertion that the real goal of science was and always had been not knowledge at all, but only power (cf. above, p. 317). Seizing only one half of Bacon's Great Instauration, Comte declared in his *Cours de Philosophie Positive* (1830), Première Leçon, that the object of science was '*savoir, pour prévoir*', in effect prediction to give control. This needed only knowledge of empirical sequences, and to ask for knowledge of the nature of things beyond this was not only useless but was to ask for something unattainable. It was to provide sure means of establishing such empirical connexions that Comte's friend John Stuart Mill developed his own systematic account of scientific method. On the opposite side, Kant's account of scientific inquiry not as a mere dissection of nature, but as a process of active questioning in the light of preconceived principles, was used by William Whewell in his emphasis, in opposition to Comte and Mill, on the role of 'ideas' and hypotheses in scientific inquiry. Harking back to the 'argomento *ex suppositione*' and the 'resolutive-compositive' method of Galileo, the same point has been

---

behaviour of both bodies and minds was determined by external events. Kant's pre-critical and critical writings both indicate a concern with the mechanism-organism problem. In the 'Critique of the Teleological Judgement' (Part 2 of the *Critique of Judgement*, 1790), a brilliant contribution to the philosophy of biology, he made a point of the impossibility in principle of explaining the facts of organic unity in mechanistic terms, even though all the parts of the unity could be analysed mechanistically. Thus he concluded that a living organism was not a mere aggregate of unrelated mechanistic constituents, but a functionally related system of parts bound together by a principle of unity. Analogously, in the *Critique of Pure Reason*, he gave to the mind a principle whereby it determined the connexions of impressions and ideas according to its own plan. In both cases the emphasis is placed on the actively controlling role of the intrinsic principle, and in this Kant reintroduces something like Aristotle's matter and form in opposition to the mechanical philosophy of the 17th century.

made by Mill's recent critics in stressing the 'hypothetico-deductive' structure of science. Twentieth-century 'conventionalism', immediately the result largely of internal developments of physics which led to the abandonment of some of Newton's basic principles and the use of non-Euclidean geometries to 'save the appearances', is likewise both an advance on the position reached by Kant and a return to an earlier position. Physics itself having disposed at least of the necessity of assuming Euclid's principles, the conception has grown, especially under the influence of Mach, Henri Poincaré and Duhem, that any theoretical system can be used to correlate experience, provided it passes the tests of logical coherence and experimental verification. Saluting the attempts made from Simplicius to Bellarmine to make sense of the state of astronomical theory before Kepler, the attemps of this school to deal with an analogous modern problem have made the choice of a system, apart from these tests, simply a matter of convenience and convention.

At the beginning of the European philosophical adventure, the search for the rational intelligibility of the world as we experience it, Hesiod's Muses announced darkly: 'We know how to tell many fictions that wear the guise of truth; but we also know how to declare the truth, when we will.' Lacking the gift of oracular understanding, the men who have in fact conducted the adventure since Greek times have themselves been able to make this philosophical distinction only by searching not only for the truth but also for principles for distinguishing truth from falsehood. Ever since the Greeks took the decisive step in cosmology of looking for explanations deductively connected with the means of prediction, the step by which they established the European scientific tradition as distinct for example from Babylonian astronomy in which there was a total logical disjunction between the highly developed technological predictions and the myths that did service for explanations, the problem of finding criteria for distinguishing true explanations from false has been a pre-eminent question in the growth of science. Seeking as they did knowledge as well as utility, the Greeks established European science as a philosophical activity different both from

Eastern technology which largely knew no science and from Western technology which is science applied.

Inevitably in such an enterprise conceptions of scientific truth have themselves undergone development under the impact both of the internal problems of science and of philosophical criticism. But through the diversity of such conceptions and of actual scientific performance, from Plato's time to the present, the philosophical policy of science has remained consistently the same. Of this there could be no more telling illustration than that provided by the period reviewed in the foregoing pages. Apparently so full of metaphysical and theological distractions, even these were turned to good account, first in the conception of a system of rational explanation as such, and eventually in the great theoretical formulations of the period of Kepler and Galileo. The creative processes of original discovery and invention, always mysterious, are as little open to direct inspection as the laws of nature themselves. It is part of the philosophical enlightenment provided by the history of science to discover that the thought of great innovators whose effectiveness we admire was organized on a pattern in many ways so utterly different from our own, that they accepted a complex of non-empirical conceptions and 'regulative beliefs' that, alien though they are to us, nevertheless gave construction and form to theories of the greatest predictive and explanatory power. But it is a further part of enlightenment to discover that in spite of immediate appearances the policy for dealing with such a pattern of thought, the criteria of verification and the objective towards which they are applied, has preserved its essential continuity throughout the whole European tradition.

Putting forward theories as true, but always submitting them to the experimental test, the intuition that has governed the scientific tradition has been characterized by Pascal in his *Pensées* (395): '*Nous avons une impuissance de prouver, invincible à tout le dogmatisme. Nous avons une idée de la vérité, invincible à tout le pyrrhonisme.*' Balanced between intuition and reason, between imagination and experiment, philosophical opinion in relation to science has oscillated between the extremes of scepticism and rationalism according to whether claims to have discovered ultimate reality, putting a stop to all further inquiry, or claims that no rational knowledge is possible at all, reducing science to an irrational technology, have presented most danger to the hopes of the moment. 'For who will prescribe bounds to man's intelligence and invention?' asked Galileo, the scientific realist, in 1615. 'Who will assert that all that is sensible and knowable in the world is already discovered and known?' It is through the development of this pragmatic policy of taking each case separately on its merits, of refusing to be bound by its own constructions, that the history of the Scientific Revolution throws its most significant light, not only on the nature of science itself, but also on all those other aspects of modern European thought that have arisen from an attitude taken to its methods and conclusions.

NOTES TO PLATES

Plate 8. The lower part of the page reads as follows:
'My first observation & others following of the new found planets about Jupiter.
1610 Syon.

1. Octob. 17. ☿ [Mercury]. ho. 12ᵃ. 1ᵃ. 2ᵃ. I saw but one, & that above.
   Blackfriers, London.
2. November. 16. ♀ [Venus]. ho. 9ᵃ. 10ᵃ. I saw one fayre 9' or 10' above, and sometimes I thought I saw an other very small betwixt them, 3' or 4' à ♃ [Jupiter]. London.
3. November. 19. ☽ [moon]. ho. 9ᵃ. one under fayre.
4. Syon. Novemb. 28. ☿ ho. 9ᵃ. one under. fayre.
5. November. 30. ♀ ho. 9ᵃ one above. fayre.
6. Decemb. 4. ☿. ho. 9ᵃ. one under. fayre.
7. Decemb. 7. ho. 9ᵃ. 9ᵃ½. I saw but one, & above.
8. Mane. ho. 17ᵃ. Two seen on the west side, a little under. Sir W. Lower also saw them here. The nerest fayrest. The farther not well seen within the reach of my instrument of 20/1 of 14' dyameter.'

Plate 12. 'The heart drives out the blood in its restraining ... This thing was ordained by Nature in order that, when the right ventricle begins to shut, the escape of the blood out of its big capacity should not suddenly cease, because some of that blood had to be given to the lung; and none of it would have been given, if the valves had prohibited the exit; (but this ventricle shut, when the lung had received its quantity of blood, and when the right ventricle could press through the pores of the medium wall into the left ventricle); and at this time the right auricle made itself the depository of the superabundance of the blood which it advances to the lung that suddenly renders it to the opening of this right ventricle, restoring itself through the blood which the liver gives it. How much blood is the liver able to give it

337

through the opening of the heart? It gives as much of it as it consumes, i.e., a minimum quantity, because in one hour about two thousand openings of the heart take place. There is great weight ... 7 ounces an hour.' (*Quaderni d'Anatomia*, ed. O. C. L. Vangensten, A. Fonahn and H. Hopstock, Christiana, 1912, vol. 2, f. 17v.)

BIBLIOGRAPHY

CHAPTER I

PHILOSOPHY AND SCIENTIFIC METHOD IN GENERAL:
For extensive studies specifically on scientific method there are Crombie, *Robert Grosseteste and the Origins of Experimental Science*, Oxford, 1971, 'Quantification in medieval physics', *Isis*, lii (1961), and J. H. Randall, jr, *The School of Padua and the Emergence of Modern Science*, Padua, 1961. Cf. P. Duhem, 'Essai sur la notion de théorie physique ...', *Annales de philosophie chrétienne*, vi (1908) and *La théorie physique*, Paris, 1914 – pioneer studies of somewhat positivist tendency. Other studies of Greek and 13th-century conceptions of method and philosophy of science are mentioned in Vol. I under Chapters 1, 2, and 3 (especially the General section, and those on Biology and Cosmology and Astronomy), and below. Cf. below under Chapter 2, Mathematics and Mechanics, Astronomy, Experimental Physiology and Conceptions of Science and Method. For further discussions, and philosophical background, there are N. Abbagnano, *Guglielmo di Ockham*, Lanciano, 1931; L. Baudry, 'Les rapports de Guillaume d'Occam et de Walter Burleigh', *Archives d'histoire doctrinale et littéraire du moyen âge*, ix (1934), *Le tractatus de principiis theologiae attribué à G. d'Occam*, (*Études de philos. médiévale*, xxiii) Paris, 1936, *Guillaume d'Occam: sa vie, ses œuvres, ses idées sociales et politiques* (*Études de philos. médiévale*, xxxix) Paris, 1949– , i– ; P. Boehner, G. E. Mohan and A. C. Pegis, several articles on William of Ockham in *Franciscan Studies*, N.S. i–xi (1941–45), *Traditio*, i–iv (1943–46), and *Speculum*, xxiii (1948); P. Boehner, *Medieval Logic. An outline of its development from 1250 to c. 1400*, Manchester, 1952; R. Carton, *L'Expérience physique chez Roger Bacon* (*Études de philos. médiévale*, ii) Paris, 1924; W. C. Curry, *Chaucer and the Medieval Sciences*, New York, 1926; Nicolas de Cués, *Œuvres* choisis, trad. de Gandillac, Paris, 1942; Nicholas Cusanus, *Of Learned Ignorance*, trans. by F. G. Heron, London and New Haven, 1954; A. Edel, *Aristotle's Theory of the Infinite*, New York, 1934;

Bibliography

M. Patronnier de Gandillac, *La philosophie de Nicolas de Cués*, Paris, 1941; N. W. Gilbert, *Renaissance Concepts of Method*, New York, 1960; E. Gilson, *The Unity of Philosophic Experience*, London, 1938, *Reason and Revelation in the Middle Ages*, New York, 1938; R. Guelluy, *philosophie et théologie chez Guillaume d'Ockham*, Louvain and Paris, 1947; W. H. Hay, 'Nicolaus Cusanus: The structure of his philosophy', *The Philosophical Review*, New York, lxi (1952); V. Heynck, 'Ockham Literatur 1919–1949', *Franziskanische Studien*, xxxii (1950); W. and M. Kneale, *The Development of Logic*, Oxford, 1962; A. Koyré, 'Les origines de la science moderne', *Diogène*, No. 16 (1956); G. de Lagarde, *La naissance de l'esprit laïque au déclin du moyen âge*, Paris, 1934–46, 6 vols.; J. Lappe, *Nicolaus von Autrecourt, Sein Leben, seine Philosophie, seine Schriften (Beitr. Ges. Philos. Mittelalt.*, vi, 2), 1908; E. Longpré, *La Philosophie du B. Duns Scotus*, Paris, 1926; R. McKeon, *Selections from Medieval Philosophers*, New York, 1930, ii, 'Aristotle's conception of the development and the nature of scientific method', *J. Hist. of Ideas*, viii (1947); A. Maier, 'Zu einigen Problemen der Ockhamforschung', *Archivum Franciscanum Historicum*, Florence, xlvi (1953); A. Mansion, 'L'induction chez Albert le Grand', *Revue néo-scolastique*, xiii (1906); G. de Mattos, 'L'intellect agent personnel dans les premiers écrits d'Albert le Grand et de Thomas d'Aquin', ibid., xliii (1940); K. Michalski, 'Les courants philosophiques à Oxford et à Paris pendant le xive siècle', *Bulletin international de l'Académie polonaise des sciences et des lettres (Cracovie)*, Classe d'hist. et de philos., 1920, 'Les sources du criticisme et du scepticisme dans la philosophie du xive siècle', ibid., 1922, 'Les courants critiques et sceptiques dans la philosophie du xive siècle', ibid., 1925; P. Minges, *Joannis Duns Scoti Doctrina Philosophica et Theologica*, Berlin, 1930, 2 vols; E. A. Moody, *The Logic of William of Ockham*, New York, 1935, *Truth and Consequence in Medieval Logic*, Amsterdam, 1953; William of Ockham, 'The *Centiloquium* attributed to ...', ed. P. Boehner, *Franciscan Studies*, N.S. i (1941), ii (1942), *Summa Logicae*, ed. P. Boehner, (Franciscan Inst. Publ., text series No. 2), St Bonaventure, N.Y., and Louvain, 1951–54, 2 vols., *Philosophical Writings*, ed. Boehner, Edinburgh, 1957; J. R. O'Donnell, edition and study of Nicholas of Autrecourt in *Mediaeval Studies*, i (1939), iv (1942); C. v. Prantl, *Geschichte der Logik im Abendlande*, Leipzig, 1855–70, 4 vols.; H. Rashdall, 'Nicholas de Ultricuria, a medieval Hume', *Proc. Aristotelian Soc.*, N.S. vii (1907); P. Rossi, *Clavis Universalis*, Milan and Naples, 1960; H. Scholz und H. Schweitzer, *Die sogenannten Definitionen durch Abstraktion (Forschungen zur Logistik und zur Grundle-*

Bibliography

gung der exakten Wissenschaften, ed. H. Scholz, Heft iii), Leipzig, 1935; L. Thorndike, *Science and Thought in the 15th Century*, New York, 1929, 'Dates in intellectual history: the 14th century', J. *History of Ideas*, vi (1945), Suppl. i; S. C. Tornay, *Ockham, Studies and Selections*, La Salle, Ill., 1938; E. Vansteenberghe, *Le Cardinal Nicolas de Cués (1401–1464)*, Paris, 1920; J. R. Weinberg, *Nicolaus of Autrecourt*, Princeton, 1948.

MATHEMATICS: Archimedes, *Works*, ed. in modern notation with introductory chapters by T. L. Heath, Cambridge, 1897, New York, 1953 (in English), *Les œuvres complètes*, traduit du grec en français avec une introduction et des notes, Paris et Bruxelles, 1921; W. W. Rouse Ball, *A Short Account of the History of Mathematics*, 3rd ed., London, 1901; G. Beaujouan, 'L'enseignement de l'arithmétique élémentaire à l'université de Paris aux xiiie et xive siècles: de l'abaque à l'algorisme', *Hommage à Millás-Vallicrosa*, Barcelona, 1956, i; O. Becker und J. E. Hofmann, *Geschichte der Mathematik*, Bonn, 1951; C. B. Boyer, *The Concepts of the Calculus*, New York, 1939, *History of Analytic Geometry*, New York, 1956 (reprint of articles in *Scripta Mathematica)*; L. Brunschwig, *Les étapes de la philosophie mathématique*, 3e éd., Paris, 1929 – fundamental for mathematical method; F. Cajori, *A History of Mathematical Notations*, London, 1929; M. Cantor, *Vorlesungen über Geschichte der Mathematik*, Leipzig, 1900, ii; E. J. Dijksterhuis, *Archimedes*, Copenhagen, 1956 – very useful; Euclid, *Elements*, English trans. and introduction by Sir T. L. Heath, Cambridge, 1926, 3 vols. – excellent for geometrical method; Sir T. L. Heath, *A History of Greek Mathematics*, 2 vols., Oxford, 1921, *Mathematics in Aristotle*, Oxford, 1949 – excellent study of method; G. F. Hill, *The Development of Arabic Numerals in Europe*, Oxford, 1915; J. E. Hofmann, *A History of Mathematics*, New York, 1957 (trans. of *Geschichte der Mathematik*, Berlin, 1953, i); G. Libri, *Histoire des science mathématiques en Italie*, Paris, 1838–41, 4 vols.; Gino Loria, *Storia delle matematice*, Turin, 1929–33, 3 vols.; P. H. Michel, *De Pythagore à Euclide. Contribution à l'histoire des mathématiques préeuclidiennes*, Paris, 1950; D. E. Smith, *History of Mathematics*, New York, 1958; Dirk J. Struik, *A Concise History of Mathematics*, New York, 1948, 2 vols.; Suter, *op. cit.* in Vol. I under Chapter 3, Cosmology and Astronomy; P. Tannery, *Mémoires scientifiques*, v. *Sciences exactes au moyen âge*, publiés par J. L. Heiberg, Toulouse et Paris, 1922; K. Vogel (editor), *Die Practica des Algorismus Ratisbonensis*, Munich, 1954; H. Wieleitner, *Geschichte der Mathematik*,

Bibliography

Berlin, 1939, 2 vols.; H. G. Zeuthen, *Histoire des mathématiques dans l'antiquité et le moyen âge*, Paris, 1902.

LATE MEDIEVAL PHYSICS: The basic studies are M. Clagett, *The Science of Mechanics in the Middle Ages*, Madison, Wisconsin, 1959; E. J. Dijksterhuis, *The Mechanization of the World Picture*, Oxford, 1961; P. Duhem, *Système du Monde* and *Études sur Léonard de Vinci*, Paris, 1906–13, 3 séries; and A. Maier, *Zwei Grundprobleme der Scholastischen Naturphilosophie*, Rome, 1951 (2nd ed. of *Das Problem der intensiven Grösse* ..., 1939, and *Die Impetustheorie* ..., 1940), *An der Grenze von Scholastik und Naturwissenschaft*, Essen, 1943, *Die Vorläufer Galileis im 14. Jahrhundert*, Rome, 1949, *Metaphysische Hintergründe der spätscholastischen Naturphilosophie*, Rome, 1955, *Zwischen Philosophie und Mechanik*, Rome, 1958. *Ausgehendes Mittelalter*, 2 vols., Rome, 1964, 1967. Also useful are J. A. Weisheipl, *The Development of Physical Theory in the Middle Ages*, London, 1959, and his articles on the Arts curriculum in *Mediaeval Studies*, xxvi–xxviii (1964–66). Cf. also Vol. I under Chapter 3, Mechanics.

MATTER, SPACE, GRAVITY, DYNAMICS: C. Bailey, *The Greek Atomists and Epicurus*, Oxford, 1928; E. Borchert, *Die Lehre von der Bewegung bei Nicolaus Oresme (Beitr. Ges. Philos. Mittelalt.*, xxxi, 3), 1934; Thomas of Bradwardine, *Tractatus de Proportionibus*, ed. with English translation and introduction by H. L. Crosby, jr, Madison, Wis., 1955; J. Bulliot, 'Jean Buridan et le mouvement de la terre. Question 22ᵉ du Second Livre du "De Coelo" ', *Revue de Philosophie*, xxv (1914); Johannes Buridanus, *Quaestiones super libris quattuor de Caelo et Mundo*, ed. E. A. Moody, Cambridge, Mass., 1942; M. D. Chenu, 'Aux origines de la science moderne', *Revue des science philosophiques et théologiques*, xxix (1940); M. Clagett, *Giovanni Marliani and the late Medieval Physics (Columbia University Studies in History, Economics and Public Law*, No. 483), New York, 1941, 'The Liber de motu of Gerard of Brussels and the origins of kinematics in the West', *Osiris*, xii (1956), 73–175; Nicolaus von Cués, *Vom Globusspiel (De Ludo Globi)*, übersetzt und mit Einführung und Anmerkungen versehen von E. von Bredow (*Schriften des Nicolaus von Cués*, ... in deutscher Übersetzung herausgegeben von E. Hofmann, xiii), Hamburg, 1952; A. G. Drachmann, *Ktesibios, Philon and Heron, A Study in Ancient Pneumatics*, Copenhagen, 1948; P. Duhem, *Le mouvement absolu et le mouvement relatif*, reprinted from *Revue de la Philosophie*, xi–xiv, (1907–9), 'Roger Bacon et l'horreur du vide', in *Roger*

Bibliography

Bacon Essays, ed. Little, Oxford, 1914; D. B. Durand, 'Nicole Oresme and the medieval origins of modern science', Speculum, xvi (1941); E. Faral, 'Jean Buridan : Notes sur les manuscrits, les éditions et le contenu de ses ouvrages', Archives d'histoire doctrinale et littéraire du moyen âge, xv (1946); J. E. Hofmann, 'Zum Gedanken an Thomas Bradwardine', Centaurus, i (1959); A. Koyré, 'Le vide et l'espace infini au xive siècle', Archives d'histoire doctrinale et littéraire du moyen âge, xxiv (1949); K. Lasswitz, Geschichte der Atomistik vom Mittelalter bis Newton, 2nd ed., Leipzig, 1926, 2 vols. – still the best history of atomism in this period; A. Maier, 'Die Anfänge des physikalischen Denkens im 14. Jahrhundert', Philosophia Naturalis, i (1950), 'Die Subjektivierung der Zeit in der scholastischen Philosophie', ibid., 'Die naturphilosophische Bedeutung der scholastischen Impetustheorie', Scholastik, xxx (1955); C. Michalski, 'La physique nouvelle et les différents courants philosophiques au xive siècle', Bull. internat. de l'Acad. polonaise des sciences et des lettres, Classe d'hist. et de philos., 1927; E. A. Moody, 'Ockham, Buridan and Nicholas of Autrecourt', Franciscan Studies, N.S. vii (1947), 'Ockham and Aegidius of Rome', ibid., ix (1949), 'Galileo and Avempace', J. History of Ideas, xii (1951); S. Moser, Grundriss der Naturphilosophie bei Wilhelm von Occham (Philosophie und Grenzwissenschaften, iv. 2–3), Innsbruck, 1932; William of Ockham, The Tractatus de Successivis, ed. P. Boehner (Franciscan Inst. Publ. i), St Bonaventure, N.Y., 1944; Nicole Oresme, Le livre du ciel et du monde, ed. A. D. Menut and A. J. Denomy, in Mediaeval Studies, iii–v (1941–43); O. Pederson, Nicole Oresme og haus naturfilosofiske system, Copenhagen, 1956; S. Pines, Beiträge zur islamischen Atomenlehre, Berlin, 1936, 'Les précurseurs musulmans de la théorie de l'impetus', Archeion, xxi (1938), 'Études sur al-Zamân Abu'l Barakât al Bahdâdî', Revue des études juives, ciii (1938); H. Shapiro, 'Motion, time and place according t William of Ockham', Franciscan Studies, xvi (1956); A. Gneipl, al Meisen, om Atomos to Atom, Pittsburgh, Pa., 1952; J. A. ; Nature and compulsory movement', ibid., xxix (1955) Concept of nature', The New Scholasticism, xxv gravitation', ibid.

ntinuity of William of Ockham',
tion to works already mentioned, T. B. Birch,
B. Boyer, 'The invention of analytic
MATHEMATICAL PHY clxxx (1949), 'Early contributions to
Mathematica, xix (1953); R. Caverni,
Philosophy of S entale in Italia, 1891–98, Florence, 5 vols.;
geometry'
analy

M. Clagett, 'Richard Swineshead and the late mediaeval physics', *Osiris*, ix (1950); J. L. Coolidge, 'Origins of analytic geometry', *Osiris*, i (1936); C. Cusanus, *The Idiot in Four Books*, London, 1650; Nicolaus de Cusa, *Idiota de staticis experimentis*, ed. L. Baur (*Opera omnia*, v), Leipzig, 1937, *De Staticis Experimentis*, trans. by Henry Viets, *Annals of Medical History*, iv (1922); S. Günther, 'Die Anfange und Entwicklungsstadien des Coordinatenprinzips', *Abhandlungen der Naturhistorischen Gesellschaft zu Nürnberg*, vi (1877); E. Hoffman, 'Das Universum des Nikolaus Cusanus', *Sitzungsberichte der Heidelberger Akademie der Wissenschaften*, Philos.-hist. Klasse, 1929–30, Heidelberg, 3 Abh.; H. P. Lattin, 'The eleventh century M S Munich 14436; its contribution to the history of co-ordinates, of logic, of German studies in France', *Isis*, xxxviii (1948); A. Maier, 'Der Funktionsbegriff in der Physik des 14. Jahrhunderts', *Divus Thomas*, Freiburg, xix (1946), 'La doctrine de Nicolas d'Oresme sur les "Configurationes intensionum"', *Revue des Sciences philosophiques et théologiques*, xxxii (1948); J. Uebinger, 'Die philosophischen Schriften des Nikolaus Cusanus', *Zeitschrift für Philosophie und philosophische Kritik*, ciii (1894), cv (1895), cvii (1896); H. Wieleitner, 'Der "Tractatus de latitudinibus formarum" des Oresme', *Bibliotheca Mathematica*, xiii (1913), 'Über den Funktionsbegriff und die graphische Darstellung bei Oresme', ibid., xiv (1914); Curtis Wilson, 'Pomponazzi's criticism of Calculator', *Isis*, xliv (1953), *William Heytesbury: Medieval Logic and the Rise of Mathematical Physics*, Madison, Wis., 1956.

SCIENCE AND THE LITERARY RENAISSANCE OF THE 15TH CENTURY: Cf. Bolgar and Sandys, *op. cit.*, Vol. I, Chapter 2, and H. Baron, 'Towards a more positive evaluation of the 15th-century Renaissance', *J. History of Ideas*, iv (1943); H. S. Bennett, *English Books and Readers, 1475–1557. Being a study in the history of the book trade from Caxton to the incorporation of the Stationers' Company*, London, 19; J. Burckardt, *The Civilization of the Renaissance in Italy*, London, 1; jr, *The Renaissance*; Cassirer, P. O. Kristeller and J. H. Randall, 'Tradition and innovation' ... *y of Man*, Chicago, 1948; D. V. Durand, (1943); W. F. Ferguson, ... *th-century Italy', J. History of Ideas*, iv bridge, Mass., 1948 – a v ... *nce in Historical Thought*, Cambridge, with a full bibliography; G. ... London, 1935; F. R. Johnson and ... ted historiographical study, guage Quarterly, ii (1941); R. F. J ... *cretius and his Influence*, Language, Stanford, 1953 – a study ... 'Science', *Modern Language* ... h of the English ... uences in the development of the vernacular in the 16th century; Pearl Kibre, *The Library of Pico della Mirandola*, New York, 1936, 'Intellectual interests reflected in libraries of the 14th and 15th centuries', *J. History of Ideas*, vii (1946); A. C. Klebs, 'Incunabula scientifica et medica', *Osiris*, iv (1937); P. O. Kristeller, *Studies in Renaissance Thought and Letters*, Rome, 1956 – very important; P. O. Kristeller and J. H. Randall, jr, 'Study of Renaissance philosophy', *J. History of Ideas*, ii (1941); G. Sarton, 'The scientific literature transmitted through the incunabula', *Osiris*, v (1938), *The Appreciation of Ancient and Medieval Science during the Renaissance (1450–1600)*, Philadelphia, 1955, *Six Wings. Men of Science in the Renaissance*, Bloomington, Ind., 1957; Lynn Thorndike, 'A highly specialized medieval library', *Scriptorium*, vii (1935); H. Weisinger, 'The idea of the Renaissance and the rise of modern science', *Lychnos* (1946–47), 'English origins of the sociological interpretation of the Renaissance', *J. History of Ideas*, xi (1950), 'English treatment of the relationship between the rise of science and the Renaissance, 1740–1840', *Annals of Science*, vii (1951); G. P. Winship, *Printing in the Fifteenth Century*, Philadelphia, 1940.

CHAPTER 2

GENERAL: For introduction there are Marie Boas, *The Scientific Renaissance 1450–1630*, London, 1962; H. Butterfield, *The Origins of Modern Science*, London, 1949; C. C. Gillispie, *The Edge of Objectivity*, Princeton, 1960; A. R. Hall, *The Scientific Revolution 1500–1800*, London, 1954, *From Galileo to Newton 1630–1720*, London, 1963; and H. T. Pledge, *Science Since 1500*, London, 1939. Much valuable information is contained in the older studies of Caverni, Libri, Montucla and Whewell, and in A. Mieli, *Panorama general de historia de la ciencia*, Buenos Aires, 1945–50, 4 vols., L. Thorndike, *History of Magic and Experimental Science*, New York, 1941–58, vols. v–viii, W. P. D. Wightman, *Science and the Renaissance*, Aberdeen (in press); and A. Wolf, *A History of Science, Technology and Philosophy in the 16th and 17th Centuries*, revised by D. McKie, London, 1951; cf. also Henry Crew, *The Rise of Modern Physics*, 2nd ed., Baltimore, 1935. Useful collections of sources and miscellaneous information, sometimes inaccurate, are R. T. Gunther, *Early Science in Oxford*, 14 vols., Oxford, 1923–45, and *Early Science in Cambridge*, Oxford, 1937. Useful for selections in English translation are the Source Books published by Harvard University Press, in *Astronomy*,

Bibliography

ed. H. Shapley and H. E. Howarth, 1929, *Mathematics*, ed. D. E. Smith, 1929, *Physics*, ed. W. F. Magie, 1935, *Geology*, ed. K. F. Mather, 1939, *Animal Biology*, ed. T. S. Hall, 1951, *Chemistry*, ed. H. M. Leicester and H. S. Klickstein, 1952, and *Psychology*, ed. R. J. Herrnstein and E. G. Boring, 1966.

SCIENTIFIC THOUGHT IN A NEW SOCIAL SETTING: P. Allen, 'Scientific studies in the English universities of the seventeenth century', *J. History of Ideas*, x (1949); J. Bertrand, *L'académie des sciences et les académiciens de 1666 à 1793*, Paris, 1869; T. Birch, *History of the Royal Society*, London, 1756, 4 vols.; H. Brown, *Scientific Organizations in Seventeenth Century France (1620–1680)*, Baltimore, 1934, 'The utilitarian motive in the age of Descartes', *Annals of Science* (1936); F. Brunot, *Histoire de la langue française*, Paris, 1930, vi. i, *Le mouvement des idées et les vocabulaires techniques* (fasc. 2, 'La langue des sciences'); J. N. D. Bush, *English Literature in the Earlier Seventeenth Century, 1600–60*, Oxford, 1945; G. N. Clark, *Science and Social Welfare in the Age of Newton*, Oxford, 1937, *The Seventeenth Century*, Oxford, 1947, *Early Modern Europe from about 1450 to about 1720*, London, 1957 – a most illuminating sketch; A. C. Crombie, *Oxford's Contribution to the Origins of Modern Science*, Oxford, 1954; F. de Dainville, 'L'enseignement des mathématiques dans les Collèges Jésuites de France du XVIᵉ au XVIIᵉ siècle', *Revue d'Histoire des Sciences*, vii (1954); A. Favaro, 'Documenti per la storia dell' Accademia dei Lincei', *Bullettino di Bibliografia e di scoria delle scienze*, xx (Rome, 1887); L. P. V. Febvre, *Le problème de l'incroyance au XVIᵉ siècle*, Paris, 1947; A. J. George, 'The genesis of the Académie des Sciences', *Annals of Science*, iii (1938); H. Grossmann, 'Die gesellschaftlichen Grundlagen der mechanistischen Philosophie und die Manufaktur', *Zeitschrift für Sozialforschungen*, iv (1935); H. Hartley, *The Royal Society: Its Origins and Founders*, London, 1960; H. Hauser, 'Science et philosophie après le concile de Trente', *Scientia*, lvii (1935); Paul Hazard, *La crise de la conscience européenne (1680–1715)*, Paris, 1935 (English translation, London, 1953); R. Hooykaas, 'Science and reformation', *Cahiers d'histoire mondiale*, iii (1956), *Humanisme, Science et Réforme. Pierre de la Ramée (1515–1572)*, Leiden, 1958; W. E. Houghton, 'The history of trades', *J. History of Ideas*, ii (1941), 'The English virtuoso in the seventeenth century', ibid., iii (1942); J. Jacquot, 'Thomas Harriot's reputation for impiety', *Notes and Records of the Royal Society*, ix (1952); F. R. Johnson, 'Gresham College: precursor of the Royal Society', *J. Hist.*

Bibliography

Ideas, i (1940); R. F. Jones, *Ancients and Moderns*, St Louis, 1961; J. E. King, *Science and Rationalism in the Government of Louis XIV, 1661–1683*, Baltimore, 1949; P. H. Kocher, *Science and Religion in Elizabethan England*, San Marino, Cal., 1953; S. F. Mason, 'The Scientific Revolution and the Protestant Reformation', *Annals of Science*, ix (1953); R. K. Merton, 'Science, technology and society in 17th-century England', *Osiris*, iv (1938); J. V. Nef, *Industry and Government in France and England, 1540–1640*, Philadelphia, 1940; L. S. Olschki, *Geschichte der neusprachlichen wissenschaftlichen Literatur*, Heidelberg, 1919–27, 3 vols.; M. Ornstein [Bronfenbrenner], *The Role of Scientific Societies in the Seventeenth Century*, Chicago, 1938 – an excellent survey; L. Pastor, *The History of the Popes*, trans. E. Graf, London, 1937, 1938, xxv, xxix; P. Smith, *A History of Modern Culture*, London, 1930–34, 2 vols.; T. Sprat, *A History of the Royal Society of London*, London, 1667; R. H. Syfret, 'The Origins of the Royal Society', *Notes and Records of the Royal Society*, v (1948), 'Some early reactions to the Royal Society', ibid., vii (1950), 'Some early critics of the Royal Society', ibid., viii (1950); H. O. Taylor, *Thought and Expression in the Sixteenth Century*, New York, 1920, 2 vols.; G. H. Turnbull, 'Samuel Hartlib's influence on the early history of the Royal Society', *Notes and Records of the Royal Society*, x (1953); J. L. Vives, *On Education*, trans. by F. Watson, Cambridge, 1913; A. von Martin, *Sociology of the Renaissance*, London, 1945; C. R. Weld, *A History of the Royal Society*, London, 1848, 2 vols.; B. Willey, *The Seventeenth Century Background*, London, 1934; Louis B. Wright, *Middle-Class Culture in Elizabethan England*, Chapel Hill, N.C., 1935; E. Zilsel, 'Problems of empiricism : experiment and manual labour', *International Encyclopaedia of Unified Science*, ed. O. Neurath, 1941, II. viii, 'The sociological roots of science', *American J. of Sociology*, xlvii (1942), 'The genesis of the concept of physical laws', *The Philosophical Review*, li (1942), 'The genesis of the concept of scientific progress', *J. History of Ideas*, vi (1945).

MATHEMATICS AND MECHANICS: An excellent survey is R. Dugas, *La mécanique au xviie siècle*, Neuchâtel, 1954. In addition, besides works mentioned under Chapter 1 and in Vol. I under Chapter 3, there are A. Armitage, 'The deviation of falling bodies', *Annals of Science*, v (1948); Isaac Beeckman (1588–1637), *Journal*, ed. Cornelius de Waard, La Haye, 1953; A. E. Bell, *Christian Huygens and the Development of Science in the Seventeenth Century*, London, 1947; S. Brodetsky, *Sir Isaac Newton*, London, 1927 – a useful summary;

Bibliography

L. Brunschwig, *Les étapes de la philosophie mathématique*, Paris, 1947; E. A. Burtt, *The Metaphysical Foundations of Modern Physical Science*, London, 1932 (reprinted New York, 1955); F. Cajori, *A History of Mathematics*, New York, 1924, *A History of Physics*, 2nd ed., New York, 1929; A. Carli and A. Favaro. *Bibliografia Galileiana* (1568–1895), Rome, 1896; E. Cassirer, 'Mathematische Mystik und mathematische Naturwissenschaft', *Lychnos* (1940), 'Galileo's Platonism', in *Studies and Essays ... offered ... to George Sarton*, ed. M. F. Ashley Montague, New York, 1944; I. B. Cohen, 'Galileo's rejection of the possibility of velocity changing uniformly with respect to distance', *Isis*, xlvii (1956); Julian L. Coolidge, *History of Geometrical Methods*, Oxford, 1940; Lane Cooper, *Aristotle, Galileo, and the Tower of Pisa*, Ithaca, N.Y., 1935; R. Depau, *Simon Stevin*, Brussels, 1942 (study and French trans. of texts); René Descartes, *Oeuvres*, ed. Ch. Adam et P. Tannery, Paris, 1897–1913, 12 vols. (English trans. of philosophical works by E. S. Haldane and G. R. T. Ross, 2nd ed., Cambridge, 1931, New York, 1955; of *The Geometry* by D. E. Smith and M. L. Latham, La Salle, Ill., 1925, New York, 1954); E. J. Dijksterhuis, *De Mechanisering van het Wereldbeeld*, Amsterdam, 1950 (English trans., Oxford, 1961); F. Enriques, *Le Matematiche nella storia e nella cultura*, Bologna, 1938; *Galilée. Aspects de sa vie et de son oeuvre*, Paris, 1968; Galileo Galilei, *Opere*, ed. naz. by A. Favaro, Florence, 1890–1909, 20 vols. (English trans. of *De motu* by I. E. Drabkin. Madison, Wisc., 1960; *The Siderial Messenger*, by E. S. Carlos, London, 1880; *Dialogue concerning the Two Principal Systems of the World*, by T. Salisbury, 1661, revised by G. de Santillana, Chicago, 1953, and by S. Drake, Berkeley, Cal., 1953; *Mathematical Discourses ... [Dialogues] Concerning Two New Sciences*, by H. Crew and A. de Salvio, New York, 1914, 1952; *Discoveries and Opinions of Galileo*, trans. with an introduction and notes by S. Drake, New York, 1957 [*Starry Messenger, Letters on Sunspots, Il Saggiatore, Letter to the Grand Duchess Christina*]); L. Geymonat, *Galileo Galilei*, English trans. by Drake, New York, 1965; A. R. Hall, *Ballistics in the Seventeenth Century*, Cambridge, 1952; L. R. Heath, *The Concept of Time*, Chicago, 1936; Christian Huygens, *Oeuvres complètes*, ed. Société hollandaise des sciences, La Haye, 1888–1950, 22 vols. (English trans. of *Treatise on Light* by S. P. Thompson, Chicago, 1945); A. Koyré, *Études galiléennes* (*Actualités scientifiques et industrielles*, nos. 852–54), Paris, 1939 – very important, 'Galileo and Plato', *J. History of Ideas*, iv (1943), 'The significance of the Newtonian Synthesis', *Archives internationales d'histoire des sciences*, xxix (1950), 'An experiment in measurement', *Proceedings of the American Philosophical Society*, xcvii (1953), 'A documentary history of the problem of fall from Kepler to Newton', *Transactions of the American Philosophical Society*, N.S. xlv. 4 (1955), 'Pour une édition critique des œuvres de Newton', *Revue d'histoire des sciences*, viii (1955), 'L'hypothèse et l'expérience chez Newton', *Bulletin de la Société française de Philosophie*, i (1956); R. Lämmel, *Galileo Galilei und sein Zeitalter*, Zürich, 1942 – an excellent study; R. Lenoble, *Marin Mersenne ou la naissance du mécanisme*, Paris, 1943; *Leonardo da Vinci et l'expérience scientifique au xvie siècle* (Colloques internationaux du Centre national de la Recherche Scientifique), Paris, 1953; W. H. Macaulay, 'Newton's theory of kinetics', *Bulletin of the American Mathematical Society*, iii (1897); E. Mach, *The Science of Mechanics*, trans. by T. J. McCormack, La Salle, Ill., 1942; E. McMullin (editor), *Galileo: Man of Science*, New York, 1967; R. Marcolongo, 'Lo sviluppo della meccanica sino ai discepoli di Galileo', *Atti della Reale Accademia dei Lincei*, xiii (1920); M. Mersenne, *Correspondance*, ed. Mme Paul Tannery, Cornelis de Waard and René Pintard, Paris, 1932– , i– ; G. Milhaud, *Descartes Savant*, Paris, 1921; A. Mieli, 'Il tricentenario dei "Discorsi e dimostrazioni matematiche" di Galileo Galilei', *Archeion*, xxi (1938) – a criticism of Duhem, etc.; P. Mouy, *Le développement de la physique cartésienne, 1646–1712*, Paris, 1934; Sir Isaac Newton, *Mathematical Principles of Natural Philosophy and his System of the World*, Motte's trans. revised by F. Cajori, Berkeley, Cal., 1946; F. Rosenburger, *Isaac Newton and Seine Physikalischen Prinzipien*, Leipzig, 1895; O. Ore, *Cardano: The Gambling Scholar*, Princeton, 1953; G. Sarton, 'Simon Stevin of Brughes', *Isis*, xxi (1934); J. F. Scott, *The Scientific Work of René Descartes*, London, 1952; W. B. Parsons, *Engineers and Engineering in the Renaissance*, Baltimore, 1939; D. E. Smith, *A History of Mathematics*, Boston, 1923–25, 2 vols.; Simon Stevin, *The Principal Works*, ed. E. J. Dijksterhuis, vol. i, 'Mechanics', Amsterdam, 1955; E. W. Strong, *Procedures and Metaphysics*, Berkeley, 1936; H. J. Webb, 'The science of gunnery in Elizabethan England', *Isis*, xlv (1954); P. P. Wiener, 'The tradition behind Galileo's methodology', *Osiris*, i (1936).

ASTRONOMY: In addition to works already mentioned in the preceding section and in Vol. I under Chapter 3, G. Abetti, *The History of Astronomy*, trans. from the Italian *Storia dell' Astronomia* by Betty Burr Abetti, New York, 1952; E. J. Aiton, 'Galileo's theory of the tides', *Annals of Science*, x (1954); D. C. Allen, *The Star-crossed*

Bibliography

Renaissance, Durham, N.C., 1941 – on astrology; A. Armitage, *Copernicus*, London, 1938, 'The cosmology of Giordano Bruno', *Annals of Science*, vi (1948), ' "Borell's hypothesis" and the rise of celestial mechanics', ibid.; C. Baumgardt, *Johannes Kepler: Life and Letters*, New York, 1952; A. Berry, *Short History of Astronomy*, London, 1896; G. Bigourdan, *L'astronomie, évolution des idées et des méthodes*, Paris, 1911; I. Bouiliau, *Astronomia Philolaica*, Paris, 1645; C. B. Boyer, 'Notes on epicycles and the ellipse from Copernicus to Lahire', *Isis*, xxxviii (1947); J. Brodrick, *The Life and Work of Blessed Robert, Cardinal Bellarmine, 1542–1621*, London, 1928; W. W. Bryant, *Kepler*, New York, 1920; Tommaso Campanella, 'The Defence of Galileo of Thomas Campanella', trans. and ed. by C. McColley, *Smith College Studies in History*, Northampton, Mass., xx (1938); Max Caspar, *Johannes Kepler*, 2nd ed., Stuttgart, 1950 (trans. C. D. Hellman, New York, 1959); Copernicus, *De Revolutionibus* ... (trans. Wallis, in *Great Books of the Western World*, Chicago, 1952 – somewhat inaccurate; trans. of preface and Book I by J. F. Dobson and S. Brodsky, *Royal Astronomical Society Occasional Notes*, ii, 1947, No. 10); A. C. Crombie, *Galilée devant les critiques de la postérité* (Les Conférences au Palais de la Découverte, Série D, No. 45) Paris, 1957; H. Dingle, essays in *The Scientific Adventure*, London, 1952; J. L. E. Dreyer, *Tycho Brahe*, Edinburgh, 1890, *History of Planetary Systems*, Cambridge, 1906 (reprinted as *A History of Astronomy* ..., New York, 1953); A. Favaro, *Galileo Galilei e l'Inquisizione. Documenti del processo Galileiano* ..., Florence, 1907; J. A. Gade, *The Life and Times of Tycho Brahe*, Princeton, 1947; K. von Gebler, *Galileo Galilei and the Roman Curia*, trans. Mrs G. Sturge, London, 1879; B. Ginsburg, 'The scientific value of the Copernican induction', *Osiris*, i (1936); E. Goldbeck, *Keplers Lehre von der Gravitation*, Halle a/d.Saale, 1896; S. Greenberg, *The Infinite in Giordano Bruno*, with a translation of his dialogue *Concerning The Cause, Principle, and One*, New York, 1950; W. Hartner, 'The Mercury horoscope of Marcantonio Michiel of Venice, A study in the history of renaissance astrology and astronomy', in *Vistas in Astronomy*, ed. A. Beer, London and New York, 1955, i; C. D. Hellman, *The Comet of 1577: its Place in the History of Astronomy*, New York, 1944; G. Holton, 'Johannes Kepler's universe: its physics and metaphysics', *American Journal of Physics*, xxiv (1956); Max Jammer, *Concepts of Space: the history of theories of space in physics*, Cambridge, Mass., 1954, *Concepts of Force*, Cambridge, Mass., 1957; F. R. Johnson, 'The influence of Thomas Digges on the progress of modern astronomy in sixteenth-century England',

Bibliography

Osiris, i (1936), *Astronomical Thought in Renaissance England*, Baltimore, 1937, 'Astronomical textbooks in the sixteenth century', in *Science, Medicine and History*, ed. E. A. Underwood, Oxford, 1953, i; F. R. Johnson and S. V. Larkey, 'Thomas Digges, the Coperican System, and the idea of the infinity of the Universe in 1576', *Huntington Library Bulletin* (San Marino, Cal.), v (1934), 'Robert Recorde's mathematical teaching and the anti-Aristotelian movement', ibid., vii (1935); C. G. Jung and W. Pauli, *The Interpretation of Nature and the Psyche*, London, 1955 (German ed., Zürich, 1952) – Pauli has an interesting essay on Kepler; *Johann Kepler, 1571–1630. A Tercentenary Commemoration of his Life and Work*, Baltimore, 1931 – with bibliography; Johannes Kepler, *Gesammelte Werke*, ed. W. von Dycht and M. Caspar, München, 1938– ; A. Koyré, *Philosophical Review*, lii (1943) – on Kepler's conception of inertia, *La Révolution astronomique: Copernic, Kepler, Borelli*, Paris, 1961 – essential, 'La gravitation universelle de Kepler à Newton', *Actes du VI Congrès international d'histoire des sciences*, Amsterdam, 1950, Paris, 1953, 'L'oeuvre astronomique de Kepler', *XVIIᵉ siècle*, Paris, No. 30 (1956); T. S. Kuhn, *The Copernican Revolution*, Cambridge, Mass., 1957 – very useful; G. McColley, 'The 17th-century doctrine of a plurality of worlds', *Annals of Science*, i (1936); A. Mercati, *Il sommario del processo di Giordano Bruno (Studi e Testi, ci)* Rome, 1942; H. Metzger, *Attraction universelle et religion naturelle chez quelques commentateurs anglais de Newton*, Paris, 1938; S. I. Mintz, 'Galileo, Hobbes, and the circle of perfection', *Isis*, xliii (1952); M. H. Nicholson, *The Breaking of the Circle. Studies in the effect of the "New Science" upon seventeenth-century poetry*, Evanston, Ill., 1950; W. Norlind, 'Copernicus and Luther: a critical study', *Isis*, xliv (1953); E. Panofsky, *Galileo as a Critic of the Arts*, The Hague, 1954 – on Galileo and Kepler; Pastor, *History of the Popes*, London, 1937, 1938, xxv, xxix; S. P. Rigaud, *Supplement to Dr Bradley's Miscellaneous Works, with an account of Harriot's astronomical papers*, Oxford, 1833; E. Rosen, *Three Copernican Treatises*, 2nd ed., New York, 1959, 'The Ramus-Rheticus Correspondence', *J. Hist. Ideas*, i (1940), 'Maurolyco's attitude toward Copernicus', *Proceedings of the American Philosophical Society*, ci (1957), *Kepler's Conversation with Galileo's Sidereal Messenger*, New York, 1965; G. de Santillana, *The Crime of Galileo*, Chicago, 1955 (*Le procès de Galilée*, Paris, 1956) – the most recent study of the affair of Galileo and the Catholic Church; D. Shapeley, 'Pre-Huygenian observations of Saturn's rings', *Isis*, xl (1949); D. W. Singer, *Giordano Bruno, His Life and Thought*, New York-London, 1950; A. J. Snow, *Matter and*

Bibliography

Gravity in Newton's Physical Philosophy, London, 1926; D. Stimpson, *The Gradual Acceptance of the Copernican Theory*, New York, 1917; James Winny (ed.), *The Frame of Order. An Outline of Elizabethan beliefs taken from treatises of the late sixteenth century*, London, 1957; E. Wohlwill,. *Galilei und sein Kampf für die kopernicanische Lehre*, Hamburg and Leipzig, 1909; R. Wolf, *Geschichte der Astronomie*, Munich, 1877; H. Zaiser, *Kepler als Philosoph*, Stuttgart, 1932; E. Zilsel, 'Copernicus and mechanics', *J. History of Ideas*, i (1940); E. Zinner, *Die Geschichte der Sternkunde*, Berlin, 1931, *Entstehung und Ausbreitung der Oppernikanischen Lehre (Sitzungsberichte der physik.-mediz. Sozietät zu Erlangen)*, Erlangen, 1943.

MAGNETISM, ELECTRICITY, AND OPTICS: In addition to the works listed in Vol. I under Chapter 3, and in the next section, C. B. Boyer, 'Kepler's explanation of the rainbow', *American Journal of Physics*, xviii (1950), 'Descartes and the radius of the rainbow', *Isis*, xliii (1952); Cajori, *A History of Physics*, New York, 1929; William Gilbert, *De Magnete Magnetisque Corporibus et de Magno Magnete Tellure*, London, 1600 (English trans. by P. F. Motteley, London, 1893); N. H. de V. Heathcote, 'Guericke's sulphur globe', *Annals of Science*, vi (1950); J. Itard, 'Les lois de la réfraction de la lumière chez Kepler', *Revue d'Histoire des Sciences*, X (1957); D. J. Korteweg, 'Descartes et les manuscrits de Snellius', *Revue de Métaphysique et de Morale*, Paris, iv (1896); P. Kramer, 'Descartes und das Brechungsgesetz des Lichtes', *Abhandlungen zur Geschichte der Mathematik*, iv (1882); G. Leisegang, *Descartes Dioptrik*, Meisenheim am Glan, 1954; J. Lohne, 'Thomas Harriot (1560–1621)', *Centaurus*, vi (1959), 'Zur Geschichte des Brechungsgesetzes', *Archiv für Geschichte der Medizin und der Naturwissenschaften,* xlvii (1963); Sir Isaac Newton, *Opticks*, 4th edition, London, 1730 (reprinted, London, 1931, New York, 1952); R. E. Ockenden, 'Marco Antonio de Dominis and his explanation of the rainbow', *Isis*, xxvi (1936); E. Panofsky, *Albrecht Dürer*, 3rd ed., Princeton, 1943, 2 vols.; C. E. Papanastassiou, *Les Théories sur la nature de la lumière de Descartes à nos jours*, Paris, 1935; M. Roberts and E. R. Thomas, *Newton and the Origin of Colours*, London, 1934 (containing a reprint of Newton's 'New Theory about Light and Colours', *Philosophical Transactions of the Royal Society*, vi, 1671–72); D. H. D. Roller, 'The *De Magnete* of William Gilbert', *Isis*, xlv (1954), D. H. D. Roller (ed.), *The Development of the Concept of the Electric Charge. Electricity from the Greeks to Coulomb (Harvard Case Histories in Experimental Science*, ed. J. B. Conant, viii), Cambridge,

Mass., 1954; V. Ronchi, *Histoire de la Lumière*, Paris, 1956, *Optics: The Science of Vision*, New York, 1957; L. Rosenfeld, 'La théorie des couleurs de Newton et ses adversaires', *Isis*, ix (1927), 'Marcus Marcis Untersuchungen über das Prisma und sein Verhältnis zu Newton's Farbentheorie', *Isis*, xvii (1932); A. I. Sabra, *Theories of Light from Descartes to Newton*, London, 1967; R. Suter, 'A biographical sketch of Dr William Gilbert of Colchester', *Osiris*, x (1952); J. A. Vollgraff, 'Snellius' notes on the reflection and refraction of rays', *Osiris*, i (1936); E. T. Whittaker, *A History of Theories of Ether and Electricity*, Edinburgh, 1951, i; E. Zilsel, 'The origins of William Gilbert's scientific method', *J. History of Ideas*, ii (1941).

SCIENTIFIC INSTRUMENTS: In addition to the works listed in the preceding section and in Vol. I under Chapter 3, Astronomy and Optics, and Chapter 4, Building, etc., and Medicine, M. K. Barnett, 'The development of thermometry and the temperature concept', *Osiris*, xii (1956); M. Bishop, *Pascal, the life of genius*, Baltimore, 1936; L. C. Bolton, *Time Measurement*, London, 1924; R. S. Clay and T. S. Court, *The History of the Microscope*, London, 1932; A. Danjon and A. Couder, *Lunettes et Télescopes*, Paris, 1935; M Daumas, *Les instruments scientifiques aux XVIIᵉ et XVIIIᵉ siècles*, Paris, 1953; C. de Waard, *L'expérience barométrique. Ses antécédents et ses explications*, Thouars, 1936; A. N. Disney, C. F. Hill and W. E. W. Baker, *Origin of the Telescope*, London, 1955; Henri Michel, 'Les tubes optiques avant le télescope', *Ciel et Terre*, Brussels, lxx (1954); J. W. Olmsted, 'The application of telescopes to astronomical instruments', *Isis*, xl (1949); L. D. Patterson, 'The Royal Society's standard thermometer', *Isis*, xliv (1953); V. Ronchi, *Galileo e il cannocchiale*, Udine, 1942, 'Du *De Refractione* au *De Telescopio* de G. B. Della Porta', *Revue d'Histoire des Sciences*, vii (1954); E. Rosen, *The Naming of the Telescope*, New York, 1947, 'When did Galileo make his first telescope?', *Centaurus*, ii (1951), 'Did Galileo claim he invented the telescope?', *Proceedings of the American Philosophical Society*, xcvii (1954); C. Singer, E. J. Holmyard, *et alii, History of Technology*, Oxford, 1957, iii, chapters by D. J. Price and H. Alan Lloyd; F. Sherwood Taylor, 'The origin of the thermometer', *Annals of Science*, v (1942); R. W. Symonds, *A History of English Clocks*, London, 1947.

NAVIGATION AND CARTOGRAPHY: In addition to works listed in Vol. I under Chapter 4, J. Delevsky, 'L'invention de la projection de

Bibliography

Mercator et les enseignements de son histoire', *Isis*, xxxiv (1942); N. H. De V. Heathcote, 'Christopher Columbus and the discovery of magnetic variation', *Science Progress* (1932), 'Early nautical charts', *Annals of Science*, i (1936); J. E. Hofmann, 'Nicolaus Mercator (Kauffmann), sein Leben und Werken, vorzugsweise als Mathematiker', *Akademie der Wissenschaften und der Literatur in Mainz*, Abh. der math.-naturwiss. Klasse, No. 3, Wiesbaden, 1950; G. H. T. Kimble, *Geography in the Middle Ages*, London, 1938; S. Lorant (ed.), *The New World. The first pictures of America made by John White and Jacques le Moyne*, New York, 1946 (contains Harriot's *Brief and True Report*); S. E. Morrison, *Admiral of the Ocean Sea, a life of Christopher Columbus*, Boston, Mass., 1942, 2 vols.; A. P. Newton, *Travel and Travellers of the Middle Ages*, London, 1926; E. G. R. Taylor, *Tudor Geography, 1485–1583*, London, 1930, *Late Tudor and Early Stuart Geography, 1583–1650*, London, 1934, *The Mathematical Practitioners of Tudor and Stuart England*, Cambridge, 1954; L. C. Wroth, *The Way of a Ship, an essay on the literature of navigation*, Portland, Me., 1937.

BIOLOGY in general: T. Ballauff, *Die Wissenschaft vom Leben*, i, Freiburg and Munich, 1954; H. Daudin, *Les méthodes de la classification et l'idée de série en botanique et en zoologie de Linné à Lamarck (1740–1790)*, Paris, 1926; P. G. Fothergill, *Historical Origins of Organic Evolution*, London, 1952; E. Guyénot, *Les sciences de la vie aux 17ᵉ et 18ᵉ siècles*, Paris, 1941; E. Nordenskiöld, *The History of Biology*, London, 1929; C. Singer, *A Short History of Biology*, 2nd ed., London, 1950.

EXPERIMENTAL PHYSIOLOGY: In addition to works listed in Vol. I under Chapter 3, Marie Thérèse d'Alverny, 'Avicenne et les médicins de Venise', *Medioevo e Rinascimento, studi in onore di Bruno Nardi*, Florence, 1955; J. P. Arcieri, *The Circulation of the Blood and Andrea Cesalpino of Brezzia*, New York, 1945; R. H. Bainton, *Michel Servet, hérétique et martyr, 1511–1553*, Genève, 1953 (English edition, Boston, 1953) – a bibliographical study; E. Bastholm, *The History of Muscle Physiology*, English trans. by W. E. Calvert, Copenhagen, 1950; H. P. Bayon, 'William Harvey, physician and biologist', *Annals of Science*, iii (1938), iv (1939) – a basic study; B. Becker (editor), *Autour de Michel Servet et de Sébastien Castellion*, Haarlem, 1953; A. G. Berthier, 'Le mécanisme cartésien et la physiologie au 17ème siècle', *Isis*, ii (1914), iii (1920); H. Brown, 'John Denis and the transfusion of blood, Paris 1667–68', *Isis*, xxxix (1938); G.

Bibliography

Canguilhem, *La formation du concept de réflexe aux xvii*[e] *et xviii*[e] *siècles*, Paris, 1955; A. Castiglioni, *The Renaissance of Medicine in Italy*, Baltimore, 1934, 'Galileo Galilei and his influence on the evolution of medical thought', *Bulletin of the History of Medicine*, xii (1942); L. Chauvois, *William Harvey*, London and Paris, 1957; L. D. Cohen, 'Descartes and Henry More on the beast-machine', *Annals of Science*, i (1936); J. E. Curtis, *Harvey's Views on the Use of the Circulation of the Blood*, New York, 1915 – an illuminating study; Franklin Fearing, *Reflex Action, a study in the history of physiological psychology*, London, 1930; D. Fleming, 'William Harvey and the pulmonary circulation', *Isis*, xlvi (1955); Sir M. Foster, *Lectures on the History of Physiology during the Sixteenth, Seventeenth and Eighteenth Centuries*, Cambridge, 1901; K. J. Franklin, *A Short History of Physiology*, London, 1933, 'A survey of the growth of knowledge about certain parts of the foetal cardio-vascular apparatus, and about the foetal circulation, in man and some other animals, Part I : Galen to Harvey', *Annals of Science*, v (1941); J. F. Fulton, *Selected Readings in the History of Physiology*, London, 1930, *Michael Servetus, humanist and martyr:* With a bibliography of his works . . . by M. E. Stanton, New York, 1953; E. Gilson, *Études sur le rôle de la pensée médiévale dans la formation du système cartésien* (*Études de la philosophie médiévale*, xiii), Paris, 1930 – fundamental; William Harvey, Works, trans. by R. Willis, London, 1847, *Prelectiones Anatomiae Universalis*, annotated trans. by D. D. O'Malley, F. N. L. Poynter and K. F. Russell, Univ. of California Press, 1961, and by G. Whitteridge, Edinburgh and London, 1964, *De Motu Cordis*, text and trans. by K. J. Franklin, Oxford, 1957, *The Circulation of the Blood*, trans. Franklin, Oxford, 1958, *De Motu Locali Animalium*, Cambridge, 1959, The William Harvey Issue, *Journal of the History of Medicine*, xii (1957), No. 2; H. E. Hoff and P. Kellaway, 'The early history of the reflex', *Journal of the History of Medicine*, vii (1952); K. D. Keele, *Leonardo da Vinci on Movement of the Heart and Blood*, Philadelphia, 1952; G. Keynes, *Blood Transfusion*, Bristol and London, 1949, *The Life of William Harvey*, Oxford, 1966; *Léonard de Vinci et l'expérience scientifique au xvi*[e] *siècle*, Paris, 1953; R. Lower, *De Corde*, trans. by K. J. Franklin in R. T. Gunther, *Early Science in Oxford*, ix; D. McKie, 'Fire and the Flamma Vitalis : Boyle, Hooke and Mayow', *Science, Medicine and History*, ed. E. A. Underwood, Oxford, 1953, i; N. S. R. Maluf, 'History of blood transfusion', *Journal of the History of Medicine*, ix (1954); M. Meyerhoff, 'Ibn An-Nafis (13th century) and his theory of the lesser circulation', *Isis*, xxiii (1935); Sir W. Osler, *The Growth of*

Bibliography

Truth as Illustrated in the Discovery of the Circulation of the Blood (Harveyan Oration), London, 1906; C. D. O'Malley, *Michael Servetus*, Philadelphia, 1953; W. Pagel, 'Religious motives in the medical biology of the XVIIth century', *Bulletin of the Institute for the History of Medicine*, iii (1935), 'William Harvey and the purpose of circulation', *Isis*, xlii (1951), 'Giordano Bruno: the philosophy of circles and the circular movement of the blood', *Journal of the History of Medicine*, vi (1951), 'The reaction to Aristotle in seventeenth-century biological thought', in *Science, Medicine and History*, ed. E. A. Underwood, Oxford, 1953, i, *William Harvey's Biological Ideas*, Basel, 1967; J. R. Partington, *op. cit.* below 'Chemistry'; D'Arcy Power, *William Harvey*, London, 1897; P. A. Robin, *The Old Physiology in English Literature*, London, 1911; Sir H. Rolleston, 'The reception of Harvey's doctrine of the circulation of the blood in England', in *Essays ... presented to Karl Sudhoff*, ed. C. Singer and H. E. Sigerist, Oxford and Zürich, 1924; K. E. Rothschuh, *Entwickelungsgeschichte physiologischer Probleme in Tabellenform*, Munich and Berlin, 1952, *Geschichte der Physiologie*, Berlin, 1953; Sir Charles Sherrington, *The Endeavour of Jean Fernel*, Cambridge, 1946; C. Singer, *The Discovery of the Circulation of the Blood*, London, 1922; N. Kemp Smith, *op. cit.* below, Philosophy of Science, etc.; Nicolaus Steno, *A Dissertation of the Anatomy of the Brain ... 1665*, Copenhagen, 1950 (reprint), *Nicolai Stenonis, Epistolae et epistolae ad eum datae*, ed. G. Scherz and J. Raeder, Hafniae, 1952, 2 vols.; W. Sterling, *Some Apostles of Physiology*, London, 1902; P. Tannery, 'Descartes physicien', *Revue de Métaphysique* (1896); O. Temkin, 'Metaphors of human biology', *Science and Civilization*, ed. R. Stauffer, Madison, Wis., 1949; J. Trueta, 'Michael Servetus and the discovery of the lesser circulation', *Yale Journal of Biology and Medicine*, xxi (1948); F. A. Willins and T. J. Dry, *History of the Heart and the Circulation*, Philadelphia, 1948.

CHEMISTRY: In addition to works listed in Vol. I under Chapters 3 and 4, E. Bloch, 'Die antike Atomistik in der neueren Geschichte der Chemie', *Isis*, i (1913–14); T. L. Davis, 'Boyle's conception of the elements compared with that of Lavoisier', *Isis*, xvi (1931); Edward Farber, *The Evolution of Chemistry*, New York, 1952; F. W. Gibbs, 'The rise of the tinplate industry', *Annals of Science*, vi (1950), vii (1951); Kurt Goldammer, *Paracelsus. Sozialethische und sozialpolitische Schriften*, Tübingen, 1952; J. C. Gregory, *Short history of Atomism from Democritus to Bohr*, London, 1931, *Combustion from Heraclitus to Lavoisier*, London, 1934; Thomas S. Kuhn, 'Robert Boyle and structural chemistry in the seventeenth century', *Isis*, xliii (1952) K. Lasswitz, *Geschichte der Atomistik vom Mittelalter bis Newton*, 2 vols., Leipzig, 1960; H. Metzger, *Les doctrines chimiques en France du début du 17ᵉ siècle à la fin du 18ᵉ siècle, Paris*, 1923; R. Multhauf, 'Medical chemistry and the "Paracelsians"', *Bulletin of the History of Medicine*, xxviii (1954); L. K. Nash (ed.), *The Atomic-Molecular Hypothesis (Harvard Case Histories in Experimental Science*, ed. J. B. Conant, iv), Cambridge, Mass., 1950, 'The origin of Dalton's chemical atomic theory', *Isis*, xlvii (1956); Henry M. Pachter, *Paracelsus, Magic into Science*, New York, 1951; W. Pagel, 'The religious and philosophical aspects of van Helmont's science and medicine', *Bull. Hist. Medicine* (1944, Suppl. 2), *Paracelsus*, Basel and New York, 1958; *Paracelsus, Selected Writings*, ed. J. Jacobi, New York, 1951; J. R. Partington, *A Short History of Chemistry*, London, 1937, 'Jean Baptista van Helmont', *Annals of Science*, i (1936), 'The origins of the atomic theory', ibid., iv (1939), 'The life and work of John Mayow (1641–1679)', *Isis*, xlvii (1956); T. S. Patterson, 'John Mayow in contemporary setting', *Isis*, xv (1931); Jean Rey, *Essays*, ed. D. McKie, London, 1951; H. E. Sigerist, *Paracelsus in the Light of Four Hundred Years*, New York, 1941; G. B. Stones, 'The atomic view of matter in the xvth, xvith and xviith centuries', *Isis*, x (1928); C. M. Taylor, *The Discovery of the Nature of Air*, London, 1923; J. H. White, *History of the Phlogiston Theory*, London, 1932.

GEOLOGY: In addition to the list in Vol. I under Chapter 3, D. R. Rome, 'Nicolas Sténon et la "Royal Society of London"', *Osiris*, xii (1956); C. Schneer, 'The rise of historical geology in the seventeenth century', *Isis*, xlv (1954); Nicholas Steno, *Prodromus ...*, English trans. by J. G. Winter, New York, 1916; H. R. Thompson, 'The geographical and geological observations of Bernard Palissy the potter', *Annals of Science*, x (1954); Karl von Zittel, *History of Geology and Palaeontology*, translated by M. M. Ogilvie-Gordon, London, 1901.

BOTANY: In addition to works listed in Vol. I under Chapter 3, A. Arber, *Herbals*, Cambridge, 1938; W. Blunt, *The Art of Botanical Illustration*, London, 1950; C. Demars, 'Rembert Dodoens, 29.6.1517–10.3.1585', *IIIᵉ Congrès National des Sciences*, Brussels, 1950; F. G. D. Drewitt, *The Romance of the Apothecaries' Garden at Chelsea*, London, 1928; Knut Hagberg, *Carl Linnaeus*, London, 1952; R. Hooke, *Micrographia*, London, 1665 (reprinted in R. T. Gunther, *Early Science in Oxford*, xiii, Oxford, 1938); C. E. Raven, *John Ray*, Cambridge,

Bibliography

1942, *English Naturalists from Neckam to Ray*, Cambridge, 1947; J. Sachs, *History of Botany, 1530–1860*, trans. by H. E. F. Garnsey and I. B. Balfour, Oxford, 1890.

ANATOMY AND ZOOLOGY: In addition to works listed in Vol. I under Chapter 3, L. Choulant, *History and Bibliography of Anatomic Illustrations*, trans. and annotated by M. Frank, New York, 1945; F. J. Cole, *A History of Comparative Anatomy*, London, 1944; H. Cushing, *A Bio-Bibliography of Andreas Vesalius*, New York, 1943; P. Delaunay, *L'Aventureuse existence de Pierre Belon de Mans*, Paris, 1926 (also *Revue du seizième siècle*, Paris, ix–xii, 1922–5); C. C. Gillispie, *Genesis and Geology*, Cambridge, Mass., 1951 – for a bibliography on evolution in the 18th century; E. W. Gudger, 'The five great naturalists of the 16th century, Belon, Rondelet, Salviani, Gesner, and Aldrovandi: a chapter in the history of ichthyology', *Isis*, xxii (1934); R. Herrlinger, *Volcher Coiter, 1534–1576* (*Beiträge zur Geschichte der medizinischen und naturwissenschaftlichen Abbildung*, i), Nuremberg, 1952; H. Hopstock, 'Leonardo as anatomist', in *Studies in the History and Method of Science*, ed. Singer, Oxford, 1921, ii; S. W. Lambert, W. Wiegand, W. M. Ivins, jr., *Three Vesalian Essays*, New York, 1952; Leonardo da Vinci, *Notebooks*, arranged, rendered into English and introduced by E. MacCurdy, London, 1938, 2 vols., *Literary Works*, ed. J. P. and I. A. Richter, Oxford, 1939, 2nd ed., 2 vols.; Willy Ley, *Konrad Gesner, Leben und Werke* (*Münchener Beiträge zur Geschichte und Literatur der Naturwissenschaften*, xv–xvi) Munich, 1929; J. P. Murrich, *Leonardo da Vinci the Anatomist*, Baltimore, 1930; C. D. O'Malley and J. B. de C. M. Saunders, *Leonardo da Vinci on the Human Body*, the anatomical, physiological and embryological drawings of Leonardo da Vinci. With translation, emendations and a biographical introduction, New York, 1952; M. F. Ashley Montagu, *Edward Tyson, M.D., F.R.S., 1650–1708, and the rise of human and comparative anatomy in England* (*Memoirs of the American Philosophical Society*, xx) Philadelphia, 1943; Vittorio Putti, *Berengario da Carpi*, Bologna, 1937; E. Radl, *Geschichte der biologischen Theorien*, Part I, Leipzig, 1905; E. S. Russell, *Form and Function*, London, 1916 – fundamental for the history of comparative anatomy; J. B. de C. M. Saunders and C. D. O'Malley, articles on Vesalius in *Studies and Essays ... offered to George Sarton*, ed. M. F. Ashley Montagu, New York, 1944, and in *Bulletin of Medical History*, xiv (1943), *The Illustrations from the Works of Andreas Vesalius of Brussels*, Cleveland and New York, 1950; C. Singer and

Bibliography

C. Rabin, *A Prelude to Modern Science*, Cambridge, 1946 – on Vesalius' *Tabulae Anatomicae Sex; Vesalius on the Human Brain*, translations by C. Singer, London, 1952.

EMBRYOLOGY AND GENETICS: In addition to the works mentioned in Vol. I under Chapter 3, H. P. Bayon, 'William Harvey (1578–1657): his application of biological experiment, clinical observation and comparative anatomy to the problems of generation', *Journal of the History of Medicine*, ii (1947); F. J. Cole, *Early Theories of Sexual Generation*, Oxford, 1930; A. C. Crombie, 'P. L. M. de Maupertius, F.R.S. (1698–1759), précurseur du transformisme', *Revue de Synthèse*, lxxviii (1957); C. Dobell, *Antony van Leeuwenhoek and his "Little Animals"*, London, 1932; *The Embryological Treatises of Hieronymus Fabricius*, ed. H. B. Adelmann, New York, 1942; A. van Leeuwenhoek, *Collected Letters*, Amsterdam, 1939– ; A. W. Meyer, *An Analysis of the De Generatione Animalium of William Harvey*, Stanford, Cal., 1936, 'Leeuwenhoek as experimental biologist', *Osiris*, iii (1937), *The Rise of Embryology*, Stanford, Cal., 1939; J. Needham, *A History of Embryology*, Cambridge, 1934; W. Pagel, 'J. B. Van Helmont, *De Tempore*, and biological time', *Osiris*, viii (1948); F. Redi, *Opere*, Naples, 1778, Milan, 1809–11 (English trans. of *Experiments on the Generation of Insects*, Chicago, 1909); C. Singer, 'The dawn of microscopical discovery', *Journal of Royal Microscopical Society*, xxxv (1915).

MEDICINE: In addition to the list in Vol. I under Chapter 4, and above, D. Campbell, 'The medical curriculum of the universities of Europe in the sixteenth century', in *Science, Medicine and History*, ed. E. A. Underwood, Oxford, 1953, i; A. Castiglioni, 'The medical school of Padua and the renaissance of medicine', *Annals of Medical History*, N.S. vii (1935); J. D. Comrie, *Selected Works of Thomas Sydenham*, with a short biography, London, 1922; P. Delaunay, *La vie médicale aux 16ème, 17ème et 18ème siècles*, Paris, 1935; John F. Fulton, *The Great Medical Bibliographers: a study in humanism*, Philadelphia, 1951; D. A. Wittop Koning (ed.), *Art and Pharmacy*, Deventer, Holland, 1950; Ambrose Paré, *The Apologie and Treatise*, trans. by T. Johnson, 1634, ed. G. Keynes, London, 1951, *Textes choisis*, présentés et commentés par L. Delarnelle et M. Sendrail, Paris, 1953; G. Sudhoff, *Aus der Frühgeschichte der Syphilis (Studien zur Geschichte der Medizin*, ix) Leipzig, 1912.

Bibliography

PHILOSOPHY OF SCIENCE AND CONCEPT OF NATURE IN
THE SCIENTIFIC REVOLUTION: Cf. works listed in Vol. I
under Chapters 1, 2, 3 (Philosophy, Astronomy, Mechanics), and in
the present volume under Chapters 1 (Philosophy, Matter and Space,
etc., Mathematical Physics) and 2 (Mechanics, Astronomy, Optics,
Physiology). For the 17th and 18th centuries there are: H. G. Alexander (editor), *The Leibniz-Clarke Correspondence*, Manchester, 1956;
F. H. Anderson, 'The influence of contemporary science on Locke's
methods and results', *University of Toronto Studies, Philosophy*, ii
(1923), *Philosophy of Francis Bacon*, Chicago, 1948; E. N. da C.
Andrade, 'Robert Hooke', *Proceedings of the Royal Society* A, cci
(1950); A. Armitage, 'René Descartes (1596–1650) and the early Royal
Society', *Notes and Records of the Royal Society*, viii (1950); Sir
Francis Bacon, *Works*, ed. J. Spedding, R. L. Ellis and D. D. Heath,
London, 1857–9, 7 vols., *Letters and Life*, ed. J. Spedding, London,
1861–4, 7 vols.; Amir Mehdi Badi', *L'Idée de la méthode des sciences*,
i, *Introduction*, Paris, 1953; W. W. Rouse Ball, *An Essay on Newton's
Principia*, London, 1893 – including Newton's correspondence with
Hooke and Halley; A. G. A. Balz, *Cartesian Studies*, New York, 1951
– very useful; M. A. Bera (ed.), *Pascal, l'homme et l'œuvre*, Paris,
1956; George Berkeley, *De Motu*, in *The Works of George Berkeley*,
ed. A. A. Luce and T. E. Jessop, iv, London and Edinburgh, 1951;
M. Boas, 'The establishment of the mechanical philsophy', *Osiris*, x
(1952) – a useful monograph, 'La méthode scientifique de Robert
Boyle', *Revue d'histoire des sciences*, ix (1956); The Honourable Robert
Boyle, *Works*, ed. Thomas Birch, London, 1744, 5 vols., 2nd ed., 1772,
6 vols.; F. Brandt, *Thomas Hobbes' Mechanical Conception of Nature*,
English trans., London, 1928; E. Bréhier, *Histoire de la philosophie*,
Paris, 1942–3, i–ii; G. S. Brett, *The Philosophy of Gassendi*, London,
1908; Sir D. Brewster, *Memoirs of the Life, Writings and Discoveries
of Sir Isaac Newton*, Edinburgh, 1855 – still useful although often
mistaken; L. Brunschwig, *Les étapes de la philosophie mathématique*,
3e éd., Paris, 1929; E. A. Burtt, *The Metaphysical Foundations of
Modern Physical Science*, 2nd ed., London, 1932 (New York, 1955) – a
basic pioneer study; E. Cassirer, *Das Erkenntnis-problem in der Philosophie und Wissenschaft der neueren Zeit*, Berlin, 1922–3, i–iii, *The
Philosophy of the Enlightenment*, English trans., Princeton, 1951 (German ed., 1932); I. B. Cohen, *Franklin and Newton* (*Memoirs of the
American Philosophical Society*, xliii), Philadelphia, 1956 – very
informative; R. G. Collingwood, *The Idea of Nature*, Oxford, 1945;
A. C. Crombie, four articles on the history of scientific method in

Bibliography

Discovery, xiii–xiv (1952–3), 'Newton's conception of scientific method', Bulletin of the Institute of Physics (1957); J. H. Dempster, 'John Locke, physician and philosopher', Annals of Medical History, iv (1932) – with bibliography; Descartes, Œuvres, ed. C. Adam and P. Tannery, Paris, 1897–1913, 12 vols.; M. 'Espinasse, Robert Hooke, London, 1956; B. Farrington, Francis Bacon, philosopher of industrial science, New York, 1949; Jeremiah S. Finch, Sir Thomas Browne, A Doctor's Life of Science and Faith, New York, 1950; H. Fisch, 'The scientist as priest : a note on Robert Boyle's natural theology', Isis, xliv (1953); J. F. Fulton, 'Robert Boyle and his influence on thought in the 17th century', Isis, xviii (1932); Galileo, Opere, ed. A. Favaro, Florence, 1890–1909, 20 vols.; Pierre Gassendi, sa vie et son œuvre, Paris, 1955: A. Gewirtz, 'Experience and the nonmathematical in the Cartesian method', J. History of Ideas, ii (1941) – an important study; W. J. Greenstreet (ed.), Isaac Newton, Memorial Volume, London, 1927 – contains valuable essays on Newton's thought; O. Hamelin, Le système de Descartes, 1911; R. Hooke, Micrographia, reprinted in R. T. Gunther, Early Science in Oxford, Oxford, 1938, xiii, Posthumous Works, ed. R. Waller, London, 1705, Diary, ed. H. W. Robinson and W. Adams, London, 1935; H. Hervey, 'Hobbes and Descartes in the light of some unpublished letters of the correspondence between Sir Charles Cavendish and Dr. John Pell', Osiris, x (1952); M. B. Hesse, Forces and Fields. A Study of Action at a Distance ..., London, 1961; W. G. Hiscock, David Gregory, Isaac Newton and their Circle, Oxford, 1937; History of Science Society, Isaac Newton, London, 1928 – essays by various writers; P. Hoenan, De origine primorum principiorum scientiae, Gregorianum, xiv (1933); J. Jacquot, 'Un amateur de science, ami de Hobbes et de Descartes, Sir Charles Cavendish (1591–1654)', Thalès, vi (1949–50), papers on Harriot and Hobbes in Notes and Records of the Royal Society, ix (1952); I. Kant, Critique of Pure Reason, 2nd ed. (1787), trans. by N. Kemp Smith, London, 1933, Critique of Judgement (1790), trans. by J. H. Bernard, 2nd ed., London, 1914; A. Koyré, op. cit. above under Mechanics and From the Closed World to the Infinite Universe, Baltimore, 1957 – important; S. P. Lamprecht, 'The role of Descartes in seventeenth-century England', Studies in the History of Ideas, iii (1935); A. Lange, Geschichte des Materialismus, 3. Aufl., Leipzig, 1873–5, 2 vols. (English trans., London, 1925); R. Lenoble, Marin Mersenne, Paris, 1943; M. Leroy, Descartes social, Paris, 1931; G. Martin, Kant's Metaphysics and Theory of Science, trans. from the German by P. G. Lucas, Manchester, 1955; G. Milhaud, Descartes Savant, Paris, 1921; L. T. More, Isaac Newton; a biography, 1642–1727,

Bibliography

New York, 1934, *Life and Works of Robert Boyle*, London, 1944; Sir Isaac Newton, *Papers and Letters on Natural Philosophy*, ed. L. B. Cohen, Cambridge, 1958, *Principia* – see above, Mathematics, *Opticks* – see above, Optics, *Opera*, ed. S. Horsley, London, 1782 (correspondence with Bentley and Boyle), *Correspondence*, ed. J. Edleston, London, 1850, *Correspondence*, ed. H. W. Turnbull, Cambridge, 1959– , *Theological Manuscripts*, selected and edited with an introduction by H. McLachlan, Liverpool, 1950, *Unpublished Scientific Papers*, ed. A. R. and M. B. Hall, Cambridge, 1962; W. J. Ong, 'System, space, and intellect in Renaissance symbolism', *Bibliothèque d'humanisme et Renaissance*, travaux et documents, xviii (1956); S. B. L. Penrose, jr., *The Reputation and Influence of Francis Bacon in the Seventeenth Century*, New York, 1934; Rohault's *System of Natural Philosophy*, illustrated with Dr. Samuel Clarke's Notes taken mostly out of Sr. Isaac Newton's Philosophy. Done into English by John Clarke, 2nd ed., London, 1728–9, 2 vols.; P. Rossi, *Francesco Bacone*, Bari, 1957; L. Roth, *Descartes' Discourse on Method*, Oxford, 1937; *Royal Society Newton Tercentenary Celebrations*, Cambridge, 1947 – valuable essays by various writers; E. Simard, *La nature et la portée de la méthode scientifique, exposé et textes choisis de philosophie des sciences*, Quebec et Paris, 1956; N. Kemp Smith, *New Studies in the Philosophy of Descartes*, London, 1952 – very important; *Descartes' Philosophical Writings*, London, 1952 – a useful selection: L. Strauss, *The Political Philosophy of Hobbes, its Basis and Genesis*, trans. from the German MS by E. M. Sinclair, Oxford, 1936; F. Ueberweg, *Grundriss der Geschichte der Philosophie*, iii, *Die Philosophie der Neuzeit bis zum Ende des 18. Jahrhunderts*, 12. Aufl., neubearbeitet von M. Frischheisen-Köhler and W. Moog, Berlin, 1924; R. S. Westfall, 'Unpublished Boyle papers relating to scientific method', *Annals of Science*, xii (1956); A. N. Whitehead, *Science and the Modern World*, Cambridge, 1926; P. P. Wiener, 'The experimental philosophy of Robert Boyle', *The Philosophical Review*, xli (1932); B. Willey, *The Seventeenth-Century Background*, London, 1934; R. M. Yost, jr., 'Sydenham's philosophy of science', *Osiris*, ix (1950), 'Locke's rejection of hypotheses about sub-microscopic events', *J. History of Ideas*, xii (1951); D. O. Zöckler, *Geschichte der Beziehungen zwischen Theologie und Naturwissenschaft, mit besonderer Rücksicht auf Schöpfungsgeschichte*, Gütersloh, 1877–9, 2 vols.

A SELECTION OF VARIOUS MODERN WORKS ON THE PHILOSOPHY OF SCIENCE IS: I. Berlin, 'Logical translation',

Bibliography

Proceedings of the Aristotelian Society, N.S. 1 (1949–50); Claude Bernard, *Introduction à l'étude de la médecine expérimentale*, Paris, 1865 (English trans. by H. C. Greene, New York, 1957); W. I. B. Beveridge, *The Art of Scientific Investigation*, London, 1950; Max Black, 'The definition of scientific method', in *Science and Civilization*, ed. R. C. Stauffer, Madison, Wis., 1949; Max Born, *Natural Philosophy of Cause and Chance*, Oxford, 1949; R. B. Braithwaite, *Scientific Explanation*, Cambridge, 1953; P. W. Bridgman, *The Logic of Modern Physics*, New York, 1928; N. R. Campbell, *Physics. The Elements*, Cambridge, 1920, *What is Science?* London, 1921, New York, 1952 – a very useful introduction; W. K. Clifford, *The Commonsense of the Exact Sciences*, 2nd ed., London, 1946; E. Nagel, *The Structure of Science*, London, 1961; A. Comte, *Cours de philosophie positive*, Paris, 1830, i; J. B. Conant, *On Understanding Science*, Oxford, 1947; H. Dingle, *The Scientific Adventure*, London, 1952; J. M. C. Duhamel, *Des méthodes dans les sciences de raisonnement*, Paris, 1865–70, 4 vols.; P. Duhem, *La théorie physique: son objet, sa structure*, Paris, 1914 (English trans. by P. P. Wiener, *The Aim and Structure of Physical Theory*, New York, 1954); F. Enriques, *Problems of Science*, trans. K. Royce, Chicago, 1924; H. Feigl and M. Brodbeck, *Readings in the Philosophy of Science*, New York, 1953 – a valuable collection of recent papers, complementary to that of Wiener, below; N. R. Hanson, *Patterns of Discovery*, Cambridge, 1958; F. A. von Hayek, 'Scientism and the study of society', *Economica*, N.S. ix (1942), x (1943), xi (1944), *The Counter-Revolution in Science*, Glencoe, Ill., 1952; A. W. Heathcote, 'William Whewell's philosophy of science', *British J. Philosophy of Science*, iv (1954); H. von Helmholtz, *Popular Lectures on Scientific Subjects*, English trans. London, 1895, 2 vols., *Schriften zur Erkenntnistheorie*, ed. P. Hertz and M. Schlick, Berlin, 1921; G. Holton and D. H. D. Roller, *Foundations of Modern Physical Science*, Reading, Mass., 1958; L. O. Kattsoff, articles on postulational method, *Philosophy of Science*, ii (1935), iii (1936); F. Kaufmann, *The Methodology of the Social Sciences*, Oxford, 1944; Hans Kelsen, *Society and Nature; a sociological enquiry*, London, 1946; V. Kraft, *Die Grundformen der wissenschaftlichen Methoden* (*Sitzb. d. Akad. d. Wissensch. in Wien*, philos.-hist. Klasse, cciii. 3) Wien and Leipzig, 1925; E. H. Madden, *The Structure of Scientific Thought*, London, 1960; F. S. Marvin, *Comte, the Founder of Sociology*, London, 1936; James Clerk Maxwell, *Scientific Papers*, ed. W. D. Niven, Cambridge, 1890, i; H. Metzger, *Les concepts scietifiques*, Paris, 1926; E. Meyerson, *De l'explication dans les sciences*, Paris, 1921, 2 vols., *Identité et réalité*,

Bibliography

3e éd., Paris, 1926 (English trans. by K. Loewenberg, *Identity and Reality*, London, 1930); J. S. Mill, *A System of Logic*, London, 1843, *Auguste Comte and Positivism*, 2nd ed., London, 1866; T. P. Nunn, *The Aim and Achievement of Scientific Method*, London, 1907; K. Pearson, *The Grammar of Science*, London, 1892; H. Poincaré, *La science et l'hypothèse*, Paris, 1920 (English trans., London, 1905), *Science et méthode*, Paris, 1927 (English trans., London, 1914); M. Polanyi, *Personal Knowledge*, London, 1958; K. R. Popper, *The Poverty of Historicism*, London, 1957; 'A note on Berkeley as precursor of Mach', *British Journal for the Philosophy of Science*, iv (1953), 'Three views concerning human knowledge', in *Contemporary British Philosophy* (Third Series), London, 1956, *The Logic of Scientific Discovery*, London, 1959 (English trans. of *Logik der Forschung*, Vienna, 1935); F. P. Ramsey, *The Foundations of Mathematics*, London, 1931; B. Russell, *Human Knowledge – Its Scope and Limits*, London, 1949; Sir C. Sherrington, *Man and his Nature*, Cambridge, 1940; P. A. Schilpp (editor), *Albert Einstein, Philosopher-Scientist*, Cambridge, Mass., 1949; L. S. Stebbing, *A Modern Introduction to Logic*, 2nd ed., London, 1933; R. Taton, *Causalité, accident et la découverte scientifique*, Paris, 1955; S. Toulmin, *Philosophy of Science*, London, 1953; W. H. Watson, *On Understanding Physics*, Cambridge, 1938; F. Waismann, 'Verifiability', *Proceedings of the Aristotelian Society*, London, Suppl. Vol. xix (1945), (reprinted in *Logic and Language*, ed. A. G. N. Flew, Oxford, 1951), 'Language strata', in *Logic and Language* (Second Series), ed. A. G. N. Flew, Oxford, 1953; W. Whewell, *Philosophy of the Inductive Sciences*, London, 1840, 1847, 2 vols., *Novum Organum Renovatum*, 3rd ed., London, 1858, On the Philosophy of Discovery, London, 1860 – an opposite point of view from Mill's; A. N. Whitehead, *Introduction to Mathematics*, London, 1911; P. P. Wiener, *Evolution and the Founders of Pragmatism*, Cambridge, Mass., 1949; P. P. Wiener (ed.), *Readings in Philosophy of Science*, New York, 1953 – see above, Feigl and Brodbeck; R. L. Wilder, *Introduction to the Foundations of Mathematics*, New York, 1952; J. O. Wisdom, *Foundations of Inference in Natural Science*, London, 1952; J. H. Woodger, *Biological Principles*, London, 1920, *Biology and Language*, Cambridge, 1952.

INDEX

Abelard, Peter, 19
Abu'l Barakat al-Baghdadi, 67
Achillini, Alexander (1463–1512), 124, 276
Adelard of Bath, 19, 20, 26, 55, 289
Agricola. *See* Bauer, Georg
Albert of Saxony (*c*. 1316–90), 22, 56, 60, 71, 75, 86–7, 96, 102, 104, 107–8, 122, 135, 137, 143, 157, 274
Alberti, Leon Battista (1404–72), 112
Albertus Magnus, 34, 38, 46, 54, 55, 57, 69, 121, 122, 123, 125, 268, 275, 280, 281, 282, 283, 286
Alchemy, 38, 111, 123, 242
Aldrovandi, Ulysses (1522–1605), 268, 282, 283
Alexandria, 25, 52 n., 64
al-Farisi, 25
Alfonsine Tables, 113, 185
Alhazen, 25, 123, 124, 125
Ali ibn Ridwan, 25
al-Katibi, 87
Alkindi, 25
Alpago, Andrea, 230 n.
Alpetragius, 74
al-Shirazi, 25, 87
Anatomy, 117, 124, 125, 227–39, 245–7, 257, 263, 267, 273–89
Anselm, St, 19
Apianus, Petrus, 251
Apollonius, 116, 138, 139, 190
Aquinas, St Thomas, 48, 55, 57, 69, 72, 79, 98, 122, 212
Arabs and Arabic science, 19, 21, 25–6, 66–8, 103, 110, 116, 117, 257, 260, 264–5, 267, 277–8
Arantius, Julius Caesar (1564), 284
Archimedes, 21, 24, 27, 29 n., 56, 70, 112, 116, 119, 132, 135–6, 138, 139, 142–3, 149, 152, 191, 292
Aristarchus of Samos, 138
Aristotle and the Aristotelian system, 24–5, 28, 36, 40–43, 49, 142, 220, 254, 261, 289, 292, 301–3, 305, 313, 314, 317, 322, 325; biology in, 117, 229, 236, 243, 245, 267, 271, 280–84, 287, 312; causation in, 36,

119; chemical elements in, 53, 260, 265, 294; cosmology in, 46–7, 50–51, 54–60, 119–20, 126, 135, 182–3, 186–7, 192, 205, 211, 214, 241; criticism of, 17–130, 134–5, 142–3, 144–5, 147–8; dynamics in, 46–7, 59, 61–97, 113, 119, 134–7, 145–54, 173–4, 189, 198, 256; induction in, 19, 25, 228, 308; mathematics and physics in, 18, 36–7, 49–50, 93, 97–8, 118–20, 129, 140–43, 144–54, 182–3, 184, 199, 203
Arnald of Villanova, 123
Art and science, 125, 229, 268–9, 273–4, 276, 278–9, 281, 284
Aselli, Gasparo (1581–1626), 285
Astrolabe, 120, 123, 252, 262
Astronomy, 19, 23, 33, 36, 38–40, 55, 60–61, 65, 68, 82, 85, 87–97, 101, 103, 110, 113–15, 116, 118, 121, 123, 124–5, 129, 135, 149, 166–71, 174–226, 250–51, 257–9, 289, 333
Atomism, 49–54, 64–5, 78, 116, 169, 265–6, 287–8, 292, 293, 313
Aubrey, John, 239
Augustine, St, and Augustinianism, 20, 48, 52 n., 169 n., 207, 216, 302, 311
Autrecourt, Nicholas of. *See* Nicholas of Autrecourt
Avempace, 67–72, 74, 79, 108
Averroës and Averroïsm, 40, 48, 57, 59, 67–9, 72, 79, 116, 122, 146, 197, 198, 257, 317
Avicebron, 53, 55, 61
Avicenna, 25, 40, 66–7, 74, 117, 230, 230 n., 275

Bacon, Francis (1561–1626), 121, 133, 146 n., 193, 209 n., 264–5, 289–90, 292–9, 310, 322, 332
Bacon, Roger, 25, 34, 38–9, 46, 53, 54–5, 57, 72–3, 75, 99–100, 122–3, 124, 169 n., 254, 257, 307, 311
Baconthorpe, John, 59
Baldi, Bernardino, 137

365

Index

Index

Index

Fallopio, Gabriel (1523–62), 280, 283, 284

Fermat, Pierre de (1601–65), 124, 139, 171

Fernel, Jean, 229–31

Ferrari, Lodovico (1522–65), 138

Fibonacci of Pisa, Leonardo, 21, 116, 123

Firearms, 134, 141–2

Fludd, Robert, 194, 240, 242, 253–4, 264

Foligno, Gentile da. *See* Gentile da Foligno

Fontana of Brescia, Nicolo. *See* Tartaglia

Forli, Jacopo da. *See* Jacopo da Forli

Foscarini, Paolo Antonio, 215, 217

Fracastoro, Girolamo (1483–1553), 175, 282, 288

Franciscus de Marchia, 73, 83

Franco, Pierre, 277

Francon of Liège, 20

Frederick II, 123

Frisius, Gemma, 252

Frobisher, Martin, 252

Fuchs, Leonard, 269, 270

Galen, 19, 25, 29, 39–40, 52 n., 100, 110, 117, 125, 228–31, 234–6, 237, 243, 245, 253, 262, 264, 278–9, 285, 287, 312

Galileo Galilei (1564–1642), 17, 35, 37, 42, 59, 65, 69, 72, 84–5, 87, 89, 93, 96, 98, 105, 107–9, 114, 121, 124–5, 127–9, 134, 137, 142, 143, 144–69, 170–74, 179, 182–3, 184–5, 189, 191–2, 199, 201–28, 248, 249–50, 253–4, 256, 258, 265, 284, 291–3, 303–6, 309, 312, 317–18, 320–26, 331–5

Garcia da Orta, 268

Gassendi, Pierre (1592–1655), 167–8, 191, 224, 246, 266, 287, 313, 315, 323

Generation, theories of, 114, 241–2, 262–3, 283, 286–8, 314, 319

Gentile da Foligno (d. 1348), 115, 123

Geography, 103, 117, 134, 250–53, 268

Geology, 120, 123, 196–7, 274–5, 282–3

Gerard of Brussels, 27, 70–71, 127

Gerard of Cremona, 175

Gerbert, 20

Gerson, Levi ben, 22, 113

Gesner, Conrad (1516–65), 267–8, 269, 281–3

Geulincx (1625–69), 315

Ghisilieri, Federico, 218

Gilbert, William (1540–1603), 123, 148–9, 162, 170, 185, 196–8, 249, 256

Giles of Rome (1247–1316), 53, 55, 69, 74, 162 n., 254

Girard, Albert (1595–1632), 138

Glanvill, Joseph, 314–15

Glisson, Francis, 288

Gravity and specific gravity, 50, 57–61, 64–5, 80–90, 97, 119, 136, 137, 141–6, 157–8, 162–3, 165–6, 167–8, 182–3, 192–3, 197–200, 204–5, 219, 233, 257, 291, 309, 313, 323–6

Greeks and Greek science, 19–21, 23–5, 56, 62, 64–5, 88, 100, 103, 105, 117–19, 132, 135, 137–9, 190–1, 253, 326, 333–4

Gregory, James, 124

Gregory of Rimini, 56, 75

Grimaldi, 124

Grosseteste, Robert (c. 1170–1253), 17, 25, 27–34, 35–8, 45, 52–3, 73, 99–100, 121, 122, 123, 124, 128, 169 n., 195, 199, 291–2, 294, 307, 311, 329

Grosseteste (pseudo-), 57

Günther, Johannes (1487–1574), 230, 277–8

Guy de Chauliac, 123

Harriot, Thomas (1560–1621), 124, 138, 252

Hartmann, Georg, 197

Harvey, William (1578–1657), 121, 227–30, 233–44, 279, 284, 285–7, 297, 311–12

Henri de Mondeville, 123

Henry of Hesse (1325–97), 114

Heraclides of Pontus, 60, 175, 180 n.

Hero of Alexandria (1st century B.C.), 21, 51–2 n., 54, 65, 137–8, 253, 255, 293, 305

Heytesbury, William of. *See* William of Heytesbury

Hindus and Hindu science, 21–2, 116

Hipparchus, 86, 158

Hippocrates, 29 n., 117, 277

Hobbes, Thomas (1588–1679), 239, 316, 323

Holbach, P. H. T. d', 316

Hood, Thomas (fl. 1582–98), 133, 252

Index

Index

Mass, 64–5, 90, 162, 170, 173–4
Massa, Nicholas, 276
Mästlin, Michael (1550–1631), 187
Mathematical method, 17–24, 36–7, 50, 62–3, 68–71, 97–116, 120–21, 131–74, 248–59, 266, 289, 292, 303, 325
Mathematics, 18–24, 26, 35–6, 50, 55–6, 62–3, 69–71, 97–119, 122–3, 125–6, 129, 131–2, 137–40, 150–51, 175–7, 183–4, 188–92, 205, 211–12, 215, 249, 257, 292, 305–6, 328
Matter, theories of, 24–30, 49–61, 111–13, 131, 168–9, 261, 262–3, 292–3, 301–2, 304–5, 309, 313–15
Maudith, John, 22
Maupertuis, P. L. M. de (1698–1759), 307, 314
Maurolico, Francesco (1494–1575), 124, 138, 176, 190, 257
Maurus, Hrabanus, 116
Mayow, John (1643–79), 240, 266
Measurements, units of, 100–101, 109–12, 119, 249–51, 253
Mechanics and mechanism, 50, 55, 57, 61–97, 101, 111–13, 114, 119–20, 131–74, 177, 182, 198, 202, 204, 222, 226–40, 241, 243–6, 247–9, 258, 264–6, 284, 288, 290, 293, 296, 300–305, 311–16, 319–23, 325–6, 332 n.
Mediavilla. See Richard of Middleton
Medicine, 19, 25, 39–41, 46, 100, 110, 112, 115, 117, 121, 123, 125, 126, 243, 254, 259–61
Medieval science, summary of contributions of, 115–30
Megenburg, Conrad von, 123
Mercator, 251
Mersenne, Marin, 137, 161, 170–71, 223, 309–11, 319
Mesue, 117
Metallurgy, 111, 133, 259–61, 263–4, 331
Meteorology, 34, 38, 124, 241, 253–4
Michelangelo (1475–1564), 273
Microscope, 239, 257, 259, 266, 285–7
Middleton, Richard of. See Richard of Middleton
Mill, John Stuart (1806–73), 46, 146, 194, 320, 321, 332–3
Mills, water- and wind-, 119, 245, 255–6
Mind and body, 42–56, 314–16, 331–2 n.
Mining, 133, 255–6, 259, 298

Mirandola, Pico della. See Pico della Mirandola
Monardes, Nicolas, 268
Mondino of Luzzi, 123, 276, 278
Montaigne, Michel E. (1533–92), 49
Morice, Peter, 255
Morrison, Dr R., 239
Moslems and Moslem science, 22, 122
Mouffet, Thomas, 282
Müller, Johannes (1436–76), 113, 116, 121, 123, 174–5, 186
Murs, Jean de, 123
Music, 86, 129, 193, 247

Napier, John (1550–1617), 191
Navigation, 134, 174, 196, 250–52
Nemorarius, Jordanus. See Jordanus Nemorarius
Neoplatonism, 37, 65–74, 118, 125, 129, 184, 194–5, 197, 200, 242, 292
Neugébauer, 183–4
Newcastle, Marquis of, 248
Newcomen, Thomas, 256
Newman, Cardinal J. H., 226
Newton, Sir Isaac (1642–1727), 64, 79, 84, 121, 127, 135, 140, 142, 162, 165, 170–74, 176, 192, 199, 201, 204–5, 215, 219–20, 222, 224–6, 249, 259, 266, 289, 291, 299, 302, 307, 311, 313, 316, 320, 322–30, 333
Nicholas of Autrecourt (d. after 1350), 47–9, 53–4, 78, 116
Nicholas of Cusa (1401–64), 49, 54, 60–61, 87, 96–7, 111–13, 121, 122, 137, 174, 183, 198, 260, 262
Nicomachus, 21
Nifo, Agostino, 41–2, 122, 214–15, 329
Nominalism and science, 37, 44–9, 54, 69, 146, 261, 330
Norden, J., 252
Norman, Robert, 196
Novara, Domenico Maria (1454–1504), 175

Ockham, William of. See William of Ockham
Oldenburg, Henry, 323
Olivi, Peter (1245/9–98), 72–3
Optics, 19, 23, 25, 39, 53, 86, 99–100, 103–4, 118, 123, 124, 129, 134, 135, 137, 149, 185–6, 191, 245, 247, 257–9, 307, 323
Oresme, Nicole (d. 1382), 22, 60, 71, 78–9, 88–96, 102, 104–5, 108–9, 114,

370

Index

Index

Index